Lecture Notes in Computer Science   9021

Commenced Publication in 1973
Founding and Former Series Editors:
Gerhard Goos, Juris Hartmanis, and Jan van Leeuwen

T0233750

More information about this series at http://www.springer.com/series/7409

Nitin Agarwal · Kevin Xu
Nathaniel Osgood (Eds.)

Social Computing, Behavioral-Cultural Modeling, and Prediction

8th International Conference, SBP 2015
Washington, DC, USA, March 31 – April 3, 2015
Proceedings

 Springer

Editors
Nitin Agarwal
University of Arkansas
Little Rock
Arkansas
USA

Nathaniel Osgood
University of Saskatchewan
Saskatoon
Saskatchewan
Canada

Kevin Xu
Technicolor Research
Los Altos
California
USA

ISSN 0302-9743
Lecture Notes in Computer Science
ISBN 978-3-319-16267-6
DOI 10.1007/978-3-319-16268-3

ISSN 1611-3349   (electronic)

ISBN 978-3-319-16268-3   (eBook)

Library of Congress Control Number: 2015933612

LNCS Sublibrary: SL3 – Information Systems and Applications, incl. Internet/Web, and HCI

Printed on acid-free paper

Springer International Publishing AG Switzerland is part of Springer Science+Business Media
(www.springer.com)

Preface

The study of human behavior in all its aspects has expanded dramatically over the last couple of decades to include not only the traditional humanities and social sciences, but also disciplines across the natural, physical, information, computer, mathematical, and health sciences. A critically important contributor to the success of this endeavor is the interdisciplinary collaboration among scientists, researchers, and scholars. The emerging fields of computational social science and social computing are prime examples of interdisciplinary or multidisciplinary efforts that hold great promise. It is the intent of this volume to further advance that effort. The volume contains papers that were presented orally or as posters at the 2015 International Conference on Social Computing, Behavioral-Cultural Modeling, and Prediction (SBP 2015). Similar to last year, the SBP 2015 was colocated with the 2015 Behavioral Representation in Modeling and Simulation (BRiMS) conference. The goal of the SBP 2015 conference was to advance our understanding of human behavior through the development and application of mathematical, computational, statistical, simulation, predictive, and other models that provide fundamental insights into factors contributing to human sociocultural dynamics.

This was the eighth year of the conference and the fifth since it merged with the International Conference on Computational Cultural Dynamics (ICCCD). In 2015, the SBP conference carried on its tradition of a high level of scholarly interest. We received a total of 118 submissions (from 21 countries), the second highest in the history of SBP (the highest being in 2013 with 137 submissions). Of the 118 submissions, 96 were complete and eligible for peer review. In all, 316 reviews were obtained on the 96 papers averaging 3.3 reviews per paper. This would not have been possible without the hard work of the members of a multidisciplinary (e.g., Military and Security; Methodology; Health Sciences; Information, Systems, and Network Science; Behavioral and Social Sciences, and Economics), and international (from 15 countries) Technical Program Committee. SBP continued to be a selective single-track conference. Twenty-four submissions were accepted as oral presentations, a 25% acceptance rate. We also accepted 36 short papers to be presented in an interactive poster session. The papers accepted for oral presentations and poster presentations are included in this volume.

The conference prides itself on its tradition of promoting multidisciplinary research and participation. Although the SBP 2015 disciplinary data on registrants was not available for inclusion in this volume, the SBP 2014 data on the discipline of attendees is illustrative. For SBP 2014, 18% of registered attendees were from the behavioral sciences, 8% from engineering, 8% from the health-related sciences, 2% from the humanities, 15% from math or computer science, 16% from computational social science, 2% from the natural sciences, 11% from the social sciences, 14% from computer and information sciences, and 6% from others. Across the five topics for SBP 2015, the overall number of submissions varied somewhat with the most papers in the category of information science (41%), followed closely by methodology (28%) and the topics of behavioral, and social sciences and economics (18%), military and security (11%), and health sciences (9%).

The conference included a variety of venues beyond the oral and poster presentations. There were four tutorial that included Text and Network Analysis Methods, Behavioral Mining, Cultural Differences and Social Network Analysis in Massive Online Games, Big Data Analytics for Behavioral Modeling, and Introduction to Spatial and Social Network Data Combining R and Python. There were also keynote addresses by distinguished scholars and scientists and round-table discussions.

A conference such as this can only succeed with the hard work of many volunteers. Members of the various committees met online throughout the year, often from far flung corners of the world, to discuss and make important decisions. The success of the conference and this volume would have been impossible without their efforts. Although too many to name here, they are all acknowledged in the section to follow. This is the third year that the conference has been held at the University of California's DC Center in downtown Washington, DC and we would like to acknowledge their hospitality and logistical support. Also, we would like to thank members of the BRiMS Organizing Committee for supporting the efforts of SBP. Last but not least, we sincerely appreciate the support from the following federal agencies: Air Force Office of Scientific Research (AFOSR), Air Force Research Laboratory (AFRL), Army Research Office (ARO), National Institute of General Medical Sciences (NIGMS) at the National Institutes of Health (NIH), National Science Foundation (NSF), and the Office of Naval Research (ONR). We also would like to thank Alfred Hofmann from Springer Publishing. We thank all for their kind help, dedication, and support to make SBP 2015 possible.

January 2015   Jeffrey C. Johnson

Organization

Conference Chair

Jeffrey C. Johnson   University of Florida, USA

Program Co-chairs

Nitin Agarwal   University of Arkansas at Little Rock, USA
Kevin Xu   Technicolor Research, USA
Nathaniel Osgood   University of Saskatchewan, Canada

Steering Committee

John Salerno   AFRL, USA
Huan Liu   Arizona State University, USA
Sun Ki Chai   University of Hawaii, USA
Patricia Mabry   National Institute of Health, USA
Dana Nau   University of Maryland, College Park, USA
V.S. Subrahmanian   University of Maryland, College Park, USA
Shanchieh (Jay) Yang   Rochester Institute of Technology, USA
Nathan D. Bos   Johns Hopkins University/Applied Physics
                               Laboratory, USA
Claudio Cioffi-Revilla   George Mason University, USA

Conference Committee

Claudio Cioffi-Revilla   George Mason University, USA
Jeffrey C. Johnson   University of Florida, USA
William G. Kennedy   George Mason University, USA
Nitin Agarwal   University of Arkansas at Little Rock, USA
Shanchieh (Jay) Yang   Rochester Institute of Technology, USA
Nathan D. Bos   Johns Hopkins University/Applied Physics
                               Laboratory, USA
Ariel M. Greenberg   Johns Hopkins University/Applied Physics
                               Laboratory, USA
John Salerno   Exelis Inc., USA
Huan Liu   Arizona State University, USA
Donald Adjeroh   West Virginia University, USA
Katherine Chuang   Soostone Inc., USA

Advisory Committee

Fahmida N. Chowdhury	National Science Foundation, USA
Rebecca Goolsby	Office of Naval Research, USA
John Lavery	Army Research Lab/Army Research Office, USA
Patricia Mabry	National Institutes of Health, USA
Tisha Wiley	National Institutes of Health, USA

Sponsorship Committee Chair

Huan Liu	Arizona State University, USA

Publicity Chair

Donald Adjeroh	West Virginia University, USA

Web Chair

Katherine Chuang	Soostone Inc., USA

Tutorial Chair

Dongwon Lee	Penn State University, USA

Challenge Problem Co-chairs

Fred Morstatter	Arizona State University, USA
Kenny Joseph	Carnegie Mellon University, USA
Kang Zhao	University of Iowa, USA

Funding Panel Workshop Chair

Fahmida N. Chowdhury	National Science Foundation, USA

Student Travel Award Co-chairs

Biru Cui	Rochester Institute of Technology, USA
Syed Ashiqur Rahman	West Virginia University, USA

Technical Program Committee

Mohammad Ali Abbasi
Myriam Abramson
Donald Adjeroh
Nitin Agarwal
Kalin Agrawal
Muhammad Ahmad
Samer Al-Khateeb
Yaniv Altshuler
Soumya Banerjee
Chitta Baral
Geoffrey Barbier
Asmeret Bier
Halil Bisgin
Lashon Booker
Nathan Bos
David Bracewell
Erica Briscoe
George Aaron Broadwell
Sun-Ki Chai
Jiangzhuo Chen
Yi Chen
Xueqi Cheng
Alvin Chin
David Chin
Michele Coscia
Peng Dai
Amitava Das
Hasan Davulcu
Yves-Alexandre de Montjoye
Bethany Deeds
Jana Diesner
Wen Dong
Daniele Durante
Koji Eguchi
Jeffrey Ellen
Yuval Elovici
Zeki Erdem
Ma Regina Justina E. Estuar
Laurie Fenstermacher
William Ferng
Michael Fire
Anthony Ford
Wai-Tat Fu
Armando Geller

Matthew Gerber
Ariel M. Greenberg
Brian Gurbaxani
Soyeon Han
Shuguang Han
Daqing He
Walter Hill
Michael Hinman
Shen-Shyang Ho
Shuyuan Mary Ho
Xia Hu
Robert Hubal
Samuel Huddleston
Edward Ip
Terresa Jackson
Lei Jiang
Chandler Johnson
Jeffrey C. Johnson
Kenneth Joseph
Ruben Juarez
Byeong-Ho Kang
Jeon-Hyung Kang
Bill Kennedy
Halimahtun Khalid
Masahiro Kimura
Shamanth Kumar
Kiran Lakkaraju
John Lavery
Dongwon Lee
Jiexun Li
Zhuoshu Li
Ee-Peng Lim
Huan Liu
Ting Liu
Xiong Liu
Corey Lofdahl
Eric Lofgren
Jonathas Magalhäes
Matteo Magnani
Masoud Makrehchi
Juan F. Mancilla-Caceres
Stephen Marcus
Mathew Mccubbins
Trevor Van Mierlo

Michael Mitchell
Sai Moturu
Marlon Mundt
Keisuke Nakao
Radoslaw Nielek
Kouzou Ohara
Brandon Oselio
Nathaniel Osgood
Alexander Outkin
Fatih Özgul
Wei Pan
Wen Pu
S.S. Ravi
David Reiter
Seyed Mussav Rizi
Nasim S. Sabounchi
Tanwistha Saha
Amit Saha
Kazumi Saito
David L. Sallach
Kaushik Sarkar
Philip Schrodt
Arun Sen
Fatih Sen
Samira Shaikh
Amy Sliva
Gita Sukthankar
Samarth Swarup
George Tadda
Venkataswamy Takumatla
Sarah Taylor

Gaurav Tuli
Craig Vineyard
Anil Kumar Vullikanti
Xiaofeng Wang
Changzhou Wang
Haiqin Wang
Zhijian Wang
Rik Warren
Aleksander Wawer
Wei Wei
Nicholas Weller
Paul Whitney
Rolf T. Wigand
Kevin Xu
Ronald Yager
S. Jay Yang
Laurence T. Yang
Christopher Yang
John Yen
Lei Yu
Bei Yu
Mo Yu
Serpil Yuce
Reza Zafarani
Laura Zavala
Mi Zhang
Qingpeng Zhang
Kang Zhao
Yanping Zhao
Inon Zuckerman

SBP 2015 Challenge Track

SBP 2015 Grand Data Challenge Finding Social Inequality to Aid the Disadvantaged

Fred Morstatter[1], Kenny Joseph[2], and Kang Zhao[3]

[1] Arizona State University
fred.morstatter@gmail.com
[2] Carnegie Mellon University
[3] University of Iowa

This year's SBP Grand Challenge problem asked participants to consider the following question: "how can we use publicly available data on the web and elsewhere to find social inequality and to aid the disadvantaged?". We invited the participants to combine the open data available on the web with the research found at the SBP conference to find ways to aid the disadvantaged.

The use of social media in understanding and mitigating social inequalities and prejudice has increased at a rapid pace, from spreading information during the Arab Spring to analyzing the social media conversations that took place during the Gamergate scandal. At the same time, data used for decades to study the ways in which social inequalities permeate every facet of social structure have become increasingly accessible. Twitter data such as that pertaining to the Ferguson protest[1], and the Gamergate scandal[2] are easy to obtain. Furthermore, governments are also opening up their data, with the United States providing census, climate, and health data, amongst others.

While many have taken advantage of these resources to produce new and interesting approaches to understanding social inequalities and ways to prevent them, there is much interesting and useful work still to be done. We invite participants to consider the following questions, which are only intended to give a rough idea of what might be an interesting topic to explore for this challenge problem:

- How are stereotypes of the disadvantaged perpetuated in social media?
- How do differing levels of Internet access affect the presence and attitude of individuals online?
- How has the distribution of poverty changed over time as American cities have grown, and how has this affected the impoverished population in a negative or positive way?

The organizers thank all who participated in this challenge. The winner of the challenge will receive a cash award, as well as a $1,000 travel stipend to present their work at the SBP 2015 conference. Second place will receive a cash award. All participants will have the opportunity for their abstract to appear on the SBP 2015 challenge web site.

---

[1] https://github.com/kennyjoseph/SBP_2015_Challenge
[2] https://medium.com/message/72-hours-of-gamergate-e00513f7cf5d

Contents

Poster Presentations

Oral Presentations

A Network-Based Model
for Predicting Hashtag Breakouts in Twitter

Sultan Alzahrani[1], Saud Alashri[1], Anvesh Reddy Koppela[1],
Hasan Davulcu[1]([✉]), and Ismail Toroslu[2]

[1] School of Computing, Informatics and Decision Systems Engineering,
Arizona State University, Tempe, AZ 85287, USA
{ssalzahr,salashri,akoppela,hdavulcu}@asu.edu
[2] Department of Computer Engineering, Middle East Technical University, Ankara,
Turkey
toroslu@ceng.metu.edu.tr

Abstract. Online information propagates differently on the web, some
of which can be viral. In this paper, first we introduce a simple standard
deviation sigma levels based Tweet volume breakout definition, then we
proceed to determine patterns of re-tweet network measures to predict
whether a hashtag volume will breakout or not. We also developed a
visualization tool to help trace the evolution of hashtag volumes, their
underlying networks and both local and global network measures. We
trained a random forest tree classifier to identify effective network measures for predicting hashtag volume breakouts. Our experiments showed
that "local" network features, based on a fixed-sized sliding window, have
an overall predictive accuracy of 76 %, where as, when we incorporate
"global" features that utilize all interactions up to the current period,
then the overall predictive accuracy of a sliding window based breakout
predictor jumps to 83 %.

Keywords: Information diffusion · Hashtag volumes · Prediction ·
Social networks · Diffusion networks

1 Introduction

Online Social Networks (OSNs) such as Twitter have emerged as popular
microblogging and interactive platforms for information sharing among people.
Twitter provides a suitable platform to investigate properties of information diffusion. Diffusion analysis can harness social media to investigate viral tweets
and trending hashtags to create early-warning solutions that can signal if a viral
hashtag started emerging in its nascent stages. In this paper, we utilize the 68-
95-99.7 rule to define a simple method of hashtag volume breakouts. In statistics,
the 68-95-99.7 rule, also known as the three-sigma rule or empirical rule, states
that nearly all values lie within three standard deviations (σ) of the mean (μ) in
a normal distribution. We utilize a fixed sized sliding window (of length 20 daily

© Springer International Publishing Switzerland 2015
N. Agarwal et al. (Eds.): SBP 2015, LNCS 9021, pp. 3–12, 2015.
DOI: 10.1007/978-3-319-16268-3_1

intervals), to compute a running average and standard deviation for each hashtag's volume distribution. Then, we identify non-overlapping *episodes* within a time-series of daily volumes for each hashtag whenever its daily volume exceeds $(\mu + 1\sigma)$ of the previous 20 day periods. We label the 20 day periods preceeding an episode as the *accumulation period* of an episode. We categorize an episode as *breaking* if the hashtag volume goes on to exceed $(\mu + 2\sigma)$ without falling below max(0, μ - 2σ), or else as a *non-breaking* episode otherwise. Next, we examine multiple network metrics associated with the accumulation period of each episode and proceed to build a classifier that aims to predict whether an episode will lead to a breakout volume or not. We employ a network based classification model and to discover latent patterns for the breakout phenomena, particularly we examine which factors contribute to make hashtag volumes breakout. We also build a visualization tool called Trending Hashtag Forecaster (THF). Our THF tool helps reveal the underlying network structures, patterns and properties that lead to breakout volumes. Our experiments showed that "local" network features during an accumulation period have an overall predictive accuracy of 76%, where as, when we incorporate "global" features that utilize measures extracted from all of the network up to the current accumulation period, then the overall predictive accuracy of the Trending Hashtag Forecaster jumps to 83%.

2   Problem Formulation

Given a set of tweets $T = t_1, t_2, t_3, ..., t_n$ where n is number of tweets in our corpus. These tweets comprise textual contents, user interactions and additional meta data. We explore and analyze both textual contents filtered by a given hashtag from hashtags set H. Then we denote tweet volume as number of tweets per day. We then compute daily means $(\mu(20))$ and standard deviation $(\sigma(20))$ for each hashtag by utilizing its volume distribution during its previous 20 days window. We experimentally determined the best window size by experimenting 10, 15, 20, 25 and 30 days windows. The 20 days window shows the best performance amongst the others.

If the hashtag frequency rises above $(\mu(20) +1\sigma(20))$, then we label that period as an episode, and we mark its previous 20 days as the accumulation period of an episode. We start observing hashtag frequency for two possible outcomes:

- a breakout if hashtag volume rises above$(\mu(20) +2\sigma(20))$, without falling below max(0, $\mu(20)$ - $2\sigma(20)$), or
- non-breakout, if hashtag volume falls below max(0, $\mu(20)$ - $2\sigma(20)$), without rising above $(\mu(20) +2\sigma(20))$

In breakout scenario for an episode no further overlapping breakouts are allowed until its volume falls below max(0, $\mu(20)$ - $2\sigma(20)$). In both scenarios, as episode begins with its accumulation period and continues until the hashtag volume dies out (i.e. it falls below max(0, $\mu(20)$ - $2\sigma(20)$)). Figure 1, shows the histograms of all daily hashtag volumes in our corpus.

Next, in Section 3 we present related work. In Section 4 we describe out Tweet corpus. In Section 5, we describe our Trending Hashtags Forecaster visualization tool. In Section 6, we introduce our network based model, local and global network features to predict hashtag episode breakouts following accumulation periods. In Section 7, we present experimental results and findings. Section 8 concludes the paper and presents the future work.

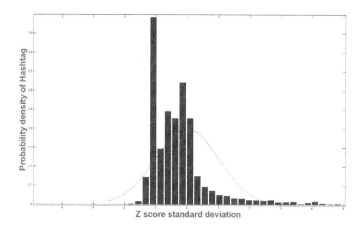

Fig. 1. Probability distribution function of all Hashtags

3   Related Work

Twitter network has more than 271 million monthly active members and 500 million tweets are generated daily [1]. The vast size and reach of Twitter enables examination of potential factors that might be correlated with breakout events and viral diffusion. We found that diffusion related studies fall into two categories. In the first category, many studies start by analyzing social networks as a graph of connected interacting nodes i.e. between users, friends or followers, and these studies investigate different factors that drive propagation and diffusion of information Arruda et al. [7] proposes that network metrics play an important role in identifying influential spreaders. They examined the role of nine centrality measures on a pair of epidemics models (i.e. disease spread on SIR model and spreading rumors on a social network). According to the authors, epidemic networks are different from social networks such that infected individuals in SIR become recovered by a probability μ while in social networks a spreader of a rumor becomes a carrier by contacts. They found centrality measures such as closeness and average neighborhood degree are strongly correlated with the outcome of spreading rumors model.

---

[1] https://about.twitter.com/company

The second category looked into the diffusion problem through content analysis by incorporating different natural language processing techniques. For instance, one study hypothesized that a specific group of words is more likely to be contained in viral tweets. Li et al. analyzed tweets in terms of emotional divergence aspects (or sentiment analysis) and they noted that highly interactive tweets tend to contain more negative emotions than other tweets [1], [8].

Weng et al. [5] investigated the prediction of viral hashtags by first defining a threshold for a hashtag to be viral, and then by examining metrics and patterns related to the community structure. They achieved a precision of 72% when threshold is set statically to 70. Romero et al studied the diffusion of information on Twitter and presented some sociological patterns that make some types of political hashtags spread more than others. Asur [11] presented factors that hinder and boost trends of topics on Twitter. They found content related to mainstream media sources tends to be main driver for trends. Trending topics are further spread by propagators who re-tweet central and influential individuals.

We propose a model that predicts hashtag breakouts thru adaptive dynamic thresholds, and by utilizing generic content-independent network measures that draws their information from (i) local networks corresponding to accumulation periods, as well as (2) from the global networks corresponding to the entire network history preceeding an accumulation period.Our experiments showed that local network features yield an overall predictive accuracy of 76%, and, global network features yield an overall predictive accuracy of 83%.

4   Data Source

The dataset we are using in this study is a collection of tweets from UK region. These tweets have been crawled based on a set of keywords with the aim to capture political groups, events, and trends in the UK. The dataset consists of more than 3 million tweets, 600K users, with more than 5.2 million interactions (both mentioning and retweeting) between users along with 1,334 hashtags.

5   Visualization Tool: Trending Hashtags Forecaster

In order to visualize and understand breaking hashtag phenomena, we built a visualization tool, depicted in Figure 2, that facilitate exploring temporal dynamics of hashtags and their underlying networks during accumulation period of each episode. Local and global network measures.are also computed and displayed as network and node features. These network measures are utilized to train and test a predictive classifier, presented in the next section.

6   Methodology

In this study, we crawled tweets containing hashtags (case insensitive) which related to political groups in UK from June, 2013 to July, 2014. After crawling,

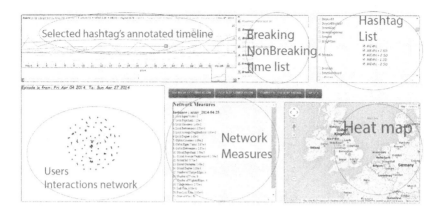

Fig. 2. THF visualization tool

we detected hashtag episodes using techniques described in Section 2. We identified the accumulation period and accumulation network of each episode, and extracted network measures corresponding to its accumulation network. Each eposide was also labeled as breaking or non-breaking based on its spread.

THF visualization tool reveals some of the discriminative patterns between breaking and non-breaking hashtags. Figure 3 shows the user interaction network for a non-breaking hashtag. User interaction network denoted by number 1 was captured during its accumulation period. Later on, this Hashtag did not breakout (i.e. did not cross its $\mu(20) + 2\sigma(20)$, but it fall back to zero volume, hence considered as a non-breaking episode. Figure 4, illustrates a breakout hashtag. Following a 20 period accumulation period, its volume exceeds $\mu(20) + 1\sigma(20)$ (denoted by network number 1), and it's volume exceeds breakout levels (by exceeding it's $\mu(20) + 2\sigma(20)$) threshold (denoted by network number 2). Network 3 shows the entire reach this episode before it's demise (i.e. by falling below $\max(0, \mu(20) - 2\sigma(20))$. An interesting observation related in the network 1 is a highly central green node, which attracts many new re-tweeters in network 2 and network 3. This observation indicates that existence of a large number of highly central nodes during the accumulation phase of an episode could be a good predictor for a following breakout. Other instances' patterns could not be cached by naked eye, yet they carry latent centrality measures correlate with our definition.

6.1   Network Based Model

In this model we investigate how users get involved in a hashtag h by mentioning, replying or retweeting. Their interactions are depicted as a directed graph G_{h_i}. We then incorporated normalized size-independent network features for directed graphs corresponding to accumulation periods of episodes. The network graph is a pair $G = (V, E)$ where V is set of vertices representing users together with a set of edges E, representing interactions between users. For instance, if a user

8   S. Alzahrani et al.

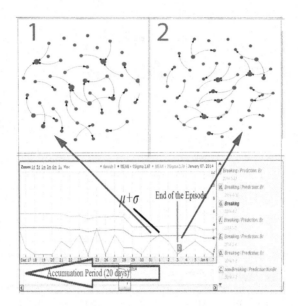

Fig. 3. Non breaking #Dawah Hashtag episode

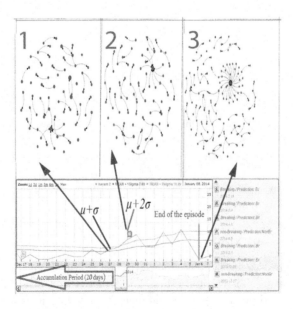

Fig. 4. Breaking #haram Hashtag episode

$u1$ mentioned, replied, or retweeted one tweet of $u2$, then a directed edge from $u1$ to $u2$ is formed.

We attempted to identify key features that contribute to the network based classification problem for breaking or non-breaking hashtags. Table 1 list all features that we used for local and global measures. Local measures are associated with user interactions during the accumulation period only, where as global measures draws their information from all interactions beginning from the start date (June 2013) until the end date of any accumulation period under consideration.

Table 1. Feature description

Feature	Description
Eigen Vector Centrality	Node's centrality depends on its neighbors centralities. If your neighbor are important you most likely are important too.
Page Rank	IVariant of Eigenvector where a node don't pass its entire centrality to its neighbors. Instead, its centrality divided into the neighbors. [3]
Closeness Centrality	A node is considered important if it is relatively close to all other nodes in the network [2].
Betweeness centrality	Measuring the importance of a node in connecting other parts of the graph [6]. This measure possesses the highest space and time complexity.
Degree centrality	It measures the number of ties a node has in undirected graph.
Indegree Centrality	It measures number of edges pointing into a node in a directed graph.
Outdegree Centrality	It is similar to the two above measure but it concerns on the number of outgoing links from a user, and it is normalized for each node.
Link Rate	Number of URLs in the tweets during the accumulation period divided by number of tweets.
Distinct Link Rate	Similar to link rate but without considering similar URLs.
Number of uninfected neighbors of early adopters	It is total number of retweets or mentioned (edges) a user has ever received globally, normalized by max-min retweets within local network in a current period being measured. [5]
Neighborhood average degree	it measures the average degree of the neighborhood of each node. [4]

7 Experiment Results and Findings

As a preprocessing step, We had 2790 for the non break out instances, while 1331 were for the break out. We sampled (without replacement) instances from both classes with oversampling for the lower represented class. We next examined the correlation between features and breaking hashtags using Principle Component Analysis (PCA). PCA is a dimensionality reduction approach that analyzes dataset to find which features give highest variance among instances and it maps the given features into lesser number of factors called components [9]. After that, in order to predict whether a given hashtag will breakout or not, we run a supervised network based learning model.

7.1 Features Correlated with Breaking Hashtags

PCA identified nine factors shown in Table 1. According to Kaiser Criterion [10], the factors to consider are the ones with eigenvalue above 1. In this study, we will focus on the first two components since they reveal interesting insights. Table 2 shows the correlation between our features and the first two components shown in Table 1. The first component is strongly correlated (negatively) with global measures, where as the second component is strongly correlated (negatively) with local measures. These two components give us a hint that global features should be grouped together and they contribute heavily (36%) to the variation in our dataset. Also, some of the local measures are also grouped together in a single factor and they somewhat contribute (21%) to the variation in our dataset.

Table 2. PCA components

Component	Eigenvalue	Variance	Cumulative Variance
1	5.79	36.16	36.17
2	3.30	20.62	56.78
3	1.669	10.43	67.21
4	1.24	7.73	74.94
5	1.01	6.31	81.24
6	0.88	5.54	86.78
7	0.663	4.14	90.92
8	0.48	3.01	93.93
9	0.42	2.65	96.57

Table 3. Correlation Between Table and Components

Feature	Component 1	Component 2	Feature	Component 1	Component 2
PageRank Local	0.14	-0.45	PageRank Global	-0.36	-0.10
Closeness Local	0.05	-0.51	Closeness Global	-0.24	0.09
Betweeness Local	0.05	-0.44	Betweeness Global	-0.35	-0.07
Avg Neighbor Degree Local	-0.11	-0.04	Avg Neighbor Degree Global	-0.3552	-0.10
Degree Cent. Local	0.11	-0.49	Degree Global	-0.3897	-0.0554
Uninfected Neighbor	0.19	0.02	In Degree Global	-0.39	-0.09
Link Rate	-0.03	0.14	Distinct Link Rate	0.0282	0.1889
Outdegree Global	-0.21	0.02	-	-	-

7.2   Network Based Model

For this model, we measured two sets of features: local and global. For local features: we have eigenvector, pagerank, closeness, betweeness, average neighborhood degree, uninfected neighbors before break out, and degree centrality. For global features we have the previous features measured globally plus in degree, out degree, and link rate. Next, we train and test a random forest classifier with 10 fold cross-validation using three approaches: prediction using all features shown in Table 3, prediction using global features that are correlated with the first factor identified by PCA shown in Table 4, and prediction using local features that are correlated with the second factor returned by PCA shown in Table 5. We achieve the highest precision of 84%, recall of 81% and F-measure of 82% for breakout prediction with the global features. We also achieve the highest precision of 82%, recall of 85% and F-measure of 84% for non-breakout prediction with the global features. On the other hand, local features archive overall lower precision and recall of roughly 76%. These findings suggest that global measures outperform local measures in predictive accuracy.

Table 4. Break out results

NETWORK	TP	FP	PRECISION	RECALL	F-MEASURE
LOCAL	0.73	0.2	0.77	0.73	0.75
GLOBAL	**0.81**	**0.15**	**0.84**	**0.81**	**0.82**
ALL FEATURES	0.8	0.16	0.83	0.8	0.81

Table 5. Non break out results

NETWORK	TP	FP	PRECISION	RECALL	F-MEASURE
LOCAL	0.79	0.27	0.75	0.79	0.77
GLOBAL	**0.85**	**0.19**	**0.82**	**0.85**	**0.84**
ALL FEATURES	0.84	0.20	0.81	0.84	0.83

8   Conclusion and Future Work

In this paper, we develop a model for predicting breaking hashtags using a content independent network model comprising both local and global network features drawn from an indicative accumulation period of hashtag volumes. For the network model, we measured and experimented with the predictive accuracies of global and local features. We also examined their importance and rankings using PCA. Global features drawn for the accumulation period network showed higher predictive accuracy compared to the local features. Network based model with global centralities for the accumulation period network can be used as a general framework to predict breaking hashtags with an overall accuracy of 82%.As future work, we propose to study the utility of content based features such as sentiment analysis, and different types of sources.

Acknowledgments. This research was supported by US DoD ONR grant N00014-14-1-0477 and USAF AFOSR grant FA9550-15-1-0004.

References

1. Li, C., Sun, A., Datta, A.: Twevent: segment-based event detection from tweets. In: Proceedings of the 21st ACM International Conference on Information and Knowledge Management, pp. 155–164. ACM (2012)
2. Newman, M.E.J. A measure of betweenness centrality based on random walks. Social networks 27.1, 39–54 (2005)
3. Brin, S., Page, L.: The anatomy of a large-scale hypertextual Web search engine. Computer Networks and ISDN Systems **30**, 107–117 (1998). doi:10.1.16/S0169-7552(98)00110-X
4. Barrat, A., Barthelemy, M., Pastor-Satorras, R., Vespignani, A.: The architecture of complex weighted networks. In: Proceedings of the National Academy of Sciences of the United States of America 101.11, PP. 3747–3752 (2004)
5. Weng, L., Menczer, F., Ahn, Y.-Y.: Virality prediction and community structure in social networks. Scientific reports 3 (2013)
6. Freeman, L.C.: A set of measures of centrality based on betweenness.Sociometry, 35–41 (1977)
7. Arruda, G., Barbieri, A., Rodrigues, F., Moreno, Y., Costa, L.: The role of centrality for the identification of influential spreaders in complex networks. Physical Review E **90**, 032812 (2014)
8. Cheng, J., Adamic, L., Dow, P.A., Kleinberg, J.M., Leskovec, J.: Can cascades be predicted?. In: Proceedings of the 23rd International Conference on World Wide Web, pp. 925–936. International World Wide Web Conferences Steering Committee (2014)
9. Pearson, K.: LIII. On lines and planes of closest fit to systems of points in space. The London, Edinburgh, and Dublin Philosophical Magazine and Journal of Science **2**(11), 559–572 (1901)
10. Bandalos, D.L., Boehm-Kaufman, M.R.: Four common misconceptions in exploratory factor analysis. Statistical and methodological myths and urban legends: Doctrine, verity and fable in the organizational and social sciences, 61–87 (2009)
11. Asur, S., et al.: Trends in social media: persistence and decay. ICWSM (2011)

Temporal Causality of Social Support in an Online Community for Cancer Survivors

Ngot Bui[✉], John Yen, and Vasant Honavar

College of Information Sciences and Technology, The Pennsylvania State University,
University Park, PA 16802, USA
{npb123,jyen,vhonavar}@ist.psu.edu

Abstract. Online health communities (OHCs) constitute a useful source of information and social support for patients. American Cancer Society's Cancer Survivor Network (CSN), a 173,000-member community, is the largest online network for cancer patients, survivors, and caregivers. A discussion thread in CSN is often initiated by a cancer survivor seeking support from other members of CSN. It captures a multi-party conversation that often serves the function of providing social support e.g., by bringing about a change of sentiment from negative to positive on the part of the thread originator. While previous studies regarding cancer survivors have shown that members of OHC derive benefits from their participation in such communities, causal accounts of the factors that contribute to the observed benefits have been lacking. This paper reports results of a study that seeks to address this gap by discovering temporal causality of the dynamics of sentiment change (on the part of the thread originators) in CSN. The resulting accounts offer new insights that the designers, managers and moderators of an online community such as CSN can utilize to facilitate and enhance the interactions so as to better meet the social support needs of the community participants. The proposed methodology also has broad applications in the discovery of temporal causality from big data.

1   Introduction

World Health Organization [1], estimated that 14.1 million new cancer cases and 8.2 million cancer-related deaths occurred worldwide in 2012. In 2014, the number of deaths due to cancer in the US was estimated to be in excess of 0.58 million and the number of new cancer cases diagnosed was estimated to be 1.66 million [2]. According to National Cancer Institute, approximately 13.7 million Americans with a history of cancer were alive on January 1, 2012 [2]. Some of these individuals were cancer free, while others still showed cancer symptoms and may have been under treatment [2].

About 72% of Internet users in the U.S. utilize the Internet for health-related purposes and 26% have read or watched someone else's experience about health or medical issues during the previous year [3]. Hence, many cancer survivors rely on an online health community (OHC) for social support. Social support

© Springer International Publishing Switzerland 2015
N. Agarwal et al. (Eds.): SBP 2015, LNCS 9021, pp. 13–23, 2015.
DOI: 10.1007/978-3-319-16268-3_2

can help cancer survivors cope better with their condition and hence improve the quality of their lives [4]. A cancer OHC that includes both survivors and their caregivers provides a forum to share experiences about their cancer, cancer treatment, and daily living issues. Through such online interaction, they support one another in ways that family members, friends or even health care providers often cannot [5]. Several studies have documented the benefits derived by cancer survivors through participation in an OHC. The benefits of OHC participation include increased social support [4,6], reduced levels of depression, stress, and psychological trauma [7], increased optimism about their life with cancer [6], increased ability to cope with their disease, and improvements on the physical and the mental aspects of their lives [4,8]. Several studies have attempted to quantify the benefits of OHCs using sentiment analysis [9,10]. Sentiment analysis and topic modeling were applied to CSN breast and colorectal cancer forums to investigate how change in sentiment of thread originators (persons who start the thread) varies by discussion topic [9]. Furthermore, sentiment analysis has been used to classify posts in an online health forum to determine whether a thread in the forum could benefit from the moderator's intervention [11]. Sentiment classification of user posts in an online cancer support community [12] has been used to determine the utility of social support, and information support to patients [13].

However, none of the preceding studies have provided a causal account of sentiment change on the part of OHC participants. Such an account could help the designers, managers and moderators of an online community such as CSN to facilitate and enhance the interactions so as to better meet the social support needs of the community participants. Against this background, we introduce a novel approach to uncover the temporal causality of the dynamics of sentiment change (on the part of the thread originators) in OHCs. The proposed approach leverages the machinery of temporal causality introduced in [14,15] to analyze the temporal ordering of sentiments of participants (i.e., thread originator and repliers) in a thread from CSN forum for the causes of the final sentiments (either positive or negative) of the thread originator.[1]

The rest of the paper is organized as follows. Section 2 introduces the key notions of temporal causality and the machinery for reasoning about causes of events. Section 3 describes our approach in temporal causality analyzing in CSN. Section 4 presents results of experiments that demonstrate the utility of the proposed approach. Section 5 concludes with a summary and a discussion and an outline of some promising directions for further research.

2 Temporal Causality

We start by reviewing a few key notions [17]: An event B is said to be a *prima facie* or potential cause of an event A iff (i) B precedes A; (ii) the probability

---

[1] Unlike the approach introduced in [16], this approach explicitly captures the temporality of the relationship between cause and effect. In addition to being able to represent properties being true for durations of time, this also allows a direct representation of the time window between cause and effect.

of B, $P(B) > 0$; and (iii) the conditional probability $P(A|B) > P(A)$. We say that B is a spurious cause of A iff B is a prima facie cause of A, and there is an event C that precedes B such that (i) $P(B,C) > 0$; (ii) $P(A|B,C) = P(A|C)$; and (iii) $P(A|B,C) \geq P(A|B)$. That is, C occurs before B and accounts for the effect A as well as B does. For example, assume that smoking and yellow finger precede the development of lung cancer and both appear to be the causes to lung cancer. However, yellow finger and lung cancer have the same common cause (i.e., smoking). A prima facie cause that is not a spurious cause is said to be a genuine cause [17]. Suppes [17] offers a method for testing whether a cause is spurious in the restricted setting where there are only two possible causes. Kleinberg [14] argued for a more stringent criterion, for a prima facie cause to be considered a genuine cause and introduced a method for assessing the causal significance of a potential cause of an effect which can be used to identify a genuine cause of an event from among a set of its potential causes.

Kleinberg's framework uses temporal logic [18] to represent and reason about events that occur in time. Temporal logic is a variant of propositional modal logic that admits the truth value of a formula, constructed from atomic propositions (sentences which are either true or false and encoded by propositional symbols) using logical connectives, (i.e., conjunction, disjunction, and negation) to be time dependent. Hence, temporal logic can be used to represent whether a property is true at some specific time. Computation Tree Logic (CTL) [19], a branching-time logic, can be used to represent the fact that a property will be true at some time in the future (e.g., at some point in the future, the train will arrive). Probabilistic Computation Tree Logic (PCTL) [20] extends CTL by specifying deadlines (requiring a property to hold before a specified window of time elapses) and quantifying the transition probability between the states in CTL. Probabilistic Kripke structures [19] can be used to represent and reason in PCTL.

Definition 1. *Probabilistic Kripke structure Let AP be a set of atomic propositions, a probabilistic Kripke structure is a four tuple: $K = \langle S, s^i, L, \mathcal{T} \rangle$, where S is a finite set of states; $s^i \in S$ is an initial state; $L : S \to 2^{AP}$ is a labeling function assigning subsets of AP to states; and $\mathcal{T} : S \times S \to [0,1]$ is a transition probability function such that $\forall s \in S : \sum_{s' \in S} \mathcal{T}\left(s, s'\right) = 1$.*

To represent the time and probability in PCTL formulas, Hansson et. al. [20] provide three types of *modal operators* which are path operators: A ("for all path") and E ("for some future path"), temporal operators: G ("holds for entire future path") and F ("finally holds"), and the "lead-to" operator. The lead-to operator, which is useful in formalizing temporal priority for causality, is defined as $f \rightsquigarrow_{\geq p}^{\leq t} g = AG\left[f \to F_{\geq p}^{\leq t}g\right]$ which means that whenever f holds there is a probability of at least p that g will hold via some series of transitions taking less than or equal to t time units. Some scenarios require the specification of a lower bound of time for g to hold. In this case, the lead-to operator can be constructed as $f \rightsquigarrow_{\geq p}^{\geq r, \leq s} g$, which denotes that g must hold in between r and s time units with probability p where $0 \leq r \leq s \leq \infty$ and $s \neq \infty$.

Armed with the machinery of PCTL, we can define a *prima facie* (or a potential) cause as follows [21]:

Definition 2. *Prima Facie Cause in PCTL* *c is a prima facie cause of e if there is a p such that all three following conditions hold. (i)* $F_{>0}^{\leq\infty}c$; *(ii)* $c \leadsto_{\geq p}^{\geq 1, \leq\infty} e$, *and (iii)* $F_{<p}^{\leq\infty}e$ *where c and e are PCTL formulas.*

For c to be a prima facie cause of an effect e, the state where c is true should be reachable with non-zero probability, and the probability of reaching a state where e is true from a state where c is true should be greater than the probability of reaching a state where e is true from the initial state of the system. This can be interpreted as requiring that c must occur at least once, and that the conditional probability of e given c is greater than the marginal probability of e.

We adopt the technique from [20] to calculate the probabilities $F^{\leq\infty}c$, $c \leadsto^{\geq 1, \leq\infty} e$, and $F^{\leq\infty}e$ where $F^{\leq\infty}e$ denotes the path probabilities summed over the set of all paths starting from the initial state of the K structure and ending at a state where e is true (means that probability of e regardless of c); $c \leadsto^{\geq 1, \leq\infty} e$ denotes the path probabilities summed over the set of all paths starting from the state where c is true and ending at a state where e is true (i.e., the probability of e given c).

3   Methodology

3.1   Learning Sentiment Classifier for Posts

Since we cannot detect the sentiment of a CSN member directly, we use the sentiment expressed in a post as a proxy for the sentiment of the CSN member at the time the post was created. As in [22], we categorize the posts as expressing either a positive sentiment or a negative sentiment.

Table 1. Features of a Post

Feature Name	Description
PosLength	The number of words in the post
PosStrength	Positive Sentiment Strength of the post
NegStrength	Negative Sentiment Strength of the post
Neg	NumberOfNeg / PostLength, where NumberOfNeg is equal to number of negative words and emotions
PosVsNeg	(NumberOfPos + 1) / (NumberOfNeg + 1), where NumberOfPos is equal to number of positive words and emotions
Name	NumberOfName / PostLength, where NumberOfName is the number of names mentioned in the post
Slang	Number of Internet slang words in the post

We used a set of 298 posts which are manually classified by two independent annotators into one of two categories: positive or negative. As in [10], we extract seven features (see Table 1) from a post to train a predictor for assigning posts to the positive or negative category. SentiStrength [22] is used to extract *PosStrength* and *NegStrength* which represent for the positive sentiment strength

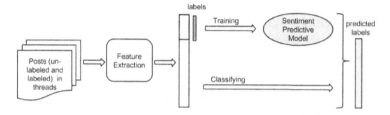

Fig. 1. Learning Sentiment Predictive Model

and negative sentiment strength of the post, respectively. We make use of the four lists[2], a list of positive and negative words [23], a list positive and negative emotions[3], a list of popular English female and male names, and a list of Internet slang words to calculate the *Neg*, *PosVsNeg*, *Name*, and *Slang* features. We use Adaboost, which has been shown to generate the best sentiment classification model [10], to classify the posts. The Adaboost sentiment classifier outputs a probability that a post expresses a positive sentiment, i.e., $\Pr(positive \,|\, post)$. If $\Pr(positive \,|\, post) > 0.5$, the post is classified as positive; otherwise, it is classified as negative. Figure 1 shows the procedure for training the post sentiment predictor and using it to classify new posts.

3.2   Cancer Survivor Network Thread as a Sequence of Sentiments

Cancer Survivor Network of American Cancer Society is an OHC with over 173,000 registered members which include cancer patients, their friends and families, and informal caregivers. In this paper, we use the CSN data set that contains all threads initiated between July 2000 and October 2010. The data set contains 48,779 discussion threads and more than 468,000 posts from 27,173 users. The data set is appropriately anonymized to protect the privacy of the CSN members.

   Our goal is to uncover the causal effect (if any) of the temporally ordered posts that make up the thread on the final sentiment of the thread originators. More specifically, we are interested in discovering causal relationship between the reply posts and the change of sentiment of those who initiate the thread. Therefore, threads used in this study need to have at least one reply and at least one self-reply (i.e., a post by the thread originator later on thread). As a result, threads that don't contain a self-reply or reply are removed from this study. The resulting data set consists of 22,854 threads.

   A thread can be represented as a temporally ordered sequence of posts $P_{o1}, P_{r1}, P_{r2}, \cdots, P_{o2}, \cdots, P_{rm}, P_{on}$ where P_{o1} is the initial post from the thread originator; P_{oi} $(i > 1)$ are self-replies; and P_{rj} are replies to the post by individuals other than the thread originator. Since we focus on the communication between two kinds of actors in a thread, the thread originator and the individuals (other than the originator) who respond to the originator's post, we

---

[2] http://sites.google.com/site/qiubaojun/psu-sentiment.zip
[3] http://en.wikipedia.org/wiki/List_of_emotions

simply compute the average probability of positive sentiment of replies between two consecutive self-replies and use it as the positive sentiment probability of the collection of replies. Formally, the average positive sentiment probability is calculated as

$$\bar{p}_{ri} = \frac{\sum \Pr\left(positive \mid Prj\right)}{|P_{rj}|}, \forall j : t_{oi} \leq t_{rj} \leq t_{o(i+1)} \tag{1}$$

where t_{oi}, $t_{o(i+1)}$, and t_{rj} are time points when posts P_{oi} and $P_{o(i+1)}$ and P_{rj} are created, respectively and $|P_{rj}|$ is the number of reply posts from t_{oi} to $t_{o(i+1)}$.

To establish a formal representation that enables temporal causality analysis, we transform the sequence of post sentiment probability in a thread to a sequence of post sentiment states as follows: [Sentiment state of initial post] [Average sentiment state of reply posts] ([Sentiment state of intermediate self-reply] [Average sentiment state of reply posts])* [Sentiment state of final self-reply] where average sentiment state of reply posts is obtained from the average sentiment probability defined in equation (1) using the threshold of sentiment state classifier described in section 3.1 (i.e., 0.5). More precisely, each sentiment state can take one of two values: positive or negative. Let b, o, r, s, and f be atomic propositions. Let b denote the beginning of a thread, o denote that the initial post sentiment is positive, r denote that the average sentiment of reply posts elicited by the initial post is positive, s denote that the sentiment of an intermediate self-reply to the initial post is positive, f denote that the sentiment of the final self-reply is positive. A thread can be represented by a sentiment state sequence $\mathbf{x} = x_0 x_1 \cdots x_n$ where $x_i \in \mathcal{X} = \{o, \neg o, r, \neg r, s, \neg s, f, \neg f\}$ where $\neg$ denotes negation of the corresponding proposition.

3.3   Probabilistic Kripke Structure in CSN

We use a probabilistic Kripke structure [19] to represent and reason about probabilistic transitions between the sentiment states of posts in a CSN. Let $\mathbf{x} = x_0 x_1 \cdots x_n$ be a sequence of post sentiments where $x_i \in \mathcal{X}$ and let X_i ($0 \leq i \leq n$) be random variable corresponding to sequence element x_i. Markov Model (MM), which captures the dependencies between the neighboring sequence elements, is used to estimate the transition probabilities between sentiment states of posts that make up a thread. In k^{th} order Markov model, the sequence element follows the Markov property: $X_i \perp\!\!\!\perp \{X_0, \cdots, X_{i-k-1}\} \mid \{X_{i-k}, \cdots, X_{i-1}\}$(i.e., X_i is conditionally independent of $X_0, \cdots, X_{i-k-1}$ given $X_{i-k}, \cdots, X_{i-1}$ for $i = k, \cdots, n$). Formally, the transition probabilities are estimated[4] over a set $\mathcal{D} = \{\mathbf{x}_l\}_{l=1}^{|\mathcal{D}|}$ of sentiment sequences as follows.

$$\hat{p}\left(X_i \mid w\right) = \left[\frac{1 + \sum_{l=1}^{|\mathcal{D}|} \#\left[w\sigma, \mathbf{x}_l\right]}{|\mathcal{X}| + \sum_{\sigma' \in \mathcal{X}} \sum_{l=1}^{|\mathcal{D}|} \#\left[w\sigma', \mathbf{x}_l\right]} \right]_{\sigma \in \mathcal{X}} \tag{2}$$

where $\#\left[w\sigma, \mathbf{x}_l\right]$ represents the number of times the symbol σ "follows" the subsequence w (of length k) in sequence $\mathbf{x}_l$ and $\hat{p}\left(X_i \mid w\right)$ is the estimate of the

---

[4] Laplace correction is used for smoothing purposes

conditional probability $P(X_i = \sigma|w)$ of sequence element X_i that "follows" the subsequence w. We use the first-order MM to determine the transition probabilities for the CSN probabilistic Kripke structure.

4 Temporal Causality in Cancer Survivor Network

4.1 Prima Facie Cause

Figure 2 shows the Probabilistic Kripke structure (K) that is constructed using the method described in previous section. The structure shows that from any state of the thread originator, i.e., $\{o, \neg o, s, \neg s\}$, there is a probability greater than 74% that it will transit to the state r. This suggests that people who reply to thread originators have a high tendency to express positive sentiment regardless the sentiment of the thread originators at the beginning and in the middle of the thread. In other words, members of CSN try to offer positive social support to others who seek support. The thread originator starts with either initial positive or negative sentiment and after the interaction with other people in the CSN forum he/she might end up with either positive or negative sentiment. Table 2 shows that about 72% of thread originators with initial negative sentiment end up with positive sentiment, and only about 24% of thread originators with initial positive sentiment end up with negative sentiment at the end.

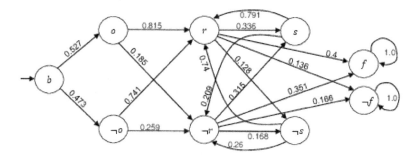

Fig. 2. Probabilistic Kripke Structure for CSN

Table 2. Sentiment Dynamics in CSN

#[o]	#[¬o]	#[¬o → f]	#[o → ¬f]
12147	10707	7512	2948

Our goal is to uncover the *prima facie* causes for final sentiment of the thread originators. Based on the definition of *prima facie* causes and the probabilistic Kripke structure K, we find that the set of prima facie causes of f and $\neg f$ are $\{r, s\}$ and $\{\neg o, \neg r, \neg s\}$, respectively.

CSN is comprised of users' data over a period of 11 years. CSN can be divided into several sub-communities (e.g., Breast Cancer, Colorectal Cancer).

Table 3. Yearly Prima Facie Causes of Final Thread Originator's Sentiment

Year	2000	2001	2002	2003	2004	2005	2006	2007	2008	2009	2010
f	o	o,r	r,s	r,s	r,s	r,s	r,s	r,s	r	r,s	r,s
$\neg f$	$\neg o$	$\neg o, \neg r, \neg s$	$\neg o, \neg r$	$\neg r, \neg s$	$\neg r, \neg s$	$\neg r, \neg s$	$\neg r, \neg s$	$\neg r, \neg s$	$\neg r$	$\neg r, \neg s$	$\neg r, \neg s$

Table 4. Communal Prima Facie Causes of Final Thread Originator's Sentiment

Community	Breast Cancer	Colorectal Cancer
f	r	r
$\neg f$	$\neg r$	$\neg r, \neg s$

Table 5. Causal Significance

$\varepsilon_{avg}(r, f)$	$\varepsilon_{avg}(s, f)$	$\varepsilon_{avg}(\neg o, \neg f)$	$\varepsilon_{avg}(\neg r, \neg f)$	$\varepsilon_{avg}(\neg s, \neg f)$
0.054	0.01	0.05	0.04	0.039

To validate the above prima facie causes, we divided CSN data set into several subsets based on the year and the sub-community.

Specifically, we group threads which were started in the same year (from 2000 to 2010) and we group threads that belong to either Breast Cancer or Colorectal Cancer community (during the period from 2005 to 2010). Surprisingly, r and $\neg r$ are consistently the prima facie causes of f and $\neg f$, respectively in both yearly and sub-community data sets. Table 3 and 4 show the prima facie causes of f and $\neg f$ in the yearly and sub-community analysis. *The results from the two tables indicate that the positive sentiment of the replies appears to causally influence the positive sentiment of the thread originator at the end of the thread; and conversely, the negative sentiment of the replies appear to causally influence the negative sentiment of the thread originators at the end of the thread.*

4.2   Assessing the significance of causes

We proceed to evaluate the significance of the prima facie causes of f and $\neg f$ using the method introduced in [21]. We calculate the significance of a cause c for an effect e as $\varepsilon_{avg}(c, e) = \frac{\sum_{x \in X \setminus c} \varepsilon_x(c,e)}{|X \setminus c|}$, where X is a set of prima facie causes of e and $\varepsilon_x(c, e) = P(e|c \wedge x) - P(e|\neg c \wedge x)$ denotes the contribution of c to the change in probability of e. Table 5 shows causal significance between causes and effects from an aggregate of data from all the years.

From the table 5, we can see that causal significance $\varepsilon_{avg}(r, f)$ is much higher than the causal significance $\varepsilon_{avg}(s, f)$ and whereas $\varepsilon_{avg}(\neg o, \neg f)$, $\varepsilon_{avg}(\neg r, \neg f)$, and $\varepsilon_{avg}(\neg s, \neg f)$ are not much different from each other.

In a similar fashion, we also examined the causal significance on data for specific years and sub-communities. Figure 3 shows the results of this analysis. Figure 3a shows that the causal significance $\varepsilon_{avg}(r, f)$ is significantly greater (pair t-test, $p < 0.01$) than the causal significance $\varepsilon_{avg}(s, f)$. However, figure 3b shows that the $\varepsilon_{avg}(\neg r, \neg f)$ and $\varepsilon_{avg}(\neg s, \neg f)$ are not significantly different

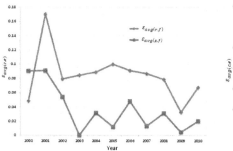 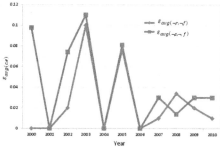

(a) Causal Significance of r and s on f **(b)** Causal Significance of $\neg r$ and $\neg s$ on $\neg f$

Fig. 3. Causal Significance According to Year

(figure 3b does not include the significance of $\neg o$ since it is not found to be a cause of $\neg f$ in most of the years (except during the first three years which account for less than 4% of the total number threads)). Our analysis of the data from the sub-communities yields a similar finding (i.e., $\varepsilon_{avg}(r, f)$ is significantly greater than $\varepsilon_{avg}(s, f)$ and $\varepsilon_{avg}(\neg r, \neg f)$ and $\varepsilon_{avg}(\neg s, \neg f)$ are not significantly different from each other).

Based on the results summarized in table 5 and figure 3, we can conclude that r causally influences f and $\{\neg r, \neg s\}$ causally affect $\neg f$. In other words, our key finding is that the positive sentiment of a reply causes the negative to positive change in the thread originator's sentiment, at least among 72% of the thread originators with initial negative sentiment. We also see that negative sentiment from a replier causes the thread originator to be left with a negative sentiment. Hence, we conclude that the sentiment of the replies drives the sentiment dynamics of the thread originator.

5   Summary and Future Work

In this work, we have introduced a framework to uncover the temporal causality of sentiment dynamics of the thread originator in the American Cancer Society's Cancer Survivor Network. To the best of our knowledge, this study is the first to uncover the factors that causally drive the sentiment dynamics in an OHC. We developed a sentiment classifier using machine learning on a training set of posts with manually labeled for their sentiment (positive versus negative). We constructed a Probabilistic Computation Tree Logic representation and a corresponding probabilistic Kripke structure to represent and reason about the transitions between sentiments of posts in a thread over time. We analyzed the Kripke structure to identify the prima facie causes of sentiment change on the part of the thread originators in the CSN forum and their significance. *Our main finding is that the positive sentiment of replies appears to causally influence the positive sentiment of the thread originator at the end of the thread; and conversely, the negative sentiment of the replies appears to causally influence the negative sentiment of the thread originators at the end of the thread.* Some promising directions for future research include: (i) exploring the causal effects of the topic being discussed on the sentiment dynamics; (ii) exploring the causal effects of (explicit as well as implicit) social relations among OHC participants on sentiment dynamics.

Acknowledgments. This research is supported by a Collaborative Agreement with American Cancer Society, which made the data of CSN available for this research. The work of Bui was supported in part by a research assistantship funded by the Edward Frymoyer Endowed Chair in Information Sciences and Technology held by Vasant Honavar. This work has benefited from discussions with Kenneth Portier, Greta Greer (both at American Cancer Society), Prasenjit Mitra (Qatar Computing Research Institute), Cornelia Caragea (University of North Texas), Kang Zhao (University of Iowa), David Reitter, and other current and former members of Cancer Informatics group at Penn State University.

References

1. Ferlay, J., et al.: Cancer incidence and mortality worldwide: sources, methods and major patterns in globocan 2012. International Journal of Cancer (2014)
2. American Cancer Society: American cancer society cancer facts and figures 2014 (2014). http://www.cancer.org/research/cancerfactsstatistics/cancerfactsfigures2014
3. Fox, S., Duggan, M.: Health online 2013. Pew Research Internet Report (2013)
4. Dunkel-Schetter, C.: Social support and cancer: Findings based on patient interviews and their implications. Journal of Social Issues **40**, 77–98 (1984)
5. Preece, J.: Empathic communities: balancing emotional and factual communication. Interacting with Computers **12**, 63–77 (1999)
6. Rodgers, S., Chen, Q.: Internet community group participation: Psychosocial benefits for women with breast cancer. Journal of Computer-Mediated Communication 10(4), July 2005
7. Beaudoin, C.E., Tao, C.C.: Modeling the impact of online cancer resources on supporters of cancer patients. New Media and Society **10**(2), 321–344 (2008)
8. Maloney-Krichmar, D., Preece, J.: A multilevel analysis of sociability, usability, and community dynamics in an online health community. ACM Trans. Comput.-Hum. Interact. 12(2), 201–232 (2005)
9. Portier, K., Greer, G.E., Rokach, L., Ofek, N., Wang, Y., Biyani, P., Yu, M., Banerjee, S., Zhao, K., Mitra, P., Yen, J.: Understanding topics and sentiment in an online cancer survivor community. Journal of the National Cancer Institute (JNCI) Monograms, 195–198 (2013)
10. Qiu, B., Zhao, K., Mitra, P., Wu, D., Caragea, C., Yen, J., Greer, G.E., Portier, K.: Get online support, feel better - sentiment analysis and dynamics in an online cancer survivor community. In: SocialCom/PASSAT, pp. 274–281 (2011)
11. Huh, J., Yetisgen-Yildiz, M., Pratt, W.: Text classification for assisting moderators in online health communities. J. of Biomedical Informatics **46**(6), 998–1005 (2013)
12. Biyani, P., Caragea, C., Mitra, P., Zhou, C., Yen, J., Greer, G.E., Portier, K.: Co-training over domain-independent and domain-dependent features for sentiment analysis of an online cancer support community. In: ASONAM 2013, pp. 413–417. ACM, New York (2013)

13. Wang, X., Zhao, K., Street, N.: Social Support and User Engagement in Online Health Communities. In: Zheng, X., Zeng, D., Chen, H., Zhang, Y., Xing, C., Neill, D.B. (eds.) ICSH 2014. LNCS, vol. 8549, pp. 97–110. Springer, Heidelberg (2014)
14. Kleinberg, S.: Causality, Probability, and Time. Cambridge University Press (2013)
15. Kleinberg, S., Hripcsak, G.: A review of causal inference for biomedical informatics. Journal of Biomedical Informatics, 1102–1112 (2011)
16. Pearl, J.: Causality: Models, Reasoning, and Inference. Cambridge University Press (2000)
17. Suppes, P.: A Probabilistic Theory of Causality. Noth-Holland, Amsterdam (1970)
18. Prior, A.: Past, Present, and Future. Clarendon Press, Oxford (1967)
19. Clarke Jr., E.M., Grumberg, O., Peled, D.A.: Model Checking. MIT Press, Cambridge (1999)
20. Hansson, H., Jonsson, B.: A logic for reasoning about time and reliability. Formal Aspects of Computing **6**, 102–111 (1994)
21. Kleinberg, S., Mishra, B.: The temporal logic of causal structures. In: Proceeding of the UAI 2009, Arlington, Virginia, United States, pp. 303–312. AUAI Press (2009)
22. Thelwall, M., Buckley, K., Paltoglou, G., Cai, D., Kappas, A.: Sentiment in short strength detection informal text. J. Am. Soc. Inf. Sci. Technol. **61**(12), 2544–2558 (2010)
23. Hu, M., Liu, B.: Mining and summarizing customer reviews. In: Proceeding of KDD 2004, pp. 168–177. ACM, New York (2004)

Are You Satisfied with Life?: Predicting Satisfaction with Life from Facebook

Susan Collins[1,2], Yizhou Sun[1], Michal Kosinski[3],
David Stillwell[3], and Natasha Markuzon[2(✉)]

[1] Northeastern University, CCIS, Boston, MA, USA
{skcoll,yzsun}@ccs.neu.edu
[2] Draper Laboratory, Boston, MA, USA
{skcollins,nmarkuzon}@draper.com
[3] Free School Lane, The Psychometrics Centre,
University of Cambridge, Cambridge CB2 3RQ, UK
{mk583,ds617}@cam.ac.uk

Abstract. Social media can be beneficial in detecting early signs of emotional difficulty. We utilized the Satisfaction with Life (SWL) index as a cognitive health measure and presented models to predict an individual's SWL. Our models considered ego, temporal, and link Facebook features collected through the myPersonality.org project. We demonstrated the strong correlation between Big 5 personality features and SWL, and we used this insight to build two-step Random Forest Regression models from ego features. As an intermediate step, the two-step model predicts Big 5 features that are later incorporated in the SWL prediction models. We showed that the two-step approach more accurately predicted SWL than one-step models. By incorporating temporal features we demonstrated that "mood swings" do not affect SWL prediction and confirmed SWL's high temporal consistency. Strong link features, such as the SWL of top friends or significant others, increased prediction accuracy. Our final model incorporated ego features, predicted personality features, and the SWL of strong links. The final model out-performed previous research on the same dataset by 45%.

Keywords: Social networking · Facebook · Satisfaction with life

1 Introduction

Have you ever Googled "happiness"? If you have, then you have noticed there are about 325 billion results and counting. This is not too surprising as most people consider happiness a desirable goal in life. Happiness has many interpretations; in this paper, we focus on satisfaction with life (SWL). SWL is a component of subjective well-being (SWB), defined by a cognitive judgmental process on how individuals evaluate their lives according to their personal criterion [8].

Since the 19th century SWL has been studied to identify and improve the quality of life for individuals and nations [21]. From an individualistic prospective, SWL has been used to gain a more robust understanding of mental illness

© Springer International Publishing Switzerland 2015
N. Agarwal et al. (Eds.): SBP 2015, LNCS 9021, pp. 24–33, 2015.
DOI: 10.1007/978-3-319-16268-3_3

by not only understanding the absence of a pathology but also the presence of happiness [9]. For instance, studies have shown that SWL can predict depression [15], occupational functioning [17], and successful interpersonal relationships [11]. From a community prospective, SWL is used to measure social progress and policy effects. In 2010, David Cameron, the prime minister of the United Kingdom, asked the Office of National Statistics to survey the nation for its life satisfaction as a part of a £2 million per year well-being project [13]. Clearly, the identification and understanding of SWL has risen to national and international attention making it a noteworthy pursuit. The question is, how can we accomplish this identification effectively and efficiently?

With the ubiquity of social media, research in data mining, natural language processing and other computational sciences has dramatically grown [16]. People are posting about their lives, family, and social interactions making sites like Twitter, Facebook, LinkedIn, etc. gold-mines for data. In other words, these users have already accomplished the tedious and resource consuming work of cataloging their interaction for us. The challenge is how to effectively transform this data into knowledge.

In our research, we developed models to predict an individual's SWL from Facebook features and identified indicators and their contributions toward predicting SWL. We took a novel approach by layering machine learning models and incorporating different types of features. We demonstrated the strong correlation between Big 5 personality features and SWL. Big 5 refers to the five broad dimensions of human personality that include: Openness, Conscientiousness, Extraversion, Agreeableness, and Neuroticism [5]. We predicted Big 5 features and used the predictions in more robust SWL models which incorporate static ego and link features. We showed that the two-step approach more accurately predicted SWL than one-step models, and outperformed previous research on the same data.

2   Related Work

In recent years, many studies have taken advantage of social media data to evaluate SWL and SWB. These studies have primarily considered ego variables (i.e. personal information such as gender, age, etc.) or link relationships (i.e. how users influence the ego) to predict SWL or SWB [2],[6],[14],[19].

In a Twitter study, researchers extracted topics and words from tweets to characterize and create a predictive model for SWL. Classic demographic features such as age, sex, and education combined with linguistic features, created the best model for prediction [19]. In another study, Facebook "likes" were used to predict a wide range of private traits and behaviors such as ethnicity, SWL, etc. [14]. This study's model accurately classified some attributes of a user, but less accurately predicted the numerical label of SWL (Pearson's Correlation R=0.17). Researches explained the less accurate results as SWL's variability caused by "mood swings" or quick changes in a user's mood.

In addition to static ego features, link features play a role in predicting SWL [2],[6]. In a recent study, Facebook researchers analyzed how emotions spread in a virtual environment. They observed that as positive posts from "friends" were reduced in news feeds, people posted fewer positive updates [6]. In another study of tweets, researchers determined if assortative mixing took place in online social networks. Assortative mixing is the tendency for individuals with similar characteristics to favor one another. The researchers concluded that Twitter is assortative, and relationships with more interconnected links were most influential [2].

In our proposed model, we combined previous SWL prediction information with a new layering technique. We incorporated linguistic features from Facebook updates to boost performance, considered temporal features to account for the "mood swing" of users, and incorporated link features by utilizing the SWL of friends as a feature.

3   Data Description

We used data collected by the myPersonality.org project [14], which contains psychometric test results and Facebook data used for social science research. The dataset contains 101,069 users with SWL scores. There are three feature types: (1) static ego, (2) temporal ego, and (3) link features described below:

3.1   Features and the Target Variable

Static Ego Features. Static ego features belong to a user but do not have timestamps associated. The following are static ego features included in the models:

- **Big 5:** The Big Five features refer to the five dimensions of human personality: Openness, Conscientiousness, Extraversion, Agreeableness, and Neuroticism. The Big 5 features values range of [1.0-5.0].
- **Age:** Reported age of a user.
- **Network Size:** Number of "friends" a user has in his network.
- **Number of Photo Tags:** Number of tagged photos of a user.
- **Relationship Status:** Categorical value representing a user's relationship status.
- **Likes:** Topical decomposition of users Like Data into 600 topics. Topics were extracted using Latent Dirichlet Allocation (LDA) [1].
- **Linguistic Inquiry Word Count (LIWC) Overall:** Linguistic Inquiry Word Count is a text analysis program that counts words into psychologically meaningful categories [20].

Temporal Ego Features. The Facebook status update feature contains temporal information. The status update is free text posted by a user. We used LIWC per post at different time frames for the temporal ego feature. We extracted features reflecting the mood of a user by calculating the LIWC of each update. Daily and weekly averages of LIWC were calculated for each user. We evaluated the potential for "mood swings" [14] by identifying how words used on a daily or weekly basis changed prediction accuracy of SWL. We defined "mood swings" as a change in word usage over time. We hypothesized that features collected closer to the SWL test would be more predictive than features collected further from the test.

Link Features. The third type of features is links associated with users, including friends and couples.

- **Friends:** We utilized the dyads table to calculate mutual friendships of users. We hypothesized that people who share a greater amount of friends were more likely to influence each other. The top 3 friends for each user were identified. Each friend's SWL score was used as a feature to determine whether a friend's "happiness" affects a user. Because not all friends had a true SWL score, we used predicted SWL from Big 5.
- **Couples:** The couples' table was utilized similarly to the friendship table; however, no calculation for rank was required. The SWL score of a significant other was used as a feature to identify how a significant others' "happiness" may influence the user. Because not all significant others had a true SWL score, we used predicted SWL from Big 5.

Target Variable: Satisfaction With Life (SWL). The SWL score was the target variable for this study. It was collected from Facebook with the Satisfaction with Life Scale – a 5 item long questionnaire designed to measure global cognitive judgments of satisfaction with one's life [8]. The SWL score is a numerical label ranging from [1.0-7.0] where 1.0 corresponds to highly unsatisfied individuals and 7.0 corresponds to highly satisfied individuals.

3.2   Sample Size

Models had variable sample sizes due to missing values for features. For example, of users with an SWL score, only 85% had Big 5 features and only 4% had LIWC. We calculated the sample size for any particular model by taking the intersection of users who contained all model's features. The sparsity in the features caused models to have drastically different sample sizes. To combat some of these small sample sizes, we chose static ego features with $n \geq 20,000$. The Static Ego models ranged from $n = 86,073$ when using Big 5 features to $n = 3,251$ when using LIWC features. Combined Static Ego models ranged from $n = 11360$ to $n = 1160$. Combined Link and Static Ego features had $n = 695$ when friends' SWL were used and $n = 171$ when a significant other's SWL was used.

4   Methods

We created data driven supervised learning methods to predict SWL from a set of features extracted from Facebook. In contrast to other models ([2],[6],[14],[19]), our models considered static ego features, temporal features, link features, and a combination approach for prediction. To reduce noise when combining high dimensional features, we employed a two-step approach of predicting Big 5 as an intermediate feature. We iteratively built prediction models starting with the most correlated features from the ego and expanded to other useful variables, such as link features.

4.1 Model Selection

Although past research [14],[19] predicting SWL used linear regression as a supervised learning model, we utilized Random Forest Regression (RFR) [3]. RFR was used for its interpretability, non-linear assumptions, efficiency, and accuracy. In this experiment, other methods such as linear regression and support vector regression were explored; however, they did not provide better prediction accuracy and afforded less interpretability than RFR. Mean square error was used as the splitting criterion [18].

Static Ego Models. We used static ego features from each dataset to train Random Forest Regression models to predict SWL. In a two-step approach, we developed models to predict Big 5 from multiple features. Predicted Big 5 scores were then incorporated as features in static ego models. The following summarizes the features for static ego models:

- **Big5:** Big 5 scores collected from a questionnaire [5].
- **FBAttrib:** Age, Network Size, Relationship Status, Number of Photo Tags
- **Likes:** "Likes" of a user as represented by 600-dimensional vector
- **LIWC:** Overall LIWC for a user represented by a 64-dimensional vector
- **Big5.FBAttrib:** Big 5 scores predicted by FBAttrib features
- **Big5.Likes:** Big 5 scores predicted by "Likes" of a user
- **Big5.LIWC:** Big 5 scores predicted from LIWC features

Combined Static Ego Models. We combined the best predictors from static ego features into one model to boost performance. When multiple Big 5 predictions were incorporated as features, we used the average for each Big 5 component as a feature. The following summarizes the features for combined static ego models:

- **Combo.Static.1:** FBAttrib features and the mean of Big5.FBAttrib, Big5. Likes, and Big5.LIWC features
- **Combo.Static.2:** FBAttrib features and the mean of Big5.Likes and Big5. LIWC features

Temporal Models. Temporal Models tested whether words expressed in Facebook statuses closer to the time of the SWL test had greater prediction accuracy than previous posts. We considered two granularities: daily and weekly statuses. The following summarizes the features for temporal models:

- **Temporal.1:** LIWC derived from Facebook statuses "n" days before SWL test, where n = [1-7]
- **Temporal.2:** LIWC derived from Facebook statues "n" weeks before SWL test, where n = [1-7]

Combined Static Ego and Link Models. Our final models merged Combo. Static.1 with two link features: top 3 friends' SWL and significant other's SWL. SWL scores of link features were predicted from the Big5 model. The following summarizes the features for combined static ego and link models:

- **FBAttrib.Big5.FriendSWL:** Combo.Static.1 features combined with top 3 friends' predicted SWL
- **FBAttrib.Big5.NoFriend:** Combo.Static.1 features of users with top 3 friends: We used this model as a Baseline to determine the lift of the top 3 friends' SWL.
- **FriendSWL:** Top 3 friends' predicted SWL: We use this model to determine the accuracy of the these features by themselves.
- **FBAttrib.Big5.OtherSWL:** Combo.Static.1 Model combined with the significant other's predicted SWL
- **FBAttrib.Big5.NoOther:** Combo.Static.1 features of users with a significant other: We used this model as a Baseline to determine the lift of a significant other's SWL.
- **OtherSWL:** Significant other's predicted SWL: We use this model to determine the accuracy of these features by themselves.

5   Experiment

5.1   Experimental Setting

To evaluate our models we used mean absolute error (MAE) measure [18]. MAE is defined as the average of the absolute errors over n samples, $e_i = |f_i - y_i|$, where f_i is the predicted value and y_i is the actual value.

$$MAE = \frac{1}{n}\sum_{i=1}^{n}|f_i - y_i| \qquad (1)$$

We evaluated our model by calculating MAE for SWL prediction and comparing it to the MAE of a random model generated from the probability distribution of a sample. The probability distribution function was estimated by interpolating over a 10-bin histogram of the labeled data. Because there are no other experiments that predict SWL from all chosen features, we found the random baseline as a naive but appropriate baseline. To make our models consistent with the qualitative interpretation of SWL scores [10], we consider models with average error rates ≤ 1.0 to be good models.

We also compared our model to a previously discussed model which used linear regression and user likes to predict SWL [14]. We replicated their methods and found that MAE=1.22 $\pm$ 0.04 (n=3,920) using the same data in the Likes model. All experimental results were based on 10-fold cross validation.

5.2   Experimental Results

Table 1 summarizes correlations between features and SWL, confirming the strong correlation between Big 5 and SWL. We observed some correlations between LIWC categories and SWL, signifying linguistic features' utility for predicting SWL. When ego features were averaged over SWL scores, we observed other highly correlated features (age, network size, relationship status, and number of tags), which were subsequently used in prediction models.

Table 1. Pearson's R Between Feature and SWL. Some features are averaged over SWL as annotated with *.

Feature	R	# of Samples
agreeableness *	0.988	86073
conscientiousness *	0.986	86073
extraversion *	0.997	86073
neuroticism *	-0.998	86073
openness *	0.901	86073
age *	0.249	42264
network size *	0.846	60863
num of tags *	0.596	23197
anger (LIWC)	-0.105	3505
body (LIWC)	-0.106	3505
negative emotion (LIWC)	-0.160	3505
swear (LIWC)	-0.148	3505

Table 2. Static and Combined Static Ego Models

Model	# of Samples	MAE	Random MAE
Big5	86073	0.97	1.58
FBAttrib	9461	1.19	1.34
Likes	3920	1.15	1.57
LIWC	3251	1.16	1.60
Big5.FBAttrib	9242	1.10	1.61
Big5.Likes	3693	1.13	1.56
Big5.LIWC	3251	1.16	1.60
Combo.Static.1	1360	1.04	1.64
Combo.Static.2	1190	1.07	1.61

Static Ego Models. After initial feature selection, we created static ego models predicting SWL. Table 2 (above) shows all models perform better than the Random Baseline. All models out-performed a more sophisticated models using "Likes" features (MAE=1.22) from a previous study [14]. The Big 5 model is the best predictor of SWL (MAE=0.97), and we utilized this insight to create layered models using Big 5 as an intermediate prediction variable. Table 2 (below) shows combination models of static ego features which included predicted Big 5 values as features. Combining ego static features yielded greater accuracy than employing ego features alone (MAE=1.04); however, the Big5 model still out-performed both combination models. This underscores the importance of Big 5 when predicting SWL.

Combined Static Ego and Link Models. Table 3 summarizes the findings when link information was added to the input feature vector. When incorporating the Top 3 Friends features, we observed a slight performance boost (MAE=0.822) in the FBAttrib.Big5.FriendSWL model over the model that did not use link features (MAE=0.827); however, we found that models incorporating link features were significantly better than our previous best model (Big5 with MAE=0.97). This may indicate that not only an individual's situation influences his "happiness" but also the "happiness" of those close to him. Perhaps another explanation for this phenomenon could be that SWL is assortative, where "happy" people gravitate toward "happy" people. When incorporating a significant other's SWL, we observed an even greater performance boost (MAE=0.670). Similar to top 3 friends, the model utilizing significant other's SWL is only slightly better than the model that did not include link features (MAE=0.681). We noted that a significant other's SWL predicted a user's SWL more accurately than his Top 3 friends. This finding is plausible since a significant other is more likely to share in daily life events and may be more influential than "friends" on Facebook.

Table 3. Combined Models of Static Ego and Link Features (above: top-3 friends; and below: significant other).

Model	# of Samples	MAE	Random MAE
FBAttrib.Big5.FriendSWL	695	0.822	1.54
FBAttrib.Big5.NoFriend	695	0.827	1.54
FriendSWL	695	0.865	1.54
FBAttrib.Big5.OtherSWL	171	0.670	1.48
FBAttrib.Big5.NoOther	171	0.681	1.48
OtherSWL	171	0.804	1.48

Temporal Model. Although the temporal models proved predictive of SWL, there was little variance over time. This suggests "mood swings" (expressed through LIWC), do not affect SWL. In particular, Temporal.1's performance showed no significant difference in prediction accuracy when using recent posts (1 Day: MAE = 1.18) versus earlier days (2 Days: 1.18, 3 Days: 1.18, 4 Days: 1.19, 5 Days: 1.19, 6 Days: 1.18, 7 Days: 1.18). Similarly, Temporal.2 showed no significant difference on a weekly scale (1 week: 1.18, 2 weeks: 1.17, 3 weeks: 1.18, 4 weeks: 1.17, 5 weeks: 1.17, 6 weeks: 1.17, 7 weeks: 1.17).

6   Conclusion and Future Work

This study showed that ego features such as network size, number of photo tags, age, relationship status, likes, and overall word usage (LIWC) can be combined to make a good predictor of SWL. Big 5 consistently predicted SWL and reducing variables from high dimensions, e.g. 600-dimensional "Likes", to highly predictive variables, e.g. Big 5, increased the performance of our model.

When using link features, we found a performance boost over the Big5 model. However, when compared to combined static ego feature models, the boost was minimal. This may be attributed to the noise of using a predicted SWL score for friends and couples. If we had the true SWL values for link relationships these models may have shown more lift.

Although LIWC is a good predictor of SWL, the temporal feature of Facebook statuses showed no improvement to our models. This may be attributed to SWL's high internal and temporal consistency as noted in previous research [8]. Because SWL measures a cognitive-judgmental process, it is plausible that "mood swings" expressed by LIWC would not be a large indicator of a user's overall SWL. Another explanation could be that the timeframes were not granular enough to capture the transient mood of a user prior to the SWL test.

Overall, when compared to the Random Baseline, all of our models out performed random prediction by at least 11%. When compared to a linear regression model that used "Likes" features [14], we found our best model was 45% more accurate. We believe that the selection of Random Forest Regression, a combination of static ego features, and "important" link features provided an increase in prediction accuracy.

Our study demonstrated how social media sites such as Facebook contain a set of features useful for predicting private traits. The ability to predict user attributes like SWL may benefit social sciences at the individual and community level. From an individual stand-point, we can create early warnings schemes to identify users in distress. For example, SWL could indicate issues like depression in students [12] or PTSD in veterans [4]. From a community stand-point, collection of SWL from social media can provide an efficient evaluation for public wellness. For government entities like the European Union, this could save efforts and costs for collecting and processing international SWB survey [13].

Sparsely recorded data presented a major limitation to the study. Several features (e.g. number of groups) correlated highly with SWL ($R = 0.678$) but were poorly populated, and therefore were not utilized in the current model. We suggest future work to focus on fine-tuning feature collection and selection. We also suggest using link relationships as predictors of SWL. In our models, we saw promising results from link features (MAE=0.67); however the sample size (n=171) was relatively small.

Acknowledgments. This research was partially supported by the Draper Laboratory internal Research and Development funding.

References

1. Blei, D.M., Ng, A.Y., Jordan, M.I.: Latent dirichlet allocation. The Journal of machine Learning Research **3**, 993–1022 (2003)
2. Bollen, J., et al.: Happiness is assortative in online social networks. Artificial Life **17**(3), 237–251 (2011)
3. Breiman, L.: Random forests. Machine Learning **45**(1), 5–32 (2001)
4. Bryant, R.A., et al.: Posttraumatic stress disorder and psychosocial functioning after severe traumatic brain injury. The Journal of Nervous and Mental Disease **189**(2), 109–113 (2001)
5. Costa Jr., P. T., McCrae, R.R.: Neo personality inventoryrevised (neo-pi-r) and neo five-factor inventory (neo-ffi) professional manual. Psychological Assessment Resources, Odessa (1992)
6. Coviello, L., et al.: Detecting Emotional Contagion in Massive Social Networks. PloS One **9**(3), e90315 (2014)
7. Diener, E.: Subjective well-being: The science of happiness and a proposal for a national index. American Psychologist **55**(1), 34 (2000)
8. Diener, E.D., et al.: The satisfaction with life scale. Journal of Personality Assessment **49**(1), 71–75 (1985)
9. Diener, E., Oishi, S., Lucas, R.E.: Personality, culture, and subjective well-being: Emotional and cognitive evaluations of life. Annual Review of Psychology **54**(1), 403–425 (2003)
10. Diener, E.: Understanding scores on the satisfaction with life scale (2009) (retrieved August 8, 2014)
11. Furr, R.M., Funder, D.C.: A multimodal analysis of personal negativity. Journal of Personality and Social Psychology **74**(6), 1580 (1998)
12. Frisch, M.B., et al.: Predictive and treatment validity of life satisfaction and the quality of life inventory. Assessment 12(1), 66–78 (2005)
13. GOV.UK. Wellbeing: Introduction to Subjective Wellbeing Datasets. Research and Analysis. Cabinet Office March 27, 2013; Web. August 27, 2014
14. Kosinski, M., Stillwell, D., Graepel, T.: Private traits and attributes are predictable from digital records of human behavior. Proceedings of the National Academy of Sciences **110**(15), 5802–5805 (2013)
15. Lewinsohn, P.M., Redner, J., Seeley, J.: The relationship between life satisfaction and psychosocial variables: New perspectives. Subjective well-being: An Interdisciplinary Perspective, pp. 141–169 (1991)
16. Manyika, J., et al.: Big data: The next frontier for innovation, competition, and productivity (2011)
17. Marks, G.N., Fleming, N.: Influences and consequences of well-being among Australian young people: 1980-1995. Social Indicators Research **46**(3), 301–323 (1999)
18. Rice, J.: Mathematical statistics and data analysis. Cengage Learning (2006)
19. Schwartz, H.A., et al.: Characterizing Geographic Variation in Well-Being Using Tweets. in: ICWSM (2013)
20. Tausczik, Y.R., Pennebaker, J.W.: The psychological meaning of words: LIWC and computerized text analysis methods. Journal of Language and Social Psychology **29**(1), 24–54 (2010)
21. Veenhoven, R.: The study of life-satisfaction (1996)

Social Computing for Impact Assessment of Social Change Projects

Jana Diesner[(✉)] and Rezvaneh Rezapour

University of Illinois Urbana Champaign, iSchool, Champaign, USA
(jdiesner,rezapou2)@illinois.edu

Abstract. One problem that both philanthropic foundations and scientific organizations have recently started to tackle more seriously is assessing the societal impact of the work they are funding by going beyond traditional methods and metrics. In collaboration with makers and funders of social justice information products, we have been leveraging social computing techniques for practical impact assessment. In this paper, we identify which of the main impact goals as defined in the social change domain can be assessed by using our computational solution, illustrate our approach with an empirical case study, and compare our findings to those that can be obtained with traditional methods. We find that our solution can complement and enhance the findings and interpretations that can be obtained with standard techniques used in the given application domain, especially when applying data mining techniques to natural language text data, such as representations of public awareness, dialogue and engagement around various issues in their cultural contexts.

Keywords: Impact assessment · Geo-cultural information · Social justice · Semantic networks · Natural language processing

1 Introduction

Philanthropic foundations give out millions of dollars each year to "work with visionaries on the frontlines of social change worldwide" (Ford Foundation[1]), create "informed and engaged communities" (Knight Foundation[2]), and "tackle critical problems" in a way that "emphasizes collaboration, innovation, risk-taking, and, most importantly, results" (Gates Foundation[3]). One common problem that foundations have been facing and recently started to address more seriously is how to measure if the above-mentioned results have been achieved [1, 2]. By results, foundations typically mean impact, i.e. change [3]. This change is often on a social level; requiring the consideration of relevant and meaningful indicators, collection and analysis of appropriate data, use of suitable methods and tools, and drawing of justified conclusions.

---

[1] http://www.fordfoundation.org/
[2] http://www.knightfoundation.org/
[3] http://www.gatesfoundation.org/what-we-do

© Springer International Publishing Switzerland 2015
N. Agarwal et al. (Eds.): SBP 2015, LNCS 9021, pp. 34–43, 2015.
DOI: 10.1007/978-3-319-16268-3_4

Prior work on impact assessment of social justice projects is limited by the comprehensiveness and scalability of theories, methods and tools [4, 5] (more on that in the background section). To address this challenge, previously and in close collaboration with the Ford Foundation, we have developed a theoretically grounded, empirical and computational methodology and pertinent technology[4] to assess the impact of social justice information products; mainly documentary films [5, 6]. In this paper, we provide an additional evaluation of our solution by comparing it to the impact goals and assessment procedures and outcomes that are used by foundations and practitioners. For this purpose, we a) identify which of those goals can be measured by our solution and if so how (methods section), b) illustrate our approach with an empirical case study and c) compare our findings to those obtained by using common (state of the art is the same as cutting edge in this domain) assessment methods (results section). We find that our approach can a) complement and enhance common practice from the given application domain by leveraging social computing techniques and b) measure the types or dimensions of impact that funders in this domain care about.

2   Background

The philanthropic sector is not the only domain where impact assessment has recently become a real-world need and heavily debated topic as foundations have started to request impact assessments from their grantees. In science and bibliometrics, impact has been traditionally measured in terms of citation counts and metrics computed over these counts, such as the h-index and i-index [7]. In recent years, altmetrics has been emerging as an initiative to introduce alternative metrics for evaluating scholarly impact, such as the sharing of raw data (e.g. datasets and databases), the number of article views and downloads from online repositories, and references to scholarly work in traditional and social media [8, 9]. Like our approach [6], altmetrics is supposed to generalize to other information products beyond articles.

The historical evolution and ongoing efforts in the foundation's sector are comparable to the scientific domain: traditionally, impact of social justice information products and initiatives has been assessed in two ways [6]: first, in a quantitative and scalable fashion by counting the number of e.g. visitors, screenings, webpage visits, click throughs and downloads. Second, in a qualitative yet less scalable way by conducting focus group interviews; comparing the perception of a topic before and after users' exposure. Impact reports, which are typically a required deliverable for grantees at the end of their funding period, often combine both strategies. A set of representative, high-quality examples are reports provided by BritDoc[5]; a main funder of social justice documentaries in the UK. It is not unimaginable that scientific funding will become subject to broader impact assessment strategies in the future as well.

---

[4] http://context.lis.illinois.edu/
[5] http://britdoc.org/real_good/evaluation

3   Method

We are using the "Women, War & Peace" series (WWP) as a case study because their defined impact objectives and evaluation methods are representative for this domain. What is WWP? This five-part TV broadcast series was originally screened by PBS during October and November of 2011[6]. Since then, the Peace is Loud (PiL) organization has made WWP available for screenings as a media kit with accompanying educational material. The theme of the series is the impact of war on women and the role of women in peace-building processes in four different geo-cultural contexts: (1) "I Came to Testify": Bosnian women who became victims of sexual abuse and brought this case to court. (2) "Pray the Devil Back to Hell": Liberian women protesting the Charles Taylor dictatorship. (3) "Peace Unveiled": Afghan women participating in peace talks and negotiations with the Taliban. (4) "The War We Are Living" Colombian women defending their gold-rich lands and resisting to become displaced. The fifth film (War Redefined) is a series of interviews with high profile individuals, e.g. Madeleine Albright and Condoleezza Rice. We disregarded the last film for this study as it is not embedded in a geo-cultural context. PiL has given us access to their impact reports [10] and film material, e.g. transcripts.

How has WWP's impact been assessed? Table 1 lists the impact goals as defined by PiL, who also measured the achievement of these goals using state of the art methods:

- Quantitative techniques and metrics: aggregated statistics, e.g., 12.57 million viewers of the series and 1,461 hostings of screenings [10].
- Qualitative techniques: (1) Surveys at screenings, which capture self-reported information on media coverage and audience demographics, engagement with the given topic and intent to further discuss the topic. (2) Listing of feedback from testimonials, press quotes, website comments and social media comments.

The quantitative indicators are easy to calculate if one has access to these data and also easy to interpret – basically, the more the better. The qualitative indicators, which in this case were thoroughly gathered and reported by PiL, are not only tedious to collect, but also require further data analysis in order to arrive at valid, meaningful and comparative conclusions and interpretations. This is where our approach to social impact assessment starts being useful and complementary to traditional techniques: in a nutshell (for methodological and technical details see [6]), we collect publicly available information from media (through LexisNexis Academic) and social media sources (Facebook, Twitter, YouTube, Amazon reviews) in a semi-automated fashion. From these data, we build a baseline model, which represents the public discourse on the main theme(s) addressed in a film (as defined by film maker) prior to film release. This model comprises semantic networks of the main issues addressed in a documentary and social networks of stakeholders (individuals and organizations) associated with these issues. Building these networks combines techniques from natural language processing (NLP) and network analysis. We also build a ground truth model (semantic network, NLP results) of information contained in the actual documentary, book etc.. This model represents the information a film can convey. We then track the a) evolution of the baseline model from before to after release and onwards and b)

---

[6] http://www.pbs.org/wnet/women-war-and-peace/

(social) media coverage of the film. We compare a) to b); looking for correlations and differences, and test if parts from the ground truth model occur in a) and/or b).

Which of the common social justice impact goals (as defined by practitioners and funders) can we assess with our given solution? Table 1 lists PiL's goals and specifies how we approach their measurement – if we do. The results section provides an example of the actual outcomes from brining our solution to this problem and series.

Table 1. Feasibility of measurement of goals with existing computational solution

Goal	Can we measure achievement?	How?
1. Build awareness for WWP	yes	Over-time, semantic
2. Spark dialogue	yes	and social networks
3. Reach and engage key constituencies	yes	from media and
4. Continued utilization of series	yes	social media data,
5. Introduce series to new, varied audience	yes	plus natural language processing
6. Increase public engagement with topic	partially (words yes, actions not)	techniques (details in [6])
7. Inform stakeholders, serve as resource for stakeholders	not yet	
8. Highlight immediacy, proximity of topic	not yet	

3.1   Data Collection and Network Construction

To collect media data, we consult with the filmmakers to identify the main themes of a production. We translate their input into key-word-based Boolean queries. This step is crucial as it generated the raw data for analysis. Table 2 lists the queries and amount of retrieved articles for the baseline model before and after film release (three years of data in each direction), and press on films. The amount of coverage of the topics does not correlate with coverage of the films; indicating that different factors affect the importance of each subject.

Table 2. Queries and amount of retrieved data

Country	Keywords (baseline: woman, women, war, war-time, peace*, <country name>)	Before	After	Press on film
Afghanistan	peace talks, Taliban	450	1,069	4
Liberia	protest*, Charles Taylor	493	605	85
Colombia	gold*, displace* (not Olympic)	80	109	3
Serbia	rape, sexual violence	54	66	22

We herein focus on semantic networks as they allow us to gain a structural look at the development of the public awareness and dialogue around an issue as well as engagement with this topic (these represent defined impact goals). The data cleaning, preprocessing, management and analysis were done in ConText. We construct two types of semantic networks based on different types of information from the articles:

meta-data networks link index terms that co-occur with at least a certain threshold value per article (from the "subject" category). Such networks provide a high-level summarization of the main themes covered in an article [11]. We also extract semantic network from the text bodies of the articles, which provide a more in-depth and culturally sensitive view [11]. In these networks, nodes represent the most salient pieces of information (based on cumulative (weighted) frequencies and tf*idf scores of terms including proper N-grams). Edges are based on term co-occurrences within a user-defined distance (we used seven words for the given corpora). The media data networks were visualized in Gephi, where node colors indicate cluster affiliation (based on modularity), node size is scaled by degree (number of direct neighbors), and tie width represents frequency. For social media data collection, analysis and visualization, we used NodeXL (http://nodexl.codeplex.com/). Since most of the films don't have their own social media presence (which is typical for umbrella campaigns), we used the WWP fanpage. To be in sync with the methods for network construction from articles, we linked salient terms (as per tf*idf) that co-occur at least twice (posts) or thrice (comments) per page. The parametric choices are based on the actual data, and similar to those from other impact assessments we have done.

4   Results

Even though the queries for all retrieved corpora weighted women as strongly as the main issue(s) per film, the networks for most films and points in time are dominated by representations of the given substantive issues, while women are positioned marginally and hardly ever tied into the main issues and respective clusters (Table 3).

Table 3. Main findings from semantic network analysis per dataset

Film	Press on theme before release		Press on theme after release		Transcript
	Main cluster(s) and key nodes	Women	Main cluster(s) and key nodes	Women	(country name excluded)
Afghanistan (Peace unveiled)	(1) war & conflict, Taliban, muslims, peace process	2nd yet smaller cluster with human rights	(1) like before, (2) peace process, talks & meetings	marginal, separated from main clusters	women, Taliban, support, war, peace, conference
Liberia (Pray the devil back to hell")	(1) war & conf., civil war, rebellion & insurg. (2) elections	very marginal, no cluster	(1) like before (2) war crimes	3rd cluster with protests & demonstrations, nobel peace prize	Leymah Gbowee, women, peace, Charles Taylor
(Colombia (War we are living)	(1) war & conflict, human rights	marginal cluster with international relations	(1) rebellion & insurgencies, war & conflicts	2nd main cluster with human rights and displaced people	war, family, land, community, government
Serbia (I came to testify)	(1) war & conflict, ethnic conflict, religion (2) international legal issues	marginal cluster with sex offenses and human rights	(1) war & conflict, ethnic conflict, human rights (2) war crimes	marginal, no cluster	rape, women, witness, war, crime, tribunal

A notable exception is the coverage of the Liberia issue after film release (Figure 1 before, Figure 2 after), where Leymah Gbowee won a shared Nobel Peace Prize; drawing attention to the role of women in this conflict and moving this theme closer to the center of the debate. Overall, for two of the four films, women became more marginalized and disjoint from the core in the networks since film release, while in the other two, they got more connected to the main issues. The observed effects are correlational. The semantic networks also reveal additional central themes that are closely tied to the query concepts (Table 3).

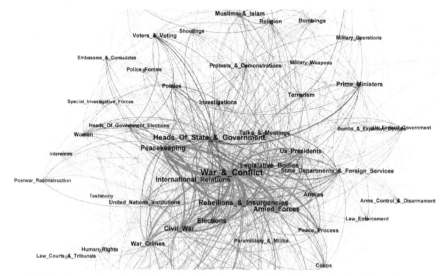

Fig. 1. Semantic network of meta-data of press on Liberia issues before film release

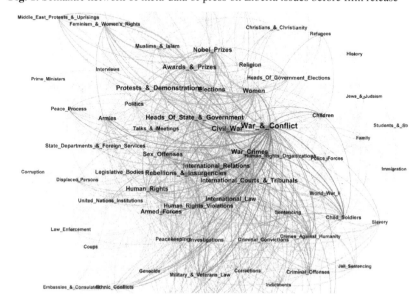

Fig. 2. Semantic network of meta-data of press on Liberia issues after film release

The ground truth model (semantic network based on film transcript) for all films feature women as a key node, followed by references to the geo-political region, main issues addressed in the film, as well as war and peace (last column in Table 3). How does this compare to the press coverage of the films? In general, in press articles about social justice documentaries, we often see a strong focus on artistic features and embedding the film in the wider context of film making, festivals, awards and screenings. While aligned with the quality standards of film making, this (journalistic) decision does not contribute much to increasing the film's impact on a given issue and is a missed opportunity for drawing attention to the film's content or problem domain. We have discussed this issue in meetings with journalists who cover this domain. For WWP, we observe mixed results: First, for all films, women are more central in the press coverage on the film than press on the topic. In "Pray the devil back to hell" (Figure 3) (Liberia, most film press), the main theme is film (making) and related festivals and awards, followed by a smaller cluster about religious issues; with the latter being more central to the content of the film. We see the opposite for "I came to testify" (Figure 4) (Serbia, 2nd most film press), where the core of the semantic networks is on international legal matters as they relate to women and violence, which is right at the heart of the film. Most articles about these films also mention the series; leading to a moderate overlap in nodes and edges of the networks for all films.

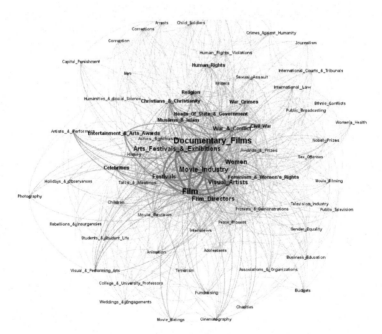

Fig. 3. Semantic network of press on film "Pray the devil back to hell"

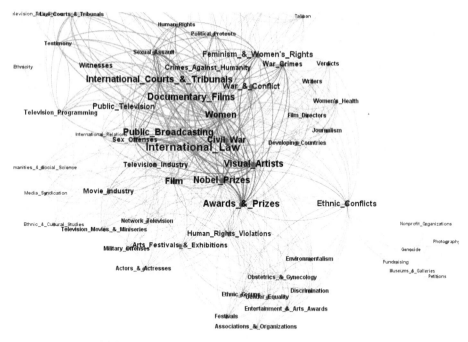

Fig. 4. Semantic network of press on film "I came to testify"

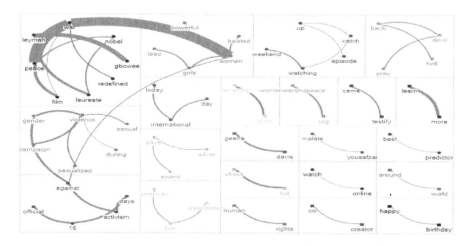

Fig. 5. Semantic network of posts on WWP Facebook fan page

How do public awareness, dialogue and engagement unfold on social media? While we have analyzed multiple platforms, we focus on Facebook here. The posts on the WWP fanpage, which are often authored by a staff member involved with the production and can be considered as a stimulus, center on three themes (Figure 5): the winning of a shared Nobel Peace Prize by one of the women in "Pray the devil back to hell", sexual violence, and empowering women and girls. This differs from the heavy focus on screening announcements that we typically see in posts and might indicate actual user contributions. How do the users react to these inputs (Figure 6)? The commenters focus on the sexual violence issue and add additional concepts to the debate (men, children), but the overall user reaction seems less diverse, thematically involved and active as we have previously observed for other productions.

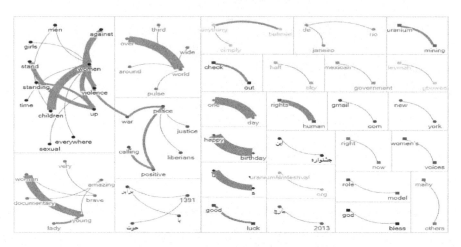

Fig. 6.   Semantic network of comments on WWP Facebook fan page

5   Discussion and Conclusion

We have shown how our assessment approach can a) measure the achievement of a large portion of the common impact goals defined by funders and evaluators in the social impact domain, and b) complement and enhance the findings and interpretations that can be obtained with standard techniques used in that field. Our solution brings social computing techniques, particularly network analysis and natural language processing to applications this domain; enabling the systematic and efficient analysis of small to large amounts of data across time and productions.

Practitioners and analysts in this domain typically collect and often only list semi-structured (key words) and unstructured (content of articles) text-based data (or cherry picked excerpts thereof if too much data), such as press coverage and social media data, in their reports. For these data, summarization and content analysis techniques – including semantic network analysis - can help to gain a more concise picture systematically and efficiently. These techniques are readily applicable to the kind of data that practical evaluators already gather, including the content of interviews with focus groups, which otherwise are aggregated into statistics that disregard the content of user statements. As academics might not have access to these data and practitioners lack the skills for analyzing them, we have been engaging in a series of collaborations with film makers and funders to realize the potential of these data and methods.

Currently, we are synthesizing the results from about a dozen social justice impact assessment studies that we have conducted into a framework for impact trajectories depending on a set of features. This work aims to lead to a theory of impact evolution as well as generalizable and practically useful guidelines for designing for impact. In future work, we plan to refine our methodology by considering prior work on causal inference in observational data and quasi-experimental research designs.

As part of this project, we are generating and continuously expanding a) a dictionary of terms, concepts and associated entity types relevant for the social impact domain, and b) a valence (aka sentiment) dictionary and classifier for this field. These resources are being made publicly available in ConText.

Acknowledgements. This work is supported by the FORD Foundation, grants 0125-6162 and 0145-0558. We thank Angie Wang from Peace is Loud for information access and sharing, Dr. Susie Pak from St. John's University for her feedback on this work, and graduate student Aseel Addawod, Informatics/UIUC, for her help with social media analysis.

References

1. Barrett, D., Leddy, S.: Assessing Creative Media's Social Impact. The Fledgling Fund (2008)
2. John, S., James, L.: Impact: A practical guide for evaluating community information projects. Knight Foundation (2011)
3. Clark, J., Abrash, B.: Social justice documentary: Designing for impact. Center for Social Media (2011)
4. Napoli, P.: Measuring media impact: An overview of the field, Media Impact Project, USC Annenberg Norman Lear Center (2014)
5. Chattoo, C.B., Das, A.: Assessing the Social Impact of Issues-Focused Documentaries: Research Methods & Future Considerations. Center for Media & Social Impact (2014)
6. Diesner, J., Kim, J., Pak, S.: Computational Impact Assessment of Social Justice Documentaries. Journal of Electronic Publishing 17(3) (2014)
7. Hirsch, J.E.: An index to quantify an individual's scientific research output. Proceedings of the National academy of Sciences of the United States of America **102**(46), 16569–16572 (2005)
8. Priem, J., Taraborelli, D., Groth, P., Neylon, C.: Altmetrics: A manifesto (2010)
9. Thelwall, M., Haustein, S., Larivière, V., Sugimoto, C.R.: Do altmetrics work? Twitter and ten other social web services. PloS One **8**(5), e64841 (2013)
10. Wang, A.: Women, War & Peace. U.S. Television Broadcast: Outreach and Audience Engagement. Impact Statement. Peace is Loud (2012)
11. Diesner, J.: From Texts to Networks: Detecting and Managing the Impact of Methodological Choices for Extracting Network Data from Text Data. Künstliche Intelligenz/ Artificial Intelligence **27**(1), 75–78 (2013)

Social Network Extraction and High Value Individual (HVI) Identification within Fused Intelligence Data

Alireza Farasat[1], Geoff Gross[1], Rakesh Nagi[2], and Alexander G. Nikolaev[1]([⊠])

[1] Department of Industrial and Systems Engineering,
University at Buffalo (SUNY), Buffalo, NY, USA
{afarasat,gagross,anikolae}@buffalo.edu
[2] Department of Industrial and Enterprise Systems Engineering,
University of Illinois at Urbana-Champaign, Champaign, IL, USA
nagi@illinois.edu

Abstract. This paper reports on the utility of social network analysis methods in the data fusion domain. Given fused data that combines multiple intelligence reports from the same environment, social network extraction and High Value Individual (HVI) identification are of interest. The research on the feasibility of such activities may help not only in methodological developments in network science, but also, in testing and evaluation of fusion quality. This paper offers a methodology to extract a social network of individuals from fused data, captured as a Cumulative Associated Data Graph (CDG), with a supervised learning approach used for parameterizing the extraction algorithm. Ordered, centrality-based HVI lists are obtained from the CDGs constructed from the Sunni Criminal Thread and Bath'est Resurgence Threads of the SYNCOIN dataset, under various fusion system settings. The reported results shed light on the sensitivity of betweenness, closeness and degree centrality metrics to fused graph inputs and the role of HVI identification as a test-and-evaluation tool for fusion process optimization.

Keywords: Social network analysis · Data fusion · Testing and evaluation · Centrality · High value individuals

1 Introduction

Traditional data fusion has transitioned from exclusively processing hard physical sensor data to fusing soft data provided by human sources as well. In the recent multi-disciplinary university data fusion research program [10,12], Natural Language Processing together with graph analytic techniques have been applied for fusion of hard and soft information so that an analyst is able to obtain situational awareness [9] (see Figure 1). The main processing steps of this effort include (1) natural language understanding and physical sensor data processing, (2) Cumulative Data Graph (CDG) construction via association of

© Springer International Publishing Switzerland 2015
N. Agarwal et al. (Eds.): SBP 2015, LNCS 9021, pp. 44–54, 2015.
DOI: 10.1007/978-3-319-16268-3_5

entities and relations in attributed graphs, (3) situation assessment via graph matching identifying graphical situations of interest within the CDG, and (4) social network extraction and High Value Individuals (HVIs) identification (see [9] for details). It is hard, however, to evaluate the success of analytic approaches within fusion systems due to the upstream propagation of errors. Social net-

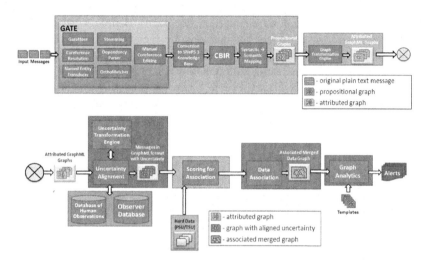

Fig. 1. Data fusion architecture in the multi-university research program

work extraction in fusion is troublesome due to complex nature of this procedure; e.g., the prediction of missing links in social network analysis (SNA) is known to be hard. When the data specifying underlying relationships is not available, inference and data mining tools are employed [21,22].

This paper reports on the use of SNA methods in the fusion domain. In evaluating a toolset for automatic processing of hard-and-soft data, one might like to reveal the relationships between the actors reported on in the data messages/signals. This paper addresses how one can assess the quality of automatic identification of HVIs, based on their structural positions in a social network [5,11,19]. The reported results shed light on the sensitivity of betweenness, closeness and degree centrality metrics to fused graph inputs and the role of HVI identification as a test-and-evaluation tool for fusion process optimization.

The paper proceeds as follows. Section 2 reviews the methodologies for extracting the social network from a CDG using a supervised learning approach for SNA of HVI identification. Section 3 provides a supervised learning approach based on Kendall Tau distance to tune our social network extractor. Section 4 discusses the results of comparison of CDG-extracted social networks (CDGSN) versus ground-truth social networks (GTSN) over multiple test datasets. Section 5 concludes the paper and discusses future research directions.

2   Social Network Extraction and HVI Modules

Social network extraction and HVI identification testing was performed within the hard-and-soft information fusion framework of the Network-based Hard+Soft Information Fusion Multidisciplinary University Research Initiative (MURI) [8–10]. To illustrate how these graph analytic tools work, the *Sunni criminal thread (SUN)* and *Bath'est Resurgence Thread (BRT)* of SYNCOIN are used [7]. Given a CDG with fused hard and soft data (e.g., locations, people, events, relationships, etc.), this section addresses the challenge of recovering the social network between person-type nodes (actors). The comparisons in recognizing HVIs are done using multiple metrics, since different metrics capture different properties of actors positions in a social network [3]. The experiments with those different metrics are conducted using supervised learning, with multiple processing modules involved in the network extraction (see Figure 2).

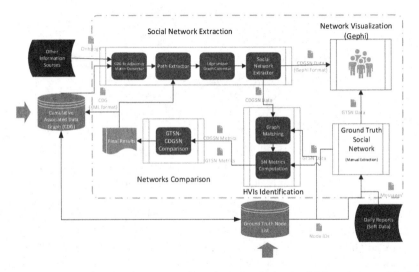

Fig. 2. Social Network Extraction and High Value Individual Identification Modules

Several methods can be used to quantify node position value in a network [1, 3, 18]; indeed, the notion of 'High Value' is application-dependent. The reported results are based on well-accepted SNA measures of centrality.

Centrality metrics quantify the prominence of actors embedded in a social network [3, 20, 23]. Based on the presumed HVI qualities and the nature of information/items exchanged between them, the betweenness, closeness or degree centrality can be adopted for HVI ranking. Degree centrality is easy to compute and reflects the number of direct connections a node has, while betweenness and closeness centralities are based on shortest paths connecting node pairs and reflect the distances between peers [4]. For the path-based centrality calculation, this study employs the very efficient algorithm of Brandes [4].

2.1   Ground Truth Social Network Extraction

For each given dataset, a corresponding Ground Truth Social Network (GTSN) is extracted by multiple analysts (those who did not participate in the dataset creation), following five steps: (1) all the raw data messages are read by each analyst, (2) all the actors mentioned, including both organization- and person-types nodes, are enumerated, (3) person-type nodes are distinguished from the complied list (as potential HVIs), (4) weights are assigned to the links connecting the persons based on the source messages, (5) the links not directly mentioned in the messages but implied are inferred (per subjective analyst judgment) by fusing the information across multiple messages. Note that controversy may arise in Steps (4-5) as the experts may not agree on the existence or the weight of a particular link. In such cases, a set of ground rules set *a-priori* by the analyst group is used for dispute resolution. To visualize the GTSN and CDGSN, an open-source software Gephi [2] is utilized (see Figure 3).

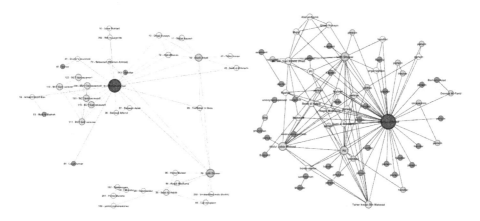

Fig. 3. Ground Truth Social Network (GTSN) (left) and Cumulative Data Graph Social Network (CDGSN) (right)

2.2   Cumulative Data Graph Social Network Extraction

The main idea employed in extracting a social network from CDG lies in traversing feasible paths between all the pairs of person-type nodes, under the feasibility and acceptance constraints. The first step is to convert a CDG XML-formatted file to an adjacency matrix of all the nodes in CDG. In extracting the feasible paths between each pair of person-type nodes, the acceptance constraints are imposed to ensure that (1) no acceptable path contains a person node, and (2) the length of an acceptable path does not exceed a pre-set threshold (T). A Depth First Search extraction procedure recovers the desired paths, with the $O(n^2T)$ time required for handling all the person node pairs.

To infer a weight of a tie between two actors (two person nodes), the number of connecting paths and the paths' lengths are taken into account. The more connections two nodes share, the higher the weight (indicating a closer relationship between the nodes). Thus, an edge between a pair of person nodes two hops away from each other has a greater weight than an edge between a pair of person nodes who are three hops away from each other. Also, if multiple paths of the lengths smaller than the hops threshold (T) exist between a pair of nodes, then the weight of the edge between those nodes also increases. More specifically, let $w(i,j)$ denote the weight of an edge between node i and node j; P_{ij} denote a set of all paths between node i and node j; p_{ij} denote a single path such that $p_{ij} \in P_{ij}$; and $h(p_{ij})$ denote the number of hops along path p_{ij}. Then,

$$w(i,j) = \frac{1}{\sum_{p_{ij} \in P_{ij}} \frac{1}{h(p_{ij})}},$$

with $h(p_{ij}) \leq T$. Note that the inverse of this weight, $1 - w(i,j)$, is interpreted as the relationship strength, thus distinguishing strong ties (e.g., close friends) from weak ties (e.g., acquaintances) [24].

2.3   Comparison Between CDGSN and GTSN

Given two networks A and B, a simple way to assess their similarity is to count the number of changes that one has to do to transform one graph into the other (this measure is known as graph edit-distance [13]). Various edit operations have been introduced to date, including edge rotation, edge addition and subtraction, vertex addition and subtraction (if the networks do not have the same number of nodes), etc.; note that it is not obvious how to weigh these changes against one another [13].

There is a wide array of other methods proposed in the literature and exploited to measure the similarities between two given networks. Spectral analysis is used to approximate the graph edit-distance by the difference in the spectrum of eigenvalues between Laplacians of the adjacency matrices [15]. Other related research introduces p^* models (now widely known as Exponential Random Graph Models [16]), graph kernels [17] and motif analysis [14]. The p^* models and motif analysis are based on the presence of small subgraphs in the compared networks [16]. Graph kernels map graph features to points in high dimensional inner product spaces [13].

However, given the objective of HVI identification, the HVI-based measures, e.g., rankings, can be directly used to evaluate the quality of the extracted network. With the ranking of aforementioned centrality metrics used to identify HVIs, Kendall Tau distance [6] can be accordingly utilized to compare the quality of CDGSNs relative to GTSNs. Moreover, Kendall Tau can be used as a rank correlation coefficient for the evaluation of the association (agreement) between two measured lists [6] (e.g., betweenness of nodes in two different networks). The normalized Kendall Tau distance is measured as follows:

$$\tau = \frac{N_D}{\binom{N}{2}},$$

where N_D is the counts of discordance pairs and $\binom{N}{2}$ is the total number of pairs. Any pair of observations (x_i, y_i) and (x_j, y_j) is termed discordant, if $x_i > x_j$ and $y_i < y_j$ or if $x_i < x_j$ and $y_i > y_j$. For instance, suppose two ranked lists, $l_1 = A, C, B$ and $l_2 = B, A, C$, contain the rankings of elements A, B and C where $A = (1, 2)$ and $B = (3, 1)$ represent the rankings of A and B in l_1 and l_2, respectively. Then, A and B are discordant because A has the higher rank in l_1 and the lower rank in l_2.

3   Hop Threshold Optimization

This section describes the design of a metric for performing systemic error trail analysis and parametric optimization of the social network extraction using a supervised learning approach. Again, the main utility of an extracted social network is assumed to lie in distinguishing HVIs. The main expected utility of the extracted social network is to allow one to correctly rank HVIs for any given dataset (due to errors both in reporting and fusion, the absolute centrality values of network actors may not be reliable).

The metric utilized for minimizing the number of the HVIs misplaced in the ranked list needs to allow for the comparison of ranks of the selected centrality metrics for each node in both GTSN and CDGSN. It should be noted that CDGSN and GTSN are not necessarily expected to have the same number of nodes due to upstream processing errors. However, most or all ground truth nodes are expected to be present in both networks; in this case, the node ranks by centrality metric values can be compared using Kendall Tau distance.

Let B_i^C and B_i^G represent the betweenness values of node i in CDGSN and GTSN, respectively. Similarly, let C_i^C and C_i^G denote the closeness of node i in CDGSN and GTSN, and D_i^C and D_i^G denote degree of node i in CDGSN and GTSN ($\forall\ i \in \Omega$ where Ω is a set of common nodes in both CDGSN and GTSN). Considering the pair (B_i^C, B_i^G), the betweenness-based Kendall Tau (τ_B) is defined as follows:

$$\tau_B = \frac{|\Omega_{D_i}^B|}{\binom{|\Omega|}{2}},$$

where $\Omega_{D_i}^B$ denotes the set of discordance pairs based on node i betweenness values and $|\Omega|$ is Ω-cardinality. Similarly, the closeness-based Kendall Tau (τ_C) and degree-based Kendall Tau (τ_D) for (C_i^C, C_i^G) and (D_i^C, D_i^G), respectively, are:

$$\tau_C = \frac{|\Omega_{D_i}^C|}{\binom{|\Omega|}{2}} \quad \& \quad \tau_D = \frac{|\Omega_{D_i}^D|}{\binom{|\Omega|}{2}}.$$

Again, $\Omega_{D_i}^C$ and $\Omega_{D_i}^D$ denote the sets of discordance pairs for closeness and degree, respectively.

The training process for detecting HVIs requires one parameter to be tuned; the parameter is called the hop threshold, $T \geq 1\ T \in \mathbb{Z}$, used in the path extraction process. The goal is to find T^* as

$$T^* = \underset{T}{\operatorname{argmin}} \left(w_B \tau_B + w_C \tau_C + w_D \tau_D \right),$$

where w_B, w_C and w_D are the weights for τ_B, τ_C and τ_D, respectively. There exists no analytical solution for this optimization problem. However, a supervised learning approach can help us tune the hops threshold parameter using training datasets.

4   Training and Test Data Set Results

Two SYNCOIN dataset [7] threads were used in this test and evaluation study: the SUN thread with 114 soft messages and four hard data reports and the BRT thread with 114 soft messages. Multiple versions of CDG were generated with various upstream fusion algorithms and parameter settings utilized: fifteen CDGs for SUN and three for BRT, all in XML format.

In order to find the optimal value(s) of T, the social network extraction algorithm of Section 2 is run for $1 \le T \le 8$ such that the weighted sum of Kendal Tau distances is minimized. Note that this operation is computationally expensive, especially with large CDGs. Figure 4 reports the weighted sums of Kendall Tau distances for four SUN Thread CDGs. Across all the CDGs obtained under

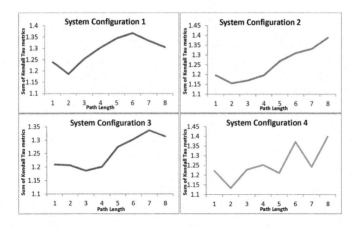

Fig. 4. Weighted Kendall Tau distance value dynamics for four SUN CDG outputs over varied values of hop threshold T

diverse fusion pipeline settings, the threshold values $T = 2$ or $T = 3$ resulted in the minimal normalized Kendall Tau distance (see Figure 4). These four configurations were chosen as exemplars demonstrating the performance under optimal data association settings for a given natural language understanding input. Naturally, the larger the value of T, the more dense the CDGSN becomes (multiple paths are found between actor pairs). Table 1 compares the top five (about top 10%) HVIs in the SUN Thread GTSN with their counterparts in CDGSN; the top 10% is a reasonable choice for the number of individuals to consider since this is reflective of the number of key actors in the SUN thread. It is found that the HVIs discovered from GTSN and CDGSN generally match. For comparing

Table 1. Top five HVIs in GTSN and CDGSN of SUN Thread dataset

Rank	Betweenness		Closeness		Degree	
	Rank in CDGSN	Rank in GTSN	Rank in CDGSN	Rank in GTSN	Rank in CDGSN	Rank in GTSN
1	Dhanun Ahmad	Dhanun Ahmad	Dhanun Ahmad	Dhanun Ahmad	Dhanun Ahmad	Dhanun Ahmad
2	Lufti Dilawar	Lufti Dilawar	Davood Alfarid	Lufti Dilawar	Lufti Dilawar	Lufti Dilawar
3	Trafficker in Dora	Jabar Wahied	Bameen Azad	Jabar Wahied	Khalil Jihad	Jabar Wahied
4	Khalil Jihad	Khalil Jihad	Trafficker in Dora	Davood Alfarid	Saiar Al Habib	Khalil Jihad
5	Saiar Al Habib	Saiar Al Habib	Doctor's assistant	Taher Iravan	Sattar Ayyash	Taher Iravan

the HVI ranks, betweennees centrality appears the most robust - this observation can inform the selection of the weights w_B, w_C and w_D.

With $T = 2$ or $T = 3$, it is also found that the Kendall Tau distance remains robust under different fusion settings, i.e., sub-optimal association settings as opposed to Fig 4 where optimal association settings are utilized (see Figure 5). The variability of Kendall Tau distance values at plus/minus 10% are commensurate with the error magnitudes observed in upstream data fusion stages [9].

The same conclusions about the robustness of the weighted Kendall Tau metric behavior were made in working with the BRT Thread data. Table 2 compares the top five HVIs in both GTSN and CDGSN of the BRT dataset. Again, most of top HVIs are common between BRT CDGSN and GTSN.

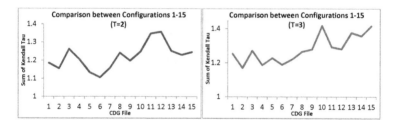

Fig. 5. Comparison of normalized Kendall Tau distance for different SUN CDG outputs

Table 2. Top five HVIs in GTSN and CDGSN of BRT dataset

Rank	Betweenness		Closeness		Degree	
	Rank in CDGSN	Rank in GTSN	Rank in CDGSN	Rank in GTSN	Rank in CDGSN	Rank in GTSN
1	Ali Tikriti	Khaari Elahi	Ali Tikriti	Khaari Elahi	Ali Tikriti	Khaari Elahi
2	Khaari Elahi	Ali Tikriti	Khaari Elahi	Ali Tikriti	Iraqi boy's mother	Ali Tikriti
3	MNCI Officer	MNCI Officer	Anwar Khan	Iraqi boy's mother	Khaari Elahi	Iraqi boy's mother
4	Ahmaad Hasseeb	Iraqi boy's mother	Ahmaad Hasseeb	Khalid Youssef	Anwar Khan	Khalid Youssef
5	Iraqi boy's mother	Suleiman	Khalid Youssef	MNCI Officer	Ahmaad Hasseeb	MNCI Officer

5   Conclusion

This paper extracts the underlying social networks of individuals from fused data, captured as Cumulative Associated Data Graphs (CDGs). Ordered, centrality-based HVI lists are obtained with CDGs constructed from Sunni Criminal Thread and Bath'est Resurgence Threads of the SYNCOIN dataset, under various fusion system settings. A supervised learning approach is used to tune the social network extractor parameter - the hop threshold. Since the rank of centrality measures can guide one to recognize HVIs, the normalized Kendall Tau distance is accordingly adopted to evaluate the identified HVI rankings against those based on the ground truth data. The employed method to set parameter T is a supervised learning approach: both the training dataset and GTSN are exploited to identify the settings that provide the minimal weighted Kendall Tau distance.

The experiments with the SUN Thread and BRT Thread datasets demonstrate that HVIs can be successfully identified in CDGSN with $T \leq 3$. We recognize that error propagated from upstream processes (e.g., natural language understanding and data association) may be responsible for variations in the Kendal Tau distances. Overall, The proposed approach is useful not only in evaluating the performance of HVI identification, but also in evaluating wider, systemic data fusion performance and error propagation across fusion processes. We plan to extend this analysis to the wider fusion system in future work. It is observed that betweenness centrality appears more robust than other centrality measures in HVI ranking.

Future work will focus on path semantics in HVI identification recognizing link types: e.g., a familial and organizational link is stronger than a locational link. Fusing alternative paths to arrive at "composite weights" is also promising.

Acknowledgments. This work has been supported by a Multidisciplinary University Research Initiative (MURI) grant (Number W911NF-09-1-0392) for Unified Research on Network-based Hard/Soft Information Fusion, issued by the US Army Research Office (ARO) under the program management of Dr. John Lavery.

References

1. Arulselvan, A., Commander, C.W., Elefteriadou, L., Pardalos, P.M.: Detecting critical nodes in sparse graphs. Computers & Operations Research **36**(7), 2193–2200 (2009)
2. Bastian, M., Heymann, S., Jacomy, M., et al.: Gephi: an open source software for exploring and manipulating networks. ICWSM **8**, 361–362 (2009)
3. Borgatti, S.P.: Identifying sets of key players in a social network. Computational & Mathematical Organization Theory **12**(1), 21–34 (2006)
4. Brandes, U.: A faster algorithm for betweenness centrality. Journal of Mathematical Sociology **25**(2), 163–177 (2001)

5. Chen, D., Lü, L., Shang, M.-S., Zhang, Y.-C., Zhou, T.: Identifying influential nodes in complex networks. Physica a: Statistical Mechanics and its Applications **391**(4), 1777–1787 (2012)
6. Fagin, R., Kumar, R., Sivakumar, D.: Comparing top k lists. SIAM Journal on Discrete Mathematics **17**(1), 134–160 (2003)
7. Graham, J.L., Hall, D.L., Rimland, J.: A synthetic dataset for evaluating soft and hard fusion algorithms. SPIE Defense, Security, and Sensing International Society for Optics and Photonics, pp. 80620F–80620F (2011)
8. Gross, G., Nagi, R., Sambhoos, K.: A fuzzy graph matching approach in intelligence analysis and maintenance of continuous situational awareness. Information Fusion **18**, 43–61 (2014)
9. Gross, G.A., Khopkar, S., Nagi, R., Sambhoos, K., et al.: Data association and graph analytical processing of hard and soft intelligence data. In: 2013 16th International Conference on Information Fusion (FUSION), pp. 404–411. IEEE (2013)
10. Gross, G.A., Nagi, R., Sambhoos, K., Schlegel, D.R., Shapiro, S.C., Tauer, G.: Towards hard+soft data fusion: Processing architecture and implementation for the joint fusion and analysis of hard and soft intelligence data. In: 2012 15th International Conference on Information Fusion (FUSION), pp. 955–962. IEEE (2012)
11. Kiss, C., Bichler, M.: Identification of influencers - measuring influence in customer networks. Decision Support Systems **46**(1), 233–253 (2008)
12. Llinas, J., Nagi, R., Hall, D., Lavery, J.: A multi-disciplinary university research initiative in hard and soft information fusion: Overview, research strategies and initial results. In: 2010 13th Conference on Information Fusion (FUSION), pp. 1–7. IEEE (2010)
13. Macindoe, O., Richards, W.: Graph comparison using fine structure analysis. In: 2010 IEEE Second International Conference on Social Computing (SocialCom), pp. 193–200. IEEE (2010)
14. Milo, R., Shen-Orr, S., Itzkovitz, S., Kashtan, N., Chklovskii, D., Alon, U.: Network motifs: simple building blocks of complex networks. Science **298**(5594), 824–827 (2002)
15. Mitchell Peabody, Finding groups of graphs in databases, Ph.D. thesis, Citeseer (2002)
16. Robins, G., Pattison, P., Kalish, Y., Lusher, D.: An introduction to exponential random graph p-star models for social networks. Social Networks **29**(2), 173–191 (2007)
17. Shervashidze, N., Borgwardt, K.M.: Fast subtree kernels on graphs. Advances in Neural Information Processing Systems, 1660–1668 (2009)
18. Shetty, J., Adibi, J.: Discovering important nodes through graph entropy the case of enron email database. In: Proceedings of the 3rd International Workshop on Link Discovery, pp. 74–81. ACM (2005)
19. Rama Suri, N., Narahari, Y.: Determining the top-k nodes in social networks using the shapley value. In: Proceedings of the 7th International Joint Conference on Autonomous Agents and Multiagent Systems, vol. 3. International Foundation for Autonomous Agents and Multiagent Systems, pp. 1509–1512 (2008)
20. Tang, J., Musolesi, M., Mascolo, C., Latora, V., Nicosia, V.: Analysing information flows and key mediators through temporal centrality metrics. In: Proceedings of the 3rd Workshop on Social Network Systems, p. 3. ACM (2010)

21. Thelwall, M., Wilkinson, D., Uppal, S.: Data mining emotion in social network communication: Gender differences in myspace. Journal of the American Society for Information Science and Technology **61**(1), 190–199 (2010)
22. van der Aalst, W.M.P., Song, M.S.: Mining Social Networks: Uncovering Interaction Patterns in Business Processes. In: Desel, J., Pernici, B., Weske, M. (eds.) BPM 2004. LNCS, vol. 3080, pp. 244–260. Springer, Heidelberg (2004)
23. Varlamis, I., Eirinaki, M., Louta, M.: A study on social network metrics and their application in trust networks. In: 2010 International Conference on Advances in Social Networks Analysis and Mining (ASONAM), pp. 168–175. IEEE (2010)
24. Xiang, R., Neville, J., Rogati, M.: Modeling relationship strength in online social networks. In: Proceedings of the 19th International Conference on World Wide Web, pp. 981–990. ACM (2010)

Structural Properties of Ego Networks

Sidharth Gupta[1], Xiaoran Yan[2(✉)], and Kristina Lerman[2]

[1] Indian Institute of Technology, Kanpur, India
[2] Information Sciences Institute, University of Southern California, Los Angeles, USA
`everyxt@gmail.com`

Abstract. The structure of real-world social networks in large part determines the evolution of social phenomena, including opinion formation, diffusion of information and influence, and the spread of disease. Globally, network structure is characterized by features such as degree distribution, degree assortativity, and clustering coefficient. However, information about global structure is usually not available to each vertex. Instead, each vertex's knowledge is generally limited to the locally observable portion of the network consisting of the subgraph over its immediate neighbors. Such subgraphs, known as *ego networks*, have properties that can differ substantially from those of the global network. In this paper, we study the structural properties of ego networks and show how they relate to the global properties of networks from which they are derived. Through empirical comparisons and mathematical derivations, we show that structural features, similar to static attributes, suffer from paradoxes. We quantify the differences between global information about network structure and local estimates. This knowledge allows us to better identify and correct the biases arising from incomplete local information.

1   Introduction

As powerful representations for complex systems, networks model entities and their interactions as vertices and edges. Over the years, different attributes characterizing real world networks have been proposed and investigated. These include features like the degree distribution, degree assortativity and clustering coefficient that describe network structure at the global level. Many efficient models and algorithms have been developed for their generation and inferences. [1,19].

Unfortunately, efficient algorithms usually rely on global knowledge of the network, which is typically not available to each vertex of the network. This is especially the case for real world social networks like the one in Milgram's "small world" experiment [17]. In social networks where vertices correspond to people, without digital bookkeeping, individuals only have access to *local information* about their immediate neighbors. Even in online networks such as Facebook, information access is restricted by privacy settings.

This work was supported in part by AFOSR (contract FA9550-10-1-0569), by NSF (grant CIF-1217605) and by DARPA (contract W911NF-12-1-0034).

N. Agarwal et al. (Eds.): SBP 2015, LNCS 9021, pp. 55–64, 2015.
DOI: 10.1007/978-3-319-16268-3_6

While small world structures can generally explain efficient decentralized navigation [12], other connections between global and local network measures are less understood. The "friendship paradox," for example, states that on average, your friends have more friends than you do [10], which can be generalized to many other attributes [11,13]. These systematic biases have been widely observed in social studies, for attributes ranging from wealth [2] to epidemic risk [7], and has been largely attributed to distribution bias in the sampling process. These paradoxes can at a local level distort our perceptions of the ground truth, resulting in inefficient policies and social consequences.

Local information about network structure from the perspective of a vertex is captured by its ego network, which is the induced subgraph over that vertex's immediate neighbors. Ego networks are considered to be the basic structure that dominates the central vertex's perspectives and activities [3,16,20,23]. In this work, we study the structural properties of ego networks and relate them to those of the global network. We hope to quantify structural biases arising from local, incomplete information, and recover accurate global estimates.

We will first review related work on the structural properties of global networks. In section 3.1, we will establish the mathematical mappings for degree distributions between the global and ego networks. In section 3.2, we will investigate degree assortativity and clustering coefficient at ego network level, by combining our knowledge of global structures and their mathematical relations.

2   Background and Related Work

With traditional independent data, the global statistics of a population remain unbiased estimates for subsets. In networks, however, the complex dependencies can skew localized statistics, leading to inhomogeneity at different scales and positions. Numerous efforts have been made to develop generative models which can reproduce realistic structure with simple local algorithms [4,9,18,22]. Unfortunately, structural features are so intertwined that preserving one often biases another. The same difficulty is also observed in graph sampling, where different sampling techniques can lead to different biases [8,14].

However, real world networks do exhibit certain patterns. We focus on comparing the collective perceptions of individuals with the global ground truth, similar to previous work for static features, where the bias in the sampling process leads to paradoxes [10,13]. Structural features, unlike their static counterparts, change values over specific subgraphs, and thus have an additional complication.

In this section, we review and organize the relevant work on structural features of networks, including degree distribution, degree assortativity and clustering coefficient. They describe network structure at the global level. However, they are by definition aggregations of local measures, and they are closely related to each other. We will focus on undirected graph $G = (V, E)$ with vertex set V, edge set E, and size $N = |V|$.

Degree distribution is one of the best studied aspects of networks. Many real world networks display "scale-free" [4,9], or power law degree distributions:

$$d_u \sim P(k) = \frac{\gamma - 1}{k_{min}} (\frac{k}{k_{min}})^{-\gamma} , \tag{1}$$

where $k = d_u$, $k_{min} = d_{min}$ and the range of the distribution is $[d_{min}, \infty]$. The exponent γ is usually in the range $[1,3]$.

While specifying individual degrees is a step forward from simple random graph models, real world networks also exhibit higher order correlations. Degree assortativity, for example, captures the pair-wise correlation between the degrees of neighboring vertices [18]. In fact, mathematical constraints alone can predict that scale-free networks with $\gamma < 3$ cannot be completely uncorrelated, leading to the phenomenon of "structural cut-off" [6] — smaller the value of γ, lower the maximum possible positive correlation between the degrees of neighbors in the network. Many real world networks are thus more disassortative than we would normally expect.

Degree correlations are fully specified by the joint distribution:$e(k, k') = \frac{E(k,k')}{\langle k \rangle N}$, where $E(k, k')$ is the number of edges between vertices of degree k and k'. A scalar aggregation of local assortativity in the way of Pearson's correlation gives us the global assortativity,

$$r_{glo} = \frac{1}{\sigma_q^2} \sum_{k,k'} kk'[e(k, k') - q(k)q(k')] , \tag{2}$$

where $q(k) = \frac{kP(k)}{\langle k \rangle} = \frac{kN(k)}{\langle k \rangle N}$ is the probability of sampling a vertex of degree k by following a randomly chosen edge and σ_q^2 is the variance of $q(k)$.

The complexity of real world networks does not stop at pair-wise correlations. Clustering coefficient goes one step further, capturing correlations among triplets of vertices [25]. The local version is defined as the probability that a third edge between two neighbors of the same vertex v would complete a triangle, with $C_v = \frac{2T_v}{d_v(d_v-1)}$, where T_v is the number triangles containing the vertex v. We can aggregate C_v over the set of vertices of a given degree $d_v = k$, and get the degree dependent clustering coefficient [24],

$$C(d_v) = C(k) = \frac{1}{N(k)} \sum_{v,d_v=k} C_v = \frac{1}{N(k)k(k-1)} \sum_{v,d_v=k} 2T_v , \tag{3}$$

where $N(k)$ is the number of vertices of degree k. In real world networks, it has been observed that $C(k)$ is also a power law function of degree, $C_{d_u} = C_0 d_u^{-\alpha}$, where α typically ranges from $[0,1]$, with networks having strong hierarchical structures corresponding to $\alpha = 1$ [21]. C_0 is a constant depending on global clustering coefficient C_{glo}. Given the degree distribution $P(k)$, we can recover $C_{glo} = \sum_{k=2}^{k_{max}} P(k)C(k)$, where we only consider vertices with $k > 1$.

Being a third order correlation measure, clustering coefficient displays dependencies on both degree distribution and degree correlations or assortativity [5].

The interplay between degree correlations and clustering is further complicated by the fact that each edge can form multiple triangles. It has been shown that negative degree correlations can limit the maximum value of C_{glo}, as triangles are less likely to appear with disassortative connections [22].

3   Structural Features of Ego Networks

An ego network is defined as the subgraph induced over vertices directly connected to a specific vertex, called an ego, but excluding the ego itself [3,23]. A toy example is given in Figure 1. Keep in mind that the removal of the ego can disconnect an ego network.

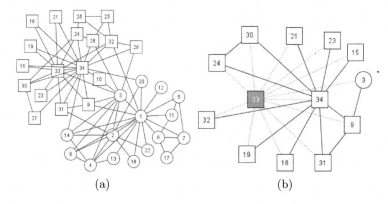

(a)   (b)

Fig. 1. A benchmark social network representing friendships among members of a karate club [26] (a) at the global level and (b) for the ego network of vertex 33.

By definition, ego network structure is closely connected to the local structure around the ego in the global network. These relationships are important to our understanding of perceptions based on local and limited information. In this section, we investigate how degree distribution, degree assortativity and clustering coefficient of ego networks depends on those of the global network. Although a full generative model that reproduces all structural features at both global and ego networks levels is difficult to build, we can however leverage our knowledge of global structures. Combining that with mathematical mappings between the two levels, we can better understand and even predict structures in ego networks.

We will approach the problem with theoretical derivations. To keep the mathematics tractable and intuitive, we will make some simplifying assumptions and educated guesses during the process. Therefore, it is very important to support our claims with empirical evidence. Our studies span a diverse range of network datasets where vertices correspond to people, such as social, coauthorship, communication and hybrid (serving social and informational purposes) networks [15], as detailed in Table 1.

Table 1. Description of the network datasets used for empirical studies

Dataset	Type	$\|V\|$	$\|E\|$	90% Eff Diameter	r_{glo}	C_{glo}
Facebook	Social	4,039	88,234	4.6	0.064	0.61
Orkut	Social	3,072,441	117,185,083	4.8	0.016	0.17
General Relativity	Coauthorship	5,242	14,496	7.9	0.66	0.53
High Energy Physics	Coauthorship	12,008	118,521	5.7	0.63	0.61
Enron email	Communication	36,692	183,831	4.7	-0.11	0.50
LiveJournal	Hybrid	3,997,962	34,681,189	6.5	0.045	0.28

In the following subsection, we will treat all ego networks as a giant disconnected graph. Here a vertex u will appear d_u times, and we index the features of each instance using a superscript. For example, the degree of vertex u in the ego network of v is denoted as d_u^v. Features without upper indices are global measures.

3.1 Degree Distribution

We start off by investigating the degree distribution in ego networks. The first simple connection to observe is that the size of the ego network of vertex v is simply its global degree d_v. The edge density of the ego network of vertex v is the local clustering coefficient C_v in the global network. The degree of vertex u in the ego network of v, or d_u^v, corresponds to the number of triangles containing the edge (u, v), which is symmetric for undirected graphs

$$d_u^v = d_v^u = m_{uv} \, , \tag{4}$$

where m_{uv} is the number of triangles sharing the edge (u, v). By summing over egos, we get the total degree of vertex u across all ego networks that it appears in, which is equal to the total degree of all vertices in the ego network of u:

$$\sum_v d_u^v = \sum_v d_v^u = 2T_u = C_u d_u (d_u - 1) \, . \tag{5}$$

The above equality gives us the average degree of vertex u across ego networks:

$$\langle d_u^v \rangle = \frac{\sum_v d_u^v}{d_u} = C_u (d_u - 1) \, , \tag{6}$$

which by symmetry is also the average degree of the ego network of vertex u. However, if we treat each instance of vertex u in all ego networks as independent variables, the average becomes:

$$\langle d_u^v \rangle_{ego} = \frac{d_u \langle d_u^v \rangle}{\langle d_u \rangle} = \frac{1}{\langle d_u \rangle} C_u d_u (d_u - 1) \, . \tag{7}$$

This over-representation of high degree vertices is the result of edge sampling. If we assume that both the global degree distribution and the degree dependent clustering follow power laws, as defined in the previous section, we get

$$d_u \sim P(x) = \frac{\gamma - 1}{x_{min}}(\frac{x}{x_{min}})^{-\gamma}, \qquad C_{d_u} = C_0 d_u^{-\alpha},$$

By a change of variables $\langle d_u^v \rangle_{ego} \approx \frac{1}{\langle d_u \rangle} C_u d_u^2 = \frac{C_0}{\langle d_u \rangle} d_u^{(2-\alpha)} = Z d_u^{(2-\alpha)}$, we have

$$\langle d_u^v \rangle_{ego} \sim P(y) = \left| \frac{1}{2-\alpha}(\frac{y}{Z})^{\frac{\alpha-1}{2-\alpha}} \frac{1}{Z} \right| (\frac{y}{Z})^{\frac{-\gamma}{2-\alpha}} x_{min}^{\gamma} \frac{\gamma-1}{x_{min}}$$

$$= \frac{x_{min}^{(\gamma-1)}(\gamma-1)}{(2-\alpha)Z}(\frac{y}{Z})^{\frac{\alpha-\gamma-1}{2-\alpha}}. \tag{8}$$

Since most real world networks have $1 \leq \gamma \leq 3$ and $0 \leq \alpha \leq 1$, the above exponent can be written as

$$\frac{\alpha-\gamma-1}{2-\alpha} = -\gamma + \frac{(\gamma-1)(1-\alpha)}{2-\alpha} \geq -\gamma,$$

which means the $\langle d_u^v \rangle_{ego}$ actually follows a power law with a heavier tail than the original degree distribution $P(k)$. In the extreme when $\alpha = 1$, as in many cases for networks with strong hierarchical structures [21], the mean degree of vertex u across ego networks and the mean degree of ego networks both become constants (uniform distribution).

The full distribution of d_u^v generally requires the complete knowledge of higher correlations. However, we do know that by definition $\langle d_u^v \rangle = E[d_u^v] = E[\langle d_u^v \rangle_{ego}]$. Assuming it also follows a power law distribution, we have

$$d_u^v \sim P(z) = \frac{\eta-1}{z_{min}}(\frac{z}{z_{min}})^{-\eta},$$

$$E[d_u^v] = z_{min}(1 + \frac{1}{\eta-2}) = y_{min}(1 + \frac{1}{\frac{-\alpha+\gamma+1}{2-\alpha} - 2}) = E[\langle d_u^v \rangle_{ego}].$$

Since smallest instances of d_u^v is smaller than its average $\langle d_u^v \rangle_{ego}$, we have $z_{min} < y_{min}$ and thus $\eta < \frac{-\alpha+\gamma+1}{2-\alpha}$, which means that the full distribution of d_u^v has a even heavier tail. Considering that $P(y)$ will be the same as $P(z)$ if all the vertex instances have the same degree, and we will underestimate the variance otherwise, we do expect $P(z)$ to have a wider spread.

Table 2. Properties of degree distributions at global and ego network levels

Network	$med(d_u)$	$\langle d_u \rangle$	$\langle d_u \rangle_{nn}$	$\langle d_u^v \rangle$	$\langle d_u^v \rangle_{nn}$	P_{glo}	$frac_{u,v}(d_u^v = 0)$
Facebook	25	43.7	106.6	54.6	95.8	1.1E-2	0.09%
Orkut	45	76.3	390.3	16.1	72.3	2.5E-5	13.64%
General Relativity	3	5.5	16.9	10.0	29.2	1.1E-3	10.99%
High Energy Physics	5	19.7	129.9	85.0	187.0	1.6E-3	2.16%
Enron email	3	10.0	140.1	11.9	34.5	2.7E-4	7.65%
LiveJournal	6	17.3	123.7	15.4	149.1	4.3E-6	16.75%

This is consistent with our empirical observations (Figure. 2). In practice, the distribution of $\langle d_u^v \rangle_{ego}$ can be empirically constructed by putting d_u copies of $C_u(d_u - 1)$ together. Independent of the shape of $P(d_u)$, our intuitions that ego networks have heavier tails holds as long as $\alpha \le 1$.

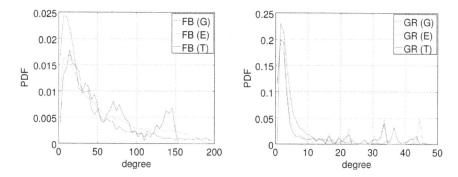

Fig. 2. Degree distributions of Facebook (left) and General Relativity (right). Red curves are for global degrees (G), green curves are for ego network degrees (E) and blue curves are for our theoretical approximation $\langle d_u^v \rangle_{ego}$ (T).

Table 2 summarizes empirical properties of degree distributions for all the data set. As predicted by our theory, $\langle d_u^v \rangle$ is usually greater than $\langle d_u \rangle$, except for two large networks Orkut and LiveJournal. Their low densities P_{glo} lead to disconnected ego networks, illustrated by high fraction of degree zero ego network instances $frac_{u,v}(d_u^v = 0)$, which breaks our approximation in Eq. 8.

In real social settings, the consistent bias of ego networks towards higher degrees can lead to wrong perceptions. For static features, over-representation of high degree hubs is identified as an important origin of "friendship paradox" and its generalizations [10]. The heavy tail degree distributions make the matter worse if the arithmetic mean is used [13]. Our analysis of degree distributions of ego networks show that both effects are still in play for structural features, leading to the surprising result $\langle d_u^v \rangle > \langle d_u \rangle$ even after the ego is taken out.

As a result, one should take extra caution when making claims about the global structure from local observations. Many popular connectivity and centrality measures for networks aggregates local structural features, they are thus potentially biased estimates of the global truth. However, our derivation of $\langle d_u^v \rangle_{ego}$ also shows that with appropriate assumptions, we can approximate global truth by its mathematical connection to local measures. In this case, we suggest using $\langle d_u^v \rangle = C_u(d_u - 1)$ to avoid over-representation, taking C_u in to account when your information is limited to u's neighborhood, and using medians instead of mean as suggested in [13].

3.2   Degree Assortativity and Clustering Coefficient

In the global network, degree correlations are heavily constrained by the degree distribution. With our understanding of the ego network degree distribution, we are ready to study its implications. According to Eq. 2, the assortativity of ego networks can be defined as

$$r_{ego} = \frac{1}{V[d_u^v]} \left(\sum_{k,k'=min(d_u^v)}^{max(d_u^v)} kk' e_{ego}(k,k') - E^2[d_u^v] \right) , \qquad (9)$$

where we plugged in ego network level features. Assortativity is largely determined by the difference between the positive terms $kk' e_{ego}(k,k')$ and the negative terms $E^2[d_u^v]$. By the results of last subsection, we know that $E^2[d_u^v]$ has generally become bigger. For the former, if we again assume that all the instances of a degree k vertex have the same degree $C(k)k$, we have

$$e_{ego}(kC(k), k'C(k')) = \frac{m(k,k')}{m_{glo}} e_{glo}(k,k') .$$

The change of the positive term depends on the details of $m(k,k')$, i.e. how edges are shared between triangles. Our empirical observations, however, confirms that degree assortativity are smaller in ego networks (see Table 3). The reduction in degree assortativity across ego networks is consistent with the argument of structural cut-offs. Since we know that ego networks generally have fatter tails and thus smaller γ, they are naturally more disassortative.

Next we analyze clustering coefficients of ego networks. Based on our knowledge of global features, clustering coefficients have very complicated dependencies with degree distributions and assortativities. However, our empirical measure reveals a very simple pattern (see Table 3).

As compared to global networks, ego networks display only slightly higher clustering coefficients. If we consider the ego network of vertex v a Erdös–Rényi random graph, then the local clustering coefficient C_v in the global network is the edge generating probability. Averaging it over all vertices we get C_{glo}. For the global network, this probability P_{glo} is orders of magnitude smaller than C_{glo}. The insignificant difference between C_{glo} and C_{ego} indicates that ego networks are much closer to random graphs than global networks. This observation confirms what Ugander et al. reported in their study of subgraph frequencies [23], where generative models with triangle closure is capable of reproducing higher order correlations observed in real world networks.

In fact, the probability of completing a triangle (i,j,k) given the edges (i,j) and (j,k), in the ego network of v, is equivalent to the probability of completing the 4-clique (i,j,k,v) in the global network, given the triangles (i,j,v) and (j,k,v). Assuming triangle completions at the global level are independent, with uniform probability C_{glo}, we can estimate C_{ego}^{rand} for ego network clustering,

$$C_{ego}^{rand} = 1 - (1 - C_v)(1 - C_u) \approx 2C_{glo} - C_{glo}^2 .$$

Table 3. Assortativities and clustering coefficients at global and ego network levels

Network	P_{glo}	C_{glo}	C_{ego}^{rand}	C_{ego}	r_{glo}	r_{ego}
Facebook	1.1E-2	0.61	0.848	0.76	0.064	-0.23
Orkut	2.5E-5	0.17	0.311	0.37	0.016	0.013
General Relativity	1.1E-3	0.53	0.779	0.63	0.66	-0.14
High Energy Physics	1.6E-3	0.61	0.848	0.85	0.63	-0.005
Enron email	2.7E-4	0.50	0.750	0.63	-0.11	-0.19
LiveJournal	4.3E-6	0.28	0.482	0.42	0.045	-0.248

Compared with the observed value C_{ego}, the constrains from degree assortativities is apparent. Ego networks with negative assortativities all have $C_{ego} < C_{ego}^{rand}$. The exception is Orkut, the only network with positive r_{ego}.

In real social networks, the bias of ego networks towards disassortativity and random triangles lead to a "flattened" view of the global world. If we all build social connections only with local information, assortative cliques are naturally formed even if we try to be open minded. Assortative communities are particularly prevalent in social networks, but this polarization effect is much harder to experience from individual perspectives. Similar lensing effect is also observed in Table 2, where the average degrees of neighbors in ego networks $\langle d_u^v \rangle_{nn}$ decrease from their global counterparts $\langle d_u \rangle_{nn}$, making the paradox seemingly weaker. While we cannot make precise corrections for degree assortativity and clustering coefficient based on local, incomplete samples, we should always remind ourselves that the former is usually underestimated and the latter actually captures correlations at a higher level than just triangles.

4   Conclusion

When only local information is available, statistical perceptions of networks structures deviates systematically from the global ground truth. In this work, we investigate the mathematical relationships between structural features at the global and ego network levels. We proposed a simple approximation of degree distributions of ego networks when the global distribution is known. Combined with empirical observations, we discovered that the heavier tailed degree distribution leads to more disassortative structures and random triangle completion at the ego network level. These insights could help us to better understand and correct for the biases arising from local and limited information in social networks, facilitating more accurate analysis of social behaviors.

References

1. Albert, R., Barabási, A.L.: Statistical mechanics of complex networks. Reviews of Modern Physics **74**(1), 47 (2002)
2. Amuedo-Dorantes, C., Mundra, K.: Social networks and their impact on the earnings of Mexican migrants. Demography **44**(4), 849–863 (2007)

64   S. Gupta et al.

3. Backstrom, L., Kleinberg, J.: Romantic partnerships and the dispersion of social ties: a network analysis of relationship status on facebook, pp. 831–841. ACM Press (2014)
4. Barabási, A.L., Albert, R.: Emergence of scaling in random networks. Science **286**(5439), 509 (1999)
5. Boguñá, M., Pastor-Satorras, R.: Class of correlated random networks with hidden variables, 68(3), 036112, September 2003
6. Boguñá, M., Pastor-Satorras, R., Vespignani, A.: Cut-offs and finite size effects in scale-free networks. The European Physical Journal B - Condensed Matter **38**(2), 205–209 (2004)
7. Christakis, N.A., Fowler, J.H.: Social network sensors for early detection of contagious outbreaks. PloS One **5**(9), e12948 (2010)
8. Clauset, A., Moore, C.: Accuracy and Scaling Phenomena in Internet Mapping (October 2004). eprint arXiv:cond-mat/0410059
9. Clauset, A., Shalizi, C.R., Newman, M.E.: Power-law distributions in empirical data (2007). Arxiv preprint arXiv:0706.1062
10. Feld, S.L.: Why Your Friends Have More Friends Than You Do. American Journal of Sociology **96**(6), 1464–1477 (1991)
11. Hodas, N.O., Kooti, F., Lerman, K.: Friendship Paradox Redux: Your Friends Are More Interesting Than You. ICWSM **13**, 8–10 (2013)
12. Kleinberg, J.: Complex networks and decentralized search algorithms. In: Proceedings of the International Congress of Mathematicians (ICM), vol. 3, pp. 1019–1044 (2006)
13. Kooti, F., Hodas, N.O., Lerman, K.: Network Weirdness: Exploring the Origins of Network Paradoxes. CoRR abs/1403.7242 (2014)
14. Leskovec, J., Faloutsos, C.: Sampling from large graphs. In: Proceedings of the 12th ACM SIGKDD International Conference on Knowledge Discovery and Data Mining, pp. 631–636. ACM (2006)
15. Leskovec, J., Krevl, A.: SNAP Datasets: Stanford Large Network Dataset Collection (June 2014). http://snap.stanford.edu/data
16. Leskovec, J., Mcauley, J.J.: Learning to discover social circles in ego networks. In: Advances in Neural Information Processing Systems, pp. 539–547 (2012)
17. Milgram, S.: The small world problem. Psychology Today **2**(1), 60–67 (1967)
18. Newman, M.E.: Mixing patterns in networks, 67(2), 026126, February 2003
19. Newman, M.E.: The structure and function of complex networks. SIAM Review **45**(2), 167–256 (2003)
20. Newman, M.E.: Ego-centered networks and the ripple effect. Social Networks **25**(1), 83–95 (2003)
21. Ravasz, E., Barabási, A.L.: Hierarchical organization in complex networks. Physical Review E 67(2), February 2003
22. Serrano, M.A., Boguná, M.: Tuning clustering in random networks with arbitrary degree distributions. Physical Review E **72**(3), 036133 (2005)
23. Ugander, J., Backstrom, L., Kleinberg, J.: Subgraph Frequencies: Mapping the Empirical and Extremal Geography of Large Graph Collections. ArXiv e-prints, April 2013
24. Vázquez, A., Pastor-Satorras, R., Vespignani, A.: Large-scale topological and dynamical properties of the Internet. Physical Review E **65**(6), 066130 (2002)
25. Watts, D.J., Strogatz, S.H.: Collective dynamics of 'small-world'networks. Nature **393**(6684), 440–442 (1998)
26. Zachary, W.W.: An information flow model for conflict and fission in small groups. Journal of Anthropological Research **33**(4), 452–473 (1977)

Conflict and Communication
in Massively-Multiplayer Online Games

Alireza Hajibagheri[1], Kiran Lakkaraju[2], Gita Sukthankar[1(✉)],
Rolf T. Wigand[3], and Nitin Agarwal[3]

[1] University of Central Florida, Orlando, FL, USA
{alireza,gitars}@eecs.ucf.edu
[2] Sandia National Labs, Albuquerque, NM, USA
klakkar@sandia.gov
[3] University of Arkansas, Little Rock, AR, USA
{rtwigand,nxagarwal}@ualr.edu

Abstract. Massively-multiplayer online games (MMOGs) can serve as a unique laboratory for studying large-scale human behaviors. However, one question that often arises is whether the observed behavior is specific to the game world and its winning conditions. This paper studies the nature of conflict and communication across two game worlds that have different game objectives. We compare and contrast the structure of attack networks with trade and communication networks. Similar to real-life, social structures play a significant role in the likelihood of inter-player conflict.

Keywords: Social network analysis · Massively-multiplayer online games

1 Introduction

Most online social media platforms are optimized to support a limited range of social interactions, primarily focusing on communication and information sharing. In contrast, relations in massively-multiplayer online games (MMOGs) are often formed during the course of gameplay and evolve as the game progresses [1]. Even though these relationships are conducted in a virtual world, they are cognitively comparable to real-world friendships or co-worker relationships [2]. The amount and richness of social intercourse makes it possible to observe a broader gamut of human experiences within MMOGs such as World of Warcraft [3], Sony EverQuest II [1,4], and Travian [5,6] than can be done with other data sources.

In particular, inter-player aggression is more openly expressed within MMOGs since combat often comprises a large portion of gameplay. This paper studies how conflict in MMOGs shapes the underlying social networks. In MMOGs, conflict and cooperation are inextricably linked since many attacks are launched by coalitions of players to gain resources, control territory, or subjugate enemies. For our analysis of conflict within MMOGs, we selected two browser-based games,

© Springer International Publishing Switzerland 2015
N. Agarwal et al. (Eds.): SBP 2015, LNCS 9021, pp. 65–74, 2015.
DOI: 10.1007/978-3-319-16268-3_7

Game X and Travian, in which there have been extensive previous studies of cooperation between players [5–10]. The two games differ in game objectives: there is no official winning condition for Game X. In contrast, Travian players develop their civilizations over a fixed period of time in order to be the first to build a magnificent Wonder of the World, a construct that can only be erected through extensive collaboration among a team of players. Constant raiding is required to amass the resources required to grow one's civilization.

To analyze the players' behavior, we constructed multiplex networks for both games, with link types for attack, communication, and trading. Our research compares and contrasts the network structure of players in both games; while there have been other studies of multiplex networks in a single game (e.g., [1,8]), there have been few studies that have looked for gameplay patterns across multiple games. For instance, griefing behavior was compared between World of Warcraft and Toontown, but not within the context of social network structures [11].

The overarching aim of our study is to understand the evolution of conflict in different game worlds. First, we discuss how differences in MMOG game objectives shape the structure of conflict. Then we study how the attack networks differ from communication and trade networks. Finally, we analyze how communication, trade, and geographic connections affect the likelihood of two players engaging in hostilities.

2   Related Work

Massively-multiplayer online games have been a fertile testing ground for many types of human studies, enabling scientists to overcome key difficulties in studying social dynamics by providing an experimental platform for collecting high resolution data over longer time period [1,3,5,6]. They have been particularly valuable for studying groups [3], teams, and organizations [5], since banding together yields economic and combat advantages in most games. Geographically-separated players must work together to achieve shared goals using a similar combination of email, chat, and videoconferencing as remote employees, hence game guilds can be viewed as analogous to virtual workplace organizations [5]. Trust between guild members is positively correlated with group performance [10], and willingness to grant shared access to property, items, or user accounts can be used to measure trust between players. Roy et al. [1] demonstrated that network structures in multiplex networks are very useful for predicting trust between MMOG players. Similarly we believe that the network structure of multiplex networks is correlated with the likelihood of conflict among players.

In real-life there are myriad potential motivations for choosing to fight. Humphreys and Weinstein [12] categorized key determinants of participation in conflicts as being long-term grievances (i.e. economic or political disenfranchisement), selective incentives (money or safety), and community cohesion. Community cohesion predicts that a person is more likely to join the conflict if they are members of a tightly-knit community and their friends have already joined.

This factor is the most relevant to fighting within MMOGs. Not only are there conflicts between guilds and alliances, but pick-up groups may spontaneously form to tackle larger challenges such as boss fights [13].

2.1   Game X

Game X is a browser-based exploration game in which players act as adventurers traveling a fictional game world in a vehicle. Similar to real-life, there is no absolute victory within the Game X world; instead players are engaged in an ongoing process of exploring the game world, mining resources, and engaging in commerce and battle with other players. Players can use resources to build factory outlets and create products that can be sold to other players. Unlike other MMOG's like World of Warcraft (WoW) and Everquest (EQ), Game X has a turn-based play system. Every day each player gets an allotment of turns. Every action (except communication) requires some number of turns to execute; example actions include: 1) move vehicle 2) mine resources 3) buy/sell resources 4) build vehicles, products, or factory outlets, 5) fight non-player characters 6) fight player characters. Players can communicate with each other through in-game personal messages, public forum posts and in chat rooms; they can also denote other players as friends or as hostiles.

In Game X fighting is one mechanism for advancement, but since fighting expends turns, the players must carefully weigh the tradeoffs between combat and pursuing an economic agenda. Players can engage in combat with non-player characters, other players, and even market centers and factory outlets. A player's skills affect the ability to attack/defend, and they can modify their vehicles to include new weaponry and defensive elements.

Game X contains four types of groups: 1) nations 2) agencies 3) races and 4) guilds. Wars in Game X can only occur between nations. There are three nations in the game, which were pre-defined by the game creators. Joining a nation provides access to restricted nation-controlled areas, quests, and vehicles. Each nation can have one of the following diplomatic relations to all others: benign, neutral, strained, or hostile. If enough members of a governing body select hostile diplomatic relations against another nation, a war is declared between the respective nations. After war has broken out, additional combat actions are available for the warring nations. In particular, war quests are available, which provide medals of valor to the players that wish to undertake and complete the quests. Any attack against the opposing nation results in accumulating a set number of war points. When the war ends, these war points determine the "winner" of the large-scale conflict. A war situation will (via the game's design) gravitate towards a state of peace. Each of the respective governing bodies must maintain a majority vote to continue the war effort. Over time, the amount of votes required to continue is increased by the game itself. Eventually, no amount of votes will suffice and the nations return to a state of peace.

Players also have a bidirectional "reputation" measure with the nations in the game. Combat with members of other nations incurs a negative penalty to this measure. During the war period this negative penalty is dropped for players of the two warring nations – allowing unrestricted combat.

2.2 Travian

Travian is a popular browser-based real-time strategy game with more than 5 million players. Games can be played in over 40 different languages on more than 300 game servers worldwide. Players start the game as chieftains of their own villages and can choose to be a member of one of three tribes (Gaul, Roman, or Teuton). Each of these three tribes has its own advantages and disadvantages. For instance, Teutons produce the cheapest military units and are the best raiders, whereas Gauls are the best at living in peace and have fast units and merchants. Players seek to improve their production capacity and construct military units in order to expand their territory through a combination of colonization and conquest. Each game cycle lasts a fixed period (a few months) during which time the players vie to create the first civilization to complete construction on one of the Wonders of the World. In the race to dominate, actors form alliances of up to 60 members under a leader or a leadership team. Alliances are equipped with a shared forum, a chat room and an in-game messaging system. Similar to the real world, teamwork and negotiation skills play a crucial role in game success.

Conflicts in Travian can be divided into two categories: attacks and raids. The goal of an attack is to destroy its target, whereas raids are meant to gather bounty and are much less vicious. The armies will do battle until at least one side is reduced in strength by 50%, and therefore the loss on both sides is usually smaller. For this paper, we construct an attack network from the raid data. A trade is an exchange of different resources (gold, wood, clay, wheat) necessary to upgrade a village's buildings. In Travian, villages may trade their resources with other villages if both villages have a marketplace. Travian has an in-game messaging system (IGM) for player communication. IGMs can also include broadcast messages, i.e. messages sent to all players by the game moderators. To study these processes we created trade and communication networks. However, in this analysis, broadcast messages were not considered as their volume could introduce bias in the results.

3 Analysis

The Game X dataset consists of data from 730 days, and the Travian data is composed of one game cycle on a high speed server in an expedited game (a period of 144 days). The experiments in this paper were conducted on a 30 day period in the middle of the Travian game cycle. This period has fewer transient bursts of activity and a more stable network than the early period (which has many less committed players who drop out) and the late period where the focus is on the Wonder of the World construction.

To analyze the relationship between conflict and communication in Game X and Travian, we identified a period of the game with comparable attack network statistics. First, for every day of the two games, we created a vector of features (the network properties from Table 1). Then, using the standard Euclidean distance measure, we find the most similar day from the Game X data for every

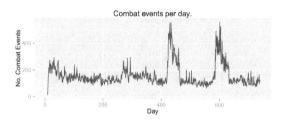

Fig. 1. Frequency of attacks in the Game X dataset (730 days). The peaks in activity represent the first and second declaration of war between nations.

Table 1. Travian and Game X attack, message, and trade network statistics

	Travian			Game X		
Parameter	Attack	Message	Trade	Attack	Message	Trade
# of Vertices	4418	3092	2649	2898	5112	5860
Frequency	633105	451669	271039	14270	907868	389629
Diameter	17	9	10	14	8	10
Avg. Path Length	5.312	3.471	2.849	4.476	3.402	4.218
Avg. Degree	7.998	14.591	32.828	8.375	27.645	25.01
Avg. Clustering Coefficient	0.065	0.319	0.154	0.012	0.137	0.117

Travian day. The features included: 1) nodes, 2) edges, 3) average path length, 4) diameter, 5) local transitivity, and 6) global transitivity. Results of this comparison reveal that the days found using this matching algorithm tend to fall in a 30 day period ([590,620]) within the second major Game X war (Figure 1). Table 1 shows the statistics for attack, trade, and message networks during the time period selected for this analysis.[1] In these networks, each node represents an individual player, and directed edges represent attacks, trades, and messages between players. Attack graphs in both Travian and Game X have a higher diameter, lower average degree, and lower clustering coefficient than either the message or trade graphs.

Although Travian and Game X possess many commonalities, an important difference between the two games is in the role of combat in gameplay. In Travian, attacking (raiding) is one of the easiest pathways for gaining the necessary resources for growing one's civilization. In Game X, attacks between players cause a loss of reputation if they are performed outside of war periods. The role of time is different in the two games. In Travian, players need to rush to grow their civilizations within a short period of time. In Game X, the players have infinite time to explore the richness of the world, but they can only take a limited number of actions per day. Each action spent fighting limits the number of actions available for economic development, hence Game X has a lower absolute frequency of attacks.

---

[1] This dataset has been made available at: http://ial.eecs.ucf.edu/travian.php

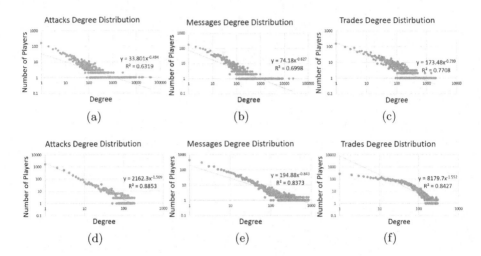

Fig. 2. Degree distribution (log-log scale) for (a) attack, (b) message, and (c) trade networks from Travian (top) and Game X (bottom)

Interestingly, the degree distributions of attack, messages, and trades in both games conform to a power law distribution (Figure 2). Clauset et al. [14] proposed a robust estimating technique to estimate the parameters of a power law; to verify the distributions, we used this method which employs a maximum likelihood estimator. This model calculates the goodness-of-fit between the data and the power law. If the resulting value is greater than 0.1 the power law is a plausible hypothesis for the data, otherwise it is rejected. Visually the trades in Game X appear to follow a double pareto lognormal distribution, with an exponential decay for higher trade values [15].

Assortativity is a preference for a network's nodes to attach to others that are similar in some way. Though the specific measure of similarity may vary, network theorists often examine assortativity in terms of a node's degree. Correlations between nodes of similar degree are found in the mixing patterns of many observable networks. For instance, in social networks, highly connected nodes tend to be connected with other high degree nodes. This tendency is referred to as assortative mixing, or assortativity. On the other hand, technological and biological networks typically show disassortative mixing, or dissortativity, as high degree nodes tend to attach to low degree nodes[16].

For Travian, as shown in Figure 3, while the message network displays disassortative mixing, attack and trade networks tend to show a non-assortative mixing. This suggests that players who send more messages are in contact with others who rarely send messages; communication in Travian often flows from alliance leaders outward to the other alliance members, reflective of a spoke-hub communication structure. In contrast, the degree of the members appears to be an unimportant consideration in dictating connectivity in the attack and trade networks. Non-assortiative networks may arise either because the networks

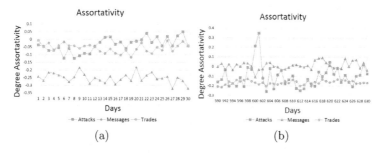

Fig. 3. Travian (a) and Game X (b) node degree assortativity

possess a balanced number of assortiative and disassortiative links or because a greater number of links in one direction is counterbalanced by a greater weight in the other [17].

For Game X, the attack and trade networks show a disassortative mixing while the message network displays non-assortative mixing. This indicates that attack and trade activity is centered around group leaders, whereas communication is more likely to be agnostic to node degree. Note that Game X has a more complicated group structure since players can belong to nations, guilds, and agencies.

Attacks in both Travian and Game X are generally inversely proportional to other types of activity. In Travian, in 41% of cases, players do not attack other players with whom they have been in contact at least once (Figure 4). On the other hand in Game X, the above statement stands for only 17% of attacks. 32% of attacks occur between players who have exchanged one message.

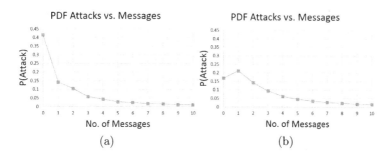

Fig. 4. Probability of attacks occurring between a pair of users vs. the number of messages they have exchanged (P(Attack and Message=x)) in Travian (a) and Game X (b)

In Travian a large number of players do not attack players with whom they have traded resources. As shown in Figure 5, 28% of the attacks in Travian occurred between two players without any trade history. However, this rate is surprisingly low in Game X; only 10% of attacks occur between players who lack a trade history. Once players have traded together, there is a sharp increase, followed by a slow decrease in attack frequency. Trading with other players indicates that they have desirable resources, making them worth attacking, and after only one trade, the players are unlikely to have established the sense of trust that may deter an attack. We believe that in some cases players who have never traded together or exchanged messages are geographically separated; hence they are less likely to attack each other because they are unaware of each other's existence.

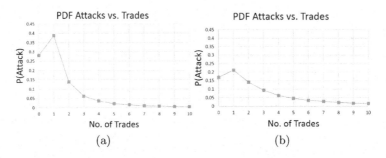

Fig. 5. Probability of attacks occurring between a pair of users vs. the number of trades they have made (P(Attack and Message=x)) in Travian (a) and Game X (b)

To test this hypothesis, we analyzed the probability of attack based on the distance between player territories in Travian (Figure 6). It was not possible to conduct a similar analysis in Game X, which has a spatial layout based on discrete tiles. To estimate distance, we calculated the territory centroids by averaging the latitudes and longitudes of the villages. Then, standard Euclidean distance was used to measure the distance between each pair of players in the attack network. Our analysis shows that attacks between immediate neighbors are rare. Attacks with close (but not immediate) neighbors are common, followed by a decay in attack activity with distance. There is a sharp peak at a distance of 100 which probably corresponds to attacks conducted at the borders of territories by players with large holdings.

Attacks are generally rare between alliance and guild members, indicating a strong level of trust in those relationships. In Travian, 4% of the attack edges are between two players within the same alliance. Surprisingly, the same rate stands for Game X, and only 4% of players attack their guild-mates.

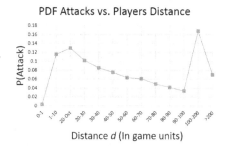

Fig. 6. Probability of attacks based on players' distance from each other

4 Discussion

In summary, our analysis reveals the following.

1. The attack networks in both Travian and Game X share a higher diameter, lower average degree, and lower clustering coefficient than either the message or the trade networks.
2. All networks in both games have similar power law degree distributions, but different degree assortativity. In Travian attack networks show non-assortative mixing, whereas in Game X they are disassortative.
3. The general trend is that attacks are inversely proportional to message frequency, trade frequency, and distance, with some specific exceptions. Players rarely attack fellow alliance or guild members in either game.

5 Conclusion and Future Work

This paper summarizes our findings across two massively multiplayer games with different game objectives, GameX and Travian. Despite the fact that Travian's game structure encourages a higher level of combat activity than Game X, attack networks in both games possess a similar network structure, and are distinctly different from message and trade networks. In future work, we hope to leverage these similarities to produce a general link prediction model for multiplex networks in MMOGs.

Acknowledgments. Research at University of Central Florida was supported by NSF award IIS-0845159. Sandia National Laboratories is a multi-program laboratory managed and operated by Sandia Corporation, a wholly owned subsidiary of Lockheed Martin Corporation, for the U.S. Department of Energy's National Nuclear Security Administration under contract DE-AC04-94AL85000. Research at the University of Arkansas at Little Rock was funded by NSF award 0838231.

References

1. Roy, A., Borbora, Z., Srivastava, J.: Socialization and trust formation: A mutual reinforcement? An exploratory analysis in an online virtual setting. In: IEEE/ACM International Conference on Advances in Social Networks Analysis and Mining, pp. 653–660 (2013)
2. Yee, N.: The labor of fun: How video games blur the boundaries of work and play. Games and Culture **1**(1), 68–71 (2006)
3. Thurau, C., Bauckhage, C.: Analyzing the evolution of social groups in World of Warcraft. In: IEEE International Conference on Computational Intelligence in Games, pp. 170–177 (2010)
4. Keegan, B., Ahmed, M., Williams, D., Srivastava, J., Contractor, N.: Dark gold: Statistical properties of clandestine networks in massively multiplayer online games. In: IEEE International Conference on Social Computing, pp. 201–208 (2010)
5. Korsgaard, M., Picot, A., Wigand, R., Welpe, I., Assmann, J.: Cooperation, coordination, and trust in virtual teams: Insights from virtual games. In: Online Worlds: Convergence of the Real and the Virtual (2010)
6. Wigand, R., Agrawal, N., Osesina, O., Hering, W., Korsgaard, M., Picot, A., Drescher, M.: Social network indices as performance predictors in a virtual organization. In: Computational Analysis of Social Networks, pp. 144–149 (2012)
7. Lakkaraju, K., Whetzel, J.: Group roles in massively multiplayer online games. In: Proceedings of the Workshop on Collaborative Online Organizations at the 14th International Conference on Autonomous Agents and Multiagent Systems (2013)
8. Lee, J., Lakkaraju, K.: Predicting guild membership in massively multiplayer online games. In: Proceedings of the International Conference on Social Computing, Behavioral-Cultural Modeling, and Prediction, Washington, D.C., April 2014
9. Alvari, H., Lakkaraju, K., Sukthankar, G., Whetzel, J.: Predicting guild membership in massively multiplayer online games. In: Proceedings of the International Conference on Social Computing, Behavioral-Cultural Modeling, and Prediction, Washington, D.C., pp. 215–222, April 2014
10. Drescher, M., Korsgaard, M., Welpe, I., Picot, A., Wigand, R.: The dynamics of shared leadership: Building trust and enhancing performance. Journal of Applied Psychology **99**(5), 771–783 (2014)
11. Warner, D., Raiter, M.: Social context in massively-multiplayer online games (MMOGs): Ethical questions in shared space. International Reviews of Information Ethics **4**, 46–52 (2005)
12. Humphreys, M., Weinstein, J.: Who fights? The determinants of participation in civil war. American Journal of Political Science **52**(2), 436–455 (2008)
13. Bennerstedt, U., Ivarsson, J., Linderoth, J.: How gamers manage aggression: Situating skills in collaborative computer games. Computer-Supported Collaborative learning **7**, 43–61 (2012)
14. Clauset, A., Shalizi, C.R., Newman, M.E.: Power-law distributions in empirical data. SIAM Review **51**(4), 661–703 (2009)
15. Seshadri, M., Machiraju, S., Sridharan, A., Bolot, J., Faloutsos, C., Leskovec, J.: Mobile call graphs: Beyond power-law and lognormal distributions. In: Proceedings of the ACM SIGKDD Conference on Knowledge Discovery and Data Mining (2008)
16. Newman, M.E.J.: Assortative mixing in networks. Physical Review Letters **89**(20), 208701 (October 2002)
17. Piraveenan, M., Chung, K.S.K., Uddin, S.: Assortativity of links in directed networks. In: Fundamentals of Computer Science (2012)

The Impact of Human Behavioral Changes in 2014 West Africa Ebola Outbreak

Kun Hu$^{(\boxtimes)}$, Simone Bianco, Stefan Edlund, and James Kaufman

Accelerate Discovery Lab, IBM Almaden Research Center, San Jose, CA, USA
{khu,sbianco,sedlund,jhkauf}@us.ibm.com

Abstract. The current outbreak of Ebola virus disease (EVD) in West Africa has caused around 23000 infections by middle of February 2015, with a death rate of 40%. The cases have been imported into developed countries, e.g., Spain and US, through travelers and returning healthcare workers. It is clear that the virus is a threat to public health worldwide. Given the absence of vaccine and effective treatment, response has focused so far on containment and education for prophylaxis. In studying the effects of human behavioral response to contain the current Ebola transmission, we built an epidemiological model in Spatio-Temporal Epidemiological Modeler (STEM), an open source platform. We simulate the course of the infection under various conditions from public available data and realistic assumptions about the disease dynamics. We ran this spatially extended simulation in three hardest-hit countries (i.e., Liberia, Sierra Leone and Guinea) in West Africa. A series of sensitivity analysis was performed to get insights of the likely human behavioral response to the change of disease epidemic, which helps understand the determinants of disease control. Our analysis suggests the reproductive number for the disease can be driven below 1.0 and effective control is possible if hospitalization occurs within 60 hours and/or if proper burials are processed within 34 hours. We also calibrated our model using processive period of reference data starting from March 14 reported by the WHO. We have an observation of gradual human behavior changes in the affected countries in response to the epidemic outbreak.

Keywords: Ebola · West Africa · Human behavioral changes · System dynamics model · Disease transmission

1 Introduction

The 2014 Ebola outbreak in West Africa has been declared by the World Health Organization (WHO) as the most serious health emergency of modern times [1]. Ebola, also known as Ebola Hemorrhagic fever (EHF) is an often fatal illness in humans. It is caused by infection with an RNA virus of the family Filoviridae, genus Ebolavirus. The virus can be spread through direct contact with infectious blood or body fluids (i.e., urine, saliva, feces, vomit, and semen), infected animals, and indirect transmission by contaminated objects (e.g., needles, bedding) [2]. The incubation period generally ranges between 2 and 21 days with average of 8-10 days [1]. Symptomatic

© Springer International Publishing Switzerland 2015
N. Agarwal et al. (Eds.): SBP 2015, LNCS 9021, pp. 75–84, 2015.
DOI: 10.1007/978-3-319-16268-3_8

infection lasts for about 2 weeks. The case fatality rate for the current outbreak is around 35% [2], which can be as high as 90% in historical outbreaks [3]. No specific vaccine or medicine has been yet approved to prevent or cure Ebola to date.

The current outbreak was triggered by a lab-confirmed case who died on Dec 6 2013 in Guinea [4]. A chain of infections in patient zero's family has taken place and many of them were dead. It is believed that the funeral served as a hub for disease dissemination to neighboring areas. Clusters of the disease soon appeared in Guinea, Liberia and Sierra Leone. The public health system was overwhelmed in the three worst-hit countries. According to the WHO latest report [1], more than 800 healthcare workers (HCWs) were infected and nearly 500 died, possibly because of shortages of personal protective equipment (PPE) or improper use, far too few medical staff for such a large outbreak and ineffective safety protocols. Public education of Ebola fell behind. Widespread distrust in the government was a more acute crisis which undermined the efforts to contain the outbreak. Many Ebola patients did not believe they contracted the deadly virus, but maybe Malaria with similar symptoms, and refused to be isolated in the treatment center. The ones that were admitted tried to escape from isolation center, even with the help of their relatives and friends. They considered isolation wards, or clinics are rather dangerous places to get infection of Ebola when mixing with other patients. Therefore, families have to care for sick relatives at home and risk infection themselves. Denial and community resistance greeted the government's efforts, meaning that the early Ebola cases in the city were difficult to identify and follow-up with. Coupled with a culture that values close interaction between mourners and the infectious deceased during a typical funeral practice in West Africa represents an important transmission venue for Ebola virus disease (EVD) in the current outbreak [1]. By the middle of January 2015, West African outbreak led to more than 23000 cases and claimed around 8500 lives [2]. Sierra Leone has been worse hit with >10000 cases and 3067 deaths by January 2015 [2]. The true toll may be three times as much because of the inaccurate reporting system [1].

2   Methods and Materials

In this paper, we proposed an epidemiological model to quantitatively assess the impact of human behavior changes on the disease progression under different conditions. In particular, we focus on effective communication and education to control the disease transmission by reducing contacts between susceptibles and Ebola patients through fast hospitalization, social distancing (e.g., isolation, self-quarantine), or cutting off interaction between relatives with deceased through prompt and safe burial.

We developed our Ebola model in an open source framework called Spatio-Temporal Epidemiological Modeler (STEM[1]). STEM can be used to quickly develop

---

[1] STEM (www.eclipse.org/STEM) is available through the Eclipse Foundation as open source software. STEM uses mathematical models of diseases (based on differential equations) to simulate the development or evolution of a disease in space and time. STEM includes a large library of reference data (including air travel) for the entire world. As an Eclipse application, every element of STEM (data, models, solvers, viewers) is reusable building blocks (Eclipse Plugins) based on the OSGI industry standard for component software.

any epidemiological model through an intuitive interface which makes use of easy model visualization. This feature allows researchers with limited programming knowledge to test interesting research hypothesis and understand the dynamics of Ebola epidemic.

2.1   Model

We extend a standard SEIR (Susceptible-Exposed-Infectious-Recovered) model with three additional compartments – death but not buried (D), hospitalization (H), and buried (B) shown in Fig.1, to properly take into account the realistic development of the Ebola infection dynamics for the current outbreak in West Africa. Infectious individuals (I) mix with susceptible people (S) in a community, potentially infecting their contacts at the rate of β_i, shown in the first expression of Eq.1. In this study, we assume that people who are exposed to EVD do not shed virus during incubation period (i.e., no asymptomatic transmission is modeled in this study shown in Eq.2). Infectious individuals may die (D), become hospitalized (H) where they are infectious at a different rate, or they may recover (R). Significant infection due to post-mortem contacts with an infectious corpse from the disease, at home or during funerals occurs a rate of β_d [5], captured by the third expression in Eq.1. The burial rate after death from the disease is modeled by the parameter δ, shown in Eq.5. Hospitalization occurs at a rate of τ, and recovery at a rate of γ_1. The rate of primary infection of clinical workers resulting from breakdown in infection control is captured by the rate of β_h^* indicated by the second expression in Eq.1 We note that the effective hospital transmission rate, $\beta_h^* = \alpha \cdot \beta_h$, is actually the product of two terms ($\alpha \cdot \beta_h$), the rate of infection in a hospital/clinical environment (β_h), and a re-scaling factor (α) that captures the fraction of the susceptible population actually exposed to contacts in a clinic (i.e., HCWs). In Liberia in 2010, there were approximately 2000 clinical workers (representing about 0.055% of the population) [6]. Infectious individuals and individuals in clinical isolation both die at rate of μ (shown in Eq.3) or recover from infection by rates of γ_1 or γ_2 respectively (see Eq.4). We assume for clinical mortalities that healthcare workers take effective measures so that no secondary infection occurs after death in a clinic. We model disease deaths outside of a clinic with the rate δ shown in Eq.5. The set of differential equations for this model (in Fig.1) are given in Eqs. (1-6).

$$\frac{dS}{dt} = -\beta_i SI - \beta_h^* SH - \beta_d SD \qquad (1)$$

$$\frac{dE}{dt} = \beta_i SI + \beta_h^* SH + \beta_d SD - \sigma E \qquad (2)$$

$$\frac{dI}{dt} = \sigma E - (\gamma_1 + \mu + \tau)I \qquad (3)$$

$$\frac{dR}{dt} = \gamma_1 I + \gamma_2 H \qquad (4)$$

$$\frac{dD}{dt} = \mu I - \delta D \tag{5}$$

$$\frac{dH}{dt} = \tau I - (\gamma_2 + \mu)H \tag{6}$$

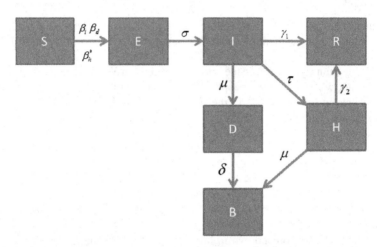

Fig. 1. Compartmental Model of Ebola in STEM is composed by seven disease transmission stages. The population is assumed to be homogenously mixing.

The basic reproductive number, R_0, is defined to be the number of secondary infections caused by one primary infection introduced to a fully susceptible population at a demographic steady state [7]. Using the Anderson and May approximation [8], it is straightforward to solve for R_0 in terms of the model parameters. The expression shown in Eq. (7) is used in the sensitivity analysis to estimate how the changes of human behavior (reflected by changing model parameters) affect R_0. The smaller R_0 becomes as a function of human behavioral control, the easier to contain the epidemic outbreak. In the case, the human behavioral intervention is so effective that R_0 is below one, the disease epidemic can be successfully contained.

$$R_0 = \frac{\beta_i + \beta_d \dfrac{\mu}{\delta} + \beta_h^* \dfrac{\tau}{\mu + \gamma_2}}{\mu + \tau + \gamma_1} \tag{7}$$

2.2   Model Calibration

We gathered daily Ebola incidence data reported for Guinea, Sierra Leone and Liberia by WHO [1] and HealthMap [11] from March 14 to October 28, 2014. The empirical data was used as reference to calibrate the Ebola model using STEM's implementation of the Nelder Mead Simplex (NMS) algorithm [12]. For this optimization, we chose a normalized mean square error (NMSE) between the aggregate incidence estimated by the model and the aggregate incidence in the reference data. We assume the transmissions happened in the hospital (isolation wards, clinic) between HCWs and patients are very limited though existing. Therefore, we focused our calibration effort on values of two parameters which we thought are important but difficult to acquire from the literature: 1) infectious transmission rate (β_i) and 2) postmortem transmission rate (β_d). The time series data of incidences plotted in Fig.2 compare the epidemic trajectory outputted from model (in green) in calibration process with empirical data (in red) that is public available from WHO. In addition, we conducted a series of calibrations using incremental period of data collected in order to evaluate the possible human behavioral changes in affected regions reflected in the empirical data. The parameter values used and obtained from the literature and calibration are shown in Table 1 with the reported range surveyed from the literature.

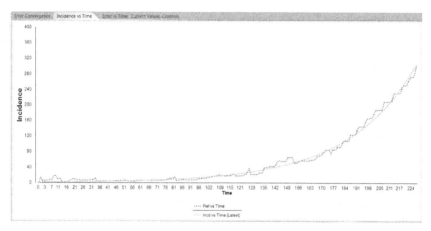

Fig. 2. Model calibration output: fitted Ebola model output (shown in green curve) vs. empirical incidence data (in red) used as reference for model calibration

Table 1. Summary of model parameters obtained from literature and model calibration

Model input / Parameter	Value	Range	Sources (ref.)	Unit
R_0, basic reproductive number	1.32	1.30-2.30	[5, 8, 9]	dimension-less
β_i, infectious transmission rate	0.29*	0.16-0.45	[8, 10]	day^{-1}
β_d, postmortem transmission rate	0.65*	0.16-0.69	[8, 10]	day^{-1}
β_h*, infectious transmission rate from admitted in hospital	0.00016	N.A.	[6, 10]	day^{-1}
σ, incubation rate	0.10	0.08-0.19	[2, 5, 9]	day^{-1}
γ_1, infectious recovery rate human in general community	0.07	0.05-0.18	[2, 5, 8, 9]	day^{-1}
γ_2, infectious recovery rate of human from hospital	0.10	0.05-0.18	[2, 5, 8, 9]	day^{-1}
μ, disease death rate	0.12	0.08-0.13	[2, 9]	day^{-1}
δ, corpse burial rate	0.33	0.22-0.50	[2, 9]	day^{-1}
τ hospital admission rate	0.20	0.20-0.31	[9]	day^{-1}

N.A.: not available

* calibrated value from the WHO public available data

2.3   Sensitivity Analysis

To measure the sensitivity of the epidemic to key model parameters, a series of experiments were run using the Runge-Kutta Cash-Karp deterministic solver. The populations of the three considered countries are based on census data in 2010. We employed the same initial condition of simulation as used in the process of model calibration. Each simulation started on Dec 1, 2013, and was seeded with two infectious cases in Guinean, five cases in Liberia and five cases in Sierra Leone. We run the simulation for 3 years and log daily results for analysis shown in Figs 3-5. For some parameter values, the epidemic does not go extinct at the end of allotted simulation time. Since the reference incidence data was collected at country level (administrative level 0) resolution, the simulations and sensitivity analysis were also done at the same administrative level. The experimental parameters, hospital admission rate (τ) and burial rate (δ) were both varied in the range of [0, 1], with steps of 0.1. Two transmission rates, infectious transmission rate (β_i) and postmortem transmission rate (β_d), are experimented in the range of [0, 2] with steps of 0.2. For each parameter value, we also report the corresponding R_0 value using the formulation reported in Equation (7). These four parameters, while not representing the whole spectrum of behavioral modifications that can be expected during a crisis, are directly linked to a number of behavioral modifications which can impact the outcome of an Ebola outbreak. More specifically, and given the state of the healthcare in the three countries, both hospital admission rate and burial rate are direct consequence of individual or community choices instead of a mere lack of health infrastructure (which, of course, has not to be underestimated). The same can be said of transmission rates, especially in the light of a disease like Ebola with a lower intrinsic transmissibility than other diseases (e.g., influenza). We gather the following metrics for each simulation: 1) time to epidemic peak (in days), and 2) epidemic peak infection (in persons) in order to evaluate specifically how the behavioral changes can alter the trajectory of epidemic progression.

3   Results

We run a series of calibrations for infectious transmission rate (β_i) and postmortem transmission rate (β_d) using partial reference data always starting from March 14 with incremental time window to examine possible human behavioral changes as time goes by in the affected regions. It is interesting to see a trend of small decline in disease transmission between susceptible and infectious populations (β_i) starting from September 5th 2014. It may indicate the onset of consistent behavioral changes, such as social distancing in the community or willingness to go to the hospital seeking for care, maybe as a consequence of a higher confidence of the general public towards public health officials triggered by intensive education campaigns. Unfortunately, the decline of infectious transmission (β_i) is not dramatic. Meanwhile, we observed a strong contribution to the epidemic curve from the direct interaction between deceased and their contacts (β_d) mainly because of the traditional burial/funeral practices in West Africa. People may not have been aware of the possibility that the Ebola infection could occur after death at the beginning of the outbreak. Or they chose to believe that their relatives died not because of Ebola.

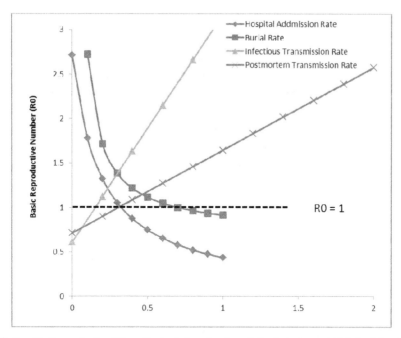

Fig. 3. Sensitivity analysis of the effect of changing hospital admission rate (τ), burial rate (δ), infectious transmission rate (β_i) and postmortem transmission rate (β_d) on containing the disease transmission in terms of R_0

We show the dependence of the reproductive number, R_0, on the changes of four experimented parameters: hospital admission rate (τ), burial rate (δ), infectious transmission rate (β_i) and postmortem transmission rate (β_d) (refer to Table 1) in Fig.3. Shown by red and blue curves, increasing either hospital admission rate or burial rate (i.e., decreasing time period), which reduces the possibility of interaction between infectious/deceased with susceptible contacts in the public community, decreases R_0. In particular, we find that R_0 falls below 1.0 if infectious individuals are willing to go to the hospital within 60 hours (2.5 days) of symptoms onset. This result suggests the importance of timely human behavioral response of new incidence in a community, and how to effectively reduce the possibility of secondary infections by altering human behavior alone. The red curve shows that burial should be performed in less than 34 hours from death by professional burial team so that the outbreak can be effectively under control. Decreasing either transmission rates also can achieve the same effect in containing the disease (decreasing R_0). Shown in green line, social distancing (isolation or quarantine) of infectious individuals could dramatically decrease the disease diffusion, especially between infectious individuals and their family members, close friends and neighbors. Similar effect is also observed from purple line for safe burial practice which suggests changing normal behavior at funeral to avoid direct contact with deceased in West Africa.

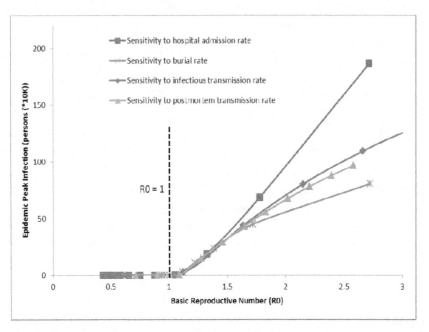

Fig. 4. Sensitivity analysis of the influence of varying hospital admission rate (τ), burial rate (δ), infectious transmission rate (β_i) and postmortem transmission rate (β_d) on the epidemic peak infection (in persons) plotted as a function of R_0

The results of sensitivity analysis are shown in Figs.4 and 5. We plotted the epidemic peak infection (numbers of cases) and the time to peak infection as a function of R_0 respectively. R_0 is varied based on the parameters shown in legend. We note that the intensity of the outbreak increases in a similar fashion when increasing either rate. Increasing the transmission rate from unburied individuals (β_d) makes the disease peak earlier with respect to an increase of infectious transmission rate (β_i). It is shown in Fig. 4 that the epidemic trajectory of peak infection is more sensitive to the hospital admission rate (τ) whereas the infectious transmission rate (β_i) primarily effects the epidemic peak time (Fig.5). While the containment of the disease when hospitalization and burial rates are increased is not surprising, it is necessary to point out the effect that the two rates have on the features of the outbreak (duration, peaks, etc.), the upper bound for the cut off times of epidemic development depending on the magnitude of the rates, as well as the fact that the analysis is performed with very high spatial resolution and accurate human mobility patterns.

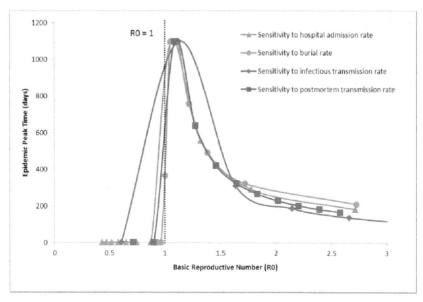

Fig. 5. Sensitivity analysis of the influence of varying hospital admission rate (τ), burial rate (δ), infectious transmission rate (β_i) and postmortem transmission rate (β_d) on the epidemic peak time (in days) plotted as a function of R_0

4   Conclusion

The 2014 Ebola outbreak has surprisingly shown a sustained prevalence in the population. At the time this is written, the disease is still spreading exponentially in the population. Sierra Leone alone has gained 30% more new cases in Dec 2014 [2]. Cases also appeared in countries, e.g., US, Spain, other than the ones where the disease has originated [2]. The general consensus about the reason for the observed pattern is behavioral rather than merely biological. Infectious tend not to report themselves or those in close proximity for fear of repercussion or social exclusion, while dangerous burial practices are still employed in several parts of the affected regions despite efforts to educate people and to deploy safe burial teams. On the other hand, the scarcity of healthcare facility and necessity is also a bottleneck to effectively control the Ebola outbreak. The WHO estimates 28 laboratories are needed, with 12 now in place, and 20,000 staff will be required to track people at risk through contact tracing. They also need 230 burial teams by December 1st but only have 140 [1]. To this extent, it is important to quantify the risk of the disease under current outbreak conditions in order to provide strategic suggestions on human behavioral modifications and effective interventions for public health officials.

We have presented an open source spatially extended mathematical model for the transmission of Ebola virus, implemented in STEM. The model explicitly considers hospital admission and transmission during burial/funerals, as well as the possibility of transmission to healthcare workers. A sensitivity analysis performed on the model shows that disease containment can be achieved through prompt hospitalization or isolation/quarantine (within 2.5 days) and prompt burial of infectious corpses (within 34 hours) by professional teams. It is also important to detect onset of infection as early as possible to accelerate isolation. Effective contact tracing is therefore essential in communities with high risk. Moreover, the outbreak profile seems to be more strongly influenced by the hospital admission rate, which calls for a concerted effort to shorten the time to admission of newly infected from the onset of disease symptoms. While more and detailed data will become available with time, we think that a proper identification of the time constraints for the disease spread is paramount towards controlling this and future outbreaks.

Reference

1. World Health Organization (WHO), Fact Sheet on Ebola Virus (2014). http://www.who.int/mediacentre/factsheets/fs103/en/
2. CDC (2014). http://www.cdc.gov/vhf/ebola/
3. Legrand, J., et al.: Understanding the dynamics of Ebola epidemics. Epidemiol Infect. **135**(4), 610–621 (2007)
4. Baize, S., et al.: Emergence of Zaire Ebola Virus Disease in Guinea. New England Journal of Medicine **371**(15), 1418–1425 (2014)
5. Towers, S., Patterson Lomba, O., Castillo Chavez, C.: Temporal Variations in the Effective Reproduction Number of the 2014 West Africa Ebola Outbreak. PLOS Currents Outbreaks September 18, 2014 (Edition 1.)
6. Ministry of Health and Social Welfare and Liberia Institute for Geo-Information Services, Government of Liberia, The national census of health workers in Liberia (2010)
7. Anderson, R., et al.: Infectious Diseases of Humans: Dynamics and Control. Oxford Science Publication (1991)
8. Althaus, C.: Estimating the Reproduction Number of Ebola Virus (EBOV) During the 2014 Outbreak in West Africa. PLOS Currents Outbreaks September 2, 2014 (Edition 1.)
9. Gomes, M.F.C., et al.: Assessing the International Spreading Risk Associated with the 2014 West African Ebola Outbreak. . PLOS Currents Outbreaks September 2, 2014 (Edition 1.)
10. Caitlin, M.R., et al.: Modeling the Impact of Interventions on an Epidemic of Ebola in Sierra Leone and Liberia. PLOS Currents Outbreaks October 16, 2014 (Edition 1.)
11. HealthMap (2014). http://healthmap.org/ebola/#timeline
12. Nocedal, J., Wright, S.J.: Numerical Optimization. Springer Series in Operations Research, ed. Glynn, P. Robinson, S.M., Springer Science & Business Media (1999)

Social Complexity in the Virtual World
Distributional Analysis of Guild Size in Massive Multiplayer Online Games

Vince Kane[(⊠)]

George Mason University, Fairfax, VA, USA
vkane2@gmu.edu

Abstract. Guild sizes for five massive multiplayer online game guilds, and their aggregate data set, were analyzed and tested for fit against four candidate distributional models: power law, linear, exponential, and log-normal. Results provide very strong support of a power law distribution, especially in the aggregated data set, for which all other distributions were rejected. This indicates that a non-equilibrium, scale-free process underlies the dynamics of guild formation. One can conclude that social processes that generate similar distributions in "real world" organizations are extendable to social life in virtual worlds.

Keywords: Social · Complexity · Virtual · Distribution · Power · Binned · Analysis · Guild · Online · Game

1 Introduction

Beginning with Pareto's studies of income, quantitative analysis of the distributional characteristics of social science data has a long tradition (the Pareto distribution is in fact a power law). Power laws take the form (most generally) $P(Event|x) = Cx^{-\alpha}$, where C is a scaling parameter and α is the scaling exponent. Over the last century, quantitative distributional analyses specific to human organizations and social institutions have been receiving increasing attention, and a number of power laws have been demonstrated here as well: rank-size, or "Zipfian", for cities; business firm sizes, (Axtell, 2001) among others; (Barabasi & Albert, 1999): degree distribution of social networks; (Liljeros, Amaral, & Stanley, 2003) on voluntary organizations. The discovery of a power law distributional model, as well as other so-called "non-equilibrium" distributions (for example exponential and log-normal), are significant and deserve special consideration because, being non-Gaussian and exhibiting the property

---

I would like to thank Claudio Cioffi-Revilla for guidance provided while instructing the graduate seminar "Computational Analysis of Social Complexity" (Spring 2014, GMU), and to classmates from the same who brought my attention to the methods of CSN 2009 and VC 2014 (referenced in the paper) for testing power law distributions and working with binned data.

© Springer International Publishing Switzerland 2015
N. Agarwal et al. (Eds.): SBP 2015, LNCS 9021, pp. 85–92, 2015.
DOI: 10.1007/978-3-319-16268-3_9

"many-some-rare" vice "rare-many-rare", they indicate an underlying complexity beyond random interaction and diffusion-type processes — thus not attributable to equilibrium processes — and explain, beyond mere statistical variation, the much higher likelihood of extreme-valued events than would be expected in a Gaussian distribution generated by an equilibrium process, or even other non-equilibrium distributions.

This paper examines the distribution of guild sizes for massive, multiplayer online (MMO) games. MMO games are characterized by a persistent virtual world in which large numbers of players may interact simultaneously. MMO guilds are selective, formalized voluntary organizations with exclusive membership (players belong to only one guild), formed for the purpose of organizing and facilitating various aspects of game play.

A natural question that the social scientist familiar with the results and methods of complexity-theoretic analysis of social science data might ask is: what does the distribution of the guild sizes look like, and its natural corollary, what could account for that distribution? This paper addresses the former question.

Discovery of a power law model of online guild sizes might indicate that theoretical explanations for generative and change mechanisms accounting for power law distributions in "real world" organizations could be applied to people's social behaviors in the "virtual" world; likewise, such explanations found in the virtual social life could serve as a model for social behavior generally – studies of virtual social behaviors are in many cases more tractable because of higher availability and better accessibility of data. Furthermore, such studies will become increasingly important as humanity's virtual presence increasingly integrates with our day-to-day, offline activities. This study and similar studies of virtual organization distributions (forum participants, news media readership, just to name a couple) would contribute to our understanding of social complexity for both the present and its evolution in the future.

2   Methodology

The methods of (Clauset, Shalizi, & Newman, 2009) (hereafter "CSN 2009") and (Virkar & Clauset, 2014) (hereafter "VC 2014") were used, with some modifications. CSN 2009 provides a rigorous algorithmic approach to testing for power laws with both continuous and discrete data. VC 2014 goes a step further by extending the methods of CSN 2009 to working with binned data – a data representation that is all too common, and unavoidable, and only binned data was available for this study.

Guild size data were obtained for five high-population MMO games: Lord of the Rings Online^{TM}(LOTRO), Everquest II^{TM}(EQ2), Eve Online^{TM}(Eve), Age of Conan^{TM}(AoC), and Guild Wars 2^{TM}(GW2). The data were manually "scraped" from the website ("MMORPG Guild List," 2014), and are shown in Table 1. The choice of games for which to analyze guild size was arbitrary, chosen for no particular reason other than that they are well-known in the industry and had a fair number of guilds (over 75 each); the analysis could be reapplied to other MMO games without loss of generality.

Table 1. Guild size bins for five MMOs, with totals. Values in parentheses are prior to elimination of duplicates.

size bin	LOTRO	EQ2	EVE	AoC	GW2	All
1-20	64	28	41	51	117	301
20-50	45	20	29	42	63	199
50-75	20	8	12	27	19	86
75-100	19	7	11	21	16	74
100-150	12	7	8	18	14	59
150-300	13	4	4	25	13	59
300-500	5	3	2	13	3	26
500-1000	0	1	1	7	2	(11) 7
1000+	7	1	1	5	8	(22) 18

The column "All" contains totals for each of the bins across all games examined. This aggregated data was modified to eliminate duplicates in the two largest bins (500-1000 and 1000+), where multiply-counting larger guilds that played multiple games would artificially – and unjustifiably – increase the weight of the upper tail. A review of the guild data did not identify duplicates in smaller guilds to be a sufficiently significant factor in the aggregate data.

The analysis was implemented in the Python programming language.[1]

As identified previously, the process recommended by VC 2014 was implemented, with two major deviations: 1) in place of the log-likelihood function for estimating the scaling parameter α, I estimated using a least-sum-of-squared-error, and 2) in place of the likelihood ratio test for comparing goodness-of-fit against alternate distributions, direct $p - value$ comparative analysis was used. These deviations will be expanded on in the process description below.

Four distributional models were tested for fit to the empirical data: power, linear, exponential, and log-normal (a visual inspection is sufficient to rule out a linear fit, but was included for completeness). For brevity, I summarize here only the main algorithmic steps of VC 2014; calculation details may be found in the reference except where I have differed, and for which I will supply the detail.

Step 1 — Estimate the scaling parameter α. VC 2014 identify the log-likelihood function as the maximum likelihood estimator for the exponent of the power law; however, there are a number of problems using this function: 1) it is only evaluable at $\alpha > 1$ (since bin sizes increase), but best estimates for a power law distribution on the data report $\alpha < 1$; 2) it does not estimate the scaling constant C of a presumed power law distribution; and 3) while it may be appropriate for estimating a presumed power law (with $\alpha > 1$), it cannot be used to estimate the parameters for alternative distributions. One of our goals is testing power law models against alternative distributions, and must of needs estimate those distributions from the same empirical data.

---

[1] The Python source code is available online at *github.com/strangeintp/mmo-guild-sizes*.

Therefore, all distributional models tested were parametrically estimated against the empirical data using Nelder-Smead optimization of the minimize() function in Python's scipy.optimize package, using sum-squared error as the objective to be minimized.

Step 2 — Estimate the lower bound of the upper tail. Based on the Hill estimation technique, this was implemented as described in VC 2014: report the K-S statistic for all potential upper tail lower bounds b_i in the range of validity and use as b_{min} the b_i that reports the lowest K-S statistic.

Step 3; Step4 — Test for goodness-of-fit; Compare against alternative distributions. My implementation combines these two steps, since p-values are used for both. The essence of the p-value test in VC 2014 and CSN 2009 is to generate random data using an estimated distributional model, perform yet another parametric estimation on that data, and compare the K-S statistic of this second estimation vs. the empirical data, against that of the first estimation vs. the empirical data. Higher K-S of second-order estimators (over many synthetic data sets) indicates that the variation of the empirical data on a presumed idealized estimated model is less than the variation generated randomly, implying a potential good fit.

In this analysis, p-values were calculated for each tested distributional model. 2500 synthetic data sets were generated for each distributional model, providing 10^{-4} resolution in p-value. In order to minimize the random variation between synthetic data sets for tested models, each time a random number was drawn for a sample in a data set, it was applied to all distributional models, so that essentially the same random number sequence was applied to all models, but resulting in different binning according to the distribution of the estimated model. Finally, VC 2014 and CSN 2009 recommend a likelihood ratio test for comparing two distributional models for fit; in this analysis, however, the p-values were computed for all candidate models and compared directly to evaluate a most likely distributional model, or reject any with low value (less than 0.1). All methods described above were tested against exact linear and log-normal distributions for verification, with positive affirmation.

3   Results

Table 2 summarizes the goodness-of-fit test results from the analysis. For two out of the five games and the aggregated data, the p-values from the analysis clearly reject other distributions besides power law. For EQ2, Eve, and AoC, a log-normal distribution is also supported.

Figure 1 shows the best parametric-estimated fits for each examined distribution to the aggregated data, on natural logarithm transformed axes.

3.1   Hill Estimation

Figure 2 shows K-S statistic results from the Hill estimation technique, with fits obtained on the aggregate data using sum squared error for solver error minimization.

Table 2. p-values for all analyzed data sets, epsilon = 0.01 (2500 synthetic data sets per column).

distribution	All	LOTRO	EQ2	Eve	AoC	GW2
power	0.471	0.6293	0.9624	0.9508	0.8952	0.2759
linear	0	0	0.3942	0.0308	0.0412	0.0144
exponential	0	0.0004	0	0	0	0
lognormal	0	0.0144	0.9812	0.8369	0.9524	0.0504

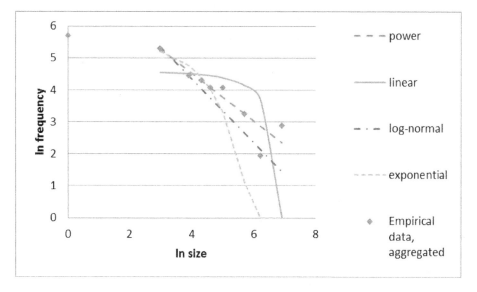

Fig. 1. Log-frequency vs. log-size, with best parametric fits against aggregated data for examined distributions, minimizing sum squared error. Note exclusion of the smallest bin (1-20) based on Hill estimation results.

Hill estimating did not use bins larger than 300-500 because of proximity to the upper tail, where sampling error is more pronounced, and in any case would have eliminated most of the useful data from the analysis.

The lowest K-S statistic for the aggregate data (from among the allowed bin size choices) obtained from a power law fit with an upper tail bin lower boundary of 20 (the "20-50" size bin), using sum squared error for optimization. The resulting power law exponent and scaling factor are $\alpha = 0.7463$ and $C = 1831.5$.

Hill estimation was similarly performed on the game-specific guild size distributions, with "best" (lowest K-S statistic) summarized in tabulated form in Table 3.

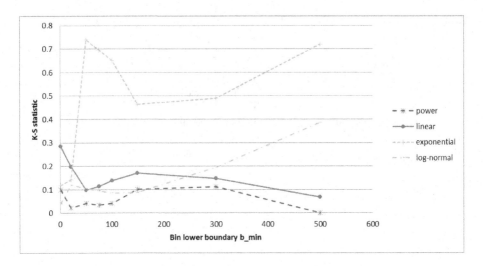

Fig. 2. Hill Estimation results on aggregate data: K-S statistics for tested distributional models to aggregated data upper tails with varying minimum bin size. The procedure reported a best (lowest) K-S statistic of 0.0222, for a power law b_{min} at 20-50.

Table 3. Results of Hill estimation for specific games: best K-S, upper tail minimum bin, and distributional model for each game's guild sizes.

Game	b_{min}	best K-S	best fit model
LOTRO	20	0.0374	power
EQ2	100	0.0268	power
Eve	75	0.0254	power
AoC	150	0.0183	power
GW2	50	0.0606	power

4   Discussion

The generated p-values, especially for those of the aggregate data, strongly support a power law distribution, indicating an underlying scale-invariant, complex process is involved in guild membership. In particular for the aggregate data, the other distributions are rejected since their p-values $\ll 0.1$ [CFN 2009]. Even visually, the power law estimate produces an astoundingly good fit with the exception of the second largest bin (500-1000) and the smallest bin (1-20). The K-S statistics as reported by the Hill estimation technique also support this conclusion.

In the cases (EQ2, Eve, and AoC) where a log-normal distribution is supported as well, the support is weak (p-values lower or not much higher than for a power law), or directly contradicted by alternate metrics of fitness (Hill estimation). Additionally, it is unlikely that one game's guild sizes would have an inherently different generative mechanism – resulting in a different distribution – than other high-membership online games.

Of note are the low bin boundaries identified for upper-tail inclusion by the Hill estimation – in fact, in most cases the power law behavior is a remarkably better fit when the lower bins are included. Additionally, it provides a consistently better fit over the whole range of bins examined, even if at certain minimum bin sizes it is slightly worse. This provides more support of scale-invariant behavior, and indicates that it extends to fairly low group sizes. Better fit with the upper tail inclusion of low bin sizes may be due to a number of factors: higher resolution in the lower bins, and truncation in the uppermost bin.

What the results also indicate universally is that scale invariance does not extend into the lowest bin, 1-20. While the phenomenon of low-size nonconformance is not uncommon in the study of power law-described behavior, it is also all too commonly dismissed — for example, as "missing" data; however, it is not apparent that explanation is justifiable here. Gaming guilds are somewhat formalized organizations of people that group together to play games, and a large motivating factor is recruitment; as such, it is unlikely that large numbers of small guilds are under-reported. On the other hand, there are certainly many small groups of people who play games together, who have no motivation or ability to form a formalized guild for recruitment and retention, or whose membership is too small to warrant the additional overhead and coordination costs of guild formation. Whether or not these small informal groups constitute "missing" data is an open question, but could possibly be answered with field work to identify how many non-guild small group teams there are, and whether that data conforms to a power law model.

An insightful extension to this work would be to examine the scaling exponent and minimum bin for upper tails for guilds in various games against the frequency and types of play activities in the game. For example, the analysis identified AoC as an outlier, having a higher minimum bin (150) for the upper tail than the other games (20 to 100). Might this indicate a causal factor in the game itself, providing less internal guild pressure to grow beyond a certain size? Social psychologists involved in game research or game business development might do well to investigate this.

A power law distributional model for online game guild sizes, as looks likely, might be generated by mechanisms similar to other voluntary social organizations. As with many organizational size distributions characterized by power laws, social networks have been found to be a crucial contributing factor. (Liljeros, 2008) describes a simulation model that exhibits scale-free characteristics due to clustered propagation processes. The well-known (Barabasi & Albert, 1999) paper shows that growth and preferential attachment are sufficient to produce scale-free social network degree distribution. If social networks are the driving force behind the size of social groupings – including voluntary organizations such as online gaming guilds – and if social networks exhibit power law size, then it is not surprising to find that social institutions also exhibit power law size.

5   Conclusion

The methods of CSN 2009 and VC 2014 were applied, with slight modification, to identify the best candidate distributional model for MMO game guild sizes. For individual games, a power law model was found to be a likely candidate fit to the empirical data with some support for log-normal. However, a power law was a clear best fit for the aggregate data, since other distributional models were rejected. Hill estimation was used to determine the most likely valid range of the upper tail, and also reported power law as the most likely distributional model.

Strong support for a Type II (absolute frequency) power law distributional model of MMO guild sizes indicates that scale-invariant social dynamics are generating and maintaining participation in
guilds. Social scientists with an interest in MMO game research and development could analyze the scaling exponent and minimum bin size for individual games to help identify potential causal differences stemming from game play.

Finally, these results indicate that scale-invariant social dynamics extend to sociality in virtual worlds — as might be expected; further related work on distributional analysis of virtual presence in other types of online social organizations would add to the body of knowledge, and could contribute to the formation of a general complexity theory for our social life, virtual or otherwise.

References

Axtell, R.L.: Zipf distribution of U.S. firm sizes. Science **293**, 1818–1820 (2001)

Barabasi, A., Albert, R.: Emergence of scaling in random networks. Science **286**(5439), 509–512 (1999)

Clauset, A., Shalizi, C.R., Newman, M.E.: Power-law distributions in empirical data. SIAM Review **51**, 661–703 (2009)

Liljeros, F.: The effect of social clustering on the size of vol- untary organizations. In: Cioffi-Revilla, C. (Ed.), Power Laws and Non-equilibrium Distributions of Complexity in the Social Sciences (pp. 66–75). Fairfax, VA: open source (2008)

Liljeros, F., Amaral, L.A., Stanley, H.E.: Universal mecha- nisms in the growth of voluntary organizations (2003). Retrieved from arxiv.org/abs/nlin/0310001

MMORPG Guild List. (2014, May). Retrieved from www.mmorpg.com/guilds.cfm/

Virkar, Y., Clauset, A.: Power-law distributions in binned empirical data. Annals of Applied Statistics **8**(1), 89–119 (2014)

A Model of Policy Formation Through Simulated Annealing: The Impact of Preference Alignment on Productivity and Satisfaction

Scott Atherley, Clarence Dillon, and Vince Kane[✉]

George Mason University, Fairfax, USA
{satherle,cdillon2,vkane2}@gmu.edu

Abstract. We are interested in how preference correlations can impact policy-maker productivity and their satisfaction with resultant policy. We applied a simulated annealing process as a model of revising draft legislation in peer and committee reviews before submission to a floor vote. Results indicate that having exogenous, common issue priorities is required for productivity but that some structures inhibit productivity, particularly where preference schedules are uncorrelated. Our model also demonstrates lower system efficiency, and lower overall satisfaction, as policy is negotiated through compromise to achieve higher production.

Keywords: Simulated annealing · Policy formation · Organizational theory · Preference alignment · Network structure

1 Introduction

It is a trope of Western democratic ideals that a democracy—governance by vote—produces the greatest happiness for its citizenry, or the least unhappiness for a majority of it. Our intuition is that diversity and divides in policy preferences can result in legislative deadlock and lower overall satisfaction with legislative outcomes. Our initial motivation for this research was consideration of the common assertion that a dysfunctional Congress (if measured by ability to pass law) is caused by tensions between the two major political parties in the U.S. – the 113th being a recent example.

Political science often models congressional ideology backward from voting decisions [9,13], party influence [1,4,5,11,12], and committee dynamics [6,14,17]. Our research complements other studies by developing a computational model of policy development through draft legislation, working forward from arbitrary

---

We wish to thank Maksim Tsvetovat who introduced us to applications of simulated annealing to Organizational Theory and inspired us to extend it, adapt it, and to think about organizations and processes where it seems to fit especially well. We also thank an anonymous reviewer for helpful feedback provided on a previous draft of this paper.

© Springer International Publishing Switzerland 2015
N. Agarwal et al. (Eds.): SBP 2015, LNCS 9021, pp. 93–100, 2015.
DOI: 10.1007/978-3-319-16268-3_10

ideological foundations. Viewing bill revision as a process that optimizes legislative body satisfaction within a system of competing legislator preferences, we judged simulated annealing (SA, hereafter) to be the non-deterministic optimization method most suited for modeling this.

As a generalized model of policy-making, the results are broadly applicable to a wide range of policy-making organizations: legislative bodies at the national or local level, regulatory bodies, standards organizations, and conference committees, for example.

Our simulation model enables quantitative analysis of the productivity and satisfaction of three typified legislatures: with unified, bi-modal (bipartisan), or completely uncorrelated sets of preferences. Results indicate that party partisanship alone does not explain a Congress's inability to produce legislation.

2   Model and Methods

2.1   Model Overview

In this section, we describe implementation of a model of policy formation through simulated annealing. We initialize each case by generating a legislature and its internal social network. Legislators have preferences on a set of issues, prioritizing some issues over others. They take turns sponsoring bills, which get reviewed and revised through two rounds of SA before the entire legislative body votes on the bill. We captured productivity, satisfaction, and amount of compromise as the relevant metrics for each simulation run. We describe the initialization and simulation processes and experimentation methods in more detail below.[1].

2.2   Model Initialization

The first step in the simulation initializes the model by generating a State object, which realizes the parameters of that run scenario: a heterogeneous set of 100 legislators, each with priority-ranked positions on a set of 75 issues, assigned stochastically but correlated through State- and party-provided seed vectors. The State object then organizes the legislators into a social network making a realistic number of connections between the most similar legislators.

Initializing the Model Environment. For our model, we assume that several core issues represent powerful, crystallizing factors that differentiate our simulated parties. Thus, party "platforms" consist of vectors of positions and priorities on a set of issues that includes both State_Priorities and a random sample of high-priority Ideology_Issues. These vectors are used as "seeds" for the stochastic generation of individual legislator preferences.

---

[1] In the interest of brevity, we have omitted some detail descriptions of the initialization and simulation; however, these details, as well as the model itself, may be found in the online supplement at https://github.com/strangeintp/garbage-can-congress/wiki

Legislator Initialization. Each legislator's issue priorities are assigned with a stochastic preferential attachment method to the seed values provided by the State object. This generates a power law distribution of priorities for that legislator and provides some correlation in legislator priorities to the extent that seed vectors are similar (as determined by state priorities and party ideology).

We assume that party-affiliated legislators adopt the positions of their party; for all other issues (and all issues for unaffiliated legislators), positions are assigned uniform randomly from the range of allowed position values (2^4 for our model). The end-result of this process is a set of legislator agents with heterogeneous but correlated policy preferences as conditioned by party ideologies and state priorities, and with the strength of correlation determined by party-alignment.

Network Generation. The final step of initialization generates a social network, using both homophily [3,15] and preferential attachment [2], based on preferences assigned in the previous step. Preferential attachment is as described in [2], with $m = 5$ new edges selected randomly from a *pdf* distribution of degree in the sub-network of potential allies of each legislator. The set of potential allies is selected using a preference-weighted likelihood over all issues. The typical outcome of this procedure is a network among legislator agents with "small-world" properties [18], consistent with existing research on social networks in the U.S. Congress [7] [2].

2.3   Simulation Algorithm Overview

Having defined a population of legislators and their relationship to each-other, we next establish a procedure for legislators to engage in the business of law-making.[3] In our model of law-making, the simulation sequentially repeats the following process for 200 proposals (or halts if all issues are passed into law):

1. Proposal:
   (a) A random legislator is chosen to sponsor a bill.
   (b) The sponsor proposes a draft bill on any issue that has not already been addressed by law. This initial draft reflects the sponsor's position on that issue.
2. Draft circulation among cosponsors:
   (a) All legislators connected to the sponsor in the social network are selected as cosponsors.
   (b) The cosponsors revise the draft using SA; new issues may be added to the bill during the revision process and solutions on existing issues may change.

---

[2] We verify "small-worldness" of the networks produced by the model using the Humphries and Gurney metrics [8].

[3] One might argue that this is a departure from realism, as the current Congress does not appear to do this. However, we are attempting to generalize a model of law-making in legislatures. Some legislatures do periodically legislate.

3. Committee review:
   (a) The draft is referred to a committee, reflecting committee agenda-setting powers [4,5]. Legislators for whom the main issue of the bill is a high priority are assigned to the relevant committee.
   (b) The committee revises the bill by SA; again, new issues may be added and existing solutions changed.
4. Floor vote:
   (a) The bill is referred to the floor for a vote.
   (b) A legislator votes 'yes' to a bill when her satisfaction with it is greater than the model parameter satisfaction_threshold.
   (c) If the bill passes by simple majority (>50% votes), the bill is made into law; *i.e.*, the solutions addressed by the bill are recorded and the issues may not be revisited for the remainder of the realization.

Simulated Annealing. Bill revision is implemented by the Metropolis algorithm for simulated annealing [10,16]. Our energy function is the cumulative dissatisfaction of all reviewers over all dimensions of the bill. Increases of 0.1 in dissatisfaction were accepted with probability $1/2$ at the maximum temperature. Higher satisfaction energy states are automatically accepted. [4]

2.4   Model Calibration

We calibrated legislators' *satisfaction thresholds*—the point at which they vote "yes" on legislation—to achieve a 4% pass-rate, comparable to passage rates in the actual US Congress (between 2% and 7% in recent history).[5]

2.5   Experiments

Table 1 identifies key parameters used in the model for the suite of experiments. The experiment suite was designed factorially, exercising all combinations of values shown in the table – excepting Green_Fraction variation when Unaffilitated_Fraction = 1.0, which is a degenerate case – resulting in 28 unique experiments. 30 simulations were realized per experiment.

For each realization, we recorded: the number of laws passed, the number of issues addressed by law (i.e. provisions), and legislative body satisfactions over all bills before and after SA revisions. To keep the data set manageable, we recorded a sample run history for only a single realization of each experiment. Aggregate statistics (averages and standard deviations) were also calculated and recorded for the output metrics of all realizations of an experiment.

---

[4] The online supplement provides more detail.
[5] Pass rates are equal to the total number of bills passed in a given Congress divided by the total counts of introduced legislation for that Congress. Data to calculate pass rates was collected from Civic Impulse LLC (http://www.govtrack.us).

Table 1. Simulation Parameter Space

Parameter	Description	Value [Variation]
Unaffilitated_Fraction	Fraction of the legislative population with no ideological party affiliation.	[0.05, 0.5, 1.0]
Green_Fraction	Fraction of the party-affiliated population belonging to the *Green* party. Remainder belong to the Yellow party.	[0.5, 0.75, 1.0]
Ideology_Issues	Ideological platform issues for the parties.	[0, 5]
State_Priorities	High-priority issues for all legislators, regardless of affiliation.	[0, 5]

3 Results and Findings

Fourteen of the 28 experiment combinations were uniformly unproductive (produced no laws), as follows:

1. All four cases with no party structure,
2. Scenarios with 50% unaffiliated, 25% Green, and 25% Yellow,
3. Scenarios with no external priorities (an additional 5), and
4. One case with 5% unaffiliated members, 50% Green members, and no ideology-based priorities.

For the remaining productive cases, Figure 1 presents run results across the three aggregate metrics captured for each run (total satisfaction and number of laws passed, plotted against number of provisions added.)

Interpreting these results, our main findings are:

1. Higher correlation of preferences results in higher productivity.
2. Higher productivity requires increased provisions.
3. Partisanship is not necessarily an impediment to productivity (see cases 2 and 4).
4. Bipartisan networks (even division of party-affiliated legislators) with more external priorities can be more productive than majorities or super-majorities (with fewer external priorities). Compare case 4 to cases 11, 18, 19, and 23.
5. Despite higher productivity, overall satisfaction decreases with increased provisions (note negative trend lines in the "Satisfaction" row of Figure 1).

4 Discussion

Based on our results and findings, we can generalize about the importance of having external priorities for self-tasking organizations, such as the U.S. Congress,

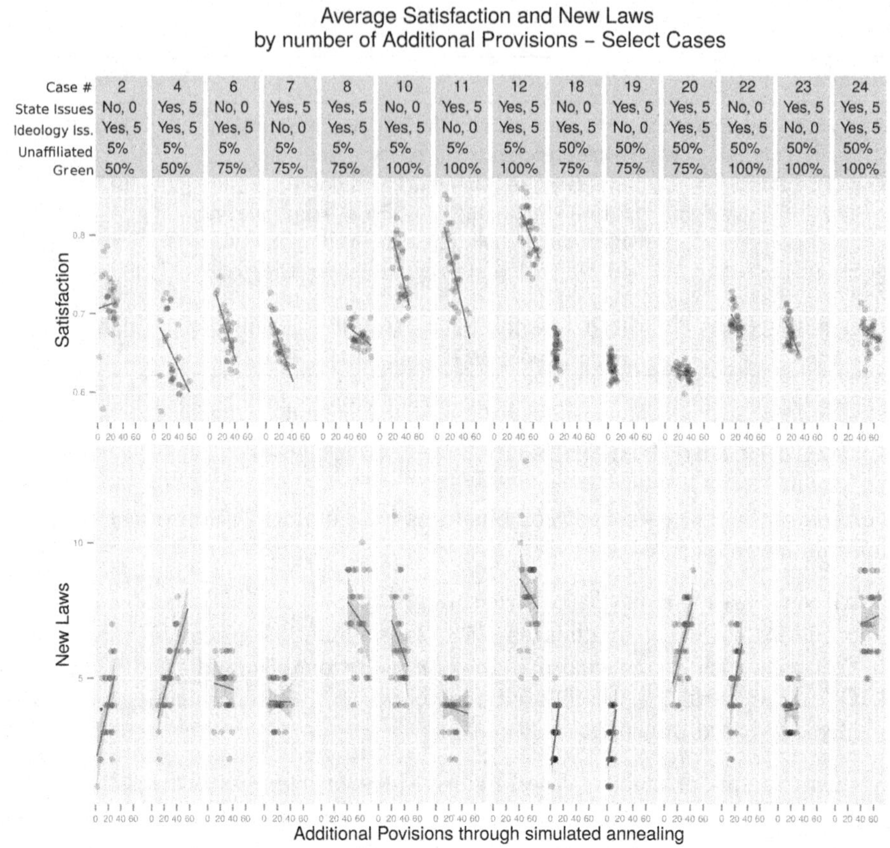

Average Satisfaction and New Laws
by number of Additional Provisions - Select Cases
Case # 2 4 6 7 8 10 11 12 18 19 20 22 23 24
State Issues No, 0 Yes, 5 No, 0 Yes, 5 Yes, 5 No, 0 Yes, 5 Yes, 5 No, 0 Yes, 5 Yes, 5 No, 0 Yes, 5 Yes, 5
Ideology Iss. Yes, 5 Yes, 5 Yes, 5 No, 0 Yes, 5 Yes, 5 No, 0 Yes, 5 Yes, 5 No, 0 Yes, 5 Yes, 5 No, 0 Yes, 5
Unaffiliated 5% 5% 5% 5% 5% 5% 5% 5% 50% 50% 50% 50% 50% 50%
Green 50% 50% 75% 75% 75% 100% 100% 100% 75% 75% 75% 100% 100% 100%
Satisfaction
New Laws
Additional Povisions through simulated annealing

Fig. 1. Experiment suite results for productive scenarios. The trend lines indicate a best fit linear correlation to the number of additional provisions.

that undertake complex tasks (law generation) requiring group approval.[6] Having at least some priorities established externally seems to be required for self-tasking organizations to be productive. As a rule, the cases with more externally-defined priorities are more productive than the cases with fewer externally-defined priorities. The cases with no externally-defined priorities were consistently unproductive. It is tempting to say that this points at the role of leadership—both from within organizations (ideology issues) and from above them (state priorities)—but this will be a question for future research.

In view of our research motivations, our results—in particular, findings 3 and 4—do not support the assertion that polarization of preferences alone leads to

[6] Note that production of an academic paper with multiple authors may fit this generalized formulation, as well.

an unproductive legislature; other factors such as manipulation of the political process are more likely causal.

Finding #2 demonstrates that higher productivity requires increased provisions. In a legislative context, these provisions (or "riders") represent effort diverted to individual interests that are not [necessarily] focused on the task at hand. These riders are commonly recognized as the "cost of doing business". Abstracting from a legislative context, we may generalize by saying that higher productivity comes with the cost of lower system efficiency.

Finally, finding #5—that satisfaction decreases with increasing provisions— captures the social aspects of negotiation and compromise. In other words, policy makers are willing to cede adverse positions on their lower-priority issues as a cost of attaining higher satisfaction with the total law, and total satisfaction decreases as these issues are added. Conversely, bills start off in a low-satisfaction state and gather provisions as a way of garnering votes. In either case, more issues are needed to retain or attain sufficient satisfaction to pass new legislation.

4.1   Implications for Future Research

Further research might vary the network-structure generating parameters with finer resolution to identify the tipping points in the outcomes: How much of a majority is needed for compromise to yield satisfaction and productivity? How much leadership intervention (ideology) is required to overcome inherently unproductive structures? How many state priorities are required to ensure sufficient preference correlation, for a given legislature?

We are also interested in understanding the effects of party affiliation on proposals and productivity. This paper reviewed aggregate results from our experiments with the model. The model could be extended to produce additional detail data about how much compromise is needed, given the characteristics of a bill's sponsor. We suspect that the sponsor's peer network, if they are in the minority, may modify an otherwise popular proposal with unpopular issues, such that it fails to garner votes at the committee or floor. If a bill's sponsor were to intentionally include members from the majority party, or conversely, exclude close connections who may be adverse to a proposal, the annealing process might produce more laws.

5   Summary

We modeled policy-making as a simulated annealing algorithm to find solutions in a complex problem space with interdependent constraints. We chose the U.S. Congress and legislative process as a case study, but this research may be applied to other policy-making organizations. Results indicate that partisanship alone is not necessarily an impediment to productivity, provided that there is sufficient alignment of priorities and preferences. However, higher productivity comes at the cost of lower satisfaction and system efficiency. We conclude that simulated annealing is a useful method for computationally modeling policy-making, and recommend it for other research projects.

References

1. Aldrich, J.: Why Parties?. University of Press, Chicago (1995)
2. Barabási, A.-L., Albert, R.: Emergence of scaling in random networks. Science **509**(5439), 509–512 (1999)
3. Bratton, K., Rouse, S.: Networks in the legislative arena: How group dynamics a ect co-sponsorship. Legislative Studies Quarterly **36**(3) (2011)
4. Cox, G.W., McCubbins, M.: Legislative Leviathan: Party Government in the House. University of California Press (1993)
5. Cox, G., McCubbins, M.: Setting the Agenda. Cambridge University Press (2005)
6. Gilligan, T., Krehbiel, K.: Asymmetric information and legislative rules with a heterogeneous committee. American Journal of Political Science **33**(2) (1989)
7. Granovetter, M.: Threshold models of collective behavior. American Journal of Sociology **83**(6) 1420–1443 (1978)
8. Humphries, M.D., Gurney, K.: Network "small-worldness": A quantitative method for determining canonical network equivalence. PLoS One **3**(4) (2008)
9. Kingdon, J.: Congressmen's Voting Decisions. University of Michigan Press (1989)
10. Kirkpatrick, S., Gelatt, C., Vecchi, M.P.: Optimization by simulated annealing. Science **220** (1983)
11. Krehbiel, K.: Information and Legislative Organization. University of Michigan Press (1991)
12. Krehbiel, K.: Pivotal Politics: A Theory of US Lawmaking. Chicago University Press (1998)
13. Mayhew, D.: Congress: The Electoral Connection (1974)
14. McKelvey, B.: Toward a complexity science of entrepreneurship. Journal of Business Venturing **19** (2004)
15. McPherson, M., Smith-Lovin, L., Cool, J.: Birds of a feather: Homophily in social networks. Annual Review of Sociology **27** (2001)
16. Metropolis, N., et al.: Equations of state calculations by fast computing machines. Journal of Chemical Physics **21**(6) (1953)
17. Shepsle, K., Weingast, B.: The institutional foundations of committee power. The American Political Science Review **81**(1) (1987)
18. Watts, D.J., Strogatz, S.H.: Collective dynamics of 'small-world' networks. Nature **393**(6684), 440–442 (1998)

VIP: Incorporating Human Cognitive Biases in a Probabilistic Model of Retweeting

Jeon-Hyung Kang[(⊠)] and Kristina Lerman

USC Information Sciences Institute, 4676 Admiralty Way,
Marina Del Rey, CA, USA
jeonhyuk@usc.edu, lerman@isi.edu

Abstract. Information spread in social media depends on a number of factors, including how the site displays information, how users navigate it to find items of interest, users' tastes, and the 'virality' of information, i.e., its propensity to be adopted, or retweeted, upon exposure. Probabilistic models can learn users' tastes from the history of their item adoptions and recommend new items to users. However, current models ignore cognitive biases that are known to affect behavior. Specifically, people pay more attention to items at the top of a list than those in lower positions. As a consequence, items near the top of a user's social media stream have higher visibility, and are more likely to be seen and adopted, than those appearing below. Another bias is due to the item's fitness: some items have a high propensity to spread upon exposure regardless of the interests of adopting users. We propose a probabilistic model that incorporates human cognitive biases and personal relevance in the generative model of information spread. We use the model to predict how messages containing URLs spread on Twitter. Our work shows that models of user behavior that account for cognitive factors can better describe and predict user behavior in social media.

Keywords: Social media · Information diffusion · Cognitive factors

1 Introduction

Online social networks can dramatically amplify the spread of information by allowing users to forward information to their followers, and those to their own followers, and so on. Predicting how people will respond to information is of immense practical and commercial interest. Prediction can guide the design of more effective marketing and public awareness campaigns, for example, those announcing the locations of clinics dispensing the flu vaccine. Researchers believe that information spread in social media is a complex process that depends on the nature of information [21], the structure of the network [2,25], the strength of social influences [1,20], as well as user interests and topic preferences [14,15]. These factors are thought to render information diffusion in social media unpredictable [1,9], although researchers have identified some features that weakly correlate with the size of information cascades [6]. On the other hand, more progress

© Springer International Publishing Switzerland 2015
N. Agarwal et al. (Eds.): SBP 2015, LNCS 9021, pp. 101–110, 2015.
DOI: 10.1007/978-3-319-16268-3_11

has been made addressing information spread in social media as a social recommendation problem. In this case, probabilistic models are used to learn users' topic preferences from the history of their item adoptions and predict what items in their social media stream users will adopt [14,17,23].

Existing models of social recommendation largely ignore cognitive factors of user behavior. One such aspect is *position bias*. Due to this cognitive bias, the amount of attention an item receives strongly depends on its position on a screen or within a list of items. Position bias is known to affect the answers people select in response to multiple-choice questions [3,18], where on the screen they look [4,7], and the links on a web page they choose to follow [8,13]. Also as a consequence of position bias, items near the top of a user's social media stream are more salient, and therefore, more likely to be viewed, than items in lower positions [16].

To distinguish position-based salience from other psychological effects, we refer to it as an item's *visibility*. After viewing the item, the user may decide to adopt it. She adopts information either because it is personally relevant to her or because it is generally interesting. To handle the former case, the model must include a hidden topic space which can be used to compute the relevance of items to users. The user may also adopt an item that is not strictly relevant, but interesting nonetheless. Such items are usually viral memes, such as breaking news, that have a high fitness, i.e., propensity to spread upon exposure regardless of the interests of adopting users.

In Section 2, we introduce a conceptually simple model of information spread that captures the factors important to information spread: item's *visibility*, *fitness*, and its *personal relevance* to user's interests. An item's visibility depends on its position in the user's social media stream. However, since position data is often not directly available, we estimate visibility from user's information load. This quantity measures the number of items a user has to inspect before finding a specific item to adopt, and it is given by the number of new messages arriving in the user's stream and the frequency the user visits the stream. The greater the number of new messages in the stream — either because the user follows more people or because she rarely visits her stream — the less visible any particular item is. Accounting for visibility allows us to learn a better model of user's interests from the history of her item adoptions. When the user does not adopt an item, the model allows us to discriminate between lack of interest and failure to see the item. While this simple model ignores some of the nuances of information spread, it has very high predictive power. In Section 3, we evaluate the proposed model on a social recommendation task using Twitter. We study the impact of visibility, item fitness, and personal relevance on information diffusion in Twitter in aggregate and through illustrative individual examples. Our study demonstrates that models of user behavior that account for cognitive biases can better describe and predict user behavior in social media, and that information spread is more predictable than previously thought.

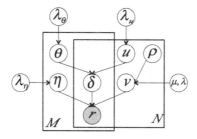

Fig. 1. The VIP model with user topic (u) and item topic (θ) profiles, personal relevance of an item to user (δ), visibility to user (v), item fitness (η), expected number of new posts user received (ρ) and item adoption (r). N is the number of users and M is the number of items.

2 The VIP Model

We describe VIP, a model that captures the three basic ingredients of information spread in social media: item's fitness and its visibility and personal relevance to the user. VIP is based on social recommendation models, whose goal is to recommend only the relevant items to users [14,15,17,23]. In social recommendation, each user is assigned a vector of topics, which serve as her interest profile, and each item also has some topics. Once these hidden vectors are learned from the history of user item adoptions, it is possible to calculate an item's *personal relevance* to the user.

Social media users adopt items even if they had not earlier demonstrated a sustained interest in their topics. This is often the case with viral, general-interest items, such as breaking news or celebrity gossip. We use the term *fitness* (or 'virality') to describe an item's propensity to be adopted upon exposure.

The key innovation of VIP is to introduce *visibility* into the generative model of item adoption. Visibility conceptually simplifies the mechanisms of information spread and explains away some of the complexity associated with it, for example, the network effects observed by [1,20,21]. Visibility explicitly takes into account the process of information discovery in social media. Online social networks are directed, with users following the activities of their friends. A user's message stream contains a list of items her friends adopted or "recommended" to her, chronologically ordered by their adoption time, with the most recent item at the top of the stream. We consider a user to be *exposed* as soon as the item enters her stream; however, exposure does not guarantee that the user will actually view the item. The probability of viewing — *visibility* — depends on the item's position in the user's stream [10]. Due to a cognitive bias known as position bias [18], a user is more likely to attend to items near the top of the screen than those deeper in the stream [4]. Below we discuss a method to quantitatively account for this visibility.

Figure 1 graphically represents the VIP model. It considers a user i with a user-topic vector u, and an item j with an item-topic vector θ. VIP generates an adoption of an item j by user i as follows:

$$r_{ij} \sim \mathcal{N}\left(v_i g_r(\delta_{ij} + \eta_j), c_{ij}^{-1}\right) \tag{1}$$

where $\eta_j \sim \mathcal{N}(0, \lambda_\eta^{-1})$, is the fitness (or interestingness) of item j, which represents the probability of adoption given the user viewed it [11,24]. The precision parameter c_{ij} serves as confidence for adoption r_{ij}. v_i represents the visibility of item to user and δ_{ij} represents user i's interest in item j. We define g_r as a linear function for simplicity. One of the key properties of VIP lies in how a user adopts items that have the same visibility. We assume that user adopts either items that are relevant to her (δ_{ij}) or interesting in general (η_j).

Users discover items by browsing through their message stream. As argued above, the position of an item in the stream determines its visibility, the likelihood to be viewed. However, an item's exact position is often not known. Instead, we estimate its average visibility from the available data. This quantity depends on user's information load [19], i.e., the flow of messages to the user's stream, and the frequency the user visits the site. The greater the number of new messages user receives between visits to the site, the less likely the user is to view any specific item. Following [12], we estimate visibility of an item to user i as:

$$v_i \sim \sum_L \left(\mathbf{G}(1/(1 + \rho_i), L)(1 - \mathbf{IG}(\mu, \lambda, L))\right) \tag{2}$$

The first factor gives the probability that L newer messages have accumulated in user i's stream since the arrival of a given item. The accumulation of items is a competition between the rates friends post new messages to the user's stream and the rate the user visits the stream to read the messages. The ratio ρ_i of these rates gives the expected number of new messages in a user i's stream since item j's arrival. Taking friends activity and user activity each to be a Poisson process, the competition gives rise to a geometric distribution with success probability $p = 1/(1 + \rho_i)$: $\mathbf{G} = (1 - p)^L p$. We will revisit how we estimate ρ_i in the section below. The second factor of Eq. 2 gives the probability that user i will navigate to at least $L + 1$'st position in her stream to view the item. This is given by the upper cumulative distribution of an inverse gaussian $\mathbf{IG}$ with mean μ and shape parameter λ and variance μ^3/λ:

$$\exp\left(\frac{-\lambda(L - \mu)^2}{2\mu^2 L}\right)\left[\frac{\lambda}{2\pi L^3}\right]^{(1/2)}. \tag{3}$$

This distribution has been used to describe the "law of surfing" [13], and it represents the probability the user will view L items on a web page before stopping. Therefore, the cumulative distribution of $\mathbf{IG}$ gives the probability the user will view at least L items, hence, navigating to $L + 1$'st position in her stream. We calculate personal relevance of the item j to user i as:

$$\delta_{ij} \sim g_\delta(u_i^T \theta_j) \tag{4}$$

where symbol T refers to the transpose operation, u_i represents topic profile of user i, θ_j represents topic profile of item j and g_δ is linear function for simplicity. We represent topic profiles of users and items in a shared low-dimensional space as follows.

$$u_i \sim \mathcal{N}(0, \lambda_u^{-1} I_K)$$
$$\theta_j \sim \mathcal{N}(0, \lambda_\theta^{-1} I_K) \tag{5}$$

where K is the number of topics. Note that if we only use personal relevance (δ) and ignore visibility and fitness, VIP model reduces to probabilistic matrix factorization (PMF) model [22] that learns latent topics from user–item adoptions.

The generative process for item adoption through a social stream can be formalized as follows:

For each user i
    Generate $u_i \sim \mathcal{N}(0, \lambda_u^{-1} I_K)$
    Generate $v_i \sim \sum_l (\mathbf{G}(1/(1+\rho_i), l)(1 - \mathbf{IG}(\mu, \lambda, l)))$
For each item j
    Generate $\theta_j \sim \mathcal{N}(0, \lambda_\theta^{-1} I_K)$
    Generate $\eta_j \sim \mathcal{N}(0, \lambda_\eta^{-1})$
For each user i
    For each recommended item j from friends
        Generate the adoption $r_{ij} \sim \mathcal{N}\left(v_i g_r(\delta_{ij} + \eta_j), c_{ij}^{-1}\right)$

Here $\delta_{ij} = u_i^T \theta_j$, $\lambda_u = \sigma_r^2/\sigma_u^2$, $\lambda_\theta = \sigma_r^2/\sigma_\theta^2$, and $\lambda_\eta = \sigma_r^2/\sigma_\eta^2$. Lack of adoption by user i of item j ($r_{ij} = 0$) can be interpreted in two ways: either user saw the item but did not like it, or user did not see the item but may have liked it had she seen it. While other models partly account for lack of knowledge about non-adoptions using smoothing [23], we properly model visibility of items to users. We set c_{ij} to a high value a^r when $r_{ij} = 1$ and a low value b^r for items recommended by friends and c^r for the rest when $r_{ij} = 0$ ($a^r > b^r > c^r > 0$). In this paper, we use the confidence parameter values, $a^r = 1.0$, $b^r = 0.03$ and $c^r = 0.01$, for c_{ij}.

2.1 Learning Parameters

To learn model parameters, we follow the approaches of [14,23] and develop coordinate ascent, an EM-style algorithm, to iteratively optimize the variables $\{u_i, \theta_j, \eta_j\}$ and calculate the maximum a posteriori estimates. MAP estimation is equivalent to maximizing the complete log likelihood (ℓ) of U, V, θ, η and R given λ_u, λ_θ, λ_η, μ, λ and ρ.

$$\ell = -\frac{\lambda_u}{2} \sum_i^N u_i^T u_i + \sum_i^N \log \left(\sum_l^L (1/\rho_i + 1)(\rho_i/\rho_i + 1)^l (1 - \mathbf{IG}(\mu, \lambda, l)) \right)$$
$$-\frac{\lambda_\theta}{2} \sum_j^M \theta_j^T \theta_j - \frac{\lambda_\eta}{2} \sum_j^M \eta_j^T \eta_j - \frac{c_{ij}}{2} \sum_i^N \sum_j^M (r_{ij} - v_i(\delta_{ij} + \eta_j))^2 \tag{6}$$

Given a current estimate, we take the gradient of ℓ with respect to u_i, θ_j, and η_j and set it to zero. The update equations are:

$$u_i \leftarrow \left(\lambda_u I_k + \Theta v_i C_i v_i \Theta^T\right)^{-1} \Theta C_i \left(v_i R_i - v_i \eta v_i\right)$$

$$\theta_j \leftarrow \left(\lambda_\theta I_k + UVC_j VU^T\right)^{-1} UC_j \left(VR_i - \eta_j VVI_N\right)$$

$$\eta_j \leftarrow \left(\lambda_\eta + v^T C_j v\right)^{-1} v^T C_j \left(R_j - VU^T \Theta_j\right)$$

where C_j is a diagonal matrix of confidence parameters c_{ij}. Item visibility to user i, v_i, is represented as a diagonal matrix V or in vector format as v. We define Θ as $K \times M$ matrix, U as $K \times N$ matrix and R_j as vector with r_{ij} values for all pairs of users i for the given item j.

2.2   Prediction

After parameters are learned, VIP can be used to predict item adoptions by a user. For user-item adoption prediction, user i's adoption of item j retweeted by a friend is obtained by point estimation with optimal variables $\{\theta^*, u^*, v^*, \eta^*\}$:

$$\mathbb{E}[r_{ij}|\mathcal{D}] \approx \mathbb{E}[v_i|\mathcal{D}]^T (\mathbb{E}[\delta_{ij}|\mathcal{D}] + \mathbb{E}[\eta_j|\mathcal{D}])$$
$$r_{ij}^* \approx v_i^* (u_i^{*T} \theta_j^* + \eta_j^*) \tag{7}$$

where $\mathcal{D}$ is the training data. The adoption probability is decided by user visibility v_i^*, user topic profile u_i^*, item topic profile v_j^*, and item fitness η_j^*.

3   Evaluation

In this section we demonstrate the utility of the VIP model by applying model to data from the social media Twitter and evaluating its performance on the prediction tasks. We collected tweets containing a *URL* to monitor information spread over the social network from Nov 2011 to Jul 2012. We start by monitoring potential seed *URL*s from streaming APIs and collected the entire history using the Twitter REST APIs to reconstruct their sharing history. This yielded 12.5M tweets with 9.5M users.

3.1   Model Selection

First, we study how parameters of VIP affect the overall performance of user-item adoption prediction using *recall*@3. We use the same "law of surfing" parameters, $\mu = 14.0$ and $\lambda = 14.0$, as [11,12] did in their study of Twitter and another social media site. The expected number of new posts including a URL user i received, ρ_i, is computed by $rate_i^{(url\ posts\ received)}/rate_i^{(visits)}$. The rate $rate_i^{(posts\ received)}$ is proportional to the number of friends ($N_{frd(i)}$) i follows and their average posting frequency [12]. To estimate posting frequency of all users, we have to track all their behaviors. Instead of tracking all users, we estimate it using the 'typical URL posting rates of users from our data: $rate_i^{(posts\ received)} = 1.4 * N_{frd(i)}$. User i visits Twitter at a rate $rate_i^{(visits)}$. This number is not available; however, we expect it to be proportional to the number we do observe: the number of posts of user i ($N_{posts(i)}$). [12] estimated that average number of visits per post was 38 for Twitter users. Also, since around 20% of tweets include a URL [5], the posting rate of user i becomes $rate_i^{(visits)} = 7.6 * N_{posts(i)}$.

For the PMF model, we vary the parameters $K \in \{10, \dots, 200\}$, λ_u and $\lambda_\theta \in \{10^{-4}, \dots, 10^4\}$ by using grid search on validation recommendations. Throughout this paper, we set parameters $K = 30$, $\lambda_u = 10^{-3}$, $\lambda_\theta = 10^{-3}$ both for PMF and VIP that performed the best for PMF. For the fitness parameter, we vary $\lambda_\eta \in \{10^{-4}, \dots, 10^4\}$, while we fix $\lambda_\theta = 10^{-3}$ and $\lambda_u = 10^{-3}$ and set $\lambda_\eta = 10^4$.

3.2   User–Item Adoption Prediction

In the prediction task, we sort the items by r_{ij}, the probability of adoption by user i, and calculate the fraction of the X top-ranked items that the user actually adopted. A user may not adopt an item either because she did not see it or because she does not like it. This makes it difficult to use precision to evaluate prediction results. Instead, we use $recall@X$ (=N(items in top X user adopted)/N(items user adopted)) to measure model's performance on the prediction task. To summarize performance of the prediction algorithm, we average recall values over all users.

We divide each user's adopted items into five folds and construct the training set and the test set. We use five-fold cross validation and compare performance of VIP to three baseline models: RANDOM, FITNESS and RELEVANCE. The RANDOM baseline chooses items at random from among the items in user i's stream, i.e., items adopted by i's friends. The baseline FITNESS uses item fitness values (η) learned by VIP to recommend X highest fitness items. The baseline RELEVANCE bases its recommendations on user-topic and item-topic vectors learned by PMF to recommend X most relevant items.

Figure 2 (a) shows the models' overall performance on the user–item adoption prediction task when we vary X, the number of recommendations made by each model. Note that a better model should provide higher $recall@X$ for different X. VIP outperforms all baselines, but the improvement is especially dramatic when the number of recommended items is small: $recall@3$ was 0.30, 0.17, 0.16, and 0.12 for VIP, RELEVANCE, FITNESS, and RANDOM models respectively. Note that as the number of recommendation (X) increases, the recall of all models improves, however, at the expense of precision.

Figure 2 (b) shows how prediction performance on the user–item recommendation task varies with user activity level, that is how the number of items adopted by the user in the training set affects $recall@3$ on the test set. The performance of the RANDOM baseline, which recommends three randomly chosen items from the user's stream, does not vary with user activity level, as expected. Similarly, FITNESS baseline does not vary significantly with activity, since it

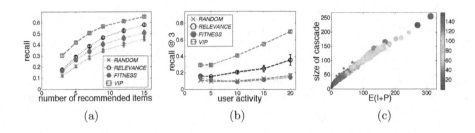

Fig. 2. Error bars are shown indicating standard deviation with the upper bar and the lower bar. (a)Recall of user–item adoption prediction with different numbers of recommended items. The number of topics was fixed at 30. (b)Average *recall@3* of user–item adoption prediction for different activity levels of users with 30 topics. (c)Cascade size vs expected values of *item fitness* plus *personal relevance* E(I+P) for all adopters. The size and color of each circle represents the expected value of that item's *visibility*.

depends only on the propensity of the item to spread. Both VIP and RELEVANCE improve with increasing user activity as they can learn better user–topic profiles with more training data. Note that for low activity users, whose interests are not well-known, recommending items based on personal RELEVANCE performs about the same as picking items based on their fitness, but as more can be learned about user's preferences, RELEVANCE outperforms picking items based on their fitness or picking them randomly from the user's stream. VIP handily outperforms baselines over all user activity levels. This shows that accounting for visibility dramatically improves predictability of user item adoptions in social media compared to using personal relevance or item fitness alone.

4   Visibility vs Item Fitness vs Personal Relevance

We analyze URL cascades on Twitter by examining how the three factors learned by the VIP model contribute to their success. Depending on the characteristics of the community that the URL has reached, fitness can vary. In our data set, URLs that have been retweeted within a community sharing a specific hobby or interest, tend to have high fitness values. Since members share common topic preferences, items received relatively high adoption rates per exposure, which also often translates into quick adoption. High fitness means high adoption rates per view with statistically significant 0.85 correlation with cascade size. However, 4% of the URLs have fitness values that are negatively correlated with cascade size and 40% of the URLs show no correlation between fitness and cascade size. Apparently fitness by itself cannot explain the spread of information, and other factors, such as visibility and personal relevance also have to be considered.

We separate the effect of item quality from its visibility to the user. We define quality very loosely as the combined effect of its fitness and relevance to adopters, and measure it by the expected value of these variables. This definition aims to make quality specific to the item itself, and separate from the details of

Table 1. Cascade size, expected values, descriptions on Youtube video URLs

Descriptions	Cascade Size	E(V)	E(I)	E(P)
Strongbow surfers Neon Night Surfing on Bondi Beach	84	49	-0.04	85.1
Jay-Z Music Video	141	50.5	-0.04	130.7
Parallels for Mac for Chrome OS and Windows 8	68	52.3	-0.05	71.6
Bahraini Activist Nabil Rajab	116	62.1	-0.03	127.6
Ellen's Swaggin' Wagon – a marriage Proposal	87	65.6	-0.13	80.2
UNICEF - Making headway toward an AIDS-free generation	102	71.7	-0.11	100.6
Whitney Houston Video	120	73.5	-0.13	127.9
Paul McCartney's message from Moscow	109	87	-0.22	102.2
Ian Somerhalder Foundation	143	98.6	-0.24	150.8

how users may discover it. Figure 2 (c) shows how the size of cascades in our data set depends on item quality and visibility. Each circle represents a URL, with its color encoding the expected visibility of the URL. Not surprisingly, higher quality URLs have larger cascades. More interestingly, some of the variance of cascade size can be explained by visibility: for URLs of similar quality, the more visible URLs spread more widely. In other words, for items cascading through a network where users have similar topic preferences, the total size of the cascade is decided by their visibility.

Next we illustrate the contributions of the three factors using specific case studies. There were 205 URLs to Youtube videos in our data set, with examples shown in Table 1. Two of the most popular URLs in our data set were *"Jay-Z Music Video"* and *"Ian Somerhalder Foundation"*, which were both adopted more than 140 times through friends' recommendation. The fitness of the *"Jay-Z Music Video"* is six times higher than that of *"Ian Somerhalder Foundation"*, but has half the expected visibility. Therefore, the high fitness value of *"Jay-Z Music Video"* makes up for the relatively low visibility and reaches a similar number of adoptions as *"Ian Somerhalder Foundation"*.

5   Conclusion

In this paper, we proposed VIP, a model that captures the mechanisms of information spread in social media. VIP can recommend items to users based on how easily users find an item in their stream, how well the item aligns with their interests, and the item's propensity to be adopted upon exposure. Prediction is surprisingly accurate, considering the crude estimates of visibility. Knowing visibility more accurately will further improve prediction performance. We plan to extend our model to take into account descriptions of items. We will further study the role of network structure and the relationship between visibility, item fitness, and personal relevance on information sharing.

Acknowledgments. This work was supported in part by AFOSR (contract FA9550-10-1-0569), by NSF (grant CIF-1217605) and by DARPA (contract W911NF-12-1-0034).

References

1. Bakshy, E., Hofman, J.M., Mason, W.A., Watts, D.J.: Everyone's an influencer: quantifying influence on twitter. In: WSDM (2011)
2. Bakshy, E., Rosenn, I., Marlow, C., Adamic, L.: The role of social networks in information diffusion. In: WWW (2012)
3. Blunch, N.J.: Position bias in multiple-choice questions. Journal of Marketing Research 21(2), 216–220 (1984)
4. Buscher, G., Cutrell, E., Morris, M.R.: What do you see when you're surfing?: using eye tracking to predict salient regions of web pages. In: SIGCHI (2009)
5. Chaudhry, A., Glodé, L.M., Gillman, M., Miller, R.S.: Trends in twitter use by physicians at the american society of clinical oncology annual meeting, 2010 and 2011. Journal of Oncology Practice 8(3), 173–178 (2012)
6. Cheng, J., Adamic, L., Dow, P.A., Kleinberg, J.M., Leskovec, J.: Can cascades be predicted? In: WWW (2014)
7. Counts, S., Fisher, K.: Taking it all in? visual attention in microblog consumption. In: ICWSM (2011)
8. Craswell, N., Zoeter, O., Taylor, M., Ramsey, B.: An experimental comparison of click position-bias models. In: WSDM (2008)
9. Goel, S., Watts, D.J., Goldstein, D.G.: The structure of online diffusion networks. In: EC (2012)
10. Hodas, N., Lerman, K.: How limited visibility and divided attention constrain social contagion. In: SocialCom (2012)
11. Hogg, T., Lerman, K.: Social dynamics of digg. EPJ Data Science 1(5) (June 2012)
12. Hogg, T., Lerman, K., Smith, L.M.: Stochastic models predict user behavior in social media. In: SocialCom (2013)
13. Huberman, B.A.: Strong Regularities in World Wide Web Surfing. Science 280(5360), 95–97 (1998)
14. Kang, J.-H., Lerman, K.: LA-CTR: A limited attention collaborative topic regression for social media. In: AAAI (2013)
15. Kang, J.-H., Lerman, K., Getoor, L.: LA-LDA: A limited attention model for social recommendation. In: SBP (2013)
16. Lerman, K., Hogg, T.: Leveraging position bias to improve peer recommendation. PLoS One 9(6), e98914 (2014)
17. Ma, H., Yang, H., Lyu, M., King, I.: Sorec: social recommendation using probabilistic matrix factorization. In: CIKM, pp. 931–940. ACM (2008)
18. Payne, S.L.: The Art of Asking Questions. Princeton University Press (1951)
19. Rodriguez, M.G., Gummadi, K., Schoelkopf, B.: Quantifying information overload in social media and its impact on social contagions. In: ICWSM (2014)
20. Romero, D.M., Galuba, W., Asur, S., Huberman, B.A.: Influence and passivity in social media. In: WWW (2011)
21. Romero, D.M., Meeder, B., Kleinberg, J.: Differences in the mechanics of information diffusion across topics: Idioms, political hashtags, and complex contagion on twitter. In: WWW (2011)
22. Salakhutdinov, R., Mnih, A.: Probabilistic matrix factorization. Advances in Neural Information Processing Systems 20, 1257–1264 (2008)
23. Wang, C., Blei, D.M.: Collaborative topic modeling for recommending scientific articles. In: KDD (2011)
24. Wang, D., Song, C., Barabási, A.-L.: Quantifying long-term scientific impact. arXiv preprint (2013). arXiv:1306.3293
25. Weng, L., Menczer, F., Ahn, Y.-Y.: Virality prediction and community structure in social networks. arXiv preprint (2013). arXiv:1306.0158

The Influence of Collaboration on Research Quality:
Social Network Analysis of Scientific Collaboration in Terrorism Studies Research Groups

Alla G. Khadka[✉] and Mike Byers

Graduate School of Public and International Affairs, University of Pittsburgh,
3601 Wesley W. Posvar Hall, Pittsburgh, PA 15260, USA
{asg38,MAB415}@pitt.edu

Abstract. Considering that counterterrorism measures are costly and cannot be afforded to fail, it is imperative that these policies are informed by evidence-based analysis. Yet the terrorism studies knowledge base continues to be dominated by conceptual pieces. To increase the number of evidence-based studies available to decision-makers, it is critical to test what kinds of research collaboration communities are the most likely to produce evidence-based research. The aim of the study is to fill this need. Drawing on models of knowledge production and social network analysis literature, normative and structural characteristics of a knowledge production system are delineated and operationalized as measurable indicators that are the most likely to correlate with the greatest amount of evidence-based studies per research community. A list of hypotheses is developed and tested in regards to terrorism research published between 1992 and 2013 by the means of regression and cluster network analyses of collaboration networks of authors and organizations contributing to terrorism research. The findings demonstrate that heterogeneous and organizationally diverse research groups are more likely to generate a greater amount of evidence-based research as compared to more homogeneous groups. Other predictors of the increase in the number of evidence-based studies produced by a research group are the group's total degree centrality and internal link count, indicating that scientific collaboration among authors and organizations is essential to producing policy research based on data analysis.

Keywords: Social network analysis · Terrorism research · Data mining · Knowledge production models

1 Introduction

Security studies are formally committed to creating knowledge that informs policy agendas and provides a basis for assessing the impact of security policies. Many of these policies are costly in material terms and they often involve the loss of human life. As a society which Machlup (1962) [18] calls a "knowledge society," we respond to this challenge by attempting to ground policy decisions in evidence-based research. In reference to evidence-based policy research, Donald T. Campbell asserts that dif-

© Springer International Publishing Switzerland 2015
N. Agarwal et al. (Eds.): SBP 2015, LNCS 9021, pp. 111–120, 2015.
DOI: 10.1007/978-3-319-16268-3_12

ferences in the social structures and processes of sciences affect the type of research produced by these sciences [3, 4]. According to the literature on knowledge production models, it is possible to increase a proportion of evidence-based studies in a knowledge base by supporting certain normative characteristics of a knowledge system, such as organizational diversity and heterogeneity. It is also possible to increase the amount of evidence-based research in a knowledge base by encouraging and funding certain types of research collaborations, including projects that engage researchers with a diverse set of skills. Furthermore, the social network analysis literature attests that if a knowledge production system is viewed as a scientific collaboration network, it is possible to determine what type of collaboration ties positively correlate with a greater amount of evidence-based studies per research community.

Considering that since September 11th there has been massive increase in the scale of government and private spending on counter-terrorism research and that this research continues to be dominated by conceptual pieces, it is critical to investigate which descriptive and structural properties of research groups in the terrorism studies knowledge base positively correlate with a greater proportion of evidence-based studies to conceptual pieces produced by a group. To this end, the descriptive characteristics serve as proxy variables for the heterogeneity and organizational diversity of a research group. Additionally, four group-level structural variables are included in the model to explore the relationship between intensity of co-authorship collaboration links in a research group and a greater proportion of evidence-based to conceptual pieces per group.

2   Theory

In the most recent systematic assessment of terrorism research, Lum et al. [16] conclude that only a small percentage of studies published between 1983 and 2004 satisfy minimal criteria for making inferences about the efficacy of different counter-terrorism strategies [16: pg.18]. Out of 15,132 articles, only 354 studies achieve sufficiently high standards of quality to be regarded as "evidence-based." The results presented by Lum et al. [16] reflect the other evaluations of terrorism literature. In examining the quality of research on terrorism in 1980s, Schmid and Jongman [24] using 1982 and 1985 survey data, concluded that approximately 80% of the literature is not evidence-based. Following Schmid and Jongman's (1988) work, Silke [25, 26] used similar methods of analysis to review terrorism research published between 1995 and 2000. Silke concluded that in the 1990s, 68% of the research was based on the literature-type studies.

While security studies research continues to be dominated by conceptual pieces, the experts agree that in security studies, the need to use evidence to inform policy processes is as, or even more, critical as in other policy issue areas [14, 16, 17, 21]. In this context, the practical implication of generating evidence-based studies and encouraging decision makers, including the military community, to use these studies in informing their strategies is especially helpful in the area of terrorism response programs. In monetary terms, in 2014-2015 these programs cost American citizens $58.6 billion [10]. Applying strategies that are predicted to reach certain results, while accounting for all the key variables, will help to minimize a potential failure.

Sociologists studying the structure of knowledge systems sought to discern how the type of generated research is affected by descriptive characteristics of knowledge systems. In this context, Merton argued that research output is affected by the norms that define how scientific enterprise functions [19]. When norms change so does the type of generated research. Merton's functionalist norms of sciences are referred to as the Mode 1 model of knowledge production [19, 20]. The four norms delineated by Merton [20, pp. 270-277] are: (1) Universalism, (2) Disinterestedness, (3) Communalism, and (4) Organized skepticism. Although Merton's norms of knowledge production were initially meant to describe the knowledge production system in the 1950s - 1970s, these norms characterize the knowledge production system that surfaced in the 20th century (professionalization and research universities) and remained prominent throughout 1990s [23].

Beginning in the 1990s research practices and values as captured by Merton underwent a number of important changes. To account for the changes in research practices engendered by the growth of information and technology, in 1994 Michael Gibbons and his colleagues develop a new framework that characterizes a contemporary structure of the knowledge production enterprise. In creating their framework, Gibbons and his co-authors contrast each attribute of the new mode of knowledge production, or **mode 2**, with **mode 1**, as delineated by Merton. The attributes are: (1) Knowledge produced in the context of application, (2) Complexity, (3) Transdisciplinarity, (4) Reflexivity, (5) Enhanced social accountability, (6) Heterogeneity, (7) Organizational diversity, and (8) Broader system of quality control [15, pp. 3-11]. This study focuses on heterogeneity and organizational diversity.

Gibbons et al. [15] introduce heterogeneity and organizational diversity as normative attributes of the knowledge structure. However, they do not provide any guidelines on how to operationalize these attributes, nor do they empirically test whether these attributes are positively correlated with the research community producing a greater proportion of evidence-based research. To address this challenge, we operationalize heterogeneity and organizational diversity in the form of measurable indicators. In view of the type of data we have in our dataset, the following hypotheses can be tested:

- **H1.** The more heterogeneous a research group is in terms of the countries it is comprised of, the more likely it is to produce a greater number of evidence-based studies and a lesser number of conceptual pieces.
- **H2.** The more diverse a research group is in terms of the types of organizations it is comprised of and the funding agency types that provide monetary support to the researchers in this cluster, the more likely it is to produce a greater number of evidence-based studies and a lesser number of conceptual studies.

Additionally, in view of the citation network analysis literature demonstrating that structure of collaboration ties in a research community affects the type of research it generates [7, 11, 27], it is reasonable to expect that:
- **H3.** A research group with higher total degree centrality and clustering coefficient is likely to generate a greater amount of empirical and case study research and a

lesser number of conceptual pieces than a group with low total degree centrality and clustering coefficient.

- **H4.** A research cluster with a lower number of internal links is likely to generate more empirical and case study research and less conceptual research than clusters with higher number of internal links as long as the number of links is not too low. Burt [2] in developing his concept of structural holes discusses a potential danger when network becomes over dense. As a network evolves over time, more and more links develop among its members, which makes it more difficult for new members with different perspectives to enter the network [28]. Thus, although an optimal research system requires some collaborative links among its members, the network should not be overly dense.

3   Method

In this study we use the Social Network Analysis (SNA) method to analyze the knowledge production system. The method produces interlinked co-authorship network structures characterized by attributes of the individual authors and organizations that constitute the system. The network analysis method used in this study is informed by citation network analysis. As introduced by Bernal [1]; Garfield [12, 13] and De Solla Price [9], the citation network analysis approach evaluates scholarly fields as structures that evolve over time, using techniques such as co-citation, co-word and co-author analyses [7, 8, 11, 27]. Linking authors and organizations co-authoring the same publication, the citation network analysis enables an assessment of bibliometric and qualitative parameters of individual authors and papers in the context of the overall network.

The cluster analysis employed in this study is informed by both bibliometric and citation network analysis in that it evaluates co-authorship networks of authors and organizations in terrorism research in terms of node-level and network-level measures. Within the broad method of dynamic network analysis [5, 6], this study employs Girvan-Newman cluster analysis algorithms to obtain frequency counts for independent and dependent variables per cluster, as well as centrality measures per cluster. The Girvan-Newman algorithm partitions a network into groups by employing edge betweenness centrality computation.   The edges with the highest betweenness centrality are eliminated from the network. The elimination of edges continues until the network is split into separate clusters with each cluster representing an interlinked community [22], which are referred to in this study as research groups. In the case of our two networks of authors and organizations the resultant groups represent communities of authors and organizations that closely collaborate with one another.

As such, if authors and organizations conduct research together, they are placed in the same group. For authors, variables per cluster include values for the number of author countries, number of funding agency types, total degree centrality, clustering coefficient, and internal link count. All of these variables are treated as independent. Additionally, the Girvan-Newman algorithm computes the percentage of evidence-based/conceptual pieces per cluster, which is treated as the dependent variable. For organizations, the independent variables are the number of organization countries, the number of organization types, the number of publication languages, total degree centrality, internal link count, and clustering coefficient per cluster. The dependent variable, as with the case of authors, is the research type.

Since the Girvan Newman algorithm breaks the networks into hundreds of clusters, we perform Pearson correlation and logistic regression to determine the statistical significance, strength, and direction of the relationship between indicators of heterogeneity, organizational diversity, and intensity of ties per cluster and whether the greater proportion of research produced by authors/organizations in the cluster is evidence-based or conceptual. The overarching aim is to determine if heterogeneity and organizational diversity of authors and organizations in research collaboration clusters and the entire system, as well as the intensity of collaboration ties, affect how much evidence-based versus conceptual research is being produced.

4   Data

This study analyzes 5121 publications on terrorism from 1992-2013. As such, the data encompasses two terms of Bill Clinton's administration, two terms of George W. Bush administration, and one term of Barack Obama's administration. When surveying the data longitudinally, the pre-9/11 period embodies a sparsely populated research collaboration network and does not exhibit any significant changes between the two presidents. However, following the September 11 terrorist attacks, the network expands disproportionately. To achieve a more accurate representation of the change between the George W. Bush and Barak Obama administrations, the four years selected to represent Obama's administration are 2008 (his campaign year), 2010, 2011 and 2012. The year 2009 was excluded, because much of the terrorism research that was published in 2009 was still framed by the political and security climate that was created under President Bush. The year 2008 is included because the presidential campaign typically reflects the incoming president's stance on security. Including 2008, 2010, 2011 and 2012, while omitting 2009, enables a more balanced representation of terrorism research influenced by and influencing the Obama administration. Since the data collection process ended at the end of November 2013, this year is not entirely representative.

All publications were gathered from the Web of Science (WoS) by means of the keyword search method. Using the HistSite function, in addition to standard bibliometric information, each paper was connected to the authors' affiliation addresses and funding agencies (in cases when the research was funded). All of the information was exported into an Excel spreadsheet.

In preparing data for network analysis, entries on authors and organizations were arranged in the form of relational matrices. There are a total of 17,996 authors in the first author column and 8,549 authors in the co-author column. When duplicates are removed there are 5,344 authors in the first-author column and 2,867 authors in the co-author column. The same method was used to arrange data on organizations. There are a total of 8,004 organizations in the first author column and 4,268 organizations in the co-author column. With duplicates removed, the number comes to 1,694 organizations in the first-author column and 946 organizations in the co-author column.

Table 1. Summary of Primary and Attribute Variables for Authors and Organizations

NAME	DESCRIPTION
Author	*Author(s) contributing to research on terrorism, 1992-2013*
Organization	*Author's primary affiliation*
Publication Title	*Title of publication*
Publication Year	*Date work is published (necessary for temporal analysis)*
Funding Agency Type (13 categories)	*Each organization is assigned to an organization type (Such as university, think tank, U.S. Federal government, military etc.*
Organization's Country	*Country author's organization is located at*
Organization Type (9 categories)	*Each organization is assigned to an organization type (Such as university, think tank, U.S. Federal government, military etc.*
Research Type	*Each publication is assigned one of the two categories based on Lum et al., 2006: (1) thought piece, (2) evidence-based research*

Organization and funding agency types were assigned manually. First, a list of types for organizations and funding agencies was created, then each relevant data point was assigned to its type. The data on the author's affiliations was obtained from information provided in the author's address. If the author listed more than one affiliation, all affiliations were included. In cases when the address was missing, it was located and ascribed manually, provided that this information is readily available. Each organization was qualified with a country variable. If an organization has multiple locations, such as the United Nations, we included all the countries where the researchers from these organizations are located based on the information captured in our dataset. Considering that the Web of Science began collecting the funding agency information in 2007, the funding agency data included in this study spans 2007-2013.

Based on the type of collected data, two one-mode networks analyzed in this study are:

(1) Authors collaboration network, and
(2) Organizations collaboration network.

Attribute variables for authors and organizations are treated as independent variables and research type is treated as the dependent variable. For the Pearson Correlation and logistical regression analyses, research type is included as a binary variable, with conceptual pieces coded 0 and evidence-based research coded 1. All data points are qualified by a publication year.

5   Analysis and Findings

Pearson Correlation is performed to assess the relationship between the number of authors' countries, funding agency types, maximum total degree centrality, clustering coefficient mean, and the number of internal links per cluster for author-author collaboration network and a greater proportion of evidence-based to conceptual studies per research group. The results indicate that all the variables have a positive statistically significant relationship with a greater rate of evidence-based to conceptual studies per cluster at p-value = 0.05, except for the funding agency type (see Table 2). The statistical insignificance of variables pertaining to funding agencies can be explained by the limitations that come with the funding agencies data. Considering that the data on funding agencies is confined to 2007-2013 period and that funding agency types and countries only appear in 64 out of 1128 clusters compared to all other values their N is too small to generate meaningful correlation results.

Table 2. Pearson Correlation Results

Variables per Cluster N=1128	Author's Country#	Funding Agency Type#	Max Total Degree	Internal Link Count	Clustering Coefficient Mean
Evidence-based Research[+]	.064**	.165	.062**	.073**	.209***
Variables per Cluster N=148	Organization Country[#]	Organization Type#	Max Total Degree	Internal Link Count	Clustering Coefficient Mean
Evidence-based Research[+]	.255**	.126	.185**	.273*	.129

[+] Binary variable: if 50% or more studies per cluster are conceptual, coded a 0; if 50% or more studies per cluster are evidence-based, coded 1
[#] Binary variable: if 1 country/organization type per cluster, coded 0; if 2+ countries/organization types per cluster coded 1
*** p-value<0.001; **p-value<0.05; *p-value<0.1

The clustering coefficient mean per cluster displays the highest correlation coefficient compared to all other variables reflecting the strength of this variable's relationship with the greater number of evidence-based studies per group. The next variable in terms of the strength of association with the evidence-based research per cluster is internal link count, followed by number of countries per cluster, followed by maximum total degree centrality per cluster.

We use binary logistic regression analysis to obtain a more nuanced understanding of predictive power for each independent variable that emerged as statistically significant in the Pearson correlation analysis (see Table 3). As with the Pearson correlation, the same binary research type variable is used as a dependent variable. The total degree centrality emerges as statistically insignificant, hence we exclude it from the results table, but all the other variables emerge as significant predictors for the probability of conceptual and evidence-based research. Considering the positive constant, the higher the probability is that a collaborating research group consists of more than one county, the higher the probability is that such group will produce a greater proportion of evidence-based studies to conceptual pieces. The same conclusion holds for the internal link count and clustering coefficient. Among all three predictors, clustering coefficient is the strongest and the number of internal links per cluster is the weakest (per B and Wald values).

Table 3. Binary Logistic Regression Results

Variables	B
Author's Country[#]	.450**
Author's Network Internal Link Count	.005**
Author's Network Clustering Coefficient	.930*
Organization's Country#	1.008*
Organization's Network Internal Link Count	.576*
Organization's Network Total Degree Centrality Max	413.768**

    [#] Binary variable: if 1 country per cluster, coded 0; if 2+ countries per cluster coded 1
    *** p-value<0.001; **p-value<0.05; *p-value<0.1

Turning to organizations, Pearson correlation results for the organizations are similar to the results we acquired for the authors. The number of organizations, countries, maximum total degree centrality, and internal link count per research group all predict a greater proportion of evidence-based research to conceptual pieces generated by a research group. The results in Table 2 further indicate the binary variable that designates 1 versus 2 or more organizations per group shows to be statistically insignificant, which appear to disprove our hypothesis in regards to organizational diversity. However, visually inspecting composition of clusters, we observe that majority of the clusters are uniform, being composed of the same organization type. Hence, what we can conclude here is that for a research group to be slightly more diverse in terms of organization types than a cluster composed of only one organizations type does not lead to a greater proportion of evidence-based studies per group.

A more detailed assessment of variables that display statistically significant relationship with a greater proportion of evidence-based research to conceptual research per cluster shows that the strongest predictor of the evidence-based research per cluster is the number of organizations' countries per cluster, followed by the number of internal links per cluster and maximum total degree centrality per cluster. It means that the greater the number of highly connected nodes there are per cluster, the higher the number of links there are per cluster, and the more likely it is for a cluster to be composed of organizations from more than one country, the greater the probability that this cluster will produce a higher rate of evidence-based to conceptual studies. The opposite relationship is true for the conceptual pieces.

6   Conclusion

In this study, we investigated whether heterogeneity, organization diversity, and level of collaboration in terrorism research communities lead to a greater proportion of evidence-based to conceptual studies per research group.

Regression results show that the research groups that are the most likely to generate a greater proportion of evidence-based to conceptual studies are those that are characterized by organizational diversity (organization types) and a range of countries. Other predictors of research quality are a group's maximum total degree centrality and internal link count. Specifically, research groups composed of more than one country are the most likely to produce a greater proportion of evidence-based to conceptual studies per group compared to those that have researchers from the same country. Moreover, a greater proportion of evidence-based research to conceptual research in a research group has a positive relationship with the number of internal links per cluster and maximum total degree centrality per cluster. This, in turn, suggests that scientific collaboration leads to a greater amount of evidence-based research in a knowledge base.

In sum, a research group that is the most likely to produce a higher proportion of evidence-based to conceptual studies is the group consisting of participants from more than one country. Moreover, to generate a greater amount of evidence-based studies, research communities must include several opinion leaders with frequent ties to and from other authors and organizations in the group as well as proclivity to frequent collaborations on numerous research projects.

References

1. Bernal, J.D.: The Social Function of Science. The Social Function of Science (1939)
2. Burt, R.S.: Structural Holes: The social structure of competition. Harvard University Press, Cambridge, MA (1992)
3. Campbell, D.T.: Science's Social System of Validity-Enhancing Collective Belief Change and the Problems of the Social Sciences. In: Fiske, D.W., Shweder, R. (eds.) Metatheory in Social Science, pp. 108–135. University of Chicago Press, Chicago, IL (1986)
4. Campbell, D.T.: Guidelines for Monitoring the Scientific Competence of Preventive Intervention Research Centers: An Exercise in the Sociology of Scientific Validity. Knowledge: Creation, Diffusion, Utilization, 8(4) (March), 389–430 (1987)
5. Carley, K.M.: A dynamic network approach to the assessment of terrorist groups and the impact of alternative courses of action. Inst of Software Research. Carnegie-Mellon University, Pittsburgh, PA (2006)
6. Carley, K.M., Pfeffer, J.: Dynamic network analysis (DNA) and ORA. Adv. Des. Cross-Cult. Act. Part (2012)
7. Chen, C.: Visualizing semantic spaces and author co-citation networks in digital libraries. Information Processing & Management 35(3), 401–420 (1999)
8. Chen, C.: Searching for intellectual turning points: Progressive knowledge domain visualization. Proceedings of the National Academy of Sciences 101(suppl 1), 5303–5310 (2004)
9. De Solla Price, D.J.: The Science of Science. Atomic Scientists, Bulletin 21(8), 2 (1965)
10. Department of Defense. Fiscal Year 2015 Budget Amendment: Overview Overseas Contingency Operations. June 2014. Office of the Under Secretary of Defense (Comptroller)/ Chief Financial Officer. For an electronic version of this document (2014). http://comptroller.defense.gov/budget2015.html
11. Estabrooks, C.A., Derksen, L., Winther, C., Lavis, J.N., Wallin, S.D.S., Profetto-McGrath, J.: The Intellectual Structure and Substance of the Knowledge Utilization Field: A longitudinal author co-citation analysis, 1945 to 2004. Implementation Science, 3 (2008)

12. Garfield, E.: Citation Indexes for Science. Science **122**(3159), 108–111 (1955)
13. Garfield, E., Sher, I.H., Torpie, R.J.: The use of citation data in writing the history of science. Philadelphia, PA: United States Air Force Office of Scientific Research. Institute for Scientific Information (1964)
14. Genovés, V.G., Farrington, D.P., Welsh, B.C.: Crime prevention: more evidence-based analysis. Psicothema **20**(1), 1–3 (2008)
15. Gibbons, M., Limoges, C., Nowotny, H., Schwartzman, S., Scott, P., Trow, M.: The New Production of Knowledge: The Dynamics of Science and Research in Contemporary Societies. Sage, London (1994)
16. Lum, C., Kennedy, L.W., Sherley, A.: Are counter-terrorism strategies effective? The results of the Campbell systematic review on counter-terrorism evaluation research. Journal of Experimental Criminology **2**(4), 489–516 (2006)
17. Lum, C., Kennedy, L.W., Sherley, A.: Is counter-terrorism policy evidence-based? What works, what harms, and what is unknown. Psicothema **20**(1) (2008)
18. Machlup, F.: The Production and Distribution of Knowledge in the United States. Princeton University Press, Princeton, NJ (1962)
19. Merton, R.K.: Social theory and social structure, enl edn. Free Press, New York, NY (1968)
20. Merton, R.K.: The sociology of science: theoretical and empirical investigations. University of Chicago Press, Chicago, IL (1973)
21. Muggah, R., Berdal, M., Torjesen, S.: Enter an evidence-based security promotion agenda. Security and Post-Conflict Reconstruction: Dealing with Fighters in the Aftermath of War, 268 (2008)
22. Rattigan, M.J., Maier, M., Jensen, D.: Graph clustering with network structure indices. In: Paper presented at the Proceedings of the 24th international conference on Machine learning (2007)
23. Rip, A.: Strategic research, post-modern universities and research training. Higher Education Policy **17**, 153–166 (2004)
24. Schmid, A.P., Jongman, A.J.: Political Terrorism: A New Guide to Actors, Authors, Concepts, Data Bases, Theories and Literature. North-Holland, Oxford (1988)
25. Silke, A.: The devil you know: Continuing problems with research on terrorism. Terrorism and Political Violence **13**(4), 1–14 (2001)
26. Silke, A.: The road less travelled: recent trends in terrorism research. Research on terrorism: trends, achievements and failures, 186–213 (2004)
27. Small, H.G.: Co-citation in the scientific literature: A new measure of the relationship between two documents. Journal of the American Society for Information Science **24**(4), 265–269 (1973)
28. Valente, T.W.: Social Networks and Health: Models, Methods, and Applications. Oxford University Press, New York, NY (2010)

Detecting Rumors Through Modeling Information Propagation Networks in a Social Media Environment

Yang Liu[1], Songhua Xu[1]($\boxtimes$), and Georgia Tourassi[2]

[1] New Jersey Insititute of Technology, University Heights, Newark, NJ 07102, USA
yl558@njit.edu
[2] Health Data Sciences Institute, Oak Ridge National Laboratory,
Biomedical Science and Engineering Center, 1 Bethel Valley Road,
Oak Ridge, TN 37830, USA

Abstract. In the midst of today's pervasive influence of social media content and activities, information credibility has increasingly become a major issue. Accordingly, identifying false information, e.g. rumors circulated in social media environments, attracts expanding research attention and growing interests. Many previous studies have exploited user-independent features for rumor detection. These prior investigations uniformly treat all users relevant to the propagation of a social media message as instances of a generic entity. Such a modeling approach usually adopts a homogeneous network to represent all users, the practice of which ignores the variety across an entire user population in a social media environment. Recognizing this limitation of modeling methodologies, this study explores user-specific features in a social media environment for rumor detection. The new approach hypothesizes that whether a user tends to spread a rumor is dependent upon specific attributes of the user in addition to content characteristics of the message itself. Under this hypothesis, information propagation patterns of rumors versus those of credible messages in a social media environment are systematically differentiable. To explore and exploit this hypothesis, we develop a new information propagation model based on a heterogeneous user representation for rumor recognition. The new approach is capable of differentiating rumors from credible messages through observing distinctions in their respective propagation patterns in social media. Experimental results show that the new information propagation model based on heterogeneous user representation can effectively distinguish rumors from credible social media content.

This manuscript has been authored by UT-Battelle, LLC under Contract No. DE-AC05-00OR22725 with the U.S. Department of Energy. The United States Government retains and the publisher, by accepting the article for publication, acknowledges that the United States Government retains a non-exclusive, paid-up, irrevocable, world-wide license to publish or reproduce the published form of this manuscript, or allow others to do so, for United States Government purposes. The Department of Energy will provide public access to these results of federally sponsored research in accordance with the DOE Public Access Plan (http://energy.gov/downloads/doe-public-access-plan).

© Springer International Publishing Switzerland 2015
N. Agarwal et al. (Eds.): SBP 2015, LNCS 9021, pp. 121–130, 2015.
DOI: 10.1007/978-3-319-16268-3_13

Keywords: Rumor detection · Heterogeneous user representation and modeling · Information propagation model · Information credibility in social media

1   Introduction

With the rapid development of social media sites and proliferation of social media content, user-generated messages can easily reach a large audience. Such a potential for rapid and far-reaching information propagation in social media brings unprecedented challenges in information quality assurance and management. Most social media sites that care about information quality currently gauge information credibility and detect rumors manually. Such practice cannot efficiently and affordably scale up to handle a large volume of messages typically seen in a popular social media environment. The incurred deficiency undesirably facilitates an easy spread of rumors to a large population at a fast pace, leading to elevated societal harm and potential damages. Consequently, automatic detection of rumors in social media is highly desirable and socially beneficial. This paper introduces a new method for automatically detecting rumors circulated in social media through observing their propagation patterns as distinct from those of credible messages. We formulate the social media rumor detection task as a microblog classification problem by introducing an information propagation model built upon a heterogeneous user representation. The derived microblog classifier follows a hypothesis that rumors and credible messages tend to propagate in a social media environment following quantitatively differentiable patterns among a heterogeneous population of human users. By extracting user context sensitive features from individual users' profiles, the new method estimates the retweeting probability of a message when it is a rumor versus a credible message respectively. Given this estimate, the method is able to classify a message as being a rumor or a truthful message by comparing the likelihood that the observed information propagation network of the message is generated by the proposed information propagation model under the rumor and credible message mode respectively.

Although many previous studies focused on Twitter, in this study, we explore the popular Chinese social media site–Sina Weibo [1]–as the experimental social media environment due to three main reasons. First, Sina Weibo is the largest microblog service site and the most popular and influential social media site in China. Like Twitter, on Sina Weibo users can set up their personal profiles and broadcast microblogs to the public. Those broadcast microblogs will appear in the timeline of a user's followers. As of today, it has 1.4 billion users registered under four general user categories, including ordinary individuals, celebrities, enterprises, and government organizations. The number of active users on the site reaches 66 million. Over 125 million microblogs are posted daily. Second, Sina Weibo users tend to share more personal information, such as their ages and education levels, than Twitter users. Being able to access such personally descriptive information is very beneficial for the proposed method. Third, Sina Weibo offers an official rumor busting service, which is not provided by many of today's social media sites, including Twitter. This service gives us a reliable collection of rumors as ground-truth data in this study. The official rumor busting expert team of Sina Weibo assesses the information credibility of suspicious microblogs reported by users through referring to third-party credible information resources and optionally consulting with external domain experts. All detected rumors are posted in a dedicated area on the homepage of Sina Weibo. According to self-released data from Sina Weibo, at least two messages are announced as rumors everyday. Those rumors usually had been retweeted thousands of times before they were officially denounced. Such a high number of retweets involved indicates that rumor spreading remains as a major problem in Sina Weibo despite the site's serious endeavor to combat the phenomenon.

The main contribution of this paper lies in its proposal of a novel information propagation model for social media environments based upon a heterogeneous user representation. The new information propagation model is capable of characterizing information propagation patterns across different user sub-populations in a social media environment. A direct application of the model is to detect and separate rumors apart from credible messages in a social media environment. We conducted comprehensive experiments using the Sina Weibo platform to explore the usefulness and advantages of applying the new information propagation model for rumor detection.

The rest of the paper is organized as follows: Section 2 briefly discusses previous studies most related to this work. Section 3 describes the proposed algorithmic method for rumor detection in a social media environment. Section 4 reports experimental results obtained using the proposed method for rumor detection. Finally, Section 5 concludes this work.

2   Related Work

Detecting rumors in social media environments has been richly studied and reported in the literature. For example, a variety of microblog classification methods has been attempted to derive features for rumor detection. In a comprehensive study [2], Castillo et al. summarized three main categories of such features, including: (1) message-based features, which focus on characterizing the content of a microblog, (2) user-based features, which consider attributes of site users, (3) propagation-based features, which encode network-based characteristics of a social media environment. In a related study, Qazvinian et al. [3] proposed three categories of features for rumor detection, including: (1) content-based features, which characterize both lexical and part-of-speech patterns of a microblog; (2) network-based features, which are extracted from a user network consisting of both users who spread rumors and those who do not participate in rumor dissemination; and (3) Twitter specific features, including Twitter hashtags and shared URLs.

It is noted that traditional content-based features generally ignore the rich information regarding human users involved in rumor spreading in a social media environment. In contrast, propagation or network-based features are extracted from a rumor propagation network. Such a network consists of users directly involved in rumor propagation. To extract propagation-based features, rumor propagation models are used. To establish these propagation models, one key task is to estimate the probability of an arbitrary user to spread a rumor. In [4], Moreno et al. modeled rumor propagation by treating a microblog propagation network as a homogeneous scale-free network. They assumed that the probability of a user to forward a microblog is affected by the number of online friends of the user who spread the microblog. Many propagation-based methods alike treat all users in a network as homogeneous instances of a common type of nodes. Their practice ignores any variance among a user population. In reality, however, even when two users have the same number of online friends, they may still respond differently to a rumor because of their personal abilities in discerning rumors. Xia and Huang argued that a person would believe a rumor if the cumulative influence from his or her online friends regarding the rumor is larger than the person's internal rumor resistance [5]. Their method only observes the number of followers to determine a user's rumor resistance threshold while ignoring many other aspects of user differentiation. Sun et al. found that besides content features, the numbers of followers and followees as well as the age of a user account all affect a user's likelihood to retweet a message [6]. Their method considers the above user-specific features in rumor detection. We recognize that in reality, factors that affect rumor spreading are far richer than the aforementioned ones. To overcome the limitations of all existing methods in comprehensively modeling and leveraging user context for rumor detection, our new method utilizes a wider spectrum of user-specific features available in Sina Weibo when constructing its information propagation model. The design of this new method is also inspired by the work of Jin and Dougherty et al. [7], which employed epidemiological models to characterize information cascades of both news and rumors in twitter for rumor detection. Their work reveals that information diffusion patterns for rumors and normal news are different. Learning from their discovery, our information propagation model captures content dissemination patterns in social media environments for rumors and credible messages respectively through two dedicated modes.

Researchers have also explored the particular problem of rumor detection on Sina Weibo. Compared with Twitter, Sina Weibo offers fewer functions to protect user privacy and anonymity yet provides richer services to encourage versatile and open-minded information sharing among users. As a result, users on Sina Weibo generally share more personal information with peers than their counterparts on Twitter. Similar to our proposed method, many studies on Sina Weibo utilize the rich spectrum of user-specific information to classify rumors from credible messages. For example, Yang et. al [1] proposed two features for rumor detection, which are respectively based on the type of computing device used for microblog posting, e.g. a mobile or PC, and the geographical location where a microblog is posted. Sun et. al [8] classified rumors on Sina Weibo that relate to social events into four categories—purely fictitious rumors, time-sensitive event rumors, rumors due to fabricated details, and rumors engineered from mismatched text and pictures. Their work particularly explored the problem of picture misuse as a source of rumors. For readers interested in more studies regarding rumor propagation and detection on Sina Weibo, they are referred to [9–12].

3   Our Method

3.1   Problem Formulation

In this study, we approach the rumor detection task as a classification problem. That is, two classes of microblogs are considered—rumors and credible messages. We consider rumors as popular microblogs primarily conveying misinformation about an event or a fact. In contrast, credible messages are microblogs not carrying such information. Under this problem formulation, the task of rumor detection is reduced to constructing a classifier capable of reliably assigning an appropriate binary class label to any given microblog.

For simplicity, we borrow the term "retweet" to refer to the message forwarding action on Sina Weibo. For each microblog, the proposed method first extracts its propagation network by tracking all its retweets and identifying all users involved in tweeting, receiving, or retweeting the microblog. We also borrow some terminologies from a previous study on rumor propagation network [13] to describe our new algorithm. We define a user's followee as a person whom the user follows. In a microblog's propagation network, we define a *spreader* as a user who receives the microblog from his/her followee and subsequently retweets the microblog. We define a *stifler* as a user who receives a microblog but doesn't retweet it further.

We hypothesize that when a user receives a microblog, the probability that the user will retweet it depends not only upon content features regarding the microblog, but also upon specific attributes of that user, such as the person's age and education background. Based on this hypothesis, we propose our new information propagation model using a heterogeneous user representation for rumor detection.

3.2   Information Propagation Model under Credible Message Mode

When constructing the new information propagation model under the credible message mode, we consider two key influence factors for a user to retweet a credible message as follows.

The first factor is the user's general tendency to retweet a microblog, For a specific user, we select two features to estimate this factor, including: (1) x_1: the number of microblogs retweeted by the user in the past three months; and (2) x_2: the number of the user's retweeted microblogs divided by the total number of microblogs tweeted or retweeted by his/her in the past three months.

Such selection of features is based upon both our empirical observations and the experimental finding in Suh's study [6], which points out that a user's past retweeting behaviors affect his or her present behaviors. In the selection process, the time window we examine spans over three months, which is of the same length as the one used in Suh's original study.

The second factor characterizes the influence of the user's followees. In the propagation network for a specific microblog, we define the influence of followees to a user u_i as $INF(u_i)$, which is abbreviated as x_3, as follows:

$$INF(u_i) = \sum_{u_j \in In(u_i)} Ret(u_j, u_i) IsF(u_j, u_i) \log |Out(u_j)| / \log |In(u_i)|; \qquad (1)$$

$$Ret(u_j, u_i) = \begin{cases} 1 & \text{if } u_j \text{ retweets that microblog to } u_i; \\ 0 & \text{otherwise} \end{cases} \qquad (2)$$

$$IsF(u_j, u_i) = \begin{cases} \alpha & \text{if } u_j \text{ and } u_i \text{ follow each other.} \\ 1 & \text{otherwise.} \end{cases} \qquad (3)$$

In (1), $In(u_i)$ is the set of followees of u_i; $|In(u_i)|$ is the number of followees of u_i; $Out(u_j)$ is the set of followers of u_j; $|Out(u_j)|$ is the number of followers of u_j. This formula is designed under the inspiration of Liu's study [13].

We assume that when the user u_i reads a microblog that is retweeted by his/her followees, the probability for u_i to retweet the message, assuming the message is a credible one, can be estimated by the following logistic function:

$$P_c(u_i|\boldsymbol{\beta_1}, \alpha) = \frac{1}{1 + e^{-\boldsymbol{\beta_1}\boldsymbol{X_1}}}, \qquad (4)$$

where $\boldsymbol{X_1} = [x_1 \ x_2 \ x_3]$ is the feature vector for u_i and $\boldsymbol{\beta_1}\boldsymbol{X_1} = \beta_1 x_1 + \beta_2 x_2 + \beta_3 x_3 + \beta_0$. In the above equation, α and $\boldsymbol{\beta}$ are model parameters.

3.3   Information Propagation Model under Rumor Mode

Besides the above general influence factors, we also assume that rumor propagation is conditioned on a user's ability to verify a message. That is, users who have a high level of verification capability, such as those who command a higher level of capability in independent thinking, a higher knowledge level and/or education level, are much less likely to believe and spread rumors than those who don't possess the above qualifications. We characterize this influence factor through the following user context sensitive features listed in Table 1. Below, we explain some features unfamiliar to Twitter users. Credit score is related to a user's behaviors of tweeting and retweeting rumors, for which Sina Weibo has an official reporting mechanism. Users reported as spreading rumors or other problematic information will be penalized in their credit scores. Personal hashtags are used to describe a user's interests, such as "travel" or "movie." Sina Weibo experts are users who hold expertise in a certain area and post a certain amount

Table 1. Description of user context sensitive features used in this study

No.	Feature	No.	Feature	No.	Feature
f_1	Age	f_2	Registration time	f_3	Number of followers
f_4	Number of followees	f_5	Number of friends	f_6	Weibo level
f_7	Active days	f_8	Credit score	f_9	Number of hashtags
f_{10}	Is VIP member	f_{11}	Is Weibo expert	f_{12}	Has college degree
f_{13}	Has personal website	f_{14}	Job description	f_{15}	Personal description

of authoritative tweets. All the above user context sensitive features are available from a user's profile in Sina Weibo. For user context sensitive features listed in Table 1, features f_1 to f_9 are integer variables; features f_{10} to f_{13} are binary variables; the last two features–f_{14}, f_{15}–regarding job and personal descriptions are represented as bags of words. We organize the aforementioned user context sensitive features as a feature vector X_1'.

The final form of feature vector for characterizing the propagation of rumorous microblogs is denoted as $X_2 = [X_1 | X_1']$, which is the concatenation of the two feature vectors X_1 and X_1' introduced in the above. The probability for a user u_i to retweet a rumorous microblog is estimated by:

$$P_r(u_i | \beta_2, \alpha) = \frac{1}{1 + e^{-\beta_2 X_2}}, \tag{5}$$

in which β_2 is another set of model parameters, whose dimensionality is higher than that of the feature vector of X_2 by one, i.e. $1 + 3 + 15 = 19$.

3.4 Microblog Classification

Given a microblog, we first extract its propagation network in which each user functions either as a spreader or a stifler. We then compute the likelihood that this propagation network is generated by the proposed propagation model under the rumor mode (P_r) and non-rumor mode (P_c) respectively. After that, we label the microblog as a rumor if $P_r > P_c$ and a credible message otherwise. The free parameters, α, β_1, and β_2, in our proposed model are optimized using the Maximum Likelihood Estimation (MAE) method.

4 Experiments

4.1 Data Collection

We collect rumors from Sina Weibo's official rumor busting account [14]. The obtained collection contains 400 rumorous microblogs published in 2013 and 2014. Those microblogs are crawled directly from the timeline of the rumor busting account, which announces all confirmed rumorous microblogs following a temporal order. We download all such rumor microblogs as our ground-truth

Table 2. High-level statistics of microblog propagation networks analyzed in our experimental study

Attributes	Rumor		Credible messages		Both	
	Mean	Std.	Mean	Std.	Mean	Std.
Total number of users in the propagation network	39466.1	1527.3	70961.4	1685.2	61860.9	1634.2
Number of spreaders	3586.7	537.5	5677.1	768.3	5258.5.8	688.4
Number of stiflers	35418.8	11369.5	64384.2	1096.7	61721.3	1329.1
Percentage of spreaders (%)	9.8	1.5	8.1	1.4	8.5	1.5
Average out degree	159.9	14.3	188.6	16.7	171.3	16.5
Depth of the network	9.5	2.1	11.3	2.9	10.1	2.7
Time span in days	7.8	1.2	10.6	1.4	9.2	1.3

rumor data. For the credible messages, we collect them from Sina Weibo's official hot topic recommendation service [15]. The service publishes up-to-date microblogs about hot topics retweeted most frequently within the previous 12 months. All those microblogs must have been reviewed by Sina Weibo's official censorship team to ensure that they don't contain any misinformation. In our experiments, we randomly select 3600 credible messages published in 2013 and 2014 from this service and label them as "credible." Table 2 shows some high-level statistics for propagation networks of individual microblogs examined in this study.

Table 3. Comparing the performance of the proposed method with that of SVM classifiers using different sets of features [2], including (1) message-based, (2) user-based, (3) propagation-based, and (4) combined, as well as two peer methods—(5) Yang's [1] and (6) Sun's [8]. "R" represents rumors; "C" represents credible messages; "W. Avg" represents weighted average of rumors and credible messages

No.	Class	Precision	Recall	F-rate	AUC	No.	Class	Precision	Recall	F-rate	AUC
(1)	R	0.708	0.636	0.672	0.659	(4)	R	0.781	0.767	0.772	0.818
	C	0.700	0.648	0.677	0.684		C	0.802	0.730	0.775	0.805
	W.Avg	0.704	0.640	0.669	0.687		W.Avg	0.761	0.782	0.776	0.812
(2)	R	0.719	0.667	0.71	0.728	(5)	R	0.732	0.741	0.735	0.749
	C	0.703	0.728	0.713	0.708		C	0.718	0.729	0.721	0.758
	W.Avg	0.711	0.697	0.710	0.713		W.Avg	0.725	0.733	0.728	0.754
(3)	R	0.687	0.606	0.654	0.685	(6)	R	0.712	0.683	0.703	0.701
	C	0.698	0.769	0.738	0.742		C	0.677	0.679	0.677	0.705
	W.Avg	0.691	0.718	0.702	0.718		W.Avg	0.705	0.680	0.688	0.702
(7)	R	0.829	0.803	0.813	0.839						
	C	0.797	0.789	0.791	0.828						
	W.Avg	**0.812**	**0.793**	**0.799**	**0.831**						

4.2 Experimental Results

To leverage the propagation network-based features for rumor detection, we train a classifier according to the description in Sec. 3. We also explored a set of popular features frequently used in previous studies for rumor detection. In the Twitter information credibility study [2], the authors summarize three categories of features used to identify rumors, which are respectively based on messages, users, and propagation patterns. Using these features, we construct a baseline method by training a SVM classifier coupled with a RBF kernel function. We also combine those three categories of features as a combined feature set. We further compare the performance of the proposed method with that of two peer methods for rumor detection on Sina Weibo, i.e. [1,8]. We use the precision, recall, F-rate, and the Area under the ROC Curve (AUC) as the evaluation metrics. Table 3 shows that the proposed method outperforms the SVM classifiers and the two peer methods. The results demonstrate the effectiveness of the proposed user context sensitive information propagation model for rumor detection.

5 Conclusion

In this paper we propose a new information propagation model based on a heterogeneous user representation for automatically differentiating rumors from credible messages in social media. We conducted a series of experiments using the popular Chinese social media site Sina Weibo to explore the effectiveness of the new method. The experimental results confirm our hypothesis that propagation patterns of rumors and credible messages in social media indeed differ and distinct from each other. Based on the captured distinction between information propagation patterns for rumors versus credible messages, the proposed method can successfully and reliably detect rumors in social media at their early stage of spreading. In principle, the proposed method can be generically applied onto other social network platforms and social media sites similar to Sina Weibo. In particular, for sites like Facebook, which support detailed user profiles, the method would work more effectively than for another group of sites, such as Twitter, where little personal information is available. In the future, we will explore the above potential of this work for application to other social media sites.

Acknowledgments. This study was performed under the Protocol, F 186-14, approved by the Institutional Review Board (HHS FWA #00003246). The study was funded in part by the National Cancer Institute (Grant number: 1R01CA170508) and the Natural Science Foundation of China (NSFC) (Grant number: 61320106008).

References

1. Yang, F., Liu, Y., Yu, X., Yang, M.: Automatic detection of rumor on sina weibo. In: Proceedings of the ACM SIGKDD Workshop on Mining Data Semantics, pp. 13:1–13:7. ACM (2012)

2. Castillo, C., Mendoza, M., Poblete, B.: Information credibility on twitter. In: Proceedings of the 20th International Conference on World Wide Web, pp. 675–684. ACM (2011)
3. Qazvinian, V., Rosengren, E., Radev, D.R., Mei, Q.Z.: Rumor has it: Identifying misinformation in microblogs. In: Proceedings of the Conference on Empirical Methods in Natural Language Processing, pp. 1589–1599. Association for Computational Linguistics (2011)
4. Moreno, Y., Nekovee, M., Pacheco, A.F.: Dynamics of rumor spreading in complex networks. Physical Review E **69**(6), 066130 (2004)
5. Xia, Z., Huang, L.L.: Emergence of social rumor: Modeling, analysis, and simulations. In: Proceedings of the 7th International Conference on Computational Science, pp. 90–97. Springer (2007)
6. Suh, B., Hong, L., Pirolli, P., Chi, E.H.: Want to be retweeted? large scale analytics on factors impacting retweet in twitter network. In: Proceedings of the 2010 IEEE Second International Conference on Social Computing (SocialCom), pp. 177–184. IEEE (2010)
7. Jin, F., Dougherty, E., Saraf, P., Cao, Y., Ramakrishnan, N: Epidemiological modeling of news and rumors on twitter. In: Proceedings of the 7th Workshop on Social Network Mining and Analysis, pp. 8:1–8:9. ACM (2013)
8. Sun, S., Liu, H., He, J., Du, X.: Detecting event rumors on sina weibo automatically. In: Ishikawa, Y., Li, J., Wang, W., Zhang, R., Zhang, W. (eds.) APWeb 2013. LNCS, vol. 7808, pp. 120–131. Springer, Heidelberg (2013)
9. Liao, Q.Y., Shi, L.: She gets a sports car from our donation: rumor transmission in a chinese microblogging community. In: Proceedings of the 2013 Conference on Computer Supported Cooperative Work, pp. 587–598. ACM (2013)
10. Lei, K., Zhang, K., Xu, K.: Understanding sina weibo online social network: A community approach. In: Proceedings of the 2013 IEEE Global Communications Conference (GLOBECOM), pp. 3114–3119. IEEE (2013)
11. Cai, G., Wu, H., Lv, R.: Rumors detection in chinese via crowd responses. In: Proceedings of the 2014 IEEE/ACM International Conference on Advances in Social Networks Analysis and Mining (ASONAM), pp. 912–917 (2014)
12. Bao, Y.Y., Yi, C.Q., Xue, Y.B., Dong, Y.F.: A new rumor propagation model and control strategy on social networks. In: Proceedings of the 2013 IEEE/ACM International Conference on Advances in Social Networks Analysis and Mining (ASONAM), pp. 1472–1473 (2013)
13. Liu, D.C., Chen, X.: Rumor propagation in online social networks like twitter-a simulation study. In: Proceedings of the Third International Conference on Multimedia Information Networking and Security (MINES), pp. 278–282. IEEE (2011)
14. Weibo Rumor Busting. http://weibo.com/weibopiyao
15. Weibo Hot Topics. http://d.weibo.com

Subjective versus Objective Questions: Perception of Question Subjectivity in Social Q&A

Zhe Liu[✉] and Bernard J. Jansen

College of Information Sciences and Technology, The Pennsylvania State
University, University Park, PA 16802, USA
zul112@ist.psu.edu, jjansen@acm.org

Abstract. Recent research has indicated that social networking sites are being adopted as venues for online information-seeking. In order to understand questioner's intention in social Q&A environments and to better facilitate such behaviors, we define two types of questions: subjective information-seeking questions and objective information seeking ones. To enable automatic detection on question subjectivity, we propose a predictive model that can accurately distinguish between the two classes of questions. By applying the classifier on a larger dataset, we present a comprehensive analysis to compare questions with subjective and objective orientations, in terms of their length, response speed, as well as the characteristics of their respondents. We find that the two types of questions exhibited very different characteristics. Also, we noticed that question subjectivity plays a significant role in attracting responses from strangers. Our results validate the expected benefits of differentiating questions according to their subjectivity orientations, and provide valuable insights for future design and development of tools that can assist the information seeking process under social context.

Keywords: Social Q&A · Social search · Information seeking · Social network · Twitter

1 Introduction

As understanding the information needs of users is crucial for designing and developing tools to support their social question and answering (social Q&A) behaviors, many of the past studies analyzed the topics and types of questions asked on social platforms [1-3]. With a similar aim in view, in this work, we also study the intentions of questioners in social Q&A, but we focus more specifically on identifying the subjectivity orientation of a question. In other words, we build a framework to differentiate the objective questions from the subjective ones. We believe this kind of subjectivity analysis can be very important in social Q&A due to several reasons: First, as previous studies suggested that both factual and recommendation/opinion seeking questions were asked on social platforms, our study allows people to automatically detect the underlying user intent behind any question, and thus provide more appropriate answers. More specifically, we assume that objective questions focus more on

© Springer International Publishing Switzerland 2015
N. Agarwal et al. (Eds.): SBP 2015, LNCS 9021, pp. 131–140, 2015.
DOI: 10.1007/978-3-319-16268-3_14

the accuracy of their responses, while subjective questions require more diverse replies that rely on opinion and experience. Second, we believe that our work can serve as the first step in implementing an automatic question routing system in social context. By automatically distinguishing subjective questions from the objective ones, we could ultimately build a question routing mechanism that can direct a question to its potential answerers according to its underlying intent. For instance, given a subjective question, we could route it to someone who shares about the same experience or knows the context well to provide more personalized responses, while for an objective question, we could contact a selected set of strangers based on their expertise or could submit it to submit it to search engines.

From the above viewpoint, we carry out our subjective analysis on Twitter. We implement and evaluate multiple classification algorithms with the combination of lexical, part-of-speech tagging, contextual and Twitter-specific features. With the classifier on question subjectivity, we also conduct a comprehensive analysis on how subjective and objective question differs in terms of their length, posting time, response speed, as well as the characteristics of their respondents. We show that subjective questions contain more contextual information, and are being asked more during working hours. Compared to the subjective information-seeking tweets, objective questions tend to experience a shorter time-lag between posting and receiving responses. Moreover, we also notice that subjective questions attract more responses from strangers than objective ones.

2   Related Work

As an emerging concept, social Q&A has been given very high expectations due to its potential as an alternative to traditional information-seeking tools. Jansen et al. [4] in their work examining Twitter as a mechanism for word-of-mouth advertising reported that 11.1% of the brand-related tweets were information-providing, while 18.1% were information-seeking . Morris et al. [1] manually labeled a set of questions posted on social networking platforms and identified 8 question types in social Q&A, including: recommendation, opinion, factual knowledge rhetorical, invitation, favor, social connection and offer. Zhao and Mei [5] classified question tweets into two categories: tweets conveying information needs and tweets not conveying information needs. Harper et al. [6] automatically classified questions into conversational and informational, and reached an accuracy of 89.7% in their experiments.

As for the task of question subjectivity identification, Li et al. [7] explored a supervised learning algorithm utilizing features from both the perspectives of questions and answers to predict the subjectivity of a question. Zhou et al. [8] automatically collect training data based on social signals, such as like, vote, answer number, etc, in CQA sites. Chen et al. [9] built a predictive model based on both textual and meta features, and co-training them to classify questions into: subjective, objective, and social. Aikawa et al. [10] employed a supervised approach in detecting Japanese subjective questions in Yahoo!Chiebukuro and evaluated the classification results using weighed accuracy which reflected the confidence of annotation.

Although a number of works exist on question subjectivity detection, none of them are conducted within social context. Considering the social nature of Q&A on SNS, we present this study, focusing on comparing objective and subjective questions in social Q&A, and propose the overarching research question of this study:

How subjective and objective information-seeking questions differ in the way they are being asked and answered?

To measure the difference, we first propose an approach which can automatically distinguish objective questions from subjective ones using machine learning techniques. In addition, we introduce metrics to examine each type of question.

3   Annotation Method

To guide the annotation process, we in this section present the annotation criteria adopted for identifying the subjective and objective questions in social context.

Subjective Information-Seeking Tweet: The intent of a subjective information-seeking tweet is to receive responses reflecting the answerer's personal opinions, advices, preferences, or experiences. A subjective information-seeking tweet is usually with a "survey" purpose, which encourages the audience to provide their personal answers.

Objective Information-Seeking Tweet: The intent of an objective information-seeking tweet is to receive answers based on some factual knowledge or common experiences. The purpose of an objective question is to receive one or more correct answers, instead of responses based on the answerer's personal experience.

Considering that not all questions on Twitter are of information-seeking purpose, in our annotation criteria we also adopted the taxonomy of information-seeking and non-information-seeking tweets from [10], although differentiating these two types are not of our interest in this study.

To better illustrate our annotation criteria used in this study, in Table 1 we listed a number of sample questions with subjective, objective or non- information-seeking intents.

Table 1. Subjectivity categories used for annotation

Question Type	Sample Questions
Subjective	• Can anyone recommend a decent electric toothbrush? • How does the rest of the first season compare to the pilot? Same? Better? Worse?
Objective	• When is the debate on UK time? • Mac question. If I want to print a doc to a color printer but in B&W how do I do it?
Non-information	• Why is school so early in the mornings? • There are 853 licensed gun dealers in Phoenix alone. Does that sound like Obama's taking away gun rights?

Given the low percentage of information-seeking questions on Twitter [11], to save our annotator's time and effort, in this study, we crawled question tweets from Replyz (www.replyz.com). Replyz is a very popular Twitter-based Q&A site, which searches through Twitter in real time looking for posts that contain questions based on their own algorithm (Replyz has been shut down on 31 July, 2014). By collecting questions through Replyz, we filtered out a large number of non-interrogative tweets.

For our data collection, we employed a snowball sampling approach. To be more specific, we started with the top 10 contributors who have signed in Replyz with their Twitter account as listed on Replyz's leaderboard. For each of these users, we crawled all the question tweets that they have answered in the past from their Replyz profile. Then, we identified the individuals who posted those collected questions and went to their profile to crawl all the interrogative tweets that they have ever responded. We repeated this process until each "seed" user yielded at least 1,000 other unique accounts. After removing non-Twitter questioners in our collection, in total, we crawled 25,697 question tweets and 271,821 answers from 10,101 unique questioners and 148,639 unique answerers.

We randomly sampled 3,000 English questions from our collection and recruited two human annotators to work on the labeling task based on our annotation criteria on subjective, objective and non-information-seeking tweets. Finally, 2,588 out of 3,000 questions (86.27%) received agreement on their subjectivity orientation from the two coders. Among the 2,588 interrogative tweets, 24 (0.93%) were labeled as with mix intent, 1,303 (50.35%) were annotated as non-information seeking, 536 (20.71%) as subjective information seeking, and the rest 725 (28.01%) as objective information seeking. Our Cohen's kappa is quite high at 0.75.

4   Question Subjectivity Detection

4.1   Feature Engineering

In this section, features extracted for the purpose of question subjectivity detection are introduced. In total, we have identified features from four different aspects, including: lexical, POS tagging, context and Twitter-specific features.

Lexical Features: we adopted word-level n-gram features. We counted the frequencies of all unigram, bigram, and trigram tokens that appeared in the training data. Before feature extraction, we lowercase and stemmed all the tokens using the Porter stemmer [12].

POS Tagging Features: In addition to the lexical features, we also believed that POS tagging can add more context to the words used in the interrogative tweets. To tag the POS of each tweet, we used the Stanford tagger [13].

Syntactic Features: The syntactic features describe the format or the context of a subjective or objective information-seeking tweet. The syntactic features that we adopted in this study include: the length of the tweet, number of clauses/sentences in the tweet, whether or not there is a question mark in the middle of the tweet, whether or not there are consecutive capital letters in the tweet.

Contextual Features: We assume that contextual features, such as URL, hashtag, etc., can provide extra signals for determining whether a question is subjective or objective. The contextual features that we adopted in this study are: whether or not a question tweet contains a hashtag, a mention, a URL, and an emoticon.

For both lexical and POS tagging features, we discarded rare terms with observed frequencies of less than 5 to reduce the sparsity of the data.

4.2 Classification Evaluation

We next built a binary classifier to automatically label subjective and objective information-seeking questions. We tested our model using a number of classification algorithms implemented in Weka, including: Naïve Bayes, LibSVM, and SMO, using 10-fold cross-validation. We only reported the best results obtained.

First, we evaluated the classification accuracies along with the number of features selected using the algorithm of information gain as mentioned above. We noticed that all three algorithms attained high accuracies when the number of features selected equaled to about 200. Next, based on the 200 features selected, we accessed the classification performances based on the evaluation metrics provided by Weka, including accuracy, precision, recall, and F-measure. The majority induction algorithm, which simply predicts the majority class, was applied to determine the baseline performance of our classifier. Table 2 demonstrated the classification results.

Table 2. Classification results using the top 500 selected features

Method	Accuracy	Precision	Recall	F1
NaïveBayes	80.12	83.46	22.16	35.02
LibSVM	76.17	90.58	43.47	58.75
SMO	81.65	87.66	26.63	40.85

5 Impact of Question Subjectivity on User Behavior

In this section, we address our research goal by understanding the impact of question subjectivity on individual's asking and answering behaviors in social Q&A. In order to do that, we first need to identify the subjectivity orientation of all 25,697 collected questions. However, as we built our classification model as a further step of providing subjectivity indication only after a question has been predetermined as informational, we can't directly apply it to the entire data set. So, to solve this challenge, we adopted the text classifier as proposed in [5] and [11] to eliminated all non-information-seeking tweets first. With the adopted method, we achieved a classification accuracy of 81.66%. We believe this result reasonable comparing to the 86.6% accuracy reported in [5], as Replyz has already removed a huge number of non-informational questions based on some obvious features, such as whether or not the question contains a linketc. We presented the overall statistics of our classified data set in Table 3.

5.1   Characterizing the Subjective and Objective Questions

Given the positive correlation reported between question length and degree of person-alization in [14], we assume that subjective information-seeking questions on Twitter are longer than the objective ones. To examine the difference, we conducted Mann–Whitney U test across the question types on character and word scales.

Table 3. Classification results using the top 500 selected features

Question Type	Non-informational	Informational	
		Subjective	Objective
Questions	15, 311	3,984	6,402
Questioners	4,762	2,267	3,072
Answers	169,690	44,636	57,495
Answerers	87,331	28,190	33,118

In our data set, information-seeking questions asked on Twitter had an average length of 81.47 characters and 14.78 words. With the empirical cumulative distribution function (ECDF) of the question length plotted in Figure 1, we noticed that both the number of characters and words differ across question subjectivity categories. Consistent with our hypothesis, in general subjective information-seeking tweets (M_c = 87, M_w = 15.95) contain more characters and words than the objective ones (M_c = 73, Mn_w = 14.05). Mann–Whitney U test further proofed our findings with statistical-ly significant p-values less than 0.05 (z_c = -17.39, p_c = 0.00 < 0.05; z_w = -15.75, p_w = 0.00 < 0.05). Through our further investigation on the content of questions, we noted that subjective questions tended to use more words to provide additional contextual information about the questioner's information needs. Examples of such questions include: *"So after listening to @wittertainment and the Herzog interview I need to see more of his work but where to start? Some help @KermodeMovie ?"*, and *"Thinking about doing a local book launch in #ymm any of my tweeps got any ideas?"*

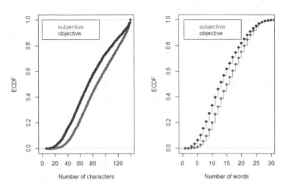

Fig. 1. Distribution of question length on character and word levels

5.2   Characterizing the Subjective and Objective Answers

So far, we have only examined the characteristics of subjective and objective information-seeking questions posted on Twitter. In this subsection, we presented how the
subjectivity orientation of a question can affect its response.

Response Speed
Considering the real time nature of social Q&A, we first looked at how quickly subjective and objective information-seeking questions receive their responses. We adopted two
metrics in this study to measure the response speed: the time elapsed until receiving the
first answer, and the time elapsed until receiving the last answer. In Figure 2, we plotted
the empirical cumulative distribution of response time in minutes using both measurements. We log transformed the response time given its logarithmic distribution.

In our data set, more than 80% of questions posted on Twitter received their first
answer in 10 minutes or less, no matter their question types (84.60% objective questions and 83.09% subjective ones). Around 95% of questions got their first answer in
an hour, and almost all questions were answered within a day. From Figure 5, we
noticed that it took slightly longer for individuals to answer subjective questions than
the objective ones. The t-test result also revealed significant difference on the arrival
time of the first answer between question types (t = -3.08, p < 0.05), with subjective
questions on average being answered in 4.60 minutes after the question was posted
and objective questions being answered in 4.24 minutes. We assumed that this might
because subjective questions were mainly posted during working hours, whereas,
respondents were more active during free time hours [14].

In addition to the first reply, we also adopted the arrival time of the last answer to
imply the temporality of each question. Define in [15], question temporality is "a
measure of how long the answers provided on a question are expected to be valuable".
Overall, 67.79% of subjective and 69.49% objective questions received their last answer in an hour. More than 96% of questions of both types closed in a day (96.68%
objective questions and 96.16% subjective ones). Again, the t-test result demonstrated
significant between-group difference on the arrival time of the last answer (t = 3.76, p
< 0.05), with subjective questions on average being last answered in 44 minutes after
the question was posted and objective questions being answered in 38 minutes. Examples of objective questions with short temporal durations include: *"Hey, does anyone know if Staples & No Frills are open today?"* and *"When is LFC v Valarenga?"*

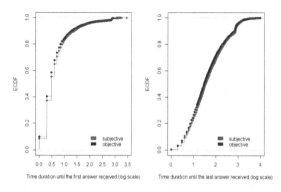

Fig. 2. Distribution of question response time in minutes

Characteristics of Respondents

In addition to the response speed, we were also interested in understanding whether the characteristics of a respondent affect his/her tendency to answer a subjective or objective question on Twitter. In order to do so, we proposed a number of profile-based factors, including: the number of followers, the number of friends, daily tweet volume, which is measured as the ratio of the total count of status to the total number of days on Twitter, and the friendship between the questioner and the respondent. Here, we only categorized questioner-answerer pairs with reciprocal follow relations as "friends", while the rest as "strangers".

We crawled the profile information of all respondent in our dataset, as well as their friendships with the corresponding questioners via Twitter API. Since our data set spanned from March 2010 to February 2014, 2,998 out of 59,856 unique users in our collection have either deleted their Twitter accounts or have their accounts set as private. So, we were only able to collect the follow relationship between 95% (78,697) of the unique questioner-answer pairs in our data set.

We used logistic regression to test whether any of our proposed factors were independently associated with the respondent's behavior of answering subjective or objective questions on Twitter. The results of our logistic regression analysis were shown in Table 4.

Table 4. Logistic regression analysis of variables associated with subjective or objective question answering behavior

Predictor	Odds Ratio	p-value
Number of followers	1.00	0.24
Number of friends	1.00	0.07
Daily tweet volume	0.99	0.00*
Friendship	1.04	0.03*

From Table 4, we noticed that among all four variables, the respondent's daily tweet volume and friendship with the questioner were significantly associated with his/her choice of answering subjective or objective questions in social Q&A. To better understand those associations, we further performed post hoc analyses on those significant factors.

First, as for the friendship between the questioner and the respondent, among all 78,697 questioner-answerer pairs in our data set, 22,220 (28.23%) of the follow relations were reciprocal, 24,601 (31.26%) were one-way and 31,871 (40.51%) were not following each other. The number of reciprocal following relations in our collection is relatively low, comparing to the 70%-80% and the 36% rates as reported in [16, 17] .We think this is because Replyz has created another venue for people to answer other's questions, even if they were not following each other on Twitter, and this enabled us to better understand how strangers in social Q&A select and answer questions.

Besides the overall patterns described, we also conducted chi-square test to examine the dependency between the questioner-respondent friendship and the answered question type. As shown in Table 5, the chi-square cross-tabulations revealed a significant trend between the two variables ($\chi^2 = 13.96$, $p = 0.00 < 0.05$). We found that in real-world settings, "strangers" were more likely to answer subjective questions than "friends". This was unexpected given previous work [1] showed that people claimed in survey that they prefer to ask subjective questions to their friends for tailored responds. One reason for this could be that compared to objective questions, subjective questions require less expertise and time investment, so that could be a better option for strangers to offer their help.

Table 5. Answered question type by questioner-answerer friendship

Question Type	Friendship Type	
	Friends	**Strangers**
Subjective	23.9% (n = 6359)	25.3% (n = 20229)
Objective	76.1% (n = 8234)	74.7% (n = 24355)

In addition in order to examine the relationship between the respondent's daily tweet volume and his/her answered question type, a Mann–Whitney U test was performed. The result was significant ($z = -7.87$, $p = 0.00 < 0.05$), with respondents to the subjective questions having more tweets posted per day ($M = 15.07$) than the respondents of the subjective questions ($M = 13.24$). This result further proved our presumption in the previous paragraph that individuals with more time spent in social platforms are more willing to answer more time consuming questions, in our case, the objective ones.

6 Discussion and Conclusion

In this work, we distinguished and analyzed 6,402 objective and 3,984 subjective questions. First, we found that contextual restrictions were imposed more often on subjective questions, and thus made them normally longer in length than the objective ones. In addition, our results revealed that subjective questions experienced longer time-lags in getting their initial answers. Furthermore, we also noticed that it took shorter time for the objective questions to receive all their responses. One interpretation of this finding could be that many of the objective questions asked on Twitter were about real-time content (e.g. when will a game start? where to watch the election debates, etc.) and were sensitive to real world events [5], so answers to those questions tended to expire in shorter durations[15]. Another possible explanation was that, since answers to the objective questions were supposed to be less diverse, individuals would quickly stop providing responses after they saw a satisfactory number of answers already exist to those questions. Of course, both speculations need support from future detailed case studies. At last, in assessing the preferences of friends and strangers on answering subjective or objective questions, we demonstrated that even though individuals prefer to ask subjective questions to their friends for tailored responds [1], it turned out that, in reality, subjective questions were being responded more by strangers. We thought this gap between the ideal and reality imposed a design challenge in maximizing the personalization benefits from strangers in social Q&A.

In terms of design implications, we believe that our work contributes to the social Q&A field in two ways: First, our predictive model on question subjectivity enables automatic detection of subjective and objective information-seeking questions posted on Twitter and can be used to facilitate future studies on large scales. Second, our analysis results allow the practitioners to understand the distinct intentions behind subjective and objective questions, and to build corresponding tools or systems to better enhance the collaboration among individuals in supporting social Q&A activities. For instance, we believe that given the survey nature of subjective questions and stranger's interests in answering them, one could develop an algorithm to route those subjective questions to appropriate respondents based on their locations and past experiences. In contrast, considering the factorial nature and short duration of objective questions, they could be routed to either search engines or individuals with equivalent expertise or availability. In summary, our work is of good value to both research community and industrial practice.

References

1. Morris, M.R., Teevan, J., Panovich, K.: What do people ask their social networks, and why?: A survey study of status message Q & A behavior. In: SIGCHI (2010)
2. Lampe, C., et al.: Help is on the way: patterns of responses to resource requests on facebook. In: CSCW (2014)
3. Liu, Z., Jansen, B.J.: Almighty twitter, what are people asking for? In: ASIS & T (2012)
4. Jansen, B.J., et al.: Twitter Power: Tweets as Electronic Word of Mouth. Journal of the American Society for Information Science and Technology 60(11), 2169–2188 (2009)
5. Zhao, Z., Mei, Q.: Questions about questions: An empirical analysis of information needs on twitter. In: WWW (2013)
6. Harper, F.M., Moy, D., Konstan, J.A.: Facts or friends?: distinguishing informational and conversational questions in social Q & A sites. In: SIGCHI (2009)
7. Li, B., et al.: Exploring question subjectivity prediction in community QA. In: SIGIR (2008)
8. Zhou, T.C., Si, X., Chang, Y.E., King, I., Lyu, M.R.: A data-driven approach to question subjectivity identification in community question answering. In: AAAI (2012)
9. Chen, L., Zhang, D., Mark, L.: Understanding user intent in community question answering. In: WWW (2012)
10. Aikawa, N., Sakai, T., Yamana, H.: Community qa question classification: Is the asker looking for subjective answers or not. IPSJ Online Transactions 4, 160–168 (2011)
11. Li, B., et al.: Question identification on twitter. In: CIKM (2011)
12. Porter, M.F.: An Algorithm for Suffix Stripping. Program: Electronic Library and Information Systems 14(3), 130–137 (1980)
13. Toutanova, K., et al.: Feature-rich part-of-speech tagging with a cyclic dependency network. In: NAACL2013-Volume 1
14. Liu, Z., Jansen, B.J.: Factors influencing the response rate in social question and answering behavior. In: CSCW (2013)
15. Pal, A., Margatan, J., Konstan, J.: Question temporality: identification and uses. In: CSCW (2012)
16. Paul, S.A., Hong, L., Chi, E.H.: Is twitter a good place for asking questions? a characterization study. In: ICWSM (2011)
17. Zhang, P.: Information seeking through microblog questions: The impact of social capital and relationships. In: ASIS & T (2012)

Analysis of Music Tagging and Listening Patterns: Do Tags Really Function as Retrieval Aids?

Jared Lorince[1]($\boxtimes$), Kenneth Joseph[2], and Peter M. Todd[1]

[1] Cognitive Science Program, Indiana University, Bloomington, IN, USA
{jlorince,pmtodd}@indiana.edu
[2] Computation, Organization, and Society Program, Carnegie Mellon University,
Pittsburgh, PA, USA
kjoseph@cs.cmu.edu

Abstract. In collaborative tagging systems, it is generally assumed that users assign tags to facilitate retrieval of content at a later time. There is, however, little behavioral evidence that tags actually serve this purpose. Using a large-scale dataset from the social music website Last.fm, we explore how patterns of music tagging and subsequent listening interact to determine if there exist measurable signals of tags functioning as retrieval aids. Specifically, we describe our methods for testing if the assignment of a tag tends to lead to an increase in listening behavior. Results suggest that tagging, on average, leads to only very small increases in listening rates, and overall the data do *not* support the assumption that tags generally serve as retrieval aids.

Keywords: Collaborative tagging · Folksonomy · Music listening · Memory cues · Retrieval aids · Personal information management

1 Introduction

In social tagging systems, users assign freeform textual labels to digital content (music, photos, web bookmarks, etc.). There are a variety of reasons for which users tag content, but it is overwhelmingly assumed that tagging for one's own future retrieval – assigning a tag to an item to facilitate re-finding it at a later time – is users' principal motivator. But is this a valid assumption?

Collaborative tagging systems are often designed, at least in part, as resource management platforms that expressly facilitate the use of tags as retrieval aids. However, the freeform, and often social, nature of tagging opens up many other possible reasons for which a user might tag a resource. While there is a significant amount of non-controversial evidence for such alternative tagging motivations (sharing resources with other users, social opinion expression, etc.), the problem with the retrieval aid assumption runs deeper than there simply existing possible

© Springer International Publishing Switzerland 2015
N. Agarwal et al. (Eds.): SBP 2015, LNCS 9021, pp. 141–152, 2015.
DOI: 10.1007/978-3-319-16268-3_15

alternatives. There is, in fact, almost no behavioral evidence that tags are ever actually used as retrieval aids. While there is much data available on user tagging habits (i.e. which terms are applied to which resources, and when), to our knowledge there is no published research providing behavioral evidence of whether or not tags, once applied to items, actually facilitate subsequent retrieval. This is an issue largely driven by a lack of data: Although a web service can in principle track a users' interaction with tags (for instance, if users use tags as search terms to find tagged content), there are no available datasets containing such information, nor can it be crawled externally by researchers.

Despite these issues, this empirical question is not intractable. While detailed information on how existing tags are utilized remains beyond our reach, an alternative approach is to examine how patterns of user interaction with tagged versus untagged content vary. If tags do serve as retrieval aids, we should expect users to be more likely to interact with a resource (e.g. visit bookmarked pages, listen to songs, view photos, etc.) once they have assigned a tag to it.

Here we test this hypothesis using a large-scale dataset consisting of complete listening and tagging histories from more than 100,000 users from the social music website Last.fm. From this dataset, we extract user-artist listening time-series, each of which represents the frequency of listening over 90 months to a particular artist by a particular user, and compare time-series in which the user has tagged the artist to those that are untagged. Specifically, we address the following two questions:

RQ1: Does tagging an artist lead to increased listening to that artist in the future, as shown by comparison of tagged versus untagged time-series?

RQ2: Are certain tags particularly associated with increases in future listening, and if so, can we identify attributes of such "retrieval-targeted" tags as opposed to others?

2   Background

Collaborative tagging has been considered one of the core technologies of "Web 2.0", and has been implemented for resources as diverse as web bookmarks (Delicious), photos (Flickr), books (LibraryThing), academic papers (Mendeley), and more. Vander Wal [15] coined the term "folksonomy" to describe the emergent semantic structure defined by the aggregation of many individual users' tagging decisions in such a system. These folksonomies have since become the target of much academic research. One of the earliest analyses of a collaborative tagging system is Golder and Huberman's [4] work on the evolution of tagging on Delicious.com. Around the same time, Hotho and colleagues [9] presented a formal definition of a folksonomy: $\mathbb{F} := (U, T, R, Y)$ [1]. The variables U, T, and R represent, respectively, the sets of users, tags, and resources in a tagging system, while Y is a ternary relation between them ($Y \subseteq U \times T \times R$).

Since 2006, an extensive literature on *how* people tag has also developed, covering topics like tagging expertise [17], mathematical [2] and multi-agent [11]

---

[1] This is a slight simplification. For details, see [9]

models of tagging choices, consensus in collaborative tagging [6], and much more. Our understanding of the dynamics of tagging behavior has greatly expanded, but the question of *why* people tag has proven much more elusive.

It is typically assumed that tags serve as retrieval aids, allowing users to re-find content to which they have applied a given tag (e.g. a user could click on or search for the tag "rock" to retrieve the songs she has previously tagged with that term). This assumption is central to Vander Wal's original defintion of a folksonomy, "the result of personal free tagging of information and objects... *for one's own retrieval*" [15, emphasis added]. This perspective is echoed in many studies of tagging patterns [3, 4, 6].

But while retrieval is the most commonly assumed motivation for tagging, other reasons certainly exist, and various researchers have developed taxonomies of tagging motivation. Among proposed motivational factors in tagging are personal information management (including but not limited to tagging for future retrieval), resource sharing, opinion expression, performance, and activism [1, 8, 18] – see [5] for a review.

While the development of motivational theories in tagging is useful, there is almost no work actually grounding them in behavioral observations. The vast majority of existing work either makes inferences about motivation based on design features of a website (e.g. social motivations in tagging require that one's tags be visible to other users, [13]), employs semantic analysis and categorization of tags (e.g. the tags "to read", "classical", and "love" can be inferred to have different uses, [18]), or directly asks users why they tag using survey methods [1]. The results of such approaches are useful contributions to the field, but few have resulted in testable behavioral hypotheses that can confirm or refute their validity.

One notable exception is work by Körner and colleagues [10]. They argue that taggers can be classified on a motivational spectrum from categorizers (who use a constrained vocabulary suitable to future browsing of their own tagged resources) to describers (who use a large, varied vocabulary to facilitate future keyword-based search with their own tags), and have developed and tested quantifiable signals of these different motivations. The main deficiency of this approach, however, is that their hypotheses are based fully on attributes of user tag vocabularies; they present no way to test whether or not describers actually use tags, once applied, for keyword-based search and categorizers use them for browsing.

Again, the problem of lack of verification arises because data on how users actually *use* existing tags is simply not available to researchers through any tagging system APIs (or through other methods) that we are aware of. Thus the existing work on tagging motivation is limited to inferring *why* people tag from *how* they tag, rather than from how they *use* their tags. In presenting our novel methods, we are aware that they still represent an inferential approach. Our approach is distinct from those just described, however, in that we test a concrete hypothesis about how tagging should affect a behavior on which we *do* have data: interaction with tagged content, in our case music listening.

3 Dataset

Last.fm incorporates two specific features that are of interest for our analysis. First, it implements a collaborative tagging system (a "broad" folksonomy, following Vander Wal's [14] terminology, meaning that multiple users tag the same, publicly available content) in which users can label artists, albums, and songs. Second, the service tracks users' listening habits both on the website itself and on media players (e.g. iTunes) via a software plugin. This tracking process is known as "scrobbling", and each timestamped instance of a user listening to a particular song is termed a "scrobble".

Here we use an expanded version of a dataset described in earlier work [11,12] that includes the full tagging histories of approximately 1.9 million Last.fm users, and full listening histories from a subset of those users (approximately 100,000) for a 90-month time window (July 2005 - December 2012, inclusive). Data were collected via a combination of the Last.fm API and direct scraping of publicly available user profile pages. For further details of the crawling process, see [11,12].

For our current purposes, we consider only those users for whom we have both tagging and listening histories. For each user, we extract one time-series for each unique artist listened to by that user. Each user-artist listening time-series consists of a given user's monthly listening frequency to a particular artist for each month in our data collection period, represented as a 90-element vector (each element of the vector represents the number of times that particular user listened to that particular artist in the particular month).

We selected a monthly timescale for listening behavior due to the fact that user tagging histories are only available at monthly time resolution. Furthermore, we perform all analyses here at the level of artists, rather than individual songs. Thus every song scrobbled is treated as a listen to the corresponding artist, and all annotations (i.e., particular instances of applying a tag to a song, album, or artist) are treated as annotations of the corresponding artist. Our choice to perform all analyses at the level of artists, rather than individual songs, is based on the facts that (a) listening and tagging data for any particular song tend to be very sparse, and (b) the number of time-series resulting from considering each unique song listened to by each user would be prohibitively large.

The 2-billion-plus individual scrobbles in our dataset generate a total of approximately 95 million user-artist listening time-series. In about 6 million of these cases, the user has assigned at least one tag to the artist (or to a song or album by that artist) within the collection period (we refer to these as tagged time-series), while in the remaining cases (approx. 89 million) the user has never tagged the artist. We summarize these high-level dataset statistics in Table 1. Comparison of these tagged and untagged listening time-series is the heart of the analyses presented in the next section.

Table 1. Dataset summary

Total users	104,829
Total scrobbles	2,089,473,214
Unique artists listened	4,444,119
Unique artists tagged	1,049,263
Total user-artist listening time-series	94,875,106
Total tagged time-series	5,930,594
Total untagged time-series	88,944,512

4   Analyses and Results

4.1   RQ1: Comparison of Tagged and untagged Time-Series

Our principal research question is whether listening patterns for tagged content are consistent with the expectation that tags serve as memory cues. If so, we would expect to see an increase in a user's listening rate to musical artists after the user has tagged them, under the assumption that a tag facilitates retrieval and increases the chances of a user listening to a tagged artist.

Unfortunately, several factors combine to make such an analysis difficult. First and foremost, the desired counterfactual of the untagged "version" of a particular tagged series, which would allow a direct testing of how tagging changes listening behavior, does not, of course, exist. We thus must utilize untagged time-series in a way that allows them to approximate what a true counterfactual might look like. In searching for such samples, a second difficulty that arises is that listening rates for tagged time-series are much greater than for untagged time-series (the average number of total listens across time-series is 16.9 when untagged and 98.9 when tagged). While suggestive of the importance of tagging, this imbalance also suggests that controls must be incorporated in both sample selection and statistical analysis to account for previous listening behavior prior to tagging. Finally, the actual point in time at which tags are expected to increase listening behavior for any given user is unknown. Thus, we must formulate our analysis to account for this uncertainty.

To alleviate issues with the non-existence of a true counterfactual, we subselect from both the tagged and untagged series using the following formal procedure. We first select only those tagged time-series that have:

- more than 25 total listens;
- a peak in listening at least 6 months from the edges of our data collection period (ensuring that the period from 6 months before to 6 months after the peak does not extend beyond the limits of our data range); and
- at least one listen in the 6 months prior to and after the peak (e.g. if the peak occurs in July, there should be at least one listen between January and June, and one between August and the following January).

We then select only those tagged time series where the tag was applied in the month of peak listening, and align all of those series at that peak point.

Constraining our time-series in this manner, we are left with a total of 206,140 tagged time-series. Next we randomly select a same-sized sample of the 4.1M untagged time-series meeting the same criteria, and also align them at the peak of listening. Where the peak was reached in multiple months in any series, we chose one of these peaks at random to align on. All results below have been verified with multiple random samplings of the untagged data.[2]

After temporally aligning the tagged and untagged samples, we limited our analysis to a 13-month period extending from 6 months prior to the peak month to 6 months after the peak. This allows us to consider a manageable variety of ways in which listening prior to the tag may affect future behavior.

In Figure 1a we plot mean listens, with 95% normal confidence intervals, for each month across all tagged and untagged time-series in the subsampled data. All values are normalized by the peak number of listens for each series, and thus values at the peak month for both the tagged and untagged lines are unity. Comparing the line heights before and after the peak, Figure 1a shows that the mean normalized listening rate increases in the months after the peak for both tagged and untagged time-series. But there is also a small but reliable difference between the tagged and untagged series: The tagged time-series show proportionally higher mean normalized listening rates after the peak month (in which the the tag was applied) as compared to untagged time-series. This is suggestive of an increase in listening as a result of tagging.[3]

While Figure 1a thus supports our hypothesis, there are two important caveats. First, as the distribution of the number of listens in any given month across all time series is heavily skewed, the mean is not fully representative of the data. Our further statistical analysis uses a log-transformed version of the listening counts to account for this. Second, the initial analysis does not control for the presumably important effect of pre-peak listening behavior on post-peak listening.

To test our hypothesis more robustly, we therefore use a regression model that incorporates previous listening behavior to predict post-peak behavior. Due to a lack of knowledge about the relationship between these variables and the volume of data we have, it was both unreasonable and unnecessary to assume a linear relationship between the dependent and independent variables. Because of this, we opted for a Generalized Additive Model (GAM, [7]) using the R package mgcv [16] applied to all of the tagged and untagged series in the selected sets described above. Our dependent variable in the regression is the logarithm of the sum of all listens in the six months after a tag has been applied, to capture the possible

---

[2] Our method thus compares tagged and untagged data aggregated over many users. While it would be preferable to perform a within-subjects analysis (i.e. comparing tagged versus untagged data for each individual user), thereby accounting for much of the variability in listening across different users, the data for any particular user tends to be too sparse, as most individuals have tagged only a few times (if ever).

[3] There is also, however, a small but reliable lower rate of listening to tagged artists versus untagged artists prior to peak listening. This may indicate that songs that "catch on" for a user more quickly (rise faster in listening from before the peak to the peak) are more likely to be tagged, a possibility to be explored in future work.

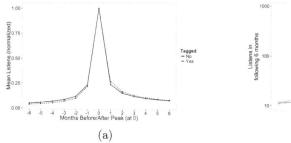
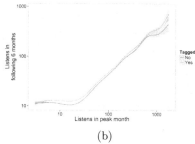

(a)   (b)

Fig. 1. Comparison of tagged and untagged listening time-series. Mean normalized listens by month (a), and regression results (b), both with 95% confidence intervals.

effect of tagging over a wide temporal window. (The results are qualitatively the same when testing listening for each individual month.) Our independent variables are a binary indicator of whether or not the time-series has been tagged, as well seven continuous-valued predictors, one each for the logarithm of total listens in the peak month and the six previous months. The regression equation is as follows, where m corresponds to the month of peak listening, L is the number of listens in any given month, T is the binary tagged/untagged indicator, and f represents the exponential-family functions calculated in the GAM (there is a unique function f for each pre-peak month):

$$\log \sum_{i=1}^{6} L_{m+i} = b_0 + b_1 T + \sum_{i=0}^{6} f(\log L_{m-i}) \tag{1}$$

The regression model, which explained approximately 30% of the variance in the data (adjusted R^2), indicated that the tagged/untagged indicator and the listening rate parameters (smoothed using thin-plate regression splines) for all seven previous months had a significant effect on post-peak listening behavior ($P \ll 0.0001$). As we cannot show the form of this effect for all model variables at once, Figure 1b instead displays the predicted difference in listening corresponding to tagging as a function of the number of peak listens, calculated with a similar model which considers only the effect of listening in the peak month on post-peak listening. This plot suggests and the full model confirms that, controlling for all previous listening behavior, a tag increases the logarithm of post-peak listens by .147 (95% CI = [.144,.150]). In other words, the effect of a tag is associated with around 1.15 more listens over six months, on average, than if it were not to have been applied. The large confidence interval on the right hand side of Figure 1b reflects the small number of userswho have extremely high listening rates for particular artists.

4.2   RQ2: Tag Analysis

To examine if and how different tags are associated with increased future listening, we ran a regression analysis similar to that described above, but with two important changes. First, instead of a single tagged/untagged indicator, we included binary (present/not present) regressors for the 2,290 unique tags that had at least five occurrences in our subsample.[4] Second, due to the data-hungry nature of the GAM and the large number of additional variables introduced by utilizing all tags as unique predictors, we chose to only control for listening in the peak month, and not the six prior months. This decision limited the computation associated with estimating a model of this size and did not appear to affect model fit substantially according to tests we ran on subsamples of the data. The same data were used as in the previous analysis (untagged time series, of course, had values of zero for all possible tags). Formally, the regression model can be represented as follows, where again m is peak month, L is the number of listens in a given month, T_i is the binary indicator for a given tag, and f is the exponential-family function calculated by the GAM:

$$\log \sum_{i=1}^{6} L_{m+i} = b_0 + f(\log L_m) + \sum_{i=1}^{2,290} b_i T_i \qquad (2)$$

After running the model, which explains approximately 28.5% of the variance in the data (adjusted R^2), 161 unique tags were statistically significant predictors at $\alpha = .001$, a threshold selected in order to account for the large number of comparisons against the null hypothesis being made in the regression model. We proceeded to examine which of these tags were relatively strong predictors in the model.

Unsurprisingly, most of the 161 tags tend to have a positive (albeit small) impact on future listening, as evidenced by positive regression coefficients and consistent with the small positive effect of tagging overall as found in the previous analysis. The most telling observation is that commonly-used genre tags (e.g. "pop", "jazz", and "hip-hop") tend to be weak positive predictors of future listening. In contrast, relatively strong predictors (both positive and negative) appear to be comparatively obscure, possibly idiosyncratic tags (e.g. "cd collection", "mymusic", "purchased 09").[5] To examine this trend quantitatively, we plot in Figure 2a the global tag usage (i.e. the total number of uses of a tag in our full dataset of approximately 50 million annotations) as a function of the tag's impact on listening indicated by its coefficient in the regression model. Similarly, we plot in Figure 2b the unique number of users utilizing the tag, again as a function of its regression coefficient. The value e^c, where c is a tag's regression coefficient, represents the number of listens we expect a user's post-listening behavior to increase or decrease by if she were to apply a (thus the strongest predictors lead to an increase of fewer than 7 listens on average). Finally, in each

---

[4] We chose a threshold of five to ensure that data was not too sparse for the regression model but was still inclusive of infrequently occurring tags. Again, qualitative results hold when using both more and less restrictive thresholds.

[5] For a full listing of the regression coefficients across all tags in the model, see https://dl.dropboxusercontent.com/u/625604/papers/lorince.joseph.todd.2015.sbp.supplemental/regression_coefficients.txt

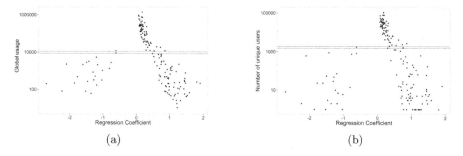

Fig. 2. In (a), each tag's global usage as a function of its regression coefficient; in (b) the number of unique users of each tag as a function of its regression coefficient

plot, the horizontal red bands mark the upper and lower limits of a bootstrapped 95% confidence interval on the popularity of the 2,129 remaining tags that were *not* significant in the regression model.

The data suggest that the most popular tags on both metrics (i.e., those above the red lines) are significant, weakly positive predictors of future listening, while relatively unpopular tags (i.e., below the red lines) tend to have relatively strong positive (and, in some cases, strong negative) impacts on listening. Tags which were not significant in the model (i.e., those that would fall between the red lines in each plot) had moderate popularity levels with respect to both metrics. The high statistical reliability but small regression coefficients for the most popular tags may be somwhat artifactual, primarily reflecting high variability in how predictive these tags are of listening across individuals. We believe this finding is still informative, however, as it indicates that popular tags are not consistently associated with future listening.

5   Conclusion

In this paper we set out to test the oft-cited assumption that tags serve as retrieval aids for individuals using collaborative tagging systems. We did so via a novel methodology, testing for evidence that tagging an artist increases a user's future listening to that artist in comparison to a carefully selected set of untagged time-series. Results suggest that tagging an artist does lead to an increase in listening, but that this increase is, on average, quite small (amounting to only 1 or 2 additional listens over a 6 month period). Given the various possible motivations for tagging, we expected only some tags to serve as retrieval cues, and thus tested the relative predictiveness of future listening for different tags. This analysis revealed systematic differences in how predictive the presence or absence of different tags was for future listening as a function of tag popularity. The data suggest that, at least for the small number of most highly significant tags that we consider, those that are globally popular have relatively little effect on future listening, and are generally associated with very small increases in post-taggging listening rates. The tags that seem to "matter" (i.e. those that are relatively strong predictors of whether or not a user will listen to an artist after tagging it) are generally much less popular. Even these stronger predictors, however, lead to relatively slight increases in listening. The strongest predictors are associated with a change of only about 7 listens over a six month period, on average.

Because we only analyzed a small sample of statistically influential tags, we are at this point tentative to make strong claims about which specific factors contribute to tags being better or worse predictors of increased listening, or even of decreased listening.[6] The evidence nevertheless suggests that relatively uncommon (and in many cases idiosyncratic) tags are most predictive of future listening behavior. The intriguing flipside is that the descriptive, popular tags that are arguably most useful to the community at large (i.e. genre labels and related tags) are not particularly associated with increases in listening for those who applied the tags, and thus are likely not functioning as memory cues.

Overall it appears that, while on average tagging an artist has a small positive effect on one's own future listening, the most common tagging activities are *not* strong predictors of future retrieval. We cannot be sure of the extent to which the many other possible tagging motivations are at play here, nor can we tell at this point if and when a tag is applied with the intention of being used for retrieval, while ultimately not being used for this purpose. That said, our results may indicate that the primary motivation for tagging on Last.fm is not for personal information management (tagging a resource for one's own retrieval), but rather may be socially or otherwise oriented, which may in turn result in tags that are useful for the community at large. This leads to the interesting possibility that a folksonomy can generate the useful, crowdsourced classification of content that proponents of collaborative tagging extol, even if this process is not strongly driven by the self-directed, retrieval-oriented tagging that is typically assumed in such systems.

While our results provide clues as to whether tags really function as retrieval aids, this remains early work addressing a hitherto unstudied research question. There is certainly room to refine and build upon the methods we present here for testing if and when tagging increases listening rates. In particular, our analysis at the level of artists (rather than the individual resources tagged) may be problematic, and we hope to develop models that operate directly at the level of the content tagged. There are also many factors we have not controlled for here that could be incorporated into future models, such as exogenous influences on listening (e.g. when an artist goes on tour or releases a new album), and we should explore alternative methods for normalizing and controlling for user listening habits beyond our approach here of simply considering raw monthly listens.

It will be critical to expand on methods for understanding which tags serve as memory cues and under what circumstances. It is clearly the case that not *all* tags function as memory cues, so more robustly identifying which tags do serve as retrieval aids is a fruitful direction for future work. Incorporating research on

---

[6] There were few enough of the latter in the current analysis that they may be largely due to statistical noise.

human memory from the cognitive sciences can also further inform hypotheses and analytic approaches to these questions, something we are actively pursuing in ongoing research. A final limitation is that we are exploring tagging in a particular collaborative tagging system, which operates in the possibly idiosyncratic domain of music. Tagging habits may vary systematically in different content domains, but until usable data becomes available, we can only speculate as to exactly how.

In closing, to address the titular question of whether or not tags function as retrieval aids, the best answer for Last.fm at least would appear to be "sometimes, but usually not". While there is much work to be done on when and why particular tags serve this function and others do not, it is clear that the overarching retrieval assumption is far from universally valid: Tags certainly do not always function as memory cues, and our results suggest that facilitating later retrieval may actually be an uncommon tagging motivation.

References

1. Ames, M., Naaman, M.: Why we tag: motivations for annotation in mobile and online media. In: Proceedings of the SIGCHI Conference on Human Factors in Computing Systems, pp. 971–980. ACM (2007)
2. Cattuto, C., Loreto, V., Pietronero, L.: Semiotic dynamics and collaborative tagging. Proceedings of the National Academy of Sciences 104(5), 1461–1464 (2007)
3. Glushko, R.J., Maglio, P.P., Matlock, T., Barsalou, L.W.: Categorization in the wild. Trends in Cognitive Sciences 12(4), 129–135 (2008)
4. Golder, S.A., Huberman, B.A.: Usage patterns of collaborative tagging systems. Journal of Information Science 32(2), 198–208 (2006)
5. Gupta, M., Li, R., Yin, Z., Han, J.: Survey on social tagging techniques. ACM SIGKDD Explorations Newsletter 12(1), 58–72 (2010)
6. Halpin, H., Robu, V., Shepherd, H.: The complex dynamics of collaborative tagging. In: Proceedings of the 16th International Conference on World Wide Web, pp. 211–220. ACM (2007)
7. Hastie, T.J., Tibshirani, R.J.: Generalized additive models (1990)
8. Heckner, M., Heilemann, M., Wolff, C.: Personal information management vs. resource sharing: towards a model of information behavior in social tagging systems. In: ICWSM (2009)
9. Hotho, A., Jäschke, R., Schmitz, C., Stumme, G.: Information retrieval in folksonomies: search and ranking. In: Sure, Y., Domingue, J. (eds.) ESWC 2006. LNCS, vol. 4011, pp. 411–426. Springer, Heidelberg (2006)
10. Körner, C., Kern, R., Grahsl, H.P., Strohmaier, M.: Of categorizers and describers: an evaluation of quantitative measures for tagging motivation. In: Proceedings of the 21st ACM Conference on Hypertext and Hypermedia, pp. 157–166. ACM (2010)
11. Lorince, J., Todd, P.M.: Can simple social copying heuristics explain tag popularity in a collaborative tagging system? In: Proceedings of the 5th Annual ACM Web Science Conference, pp. 215–224. ACM (2013)
12. Lorince, J., Zorowitz, S., Murdock, J., Todd, P.M.: "Supertagger" behavior in building folksonomies. In: Proceedings of the 6th Annual ACM Web Science Conference, pp. 129–138. ACM (2014)

13. Marlow, C., Naaman, M., Boyd, D., Davis, M.: HT 2006, tagging paper, taxonomy, flickr, academic article, to read. In: Proceedings of the Seventeenth Conference on Hypertext and Hypermedia, pp. 31–40. ACM (2006)

14. Vander Wal, T.: Explaining and Showing Broad and Narrow Folksonomies (2005). http://www.vanderwal.net/random/entrysel.php?blog=1635

15. Vander Wal, T.: Folksonomy Coinage and Definition (2007). http://vanderwal.net/folksonomy.html

16. Wood, S.N.: mgcv: GAMs and generalized ridge regression for R. R News 1(2), 20–25 (2001)

17. Yeung, C.M.A., Noll, M.G., Gibbins, N., Meinel, C., Shadbolt, N.: SPEAR: Spamming-Resistant Expertise Analysis and Ranking in Collaborative Tagging Systems. Computational Intelligence 27(3), 458–488 (2011)

18. Zollers, A.: Emerging motivations for tagging: expression, performance, and activism. In: Workshop on Tagging and Metadata for Social Information Organization, held at the 16th International World Wide Web Conference (2007)

A Novel Mental State Model of Social Perception of Brand Crisis from an Entertainment Perspective

Rungsiman Nararatwong[1,3(✉)], Roberto Legaspi[2], Hitoshi Okada[3], and Hiroshi Maruyama[2]

[1] The Graduate University for Advanced Studies, Tokyo, Japan
`rungsiman@nii.ac.jp`
[2] Research Organization of Information and Systems,
Transdisciplinary Research Integration Center,
Research Organization of Information and Systems, Tokyo, Japan
`{legaspi.roberto,hm2}@ism.ac.jp`
[3] National Institute of Informatics, Tokyo, Japan
`okada@nii.ac.jp`

Abstract. Marketers have devoted their efforts to comprehend the essence of negative electronic word-of-mouth (eWOM) since it was first introduced to understand brand crisis and the fact that social media facilitates rapid consumer communications that exacerbate the situation. This motivates marketers and researchers to investigate the complexities behind negative eWOM in order to contain potentially large-scale brand crisis. Our study proposes a novel mental state model based on existing theories of individual intrinsic motivation to engage in negative eWOM as derived from studies in the psychology of entertainment. Our model explains how individual perception of the brand may change from the initial exposure to negative contents to either feeling relieved or losing interest. We demonstrate the plausibility of our model by performing multi-way parameter sensitivity analyses with an agent-based simulation of the diffusion of brand crisis in a social network and measuring changes in the agents' social perceptions over time.

Keywords: Brand crisis · Brand perception · Social media · Entertainment

1 Introduction

Upon realizing how negative electronic word-of-mouth (eWOM) can potentially wreak large-scale damage due to the fact that social media facilitates rapid consumer communications that exacerbate the situation, both marketers and researchers started conducting systematic studies to understand this phenomenon and develop effective counter-strategies [1]. As these studies were implemented, what was observed together with frequent and considerable number of brand crises are the increasing effective reactions from organizations [2]. However, this is not always the case since other results were devastating. Hence, we believe that there is still the need to provide alternative prespectives that are novel and strategies that are more comprehensive.

© Springer International Publishing Switzerland 2015
N. Agarwal et al. (Eds.): SBP 2015, LNCS 9021, pp. 153–162, 2015.
DOI: 10.1007/978-3-319-16268-3_16

Although entertainment theories themselves have been studied for a long time, recent studies have begun to associate entertainment theories to the study of user behavior in social media [3]. What we are proposing is a novel mental state model based on existing theories of individual intrinsic motivation to engage in negative eWOM as derived from studies in the psychology of entertainment. Our model explains how individual perceptions of the brand in crisis may change from the time an individual was first exposed to negative contents about the brand while engaged in social media until the individual either experienced relief or disengaged. In addtion, since theories in the psychology of entertainment have never before been applied to the study of brand crisis in social media, we expect that these would reveal new perspectives on, and eventually help improve the understanding of, brand crisis.

In the following sections, we survey the literature on brand crisis and the psychology of entertainment. Afterwards, we present our mental state model, and then show how people's perceptions, as per this model, may dynamically change in a social network when a brand crisis persists. We show the latter through our multi-agent simulation model that involves stochastic processes and sensitivity analyses.

2   Brand Crisis in Social Media

Basically, crisis is an unpredictable event that may negatively impact stakeholder expectancies and organizational performance [4]. Although the probability of such an event is low and mostly unexpected, the damage can be acute [5,6,7]. Reputational threats to a brand include product failure and corporate misbehavior, among many other factors [8]. Damaged brand image could generate serious consequences on brand equity and trust in the brand [10]. Coombs suggested that the organization's response is extremely important to its reputation [9].

In today's hyper-connected world, social media has become the effective environment for the diffusion of brand crisis as well as a compelling opportunity for crisis communication [12]. Word-of-mouth (WOM) news is perceived as more trustworthy than mainstream media in some occasions [13]. In fact, Veil et al. suggested that the best communication channels for the public – offline, online, or in the community – should be incorporated in crisis communication. Also, in their literature review on social media in risk and crisis communication, they concluded with the best practices that are also applicable to brand crisis [14]. Moreover, Marken found that people in general trust social media and use it to search for information [15]. Yap et al., however, also raised several reasons why social media users are motivated to engage in eWOM [11]. Thus, organizations are suggested to develop a plan to assess the efficacy of using social media in crisis communication [16].

With our proposed mental state model, we aim to help practitioners understand distinctive characteristics of individuals, and the public's reaction in general, as they engage in social media. Learning the characteristics of social media users is vital to understanding how their perceptions change as they receive negative information from the public about certain brands. Practitioners should be able to strategize how to improve their communication vis-a-vis such user characteristics through our model.

3   Theories in the Psychology of Entertainment

We highlight below specific entertainment theories and elucidate how they are related to brand crisis in social media and especially to our model.

1. **Intrinsic Motivation.** According to Ryan et al., activities that hold people's intrinsic interest must have the appeal of novelty [18,19], challenge [20], or aesthetic value [21]. Intrinsic motivation "describes this natural inclination toward assimilation, mastery, spontaneous interest, and exploration" [21,22]. This is a cognitive evaluation theory that is based on the assumption that an individual has three fundamental needs not only for intrinsic motivation but also for well-being [23], namely, competence, autonomy and relatedness. To feel competent, the individual shall engage in an activity that is either more challenging or only with a level of challenge that he/she can still handle successfully. Brand crisis itself is novel and people are challenged to contemplate about what is supposed to be the result. For autonomy, it is necessary that individuals perceive their behavior to be self-determined in order to maintain intrinsic motivation [24]. This is what people perceived in social media, even though in fact they were many times influenced by those contents. Indeed, media users typically overestimate their independence and think of their actions as their own preferences [17]. Finally, relatedness is perfectly consistent with the nature of social media. All of this demonstrates that people are intrinsically motivated to engage in brand crisis in social media. It is actually the same motivation that draws their attention to the entertainment media.

2. **Unpleasant Distress.** In traditional media, conflict or obstruction between a protagonist's current state and his/her goal captures audience attention. Naturally, conflict is an unpleasant experience for the audience and we should expect them to be in a state of unpleasant distress [17]. Importantly, excitation transfer theory sheds light on how people who are in a state of unpleasant distress associate that experience with positive emotional results [25,26,27]. An unpleasantness induced by the experience of conflict escalates the feeling of relief when the conflict is solved. This positive feeling is eventually misattributed to be the enjoyment of the experience. While people are exposed in social media to bad news regarding a brand, they may experience an unpleasant distress when some conflicts arise between what they believe to be a decent thing and the negative information that they receive. Opinions from other users that contradict one's own can also amplify the conflict and the unpleasant distress. In our model, we illustrate a process from the state of unpleasant distress to state of relief based on excitation transfer theory.

3. **Disposition-based Theory.** Moral judgement is explained by the disposition theory [28,29,30]. Normally, the audience make moral judgments about the characters in a movie, which consequently elicits expectations within them as to the outcome of the movie. In the very same way, we propose that when individuals make moral judgments after they are exposed to a brand crisis, they develop in them expectations as to the resulting situation. In the unfolding part of a movie, enjoyment is a function of audience disposition toward the characters. When liked characters experienced positive outcomes or disliked characters experienced

negative outcomes, audience enjoyment increased [31]. This theory gives a clue on how entertainment takes part in brand crisis. Moreover, in crime-punishment entertainment, Raney et al. found that the outcome of moral judgment for both affective and cognitive variables can predict overall enjoyment [30]. If the outcome is the brand being punished, we should expect people to experience enjoyment.

4. **Justice.** Justice themes in media content can easily capture attention and have a potential to arouse emotion and suspense [27,32,33]. Just behavior that received reward and unjust behavior that received punishment are intrinsically satisfying [30,34]. As we can expect satisfaction from vindication to be a powerful entertainment tool, news on injustice should also be influential in capturing public attention, and people may not be consciously aware of justice issues [17].

4   Mental State Model

This model is based on the above theories in the psychology of entertainment. Each node in the model indicates an individual's current mental state and perception of the brand and may transition to the next state depending on how the events unfold. The process begins with the individual being exposed to the negative content shared on social media. Because it is negative and opposed to what the individual believes to be the decent thing the company should do, it should lead him/her to the state of unpleasant distress [17]. Take for example the infamous picture of an employee licking food that is supposed to be sold shortly afterwards [2]. Shortly after one employee posted it on social media, a storm of anger exploded out of the blue that put the company's reputation on the edge and challenged its ability to handle the situation. In the midst of uncertainty, people began to expect and wanted to see the conclusion, in this case, justice. At the end of the day, people might be satisfied if the company fired the employee, got angrier if the company's reaction was inappropriate, or lost their attention as time passed with nothing new. In general, people might just keep looking for new information until they are satisfied or simply lost interest.

Fig. 1 illustrates our mental state model. The gray circle at the top-left represents individuals who have neutral perception prior to the exposure. They may not be aware of, or have no emotional attachment to, the brand prior to the crisis. Once exposed to the negative news, however, their perception changes negatively due to a feeling of unpleasant distress [17]. This is followed by moral judgment that leads to expectations [28,29,30] and can result in one of three possible outcomes. In case the situation does not match their expectation, perhaps due to inappropriate, or lack of, response by the company, distress persists and consequently generates a loop of exasperation. Contradicting opinions in the social media may also bolster the irritation. Eventually, if the situation does not change, people whose mental state swirls within a loop of unpleasant distress and expectation will lose their attention and pull themselves off the issue. However, the result can also be satisfactory, e.g., the employee at fault was fired or the company made explicit pertinent stricter regulations. Thus, brand perception in 'relief' and 'lose attention' states can be negative or positive depending on the outcomes of the situation.

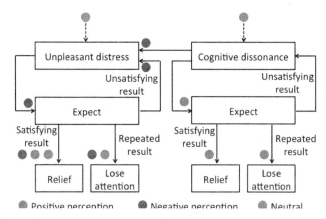

Fig. 1. Perception states and transitions as per our mental state model

At the top-right, people who already have a positive perception prior to the exposure are likely to encounter cognitive dissonance because they simultaneously hold two psychologically inconsistent cognitions [35,36], i.e., they like the brand or are loyal customers, or they might have their own reason to still think positively of the brand. Using the same example as above, when people see that picture, two cognitions become antithetical: 'I like this restaurant but what its employee did in that picture is completely unacceptable.' Naturally, cognitive dissonance is unpleasant and people are motivated to reduce it. The fact that one employee violated the company's rule does not mean that everyone licked every dish before they were served. Therefore, the company did nothing seriously wrong and some punishment on that employee would be adequate. On the other hand, it is obvious that the company neglected to rigorously keep its employees in line, which could put customer health in serious risk. The company must therefore take full responsibility if it wants consumer trust back. In the end, an individual will be motivated to choose one of these two to reduce cognitive dissonance. If one thinks that it is only a matter of that employee, then the right loop applies wherein he/she will continue to hold a positive perception, and possibly a bias towards the company. Accordingly, the company's success in overcoming the problem will bring him to a relief state. If nothing new happens over time, the individual loses attention, albeit still holds a positive perception. However, new information against the brand will again arouse cognitive dissonance. If again in the dissonance reduction process the individual this time opts to reduce dissonance by acting against the company, his perception will move to the unpleasant distress state.

5   Multi-agent Simulation Model

To give us an idea of how people's perceptions, as per our mental state model, may change dynamically in a social network with an on-going brand crisis, we constructed an agent-based simulation environment. Each agent represents an active individual in the social network that is sensitive to the on-going brand crisis and WOMs crawling in the network. Our simulation parameters are as follows:

- Total population (*TP*) of agents;
- Percentages of *TP* that began with neutral (*NP*) and positive (*PP*) perceptions;
- Agent's strength of influence (*SI*) with, as well as vulnerability (*V*) to, WOMs;
- Average number of contacts (*C*) for each agent;
- Duration of attention (*D*) of each agent in model time units; and
- Brand's positive (*BPI*) and negative (*BNI*) impact rates – the brand in crisis may or may not try, respectively, to save its reputation and recover from its current crisis.

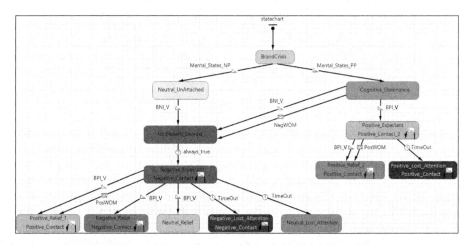

Fig. 2. Agent-based model representation of our mental state model

Fig. 2 shows our mental state model representation in the agent-based environment using the tools of the simulation software[1] we used. As brand crisis happens, agents will be randomly categorized to either have a neutral, i.e., *Mental_State_NP=NP×TP*, or positive, i.e., *Mental_State_PP=PP×TP*, perception of the brand. The nodes represent mental states, and the arrows indicate transition-triggers defined as follows:

- *BNI_V*: computed as *BNI×V×agents_in_state$_i$*, which refers to the number of agents with current mental *state$_i$* that will transition to the next state due to *BNI* and *V*. If for example *BNI*=0.1, *V*=0.4, and there are 200 agents in *state$_{Neutral\_Unattached}$*, then eight (8) agents will be randomly selected to transition to *state$_{Unpleasant\_Distress}$*.
- *BPI_ V*: computed as *BPI×V×agents_in_state$_i$*, which refers to the number of agents that will transition from the current *state$_i$* due to *V* and, this time, *BPI*.
- *Negative_Contact* and *Positive_Contact*: both are computed as *SI×C*, the difference is that the WOMs that will be delievered to the randomly selected contacts are negative (*NegWOM*) and positive (*PosWOM*), respectively.
- *NegWOM* and *PosWOM*: agents transition to the next state once these messages are received. Agents who receive such messages immediately become susceptible to WOMs due to the strength of influence of the agent that delivered the message.
- *TimeOut*: this transition is triggered each time the duration of the agent's attention, *D*, is reached, which means that the agent has lost interest.

---

[1] AnyLogic Researcher/Educational 7.1.2 multi-method simulation software.

We can see in Fig. 2 how the transition-triggers help distinguish the mental states from each other. Notice as well that the cycles in the mental state model are implemented in the agent-based model by letting the agents stay in the same state until a trigger changes their mental state. We set all model changes, e.g., initialization of agents per category and mental state transtions, to occur per model time step.

For us to obtain meaningful analyses, we accounted for dynamic uncertainties and stochastic processes in the model, i.e., the behaviors demonstrated by the model vary significantly across different runs even with the same set of parameter values. We employed our software to perform Monte Carlo simulations with our parameters, i.e., it generated inputs randomly from a probability distribution over a domain of possible inputs that we defined and it approximated the probability of simulation outcomes as it replicated each run for several times with different random seed values. In effect, our Monte Carlo experiments obtained and displayed a collection of simulation results for our model with stochastically varied parameters.

Furthermore, we carried out multi-way parameter sensitivity analyses, i.e, we accounted not just for stochastic processes, but also the sensitivity of the quantities of interest to changes of model parameters. We performed two sensitivity analyses:

- *Case-1*: We varied three parameters, namely, *BNI*, *BPI* and *V*, and drew each of their values from a uniform distribution of 0.1 to 0.5. We therefore observed the negative and positive impacts of the brand to the vulnerability of the agents.
- *Case-2*: Apart from the parameters in *Case-1*, we also activated the *NegWOM* and *PosWOM* triggers and varied two more parameters, *SI* and *C*, and drew their values from uniform distibutions of 0.1 to 0.5 and 1 to 60, respectively. Hence, here, we also looked at the impact of both positive and negative WOMs on the population depending on the strength of influence and the number of contacts the agents had.

We also varied in each case the *NP* and *PP* and drew their values from a uniform distribution of 0.01 to 0.1 while *TP* is fixed to 1,000. We ran each simulation for 1,000 times, with each simulation run lasting for 500 time units. We replicated each of the 1,000 iterations 10 to 50 times, stopping at the minimum replication when an 80% confidence is reached. The purpose of varying the number of iterations and the confidence level is to make certain the statistical significance between the current objective mean value and the best mean found in the previous iterations.

Fig. 3 shows the end-results of our multi-way parameter sensitivity analyses of both cases. Each chart is a Monte Carlo 2D histogram (a built-in software feature) wherein each cell summarizes the number of trajectories for the quantity of interest, i.e., a final mental state, that fell within a specific time interval. The color intensity of each cell corresponds to the size of the corresponding resulting 2D histogram bin.

The charts for *Case-1* (left column) indicate that the agents have equal tendencies to be positively relieved after an unpleasant distress, negatively relieved, or neutrally relieved, but with a high tendency of positive relief after a cognitive dissonance. If we are the brand in crisis, then this can be good news for us since there is higher probability that we can gain back public trust. However, this also means to continue exerting recovery efforts so as to lessen the number of negatively relieved individuals since they still retain their negative views and could influence their contacts.

The results for *Case-2* indicate changes in population behavior due to the strength of influence while using WOM and the variations of contact rates. There is an in-crease-shift (see dotted arrows from left to right charts) in the number of agents who experienced positive relief after their unpleasant distress, and decreased gradients in the negative and neutral relief charts indicate decrease in affected agents. A strategy therefore is for the management to situate more individuals who have strong enough influence to promote positive WOMs about the brand to as many contacts as possible.

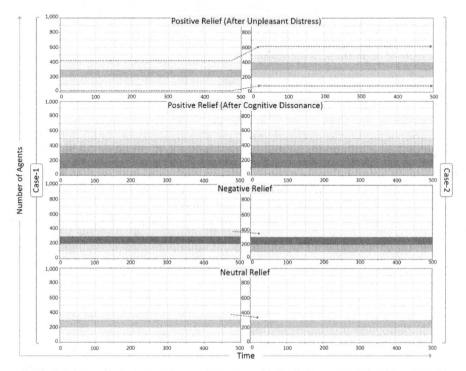

Fig. 3. Result charts of our multi-way sensitivity analyses for *Case-1* (left) and *Case-2* (right)

We are cognizant of the fact that there are numerous factors that can influence user perceptions (e.g., refer to our other current work [37]). However, having few parame-ters in our model does not mean that it is not at all viable since we based our model on sound theories. In the general sense, our model presents meaningful insights on percep-tion dynamics. What we need to do next, however, is to further its elucidation by add-ing parameters and fine-grained descriptions (e.g., levels per mental state).

Furthermore, it may be that the outcomes we obtained are merely artifacts of the modeling procedure and original assumptions. We also believe that we have yet to ground our parameter values to more realistic or proven settings. However, what we are forwarding for consideration is that our simulation environment and its procedures (e.g., stochastic processes and sensitivity analyses) can demonstrate how our mental state model may be used to predict social perception behaviors as brand crisis happens, and how to obtain and analyze results that the brand can leverage to gain public trust.

6   Conclusion

We elucidated in this paper our proposed novel mental state model that is based on existing theories in the psychology of entertainment, which may help explain social perception changes due to an on-going brand crisis and management. Not only is this understanding beneficial to researchers, by utilizing this model in brand crisis management, marketers may adjust their communication strategies with the online community to ensure positive perception at the end of the crisis. The implication of this model reveals the possibility of the increase of positive brand awareness. This is possible because during the crisis conflict is very effective in drawing people's attention. The company's appropriate reactions while it is the center of attention should therefore create a positive perception to the wider audience. Moreover, the model also emphasizes the importance of supporting a group of loyal customers as their perception is less likely to become negative. These people are especially important because in social media, their opinions can create a significant impact to public perception.

References

1. Verhagen, T., Nauta, A., Feldberg, F.: Negative online word-of-mouth: Behavioral indicator or emotional release? Computers in Human Behavior. **29**(4), 1430–1440 (2013)
2. Marzilli, T.: Taco Bell Handling of 'Taco Licker' Seems To Reassure Consumers. http://www.forbes.com/sites/brandindex/2013/06/13/taco-bell-handling-of-taco-licker-seems-to-reassure-consumers/
3. Reinecke, L., Vorderer, P., Knop, K.: Entertainment 2.0? The Role of Intrinsic and Extrinsic Need Satisfaction for the Enjoyment of Facebook Use. Journal of Communication **64**, 417–438 (2014)
4. Coombs, W.: Ongoing Crisis Communication: Planning, Managing, and Responding, 2nd edn. Sage, Thousand Oaks (2007)
5. Barton, L.: Crisis in Organizations: Managing and Communicating in the Midst of Chaos. Southwestern Publishing Company, Cincinnati (1993)
6. Pearson, C.M., Clair, J.A.: Reframing Crisis Management. Academy of Management Review **23**, 59–76 (1998)
7. Sellnow, T., Seeger, M.: Exploring the Boundaries of Crisis Communication: The Case of the 1997 Red River Valley Flood. Communication Studies **52**(2), 153–167 (2001)
8. Greyser, S.A.: Corporate brand reputation and brand crisis management. Management Decision **47**(4), 590–602 (2009)
9. Coombs, W.T., Holladay, S.J.: Helping Crisis Managers Protect Reputational Assets. Management Communication Quarterly **16**, 165–186 (2002)
10. Dawar, N., Pillutla, M.: Impact of product-harm crises on brand equity: The moderating role of consumer expectations. Journal of Marketing Research **37**(2), 215–226 (2000)
11. Yap, K.B., Soetarto, B., Sweeney, J.C.: The relationship between electronic word-of-mouth motivations and message characteristics: The sender's perspective. Australasian Marketing Journal **21**(1), 66–74 (2013)
12. Colley, K.L., Collier, A.: An Overlooked Social Media Tool? Making a Case for Wikis. Public Relations Strategist, 34–35 (2009)
13. Seo, H., Kim, J.Y., Yang, S.U.: Global Activism and New Media: A Study of Transnational NGO's Online Public Relations. Public Relations Review **35**(2), 123–126 (2009)

14. Veil, S.R., Buehner, T., Palenchar, M.J.: A Work-In-Process Literature Review: Incorporating Social Media in Risk and Crisis Communication. Journal of Contingencies and Crisis Management **19**(2), 110–112 (2011)
15. Marken, G.A.: Social Media … The Hunted can Become the Hunter. Public Relations Quarterly **52**(4), 9–12 (2007)
16. Institute of Management and Administration: How Social Media are Changing Crisis Communications – For Better and Worse. Security Directors Report, 2-5 (2009)
17. Bryant, J., Vorderer, P.: Psychology of entertainment. Lawrence Erlbaum Associates, Mahwah (2006)
18. Berlyne, D.E.: Aesthetics and psychobiology. Appleton-Century-Crofts, New York (1971)
19. Berlyne, D.E.: The new experimental aesthetics: Steps toward an objective psychology of aesthetic appreciation. Hemisphere, Washington, D.C. (1974)
20. Csikszentmihalyi, M.: Beyond boredom and anxiety. Jossey-Bass, San Francisco (1975)
21. Ryan, R.M., Deci, E.L.: Self-determination theory and the facilitation of intrinsic motivations, Social development and well-being. American Psychologist **1**, 68–78 (2000)
22. Ryan, R.M.: Psychological needs and the facilitation of integrative process. Journal of Personality **63**, 397–427 (1995)
23. Reeve, J.: Motivating others. Allyn & Bacon, Needham Heights (1996)
24. Higgins, E.T.: Knowledge activation: Accessibility, applicability, and salience. In: Higgins, E.T., Kruglanski, A.W. (eds.) Social Psychology: Handbook of Basic Principles, pp. 133–168. Guilford Press, New York (1996)
25. Zillmann, D.: Excitation transfer in communication-mediated aggressive behavior. Journal of Experimental Social Psychology **7**, 419–434 (1971)
26. Bryant, J., Miron, D.: Excitation-transfer theory and three-factor theory of emotion. In: Bryant, J., Roskos-Ewoldsen, D., Cantor, J. (eds.) Communication and Emotion: Essays in Honor of Dolf Zillmann, pp. 31–59. Lawrence Erlbaum Associates, Mahwah (2003)
27. Zillmann, D.: Theory of affective dynamics: Emotions and moods. In: Bryant, J., Roskos-Ewoldsen, D., Cantor, J. (eds.) Communication and Emotion: Essays in Honor of Dolf Zillmann, pp. 553–567. Lawrence Erlbaum Associates, Mahwah (2003)
28. Raney, A.A.: Disposition-based theories of enjoyment. In: Bryant, J., Roskos-Ewoldsen, D., Cantor, J. (eds.) Communication and Emotion: Essays in Honor of Dolf Zillmann, pp. 61–84. Lawrence Erlbaum Associates, Mahwah (2003)
29. Raney, A.A.: Expanding disposition theory: Reconsidering character liking, moral evaluations, and enjoyment. Communication Theory **14**, 348–369 (2004)
30. Raney, A.A., Bryant, J.: Moral judgment and crime drama: An integrated theory of enjoyment. Journal of Communication **52**, 402–415 (2002)
31. Zillmann, D., Cantor, J.: A disposition theory of humor and mirth. In: Chapman, T., Foot, H. (eds.) Humor and Laughter: Theory, Research, and Application, pp. 93–115. Wiley, London (1976)
32. Miluka, G., Scherer, K.R.: The role of injustice in the elicitation of differential emotional reactions. Personality and Social Psychology Bulletin **24**, 133–149 (1990)
33. Montada, L.: Understanding oughts by assessing moral reasoning and moral emotions. In: Noam, G., Wren, T. (eds.) The Moral Self, pp. 292–309. MIT-Press, Boston (1993)
34. Zillmann, D.: Basal morality in drama appreciation. In: Bondebjerg, I. (ed.) Moving Images, Culture, and the Mind, pp. 53–63. Univ. of Luton Press, Luton (2000)
35. Festinger, L.: A theory of cognitive dissonance. Stanford Univ. Press, Stanford (1957)
36. Festinger, L.: Cognitive dissonance. Scientific American **207**(4), 93–107 (1962)
37. Legaspi, R., Maruyama, H., Nararatwong, R., Okada, H.: Perception-based Resilience: Accounting for the impact of human perception on resilience thinking. In: Proc. Seventh IEEE International Conference on Social Computing and Networking, pp. 547–554 (2014)

Network-Based Group Account Classification

Patrick S. Park[1,2(✉)], Ryan F. Compton[2], and Tsai-Ching Lu[2]

[1] Cornell University, Ithaca, NY, USA
pp286@cornell.edu
[2] HRL Laboratories, Malibu, CA, USA
{rfcompton,tlu}@hrl.com

Abstract. We propose a classification method for group vs. individual accounts on Twitter, based solely on communication network characteristics. While such a language-agnostic, network-based approach has been used in the past, this paper motivates the task from firmly established theories of human interactional constraints from cognitive science to sociology. Time, cognitive, and social role constraints limit the extent to which individuals can maintain social ties. These constraints are expressed in observable network metrics at the node (i.e. account) level which we identify and exploit for inferring group accounts.

Keywords: Network-based classification · Organization prediction · Twitter · Cognitive constraint · Communication signature

1   Introduction

User accounts in social media platforms (e.g. Twitter and Facebook) are constituted by individuals, celebrities, and groups from all over the globe, each of which appropriates social media in different ways with messy behavioral traces. This heterogeneity in the user population poses a considerable challenge to the classification of non-individual users as it is difficult to invoke "domain specific" ethnographic knowledge for all users across all countries. Accurately classifying individual vs. non-individual entities on a given social media platform is important for a range of applications. In social network research, for example, the ability to accurately filter out non-individual entities directly affects the network properties the researcher observes, on which her conclusions about the network rest. In online targeted marketing where the targets are, for example, small, local businesses, these small scale business entities often do not register their accounts on central online databases such as twellow.com. Campaigns targeting such small business-es would benefit from an effective classification tool.

A number of methods for identifying non-individual group accounts exist in the literature, from early attempts to identify organizations, celebrities, and media outlets using Twitter's "list" function [1,2] to those leveraging the temporal signatures in the tweets of a given user account [3]. The state-of-the-art language-based classification approaches perform with high accuracy and variants of them which combine textual information with basic network measures (e.g. follower in-degree) have proven to be promising [4,5]. Nevertheless, the successful application of these language-based

© Springer International Publishing Switzerland 2015
N. Agarwal et al. (Eds.): SBP 2015, LNCS 9021, pp. 163–172, 2015.
DOI: 10.1007/978-3-319-16268-3_17

approaches depend on the NLP infrastructure developed for each language where considerable variation exists in the sophistication and reliability of tool-kits across languages. Furthermore, given that languages vary in the amount of information that can be packed into 140 characters per tweet (e.g. Chinese vs. English), the performance of a language-based classifier is prone to vary even more.

As a complementary approach to the language-based approach, we develop a purely network-based classifier to identify groups vs. individuals. The underlying cognitive and sociological theories [6,8,9,10,11,12,13] for the proposed method postulate that due to time, cognitive, and resource constraints, (a) the number of interactions an individual can maintain at any given time period is typically limited to a few hundred alters and (b) an individual typically exhibits a high concentration of communication with only a handful of individuals while communicating intermittently with others. These constraints affect the ways in which individuals perceive of and construct social ties. We apply these broadly documented observations regarding limited attention and skewed distribution of social interactions to classify group accounts on Twitter. The advantage of such a language-agnostic, network-based classification strategy is that it can be applied to any user regardless of language to the extent that users from different cultures, language traditions, and countries face similar resource and cognitive constraints in maintaining social ties.

2   Theoretical Basis

Our proposed method focuses on the limitations in maintaining social relationships, which have been broadly identified across the cognitive and social sciences. At the cognitive level, research shows that the human capacity to remember social relationships around one's immediate social circles, process that information, and adjust behavior accordingly is biologically limited [10,12,13,15]. At the societal level, maintaining a large number of social relationships implies that an individual bears a heavier burden of maintaining diverse role relationships which often prescribe contradicting normative demands upon the individual, causing role strain [11]. A consequence of these cognitive and social costs is that the number of social relationships maintainable at any given period in time will be limited. Another postulated consequence is that humans develop a hierarchical perception of group structure of their affiliated groups from the emotionally closest to the most distant [16] and that such a hierarchical group structure is correlated with the heavy skew in the distribution of communication to an individual's alters [6].

In contrast to individuals, human aggregates, from small groups to organizations, are not subject to the same level of cognitive and role constraints as individuals. With more than one person to manage communication ties, aggregate social entities can collectively wield more cognitive resources and overcome individual level time and cognitive constraints to expand the breadth of social relationships. In the context of Twitter, more than one group member can work on maintaining a group account, which could increase the number and diversity of communication ties.

3   Methods

3.1   Data Source and Preprocessing

We use a 10% real-time tweet stream (a.k.a Decahose) collected from April 2012 to April 2014. From this corpus, we first geolocate the users by their country using a novel label-propagation method which performs with high coverage and accuracy: labeling over 100M users at a median error of 6.33 km [20,21]. Using the inferred country labels of the user accounts, we construct within-country @mention networks as described below. By constructing within-country @mention networks, we recognize the fact that the vast majority of @mentions occur among users in the same country and that the overall level of mentioning, which affects network structure, is partly determined by Twitter penetration and the baseline propensity of @mention tweets at the country level [28].

3.2   Directed @mention Network

To leverage the widely observed constraint on human social relationships for the purpose of classifying group accounts, we extract information from the @mention tweets which capture social actions among Twitter users. Here, we adopt Max Weber's classical sociological definition of social action where an "action is 'social' if the acting individual takes account of the behavior of others and is thereby oriented in its course [17]." Since @mention tweets are explicitly oriented to other users, often for conversational purposes [7,18], we maintain that @mentions constitute a less noisy and stronger indicator of interpersonal relationships than follower ties that have been widely examined in previous Twitter studies [24,25,26].

Figure 1 shows the overall framework of the method. The procedure starts from extracting the @mentions from tweets in each country. These @mention records are then aggregated while preserving directionality (e.g. user A mentions user B X times) and are used to construct a directed, weighted @mention network for each country.

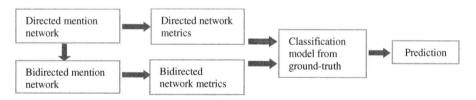

Fig. 1. Overall framework of the network-based classification for each country

3.3   Bidirected @mention Network

To incorporate more solid conversational ties, we construct the bidirected @mention network for each country which is a subset of the directed network where a tie exists by definition only if both user A and user B mutually @mention each other at least once. In addition, if account A has no two-way @mentions with any other account (i.e. network isolate), account A is removed from the bidirected network. The underlying assumption for this treatment is based on the observation that meaningful social ties are typically reciprocal [14]. The out-mentions and in-mentions within the bidirected network, then, indicate the breadth and depth of communication for each account with regard to its socially relevant communication partners. Imposing this bidirectionality constraint filters out unidirected @mention dyads where only one party recognizes the other (e.g. John Doe to Justin Bieber). Another advantage of the bidirectionality constraint is that it effectively filters out bots and spam accounts which typically never form reciprocal relationships with ordinary users [19,27]. Note that it is possible that spam and bot accounts @mention one another to outsmart Twitter's spam filtering algorithm. To guard against such behavior, we further remove all accounts in the bidirected network that are not in the largest connected component. At the end of these preprocessing steps, there are 84.3M Twitter accounts in the largest bidirected network components across 218 countries.

3.4   Network Metrics

Using the directed and bidirected @mention networks, we derive a set of metrics to be used as predictors of individual vs. group labels. The directed network metrics represent the overall spectrum of communications involving a given user account, which shows its total communication capacity, irrespective of account type (bot, spam, individual, or group) or the account type of the @mentioned account. While these directed network metrics reveal the observed total amount of communications in the data and, therefore, better approximate the cognitive and social constraints as discussed above, they are impure in the sense that not all incoming or outgoing mentions are conversational or "social" in the Weberian sense; mere name dropping of a celebrity figure in one's tweet does not constitute a meaningful social discourse.

Directed Network Metrics

Log (in-degree)
From the directed network, we measure the number of alters who @mentioned ego at least once and take the logarithm of that quantity to address the heavy skew in the distribution.

Log (out-degree to in-degree ratio)
Individual users tend to have a balanced ratio of out-degree to in-degree due to the combined effects of the norms of reciprocity in human interaction [22] and the cognitive, time-bound, and social role constraints outlined above. Specifically, since an individual's out-degree has some limit due to cognitive and time constraints, her in-degree will also be somewhat limited by the norms of reciprocity. If an individual receives mentions from more than 100 alters, but can maintain communication with only 10 of them, the other 90 who could not engage in conversations with the focal individual would be more likely to reduce communication and, at some point, cease mentioning that individual altogether. On the other hand, groups and celebrities may be less subject to such norms such that those who mention group accounts simply may not hold the same expectations of reciprocity as they would toward individuals.

Log (out-mention to in-mention ratio)
Similar to the out- to in-degree ratio, similar norms of reciprocity and constraints may apply at the mention level for individuals. Here again, we predict that a group account will exhibit lower levels of aggregate out- to in-mention ratios than an individual.

Gini coefficients of out- and in-mention signatures
We extract the out-mention and in-mention distributions of each account from the directed network and measure their skew, which represents the cognitive, time, and social role constraints associated with the hierarchical structuring of one's ego network. Theory predicts that, other factors being equal, less skew should be observed in non-individual accounts as those accounts are less limited by cognitive and time constraints than individuals. The skew of an account's mention distribution can be summarized as the well-known Gini coefficient.

$$G = 1 - \frac{\sum_{i=1}^{n} f(y_i)(S_{i-1}+S_i)}{S_n} \tag{1}$$

In equation (1), y_i is the relative out- or in-mention frequency of alter i who is connected with ego through either out- or in-mentions ($y_i < y_{i+1}$), $f(y)$ is the probability mass function, and $S_i = \sum_{j=1}^{i} f(y_i)y_j$ and $S_0 = 0$.

Bidirected Network Metrics

Log (page rank)
Previous work on spam detection reports the use of Infochimp's "trust quotient" which is a variant of page rank [27]. We argue that the page rank which is a measure of popularity can be a useful indicator of group accounts to the extent that groups are more systematically driven than ordinary individuals by the pursuit of exposure and influence on Twitter.

Log (alter's mean degree to ego's degree ratio)
Human social networks are characterized by assortative mixing [23] where connected individuals tend to have similar degrees (i.e. positive degree correlation). While evidence is sparse whether assortative mixing applies to group entities as well, there is reason to believe that disassortative mixing should be more prevalent for group accounts on Twitter. Since group accounts use Twitter primarily as a platform to engage with individual users (e.g. individual consumers) rather than with other group entities and since groups tend to have higher degree than individual users, the ratio of the alter's mean degree to ego's degree should be lower for group accounts compared to individual accounts.

Log (directed network in-mention to bidirected network in-mention ratio)
The directed network in-mention to the bidirected network in-mention ratio captures the extent to which the account receives interactive or conversational @mentions relative to the total one-way @mentions. We also add a squared term in the logistic regression classifier to capture possible non-linear associations.

3.5   Ground-Truth Labels

"Group" accounts in the current context are synonymous to "managed" accounts where more than one individual manages a given Twitter account. By this definition, a celebrity account which is not likely to be managed solely by the individual celebrity is also labeled as a group account. For the ground-truth labels, we construct a stratified user account sample based on indegree to outdegree ratio from the largest bidirected @mention network components of the UK and South Korea (272 users). In addition, we merge these hand-labeled user data with ground-truth organization labels from a previous study which consists of user accounts whose tweets are predominantly in English (103831 users) and in Spanish (118367 users), respectively [5]. All spam and bot accounts are deleted in this process. In sum, the ground-truth data consist of 177 group accounts and 222,293 individual accounts.

Table 1. Descriptive Statistics of Network Metrics

	Mean	St.D	Min	Max
Log(bidirected network pagerank)	-14.23	1.78	-18.61	-3.08
Log(directed network indegree)	4.03	1.16	0	13.96
Log(directed network out- to in- degree ratio)	0.15	0.44	-9.95	5.38
Log(directed network out- to in-mention ratio)	0.12	0.5	-10.37	7.36
Log(bidirected network neighbor degree to self degree ratio)	0.46	1.01	-5.43	8.88
Log(directed network inmention to bidirected network inmention ratio)	0.24	0.33	0	12.78
Log(directed network inmention to bidirected network inmention ratio) squared	0.16	1.09	0	163.23
Directed network outmention gini coefficient	0.54	0.14	0	0.96
Directed network inmention gini coefficient	0.54	0.14	0	0.96

4   Evaluation

Table 1 shows the descriptive statistics of the network metrics of the ground-truth dataset. We first performed a dimension reduction on our labeled data via t-SNE [29] and plotted the results in Figure 2 to provide a graphical depiction of the discriminatory power of the network metrics. In brief, the t-SNE algorithm maps pairwise distances in the high-dimensional space to distances in a low-dimensional embedding by equating "distance" with a joint probability and learning low-dimensional joint probabilities which are close (in the sense of Kullback-Leibler divergence) to the high-dimensional joint probabilities. Since 99% of our training data are individuals, we randomly selected a subset of individuals (the same as the number of group accounts) before visualization. Concentration of group labels in the upper left side and individual labels in the lower right side suggest reasonable discriminatory power.

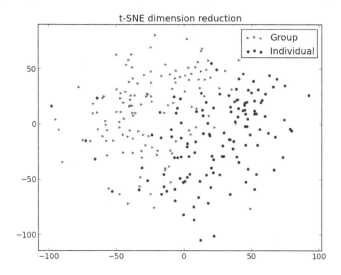

Fig. 2. Sample of ground-truth labels plotted using t-SNE dimension reduction

Next, using the network predictors with the ground-truth labels, we train K-nearest neighbor (KNN) and logistic regression classifiers. 80% of the labels are used for training and the other 20% for evaluation, reflecting the imbalance in composition between group and individual labels. Table 2 presents the aggregate performance of the two classifiers over 30 iterations for both KNN and logistic regression classifiers with the first set of rows reporting the result from models that exclude the Gini coefficients of in- and out-mention signatures and the second set of rows reporting the results from the models that include all metrics. All models achieve above 98% accuracy. However, the high accuracy is illusive since the ground-truth labels are not balanced (i.e. 222293 individuals vs. 177 groups). One would achieve a 99.92% mean accuracy by simply classifying all accounts as individual (222293/(222293+177) = 0.9992). For this reason, the more informative performance measures are precision (positive predictive value) and recall (true positive rate) shown in Table 2. Under fair random guessing ($\rho = 0.5$), the expected precision and recall are 0.000796 and 0.5, respectively. For the KNN classifier using all network predictors, the mean precision of 0.770 is 967 times higher and the mean recall of 0.568 is 1.13 times higher than random guessing, respectively. For the logistic regression classifier using all network predictors, the mean precision of 0.048 is 60 times higher and the mean recall of 0.976 is 1.95 times higher than random guessing, respectively.

We find stark differences between the KNN and logistic regression classifiers in terms of precision and recall where the former achieves higher precision (i.e. higher mean average precision (MAP)) than the latter while the latter shows higher recall than the former (i.e. higher area under the curve (AUC)). The models including the in- and out-mention Gini coefficients marginally outperform the models which exclude them, lending partial support for the cognitive constraint hypothesis.

Table 2. Overall performance of the K-nearest neighbor and logistic regression classifiers over 30 iterations

		KNN		Logistic Regression	
		mean	std	mean	std
No Gini	Precision	0.746	0.090	0.057	0.008
	Recall	0.567	0.082	0.976	0.025
	Accuracy	1.000	0.000	0.987	0.001
	MAP	0.657	0.070	0.516	0.012
	AUC	0.783	0.041	0.981	0.012
All Metrics	Precision	0.770	0.063	0.048	0.009
	Recall	0.568	0.046	0.976	0.026
	Accuracy	1.000	0.000	0.986	0.001
	MAP	0.669	0.041	0.512	0.014
	AUC	0.784	0.023	0.981	0.013

5   Discussion

Drawing from a number of theories in cognitive science and sociology, which identify sources of interactional constraints at different levels (i.e. time, cognitive, and social role), we introduced a set of language-agnostic network-based models for Twitter group account classification. The KNN classifier exhibited a high level of precision (small false negatives), but a mediocre level of recall (relatively large false positives). On the other hand, the logistic regression classifier identified nearly all group accounts, but at the expense of a high rate of false positives. Manual inspection of these false positive cases suggest that they tend to be active individuals who interact with a rather large number of other Twitter users or are local celebrities who come close to being classified as group account by our manual labeling criteria. Since the vast majority of Twitter users are individuals, the higher rate of false positives in the logistic regression classifier constitute only about 1.4% of all individual accounts.

We maintain that the focus of the specific application should dictate the choice between the KNN and the logistic regression classifiers. For example, if the group account classification is used as a filtering step in a network analysis of individual Twitter users, the KNN classifier may be a more sensible choice, given that it is less prone to misclassifying high-degree, "influential" individuals whose presence or absence will affect the observed network structure significantly. On the other hand, if the objective is to discover as many potential group accounts as possible as an intermediate step, for example, in identifying target accounts for a B2B advertisement campaign, the logistic regression classification may prove to be more useful.

A potentially fruitful future direction would combine the network metrics we constructed in this paper with the temporal signatures inscribed in users' tweets [3]. Since temporal signatures and network signatures are both language-agnostic, the potential applicability of the combined classification could be extended to language communities which do not yet possess reliable computational tools for accurate language-based classification.

References

1. Wu, S., Hofman, J.M., Mason, W.A., Watts, D.J.: Who says what to whom on twitter. In: Proceedings of the 20th International Conference on World Wide Web, WWW 2011, pp. 705–714. ACM, New York (2011)
2. Sharma, N., Ghosh, S., Benevenuto, F., Ganguly, N., Gummadi, K.P.: Inferring who-is-who in the twitter social network. In: Proceedings of the 2012 ACM Workshop on Online Social Networks, WOSN, pp. 55–60. ACM, New York (2012)
3. Tavares, G., Faisal, A.: Scaling-laws of human broadcast communication enable distinction between human, corporate and robot twitter users. PLoS One **8**(7), e65774 (2013)
4. De Choudhury, M., Diakopoulos, N., Naaman, M.: Unfolding the event landscape of twitter: Classification and exploration of user categories. In: Proceedings of the ACM 2012 Conference on Computer Supported Cooperative Work, CSCW, pp. 241-244. ACM, New York (2012)
5. De Silva, L., Riloff, E.: User type classification of tweets with implications for event recognition. In: Proceedings of the Joint Workshop on Social Dynamics and Personal Attributes in Social Media, pp. 98–108. Association for Computational Linguistics, Baltimore (2014)
6. Saramaki, J., Leicht, E.A., Lopez, E., Roberts, S.G.B., Reed-Tsochas, F., Dunbar, R.I.M.: Persistence of social signatures in human communication. PNAS **111**(3), 942–947 (2014)
7. Honeycutt, C., Herring, S.: Beyond microblogging: Conversation and collaboration via twitter. In: Proceedings of the 42nd Hawaii International Conference on System Sciences, HICSS 2009, pp. 1–10. IEEE (2009)
8. Brewer, D.D.: The social structural basis of the organization of persons in memory. Human Nature **6**, 379–403 (1995)
9. DeScioli, P., Kurzban, R.: The alliance hypothesis for human friendship. PLoS One **4**(6), e5802 (2009)
10. Dunbar, R.I.M.: Neocortex size and group size in primates: A test of the hypothesis. Journal of Human Evolution **28**, 287–296 (1995)
11. Goode, W.J.: A theory of role strain. American Sociological Review **25**(4), 483–496 (1960)
12. Miller, G.: The magical seven plus or minus two: Some limits on our capacity for processing information. Psychological Review **63**, 81–97 (1956)
13. Roberts, S., Dunbar, R.I.M., Pollet, T.V., Kuppens, T.: Exploring variation in active network size: Constraints and ego characteristics. Social Networks **31**(2), 138–146 (2009)
14. Blau, P.: Exchange and Power in Social Life. John Wiley and Sons, New York (1964)
15. Powell, J., Lewis, P.A., Roberts, N., Garcia-Finana, M., Dunbar, R.I.M.: Orbital prefrontal cortex volume predicts social network size: An imaging study of individual differences in humans. Proceedings of the Royal Society B **279**(1736), 2157–2162 (2012)
16. Zhou, W.X., Sornette, D., Hill, R.A., Dunbar, R.I.M.: Discrete hierarchical organization of social group sizes. Proceedings of the Royal Society B **272**(1561), 439–444 (2005)
17. Weber, M.: Economy and Society. University of California Press (1978)

18. Sousa, D., Sarmento, L., Rodrigues, E.M.: Characterization of the twitter @replies network: are user ties social or topical? In: Proceedings of the 2nd International Workshop on Search and Mining User-Generated Contents, SMUC 2010, pp. 63–70. ACM, New York (2010)

19. Thomas, K., Grier, C., Paxson, V., Song, D.: Suspended accounts in retrospect: An analysis of twitter spam. Proceedings of the 2011 ACM SIGCOMM Conference on Internet Measurement Conference, IMC 2011, pp. 243–258. ACM, New York (2011)

20. Jurgens, D.: That's what friends are for: Inferring location in online social media platforms based on social relationships. In: The 7th International AAAI Conference on Weblogs and Social Media, ICWSM 2013, pp. 273–282 (2013)

21. Compton, R., Jurgens, D., Allen, D.: Geotagging one hundred million twitter accounts with total variation minimization. arXiv (2014) http://arxiv.org/abs/1404.7152

22. Gouldner, A.W.: The norm of reciprocity: A preliminary statement. American Journal of Sociology **25**(2), 161–178 (1960)

23. Newman, M.E.J.: Assortative Mixing in Networks. Physical Review Letters **89**(20), 208701 (2002)

24. Java, A., Song, X., Finin, T., Tseng, B.: Why we twitter: understanding microblogging usage and communities. In: Proceedings of the 9th WebKDD and 1st SNA-KDD 2007 Workshop on Web Mining and Social Network Analysis, pp. 56–65 (2007)

25. Poblete, B., Garcia, R., Mendoza, M., Jaimes, A.: Do all birds tweet the same? characterizing twitter around the world. In: Proceedings of the 20th ACM International Conference on Information and Knowledge Management, CIKM 2011, pp. 1025–1030. ACM, New York (2011)

26. Kulshrestha, J., Kooti, F., Ashkan, N., Krishna, P.G.: Geographic dissection of the twitter network. In: The 6th International AAAI Conference on Weblogs and Social Media, ICWSM 2012, pp. 202–209 (2012)

27. Quercia, D., Capra, L., Crowcroft, J.: The social world of twitter: topics, geography, and emotions. In: The 6th International AAAI Conference on Weblogs and Social Media, ICWSM 2012, pp. 298–305 (2012)

28. Hong, L., Convertino, G., Chi, E.H.: Language matters in twitter. In: The 5th International AAAI Conference on Weblogs and Social Media, ICWSM 2011, pp. 518–521 (2011)

29. van der Maaten, L., Hinton, G.: Visualizing Data using t-SNE. Journal of Machine Learning Research **9**, 2579–2605 (2008)

Persistence in Voting Behavior: Stronghold Dynamics in Elections

Toni Pérez[(✉)], Juan Fernández-Gracia,
Jose J. Ramasco, and Víctor M. Eguíluz

Instituto de Física Interdisciplinar y Sistemas Complejos (IFISC), Majorca, Spain
toni@ifisc.uib-csic.es
http://www.ifisc.uib-csic.es

Abstract. Influence among individuals is at the core of collective social phenomena such as the dissemination of ideas, beliefs or behaviors, social learning and the diffusion of innovations. Different mechanisms have been proposed to implement inter-agent influence in social models from the voter model, to majority rules, to the Granoveter model. Here we advance in this direction by confronting the recently introduced Social Influence and Recurrent Mobility (SIRM) model, that reproduces generic features of vote-shares at different geographical levels, with data in the US presidential elections. Our approach incorporates spatial and population diversity as inputs for the opinion dynamics while individuals' mobility provides a proxy for social context, and peer imitation accounts for social influence. The model captures the observed stationary background fluctuations in the vote-shares across counties. We study the so-called political strongholds, i.e., locations where the votes-shares for a party are systematically higher than average. A quantitative definition of a stronghold by means of persistence in time of fluctuations in the voting spatial distribution is introduced, and results from the US Presidential Elections during the period 1980-2012 are analyzed within this framework. We compare electoral results with simulations obtained with the SIRM model finding a good agreement both in terms of the number and the location of strongholds. The strongholds duration is also systematically characterized in the SIRM model. The results compare well with the electoral results data revealing an exponential decay in the persistence of the strongholds with time.

1 Introduction

People making a decision in a ballot are expected to follow a rational behavior. Rational arguments based on utility functions (payoff) have been considered in the literature regarding vote modeling [1,2]. The rational hypothesis, however, tends to consider the individuals as isolated entities. This might actually be the reason why it fails to account for relatively high turnout rates in elections [3–5]. The presence of a social context increases the incentive for a voter to actually vote, as he or she can influence several other individuals towards the same option [6,11]. This effect is not only restricted to turnout, but also applies to

© Springer International Publishing Switzerland 2015
N. Agarwal et al. (Eds.): SBP 2015, LNCS 9021, pp. 173–181, 2015.
DOI: 10.1007/978-3-319-16268-3_18

the choices expressed in the election [12]. It is easy to find examples of people showing their electoral preferences in public in the hope of influencing their peers. Still social influence can also act in more subtle ways, without the explicit intention of the involved agents to influence each other. The collective dynamics of social groups notably differs from the one observed from simply aggregating independent individuals [13]. Social influence is thus an important ingredient for modeling opinion dynamics, but it requires as well the inclusion of a social context for the individuals [6,11,14–19].

Even though nowadays the pervasive presence of new information technologies has the potential to change the relation between distance and social contacts, we assume that daily mobility still determines social exchanges to a large extent. Human mobility has been studied in recent years with relatively indirect techniques such as tracking bank notes [20] or with more direct methods such as tracking cell phone communications [21,22]. A more classical source of information in this issue is the census. Among other data, respondents are requested by the census officers their place of residence and work. Census information is less detailed when considered at the individual level, but it has the advantage of covering a significant part of the population of full countries. Recent works analyzing mobile phone records have shown that people spend most of their time in a few locations [22]. These locations are likely to be those registered in the census and, indeed, census-based information has been also used recently to forecast the propagation patterns of infectious diseases such as the latter influenza pandemic [23–25].

In this work, we follow a similar approach and use the recurrent mobility information collected in the US census as a proxy for individual social context. This localized environment for each individual accounts mostly for face-to-face interactions and leaves aside other factors, global in nature, such as information coming from online media, radio and TV. Our results show that a model implementing face-to-face contacts through recurrent mobility and influence as imperfect random imitation is able to reproduce geographical and temporal patterns for the fluctuations in electoral results at different scales.

2   Definitions and Methods

We have investigated the voting patterns in the US on the county level. We have used the votes for presidential elections in years 1980-2012. For each county $i = 1, \ldots, I$, we have data about the county geographic position, area, adjacency with other counties as well as data regarding population N_{iy}, and number of voters V_{iyp} for each party p for every election year $y = 1, \ldots, Y$ (note that $\sum_p V_{iyp} \neq N_{iy}$, since not everybody is entitled to vote). Raw vote counts are not very useful for comparing the counties as populations are distributed heterogeneously. Therefore we switch from vote counts to vote shares

$$v_{iyp} = V_{iyp} / \sum_p V_{iyp} . \tag{1}$$

Since the votes received by parties others than Republicans and Democrats are minority, we have focused on the two main parties. We have further considered mostly relative voteshares r_{iyp}, which are absolute vote shares minus the national average for a given party every electoral year

$$r_{iyp} = v_{iyp} - a_{yp} \,, \tag{2}$$

$$a_{yp} = \sum_i v_{iyp}/I \,, \tag{3}$$

where I stands for the number of counties.

The relative voteshares show how much above (positive) or below (negative) the national average are the results in given county.

Our main focus is on the persistence of voting patterns. We define a *stronghold* to be a county which relative voteshare remains systematically positive (or negative).

2.1 Model Definition and Analytical Description

In the SIRM model N agents live in a spatial system divided in non-overlapping cells. The N agents are distributed among the different cells according to their residence cell. The number of residents in a particular cell i will be called N_i. While many of these individuals may work at i, some others will work at different cells. This defines the fluxes N_{ij} of residents of i recurrently moving to j for work. By consistency, $N_i = \sum_j N_{ij}$. The working population at cell i is $N_i' = \sum_j N_{ji}$ and the total population in the system (country) is $N = \sum_{ij} N_{ij}$. In this work, the spatial units correspond to the US counties and the population levels, N_i, and commuting flows, N_{ij}, are directly obtained from the 2000 census.

We describe agents' opinion by a binary variable with possible values $+1$ or -1. The main variables are the number of individuals V_{ij} holding opinion $+1$, living in county i and working at j. Correspondingly, $V_i = \sum_l V_{il}$ stands for the number of voters living in i holding opinion $+1$ and $V_j' = \sum_l V_{lj}$ for the number of voters working at j holding opinion $+1$. We assume that each individual interacts with people living in her own location (family, friends, neighbors) with a probability α, while with probability $1 - \alpha$ she does so with individuals of her work place. Once an individual interacts with others, its opinion is updated following a noisy voter model [6–10]: an interaction partner is chosen and the original agent copies her opinion imperfectly (with a certain probability of making mistakes). A more detailed description of the SIRM model can be found in [26].

3   Results

Pursuing the topic of persistence and changes in opinion, as expressed by voting results, we have investigated the *strongholds* of both parties in USA. Figure 1 shows the spatial arrangement and the duration of the strongholds (measured in elections) for data of the US Presidential Elections during the period 1980-2012 and simulations using the electoral results of 1980 as initial condition.

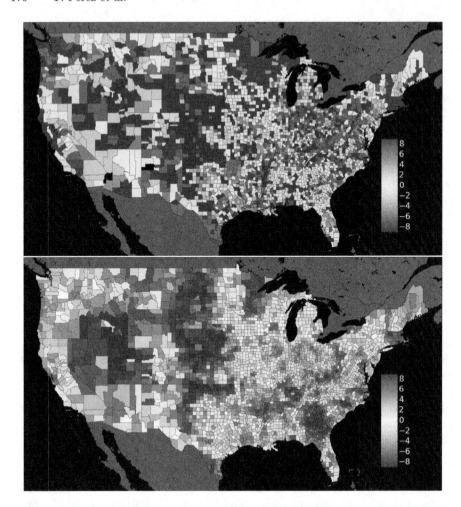

Fig. 1. Spatial distribution of the strongholds (red shows republicans while blue shows democrats). Codified by color the stronghold time measured in elections units (republicans stronghold times are given by the absolute value of the color bar). No county is a stronghold for both parties. Upper panel: US Presidential Elections data during the period 1980-2012. Bottom panel: model simulations for 9 elections using the electoral results from 1980 (at county level) as initial condition.

It can be seen at a glance, that the strongholds are not randomly distributed across the country, but clustered. This indicates that some form of correlation is present between the voting patterns. The republican strongholds seem to be concentrated mostly in the central-west, while democrat strongholds are dispersed mostly through the eastern parts, including urbanized areas. This is in agreement with the population distributions, republican strongholds being mostly lower populated counties, while democrats strongholds include some significant cities.

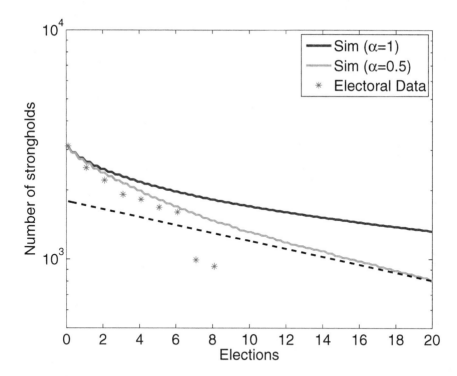

Fig. 2. Temporal evolution of the number of strongholds for electoral data and simulation results. Two different simulation results are presented, for $\alpha = 0.5$ the individuals have the same probability of interacting with individuals from other county as with individuals of his own county, while for $\alpha = 1$ interactions only within the same county would happen. Dashed line fits $f(t) = \frac{N_c}{2}\exp(-t/b)$ with $b = 25$ elections.

We quantify in Figure 2 the temporal evolution of the strongholds of the election data for the period 1980-2012 as well as for the strongholds forecasted by simulations after 9 elections. The influence of the commuting network and their interactions was revealed by contrasting the simulations results with and without network interaction. The evolution of the number of strongholds observed in the data at early stages is well described by the model with commuting interactions. For longer times, the model predicts that the number of strongholds will decay with time following an exponential law as shown in Figure 2. In the absence of commuting interaction, the number of strongholds decreases at a slower rate. Furthermore, the model overestimate the number of strongholds in this case since no other mechanisms than internal fluctuations act driving the county relative voteshare r_{iyp} towards the average. Counties set as strongholds at the beginning of the simulation remains strongholds for a longer time.

We further test the accuracy of our model by computing the percentage of strongholds accurately predicted. As Figure 3 shows, the model with no commuting interactions reproduces a higher percentage of strongholds, however, it

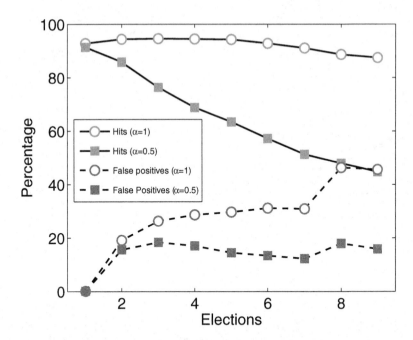

Fig. 3. Variation of the percentage of strongholds accurately predicted by the model (hits) for several elections. The number of false positives, i.e., counties identified as strongholds in the simulations without stronghold correspondence in the data is also shown.

also gives higher and increasing number of false positives. On the contrary, the model with commuting interactions maintain a flat rate of false positive below 20%. The accuracy of the model in this case remain higher than 50% after 7 elections.

Another question is what determines how long a county would be a stronghold for? We try to answer this question for the model by looking into the dependence of the strongholds duration with respect to the initial voteshare of the counties. As Figure 4 shows, there is a linear dependence for the duration of being a stronghold with the distance to the mean voteshare. Counties with initial larger deviation from the mean voteshare tend to be strongholds longer time. The fitting reveals that every tenth of relative voteshare corresponds on average to a stronghold duration of 5 elections.

4   Discussion

We have studied the persistence on the electoral system using the recently introduced Social Influence and Recurrent Mobility (SIRM) model for opinion

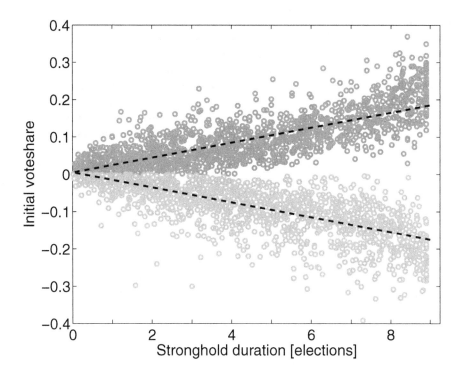

Fig. 4. Scatter plot of the initial voteshare versus the stronghold duration for model results. Red symbols shows republicans while blue symbols shows democrats. Dashed line correspond to the least square fitting revealing a symmetric linear behavior $y = mx + n$ with $m = \pm 0.02$ and $n = \pm 0.005$.

dynamics which includes social influence with random fluctuations, mobility and population heterogeneities across the U.S. The model accurately predicts generic features of the background fluctuations of evolution of vote-share fluctuations at different geographical scales (from the county to the state level), but it does not aim at reproducing the evolution of the average vote-share.

We have contrasted the evolution of the number of strongholds of the election data for the period 1980-2012 with the strongholds forecasted by simulations finding a good agreement between them. The evolution of the number of strongholds observed in the electoral data at early stages is well described by the model with commuting interactions. However, the number of strongholds observed in the data changes abruptly after the presidential election of 2008. Our model is not able/does not intend to reproduce this behavior since it involves external driving forces not included in the model.

Our results also show a good agreement between data and simulations for the location and duration of the strongholds. Strongholds are not randomly distributed across the country, but clustered. This indicates that some form of correlation is present between the voting patterns. The model reproduces nicely the spatial concentration of the both types of strongholds. Republican strongholds are mostly concentrated in the central-west, while democrat strongholds are dispersed mostly through the eastern parts, including urbanized areas.

As for the duration of the strongholds, we have found a linear dependence for the duration of being a stronghold with the distance to the mean voteshare. Counties with initial larger deviation from the mean voteshare tend to be strongholds for a longer time, on average, 5 elections for each tenth of relative voteshare. When commuting interactions are taken into consideration, our model exhibits an accuracy higher than 50% for up to 7 elections. The lack of this interactions causes an overestimation of the number of strongholds with an associated increase of the number of false positive decreasing the prediction accuracy of the model.

Our contribution sets the ground to include other important aspects of voting behavior and demand further investigation of the role played by heterogeneities in the micro-macro connection. Further elements will have to be included in order to produce predictions mimicking more accurately real electoral results. Some examples are the effects of social and communication media or the erosion of the governing party. The use of alternative communication channels is expected to affect voting behavior.

Acknowledgments. The authors acknowledge support from project MODASS (FIS2011-24785). TP acknowledges support from the program Juan de la Cierva of the Spanish Ministry of Economy and Competitiveness.

References

1. Riker, W.H., Ordeshook, P.C.: A theory of the calculus of voting. Am. Polit. Sci. Rev. **62**, 25–42 (1968)
2. Gelman, A., et al.: Estimating the probability of events that have never occurred: When is your vote decisive? Journal of the American Statistical Association **93**, 1–9 (2012)
3. Straits, B.C.: Public Opin. Q. **54**, 64–73 (1990)
4. Kenny, C.B.: Am. J. Pol. Sci. **36**, 259–267 (1992)
5. Allen, P.A., Dalton, R.J., Greene, S., Huckfeldt, R.: Am. Polit. Sci. Rev. **96**, 57–73 (2002)
6. Fowler, J.H.: Turnout in a small world in Social Logic of Politics, ed. A. Zuckerman (Temple University Press), pp 269–287 (2005)
7. Holley, R., Liggett, T.M.: Ergodic theorems for weakly interacting infinite systems and the voter model. Annals of Probability **3**, 643–663 (1975)
8. Vazquez, F., Eguíluz, V.M.: Analytical solution of the voter model on uncorrelated networks. New J. Phys. **10**, 063011 (2008)
9. Sood, V., Redner, S.: Voter model on heterogeneous graph. Phys. Rev. Lett. **94**, 178701 (2005)
10. Krapivsky, P.L., Redner, S., Ben-Naim, E.: A Kinetic View of Statistical Physics (Cambridge University Press) (2010)
11. Bond, R.M., Fariss, C.J., Jones, J.J., Kramer, A.D.I., Marlow, C., Settle, J.E., Fowler, J.H.: Nature **489**, 295–298 (2012)

12. Zuckerman, A.S.: The Social Logic Of Politics: Personal Networks As Contexts For Political Behavior (Temple University Press) (2005)
13. Lorenz, J., Rauhut, H., Schweitzer, F., Helbing, D.: Proc. Natl. Acad. Sci. USA **108**, 9020–9025 (2011)
14. Rogers, E.M.: Diffusion of Innovations. Free Press, New York (1995)
15. Castellano, C., Fortunato, S., Loreto, V.: Rev. Mod. Phys. **81**, 591–646 (2009)
16. Christakis, N.A., Fowler, J.H.: N. Engl. J. Med. **357**, 370–379 (2007)
17. Hill, A.L., Rand, D.G., Nowak, M.A., Christakis, N.A.: Proc. Biol. Sci. **277**, 3827–3835 (2010)
18. Rendell, L., Boyd, R., Cownden, D., Enquist, M., Eriksson, K., Feldman, M.W., Fogarty, L., Ghirlanda, S., Lillicrap, T., Laland, K.N.: Science **328**, 208–213 (2010)
19. Centola, D.: Science **329**, 1194–1197 (2010)
20. Brockmann, D., Hufnagel, L., Geisel, T.: The scaling laws of human travel. Nature **439**, 462–465 (2006)
21. González, M.C., Hidalgo, C., Barabási, A.-L.: Nature **453**, 779–782 (2008)
22. Song, C., Qu, Z., Blumm, N., Barabási, A.-L.: Science **327**, 1018–1021 (2010)
23. Balcan, D., Colizza, V., Gonçalves, B., Hu, H., Ramasco, J.J., Vespignani, A.: Proc. Natl. Acad. Sci. USA **106**, 21484–21489 (2009)
24. Tizzoni, M., Bajardi, P., Poletto, C., Ramasco, J.J., Balcan, D., Gonçalves, B., Perra, N., Colizza, V., Vespignani, A.: BMC Medicine **10**, 165 (2012)
25. Sattenspiel, L., Dietz, K.: A structured epidemic model incorporating geographic mobility among regions. Math Biosci **128**, 71–91 (1995)
26. Fernández-Gracia, J., Suchecki, K., Ramasco, J.J., Miguel, M.S., Eguíluz, V.M.: Phys. Rev. Lett. **112**, 158701 (2014)

Efficient Learning of User Conformity on Review Score

Kazumi Saito[1], Kouzou Ohara[2]([⊠]), Masahiro Kimura[3],
and Hiroshi Motoda[4,5]

[1] School of Administration and Informatics, University of Shizuoka, Shizuoka, Japan
k-saito@u-shizuoka-ken.ac.jp
[2] Department of Integrated Information Technology, Aoyama Gakuin University,
Kanagawa, Japan
ohara@it.aoyama.ac.jp
[3] Department of Electronics and Informatics, Ryukoku University, Shiga, Japan
kimura@rins.ryukoku.ac.jp
[4] Institute of Scientific and Industrial Research, Osaka University, Osaka, Japan
motoda@ar.sanken.osaka-u.ac.jp
[5] School of Computing and Information Systems, University of Tasmania,
Hobart, Australia

Abstract. We propose a simple and efficient method that learns and assesses the conformity of each user of an online review system from the observed review score record. The model we use is a modified Voter model that takes account of the conformity of each user. Conformity is learnable quite efficiently with a few tens of iterations by maximizing the log-likelihood given the observed data. The proposed method was evaluated and confirmed effective by two review datasets. It could identify both high and low conformity users. Users with high conformity are not necessarily early adopters. Their scores are influential to drive the consensus score. The user ranking of conformity was compared with Page Rank and HITS in which user network was roughly approximated by the directed graph induced by the observed data. The proposed method gives more interpretable ranking, and the global property of high conformity users was identified.

Keywords: Social media · Conformity · Review score · Learning

1 Introduction

The emergence of Social Media has provided us with the opportunity to collect a large number of user reviews for various items, e.g., products and movies. People now read these reviews and make decisions about their actions, e.g. buy the product, see the movie. It is thus important to be able to assess the review influence of each user who posted a review. As such influence, we focus on the conformity defined as a type of social influence involving a change in belief or behavior in order to fit in with a group. Analyzing review in depth needs natural

© Springer International Publishing Switzerland 2015
N. Agarwal et al. (Eds.): SBP 2015, LNCS 9021, pp. 182–192, 2015.
DOI: 10.1007/978-3-319-16268-3_19

language processing. Fortunately many social media sites offer review scores, i.e. rating score which is a numeric value, for individual items as well as the user reviews. We use this review score instead of review itself.

A score is defined as a rating given by a user[1] and their values vary across users and items, say, tens of thousands of users and items. We want to identify high-quality items in a given category in an efficient way from these review scores. Naive ranking would be simply to rank items according to the number of reviews or the average review score. There the emphasis is more on the statistical reliability and not on the quality and reputation of each review. They do not account for the review influence of each user that rated a specific item. Several researchers [3,6–8] incorporated the information of trust relationships with trust strengths (i.e., the local trust-value information) into low-rank matrix factorization techniques [5,10], and improved the performance in rating prediction. Tang et al. [12] proposed a method of incorporating the information of trustworthiness of users (i.e., the global trust-value information) into this framework, and further improved the performance in rating prediction. They measured the trustworthiness of a user by applying the PageRank algorithm [1] to the trust network.

However, these existing techniques assume a separate source of information, often represented by a trust network, which is not available except for a relatively limited number of review sites. In addition, it is quite difficult to precisely obtain the trust network due to its intrinsic time varying nature. In contrast, we can easily observe the phenomena related to conformity (or herding) of people [9], which gives an influence on other users and the corresponding ratings, and it is possible to estimate the conformity metric of each user based on time series data of review scores for items given by users, that is, the data consisting of 4-tuple (u, i, t, s) which means user u rated item i at time t and gave score s. The conformity metric proposed in this paper is so designed that users with high conformity tend to lead users' rating behavior for those items they rate, and users with low conformity tend to give unusual ratings for them.

We propose a simple and efficient method that learns and assesses the conformity of each user from the observed record of review scores without a need of separate trust network. The model we use is a modified version of Voter model in which we introduce the conformity of each user as a new variable to learn under the assumption that each user rates an item only once and the score is approximated by a multinomial distribution. The Voter model is one of the simplest models of opinion formation and propagation [2,11]. It assumes that a user updates his/her opinion based on his/her own and his/her direct neighbors' opinions, i.e. following the majority similarly to conformity. We assume that a user rates an item only once and never updates the value, which is different from the basic assumption of Voter model, but we borrows the idea that a user read other users' reviews and makes her choice probabilistically.

The conformity is learned such that the users' review score distributions predicted by the generative modified Voter model best match the observed score

---

[1] A user means a reviewer in this paper.

distributions, i.e. by maximizing the log-likelihood. The learning uses iterative scheme and it is very efficient taking full advantage of the convexity of the auxiliary objective function. Data is divided into half and the former half is used for learning the parameters of the model, i.e. conformity metrics, and generalization capability of the learned model is evaluated by the likelihood estimated by the unseen latter half data to determine the optimal value of the regularization factor that is introduced to avoid overfitting. Once the regularization factor is fixed, all the data are used to relearn the model.

We tested the proposed method by applying it to two review systems, Cosmetics review dataset[2] and Anime review dataset[3]. The conformity metric distribution of users depends on the characteristics of dataset used. What the proposed method found is that in the cases of adequate regularization factors, the majority of the people have the average conformity metric and only a small fraction of people have high or low conformity metrics. Thus, the method can identify two interesting groups of people that is worth paying attention to, one with high conformity metrics and the other with low conformity metrics. Users with high conformity metrics are not necessarily early adopters. Their scores are influential to drive the expected consensus score, i.e. lead global behaviors. Users with low conformity metrics deviates from the average behavior. High conformity user has the following properties: 1) she rates many items, 2) there are many followers of her who rate the same items that she rated and 3) the scores of the followers are similar to her scores. These properties are quite natural and the proposed method confirmed them. From an analogy that the trustworthiness of a user is estimated by applying the PageRank algorithm to the trust network, we have compared the ranking results of PageRank and HITS algorithms if they have the same properties. To run these algorithms, we constructed the user network by creating a directed network for each item based on the time stamp information of the observed data. The results of Page Rank and HITS are not as clear as the proposed method.

2   Model

We denote the sets of users and items by $V = \{u, v, w, \cdots\}$ and $I = \{i, \cdots\}$, respectively. When a user $v \in V$ reviewed an item $i \in I$, we denote its timestamp and score by $t_{v,i}$ and $s_{v,i}$, respectively, where each score $s_{v,i}$ is denoted by a positive integer in $S = \{1, \cdots, |S|\}$, and $|S|$ stands for the number of elements in S. Then, we can express our observed data set as $D = \{\cdots, (v, i, t_{v,i}, s_{v,i}), \cdots\}$. Hereafter, let $V(i) = \{v \mid (v, i, t_{v,i}, s_{v,i}) \in D\}$ be a set of users who reviewed an item i. For users in $V(i)$, let $U(i, t) = \{u \in V(i) \mid t_{u,i} < t\}$ be the set of users whose review times are before t, and $U(i, t, s) = \{u \in U(i, t) \mid s_{u,i} = s\}$ the set of those users whose review score is s.

As mentioned earlier, users may decide their review scores of each item by taking account not only of their own evaluations, but also of past majority scores

---

or those submitted by high conformity users. In order to stochastically cope with the opinion decision problem affected by majority scores, we can employ the basic voter model, and define the probability that a user v gives a score s to an item i at time t as $P_0(s \mid i,t) = (1 + |U(i,t,s)|)/(|S| + |U(i,t)|)$, where we employed a Bayesian prior known as the Laplace smoothing. Here we note that the Laplace smoothing corresponds to the assumption that each node initially holds one of the $|S|$ scores with equal probability. Note also that the Laplace smoothing corresponds to a special case of Dirichlet distributions that are very often used as prior distributions in Bayesian statistics.

Thus far, we assumed that all the past user scores are equally weighted. However, it is naturally conceivable that some high conformity users should have larger weights. In order to reflect this kind of effects into the model, we consider introducing a positive conformity metric $\exp(\theta_u)$ to each user u, where θ_u is a parameter. Hereafter, we denote the vector consisting of these parameters by $\boldsymbol{\theta} = (\cdots, \theta_u, \cdots)$. Then, we can extend the basic Voter model $P_0(s \mid i,t)$ and build a generative model in which user v gives score s for item i at time t with the following probability.

$$P(s \mid i,t; \boldsymbol{\theta}) = \frac{1 + \sum_{u \in U(i,t,s)} \exp(\theta_u)}{|S| + \sum_{u \in U(i,t)} \exp(\theta_u)}. \tag{1}$$

In this paper, in order to estimate $\boldsymbol{\theta}$ from the observed data set D, we consider maximizing the following logarithmic likelihood function based on Eq. (1):

$$L(D; \boldsymbol{\theta}) = \sum_{i \in I} \sum_{v \in V(i)} \log P(s_{v,i} \mid i, t_{v,i}; \boldsymbol{\theta}). \tag{2}$$

Here each review score s is replaced by the observed pair of $t_{v,i}$ and $s_{v,i}$ in Eq. (2).

3   Learning Algorithm

The number of users, which corresponds to the dimensionality of $\boldsymbol{\theta}$, easily becomes quite large, say tens of thousands. Thus, in order to avoid an overfitting problem, we consider minimizing the objective function with a standard regularization term defined by $J(\boldsymbol{\theta}) = -L(D ; \boldsymbol{\theta}) + \frac{\eta}{2}\|\boldsymbol{\theta}\|^2$, where $\|\boldsymbol{\theta}\|^2 = \sum_{u \in V} \theta_u^2$, and η stands for a regularization factor. In order to minimize $J(\boldsymbol{\theta})$ with respect to $\boldsymbol{\theta}$, we propose a learning algorithm based on the gradient descent method equipped with the second-order optimal step-length calculation for an auxiliary objective function of $J(\boldsymbol{\theta})$.

Now, let $I(v) = \{i \mid (v, i, t_{v,i}, s_{v,i}) \in D\}$ be a set of items which were reviewed by a user v. For users $V(i)$, let $W(i,t) = \{w \in V(i) \mid t_{w,i} > t\}$ be the set of users whose review time is after t and $W(i,t,s) = \{w \in W(i,t) \mid s_{w,i} = s\}$ be the set of those users whose review score is s. Moreover, we define two terms, $q_{v,i,t}(\boldsymbol{\theta}) = \exp(\theta_v)/(|S| + \sum_{u \in U(i,t)} \exp(\theta_u))$ and $q_{v,i,t,s}(\boldsymbol{\theta}) = \exp(\theta_v)/(1 + \sum_{u \in U(i,t,s)} \exp(\theta_u))$, just like posterior probabilities typically used in the EM algorithm. Then, by setting the search direction $\boldsymbol{\delta} = (\cdots, \delta_v, \cdots)$ to the negative gradient, i.e., $\delta_v = -\partial J(\boldsymbol{\theta})/\partial\theta_v$, we can calculate δ_v as follows:

$$\delta_v = \sum_{i \in I(v)} \sum_{w \in W(i, t_{v,i}, s_{v,i})} q_{v,i,t_{w,i},s_{v,i}}(\boldsymbol{\theta}) - \sum_{i \in I(v)} \sum_{w \in W(i, t_{v,i})} q_{v,i,t_{w,i}}(\boldsymbol{\theta}) + \eta\theta_v. \quad (3)$$

From the first term of the right-hand-side in Eq. (3), we can easily see that θ_v of user v tends to increase if $|W(i, t_{v,i}, s_{v,i})|$ is relatively large for many items i.

By using two terms defined by $N(v, i;\ \boldsymbol{\theta}) = \log(1 + \sum_{u \in U(i, t_{v,i}, s_{v,i})} \exp(\theta_u))$ and $D(v, i;\ \boldsymbol{\theta}) = \log(|S| + \sum_{u \in U(i, t_{v,i})} \exp(\theta_u))$, we can rewrite our objective $J(\boldsymbol{\theta})$ as follows:

$$J(\boldsymbol{\theta}) = -\sum_{i \in I} \sum_{v \in V(i)} N(v, i;\ \boldsymbol{\theta}) + \sum_{i \in I} \sum_{v \in V(i)} D(v, i;\ \boldsymbol{\theta}) + \frac{\eta}{2}\sum_{v \in V}\theta_v^2.$$

By using a term defined by $M(v, i;\ \boldsymbol{\theta} \mid \bar{\boldsymbol{\theta}}) = \sum_{u \in U(i, t_{v,i}, s_{v,i})} \theta_u\ q_{u,i,t_{v,i},s_{v,i}}(\bar{\boldsymbol{\theta}})$, we consider the following auxiliary objective function $Q(\boldsymbol{\theta} \mid \bar{\boldsymbol{\theta}})$ to describe our method for stably calculating the step-length λ with respect to $\boldsymbol{\delta}$.

$$Q(\boldsymbol{\theta} \mid \bar{\boldsymbol{\theta}}) = -\sum_{i \in I} \sum_{v \in V(i)} M(v, i;\ \boldsymbol{\theta} \mid \bar{\boldsymbol{\theta}}) + \sum_{i \in I} \sum_{v \in V(i)} D(v, i;\ \boldsymbol{\theta}) + \frac{\eta}{2}\sum_{v \in V}\theta_v^2.$$

where $\bar{\boldsymbol{\theta}}$ stands for the current estimate of $\boldsymbol{\theta}$. Here, by introducing a virtual node $z \notin V$ whose parameter value is fixed at zero, i.e., $\theta_z = 0$, an augmented set defined by $UA(i, s, t) = U(i, s, t) \cup \{z\}$ for each pair of item i and score s, and a cross-entropy term, $H(v, i;\ \boldsymbol{\theta} \mid \bar{\boldsymbol{\theta}}) = -\sum_{u \in UA(i, t_{v,i}, s_{v,i})} q_{v,i,t_{u,i},s_{u,i}}(\bar{\boldsymbol{\theta}})$ $\log q_{v,i,t_{u,i},s_{u,i}}(\boldsymbol{\theta})$, we can obtain $N(v, i;\ \boldsymbol{\theta}) = M(v, i;\ \boldsymbol{\theta} \mid \bar{\boldsymbol{\theta}}) + H(v, i;\ \boldsymbol{\theta} \mid \bar{\boldsymbol{\theta}})$. Thus, we can see that the objective function $J(\boldsymbol{\theta})$ is expressed as $J(\boldsymbol{\theta}) = Q(\boldsymbol{\theta} \mid \bar{\boldsymbol{\theta}}) - \sum_{i \in I}\sum_{v \in V(i)} H(v, i;\ \boldsymbol{\theta} \mid \bar{\boldsymbol{\theta}})$. Therefore, since $H(v, i;\ \boldsymbol{\theta} \mid \bar{\boldsymbol{\theta}})$ is minimized at $\boldsymbol{\theta} = \bar{\boldsymbol{\theta}}$, the objective function $J(\boldsymbol{\theta})$ is optimized by minimizing the auxiliary objective function $Q(\boldsymbol{\theta} \mid \bar{\boldsymbol{\theta}})$ with respect to $\boldsymbol{\theta}$.

By considering a univariate objective function defined as $F(\lambda) = Q(\bar{\boldsymbol{\theta}} + \lambda\boldsymbol{\delta} \mid \bar{\boldsymbol{\theta}})$, we can calculate the second-order optimal step-length λ as $\lambda = -F'(0)/F''(0)$, where $F'(0)$ and $F''(0)$ mean the first- and second-order derivatives of $F(\lambda)$ with respect to λ at $\lambda = 0$. Note that $F'(0) = -\|\boldsymbol{\delta}\|^2$, and we can efficiently calculate $F''(0)$ as follows:

$$F''(0) = \sum_{i \in I} \sum_{v \in V(i)} \sum_{u \in U(i, t_{v,i})} \delta_u^2\ q_{u,i,t_{v,i}}(\bar{\boldsymbol{\theta}}) - \sum_{i \in I} \sum_{v \in V(i)} \left(\sum_{u \in U(i, t_{v,i})} \delta_u\ q_{u,i,t_{v,i}}(\bar{\boldsymbol{\theta}})\right)^2 + \eta\|\boldsymbol{\delta}\|^2.$$

We can easily see that $F''(0) > 0$, $\lambda > 0$ and $F(\lambda)$ has a unique global optimal solution due to $F''(0) > 0$. Namely, we can stably calculate the step-length without using a technique like safeguarded Newton's method. Note that straightforward univariate objective function defined by $G(\lambda) = J(\bar{\boldsymbol{\theta}} + \lambda\boldsymbol{\delta})$ does not hold this kind of nice properties.

4   Experiments

We collected review score records from two famous review sites in Japan and constructed two datasets for this experiment. One consists of review scores for cosmetics extracted from "@cosme" which is a Japanese word-of-mouth communication site for cosmetics. We refer to this dataset as the Cosmetics review dataset. The other one is composed of review records collected from "anikore", a ranking and review site for anime, which is referred to as the Anime review dataset. In both the datasets, each record has 4-tuple (u, i, s, t) as mentioned above, which means user u gives a score s to item i at time t. The Cosmetics review dataset has $297,453$ review records by $10,403$ users for $46,398$ items from $2008/12/07$ to $2009/12/09$, while the Anime review dataset has $300,327$ records by $13,112$ users for $1,790$ items from $2010/8/01$ to $2012/8/08$. Thus, the average numbers of reviews per user and item were 28.6 and 6.4 in the Cosmetics dataset, and 22.9 and 167.8 in the Anime dataset. The score is an integer value ranging from 1 to 7 and its average of overall ratings was 4.4 in the Cosmetics dataset, and from 1 to 5 and 3.9 in the Anime dataset.

4.1   Learning Results

First of all, we evaluated how the regularization factor η affects the learning performance of the proposed method. To this end, we divided each dataset in half and learned the conformity metric of each user from the former half (training data) varying the value of η, and then evaluated the generalization capability of the learned model by using the latter half (test data) in terms of perplexity defined as $exp(-L(D_{test}; \hat{\theta}))$ where D_{test} means the test data and $\hat{\theta}$ a parameter vector learned from the training data. Table 1 shows the resulting values of the perplexity for different regularization factor η. Here, note that the smaller the value of perplexity is, the better a learned model is in terms of the generalization capability. From this result, we see that the resulting perplexity becomes smaller as η becomes larger, and levels off for η greater than a certain threshold, say 10 in these datasets. This means that the proposed method is robust for a wide rage of η and can achieve good generalization capability unless η is too small.

Next, we investigated how the value of η affects the efficiency of the proposed method by comparing the number of iterations spent by the proposed method until the norm of the gradient vector $||\boldsymbol{\delta}||$ converges to nearly 0 for different values of η. The results are shown in Figure 1, from which we can say that the conformity metrics can be learned quite efficiently if η is set to a value not too small. In this experiment, for η that is equal to or greater than 10, around 10 iterations and a few tens of iterations are enough for the Cosme and Anime review datasets, respectively.

We further investigated the distribution of the resulting conformity metrics and plotted them for each value of η in Figure 2. It is clear that the trends observed from these figures are quite similar for both the datasets. Namely, with adequate values of η, the majority of the users have the average value, i.e., 1.0

Table 1. Fluctuation of perplexity for the test data as a function of η

η	0.1	1.0	10	20	30
Cosme	6.2073	6.0730	6.0404	6.0403	6.0408
Anime	2.8118	2.7685	2.7669	2.7678	2.7685

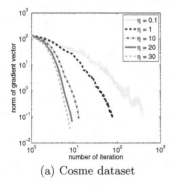

(a) Cosme dataset

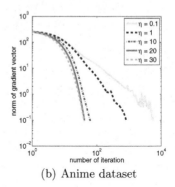

(b) Anime dataset

Fig. 1. Number of iterations spent by the proposed algorithm for different values of η

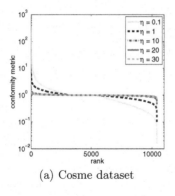

(a) Cosme dataset

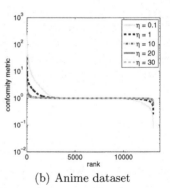

(b) Anime dataset

Fig. 2. Distributions of the resulting conformity metrics

and only a small fraction of them have high or low conformity metric[4]. These characteristic users are worthy of attention. It is expected that the user with high conformity metric provides a consensus score early on[5]. Thus, such scores are worth considering. On the other hand, those who have low conformity metrics tend to rate an item in a different way from the majorities. Thus, their scores may be useful for a particular user with similar interests. Further, knowing these low scores help preventing other users from being confused by such unusual ratings.

---

[4] We did not place a constraint that the average of the conformity metric is 1.0, but the results are indeed very close to 1.0 for each η.

[5] Later, we see that this is not necessarily true.

Moreover, we should emphasize that in the case of the Cosmetics dataset in which each user page can have fan (or reader) links, the numbers of links were generally large for users with high conformity metrics, but small for those with low metrics.

4.2   Evaluation of Conformity Metrics

In this section, we investigate how a user with high conformity metric can be characterized by other naive measurements. For this purpose, we focus on the following three properties which are naturally considered as the necessary conditions for a user to be high conformity metric: 1) she rates many items compared to other users (number of items), 2) she has many followers who rate the same items that she rated (number of successors), and 3) the scores of the followers are similar to those she rated (rating similarity). We first ranked users in each dataset according to i) these three naive measurements and ii) the conformity metrics returned by our proposed method, and investigated how top-K ranked users for each metric perform for other metrics, e.g., Do the top-K ranked users of "rating similarity" perform good or bad for "number of successors" and "number of items"? The number of successors of user v is given by $(1/|I(v)|)\sum_{i \in I(v)}|W(i,t_{v,i})|$, while the rating similarity between the score of user v and ones given by its followers is defined as follows:

$$1 - \frac{1}{|I(v)|} \sum_{i \in I(v)} \frac{\left|s_{v,i} - \frac{1}{|W(i,t_{v,i})|}\sum_{u \in W(i,t_{v,i})} s_{u,i}\right|}{|S|-1} \tag{4}$$

Hereafter, we refer to each ranking method as "#Items", "#Successors", "Rating-sim", and "Proposed", respectively according to the measurement metric used.

We, second, compared the results by the above ranking methods with those obtained by PageRank [1] and HITS [4]. These two algorithms rank nodes in a network. Tang J. et.al. (2013) [12] used a trust network and ranked nodes by PageRank and used this information to place a weight on each node to reflect conformity metric of each node. The trust network is a different source of information, i.e., social relations, and no such information is available to us for the two review systems we are using. Thus, we approximately induced a trust network from the observed score records by considering users as nodes and linking user u and v with a directed link (u,v) (link from u to v) if user v has rated an item i before user u rates it. This linking method shares, with our proposed method, the idea that a user rates an item considering scores already given to the same item by others. We set the teleportation probability of the PageRank algorithm to a typical value, i.e., 0.15, and used the authority score to rank nodes for the HITS algorithm. We call these ranking methods as "PageRank" and "HITS" in what follows.

The results for the Cosme and Anime review datasets are shown in Figures 3 and 4, respectively. Each line depicts the average score of the mentioned metric (see the title) for the corresponding ranking method in the caption box (one of

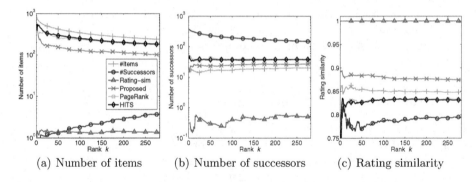

(a) Number of items   (b) Number of successors   (c) Rating similarity

Fig. 3. Comparison of the top 282 users derived from the Cosme review dataset by using the six methods in the three measures

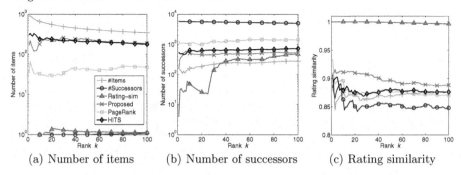

(a) Number of items   (b) Number of successors   (c) Rating similarity

Fig. 4. Comparison of the top 100 users derived from the Anime review dataset by using the six methods in the three measures

#Items, #Successors, Rating-sim, Proposed, PageRank, and HITS) over the top k users. Thus, the mentioned method always gives the best result in Figures 3 and 4. In this experiment, we adopted $K = 100$. But, for the Cosme review dataset, we actually considered the top 282 users because rating similarity score is the same for the top 282 users, which is 1.0. We set η to 20 for the Cosme dataset, and 10 for the Anime dataset, respectively in the proposed method because they achieved the best (lowest) perplexity in the previous experiment.

In Figures 3(a) and 4(a), the number of items by Proposed is much larger than those by #Successors and Rating-sim. It is noted that, in an extreme case, even if user v rated only one item, v can achieve a high score in the number of successors if many users rate the item after she did. Similarly, her rating similarity can be large if her followers give the same score as she did even if the number of items she rated is only one. This is the reason why the resulting values by #Successors and Rating-sim tend to be quite small in these figures. This implies that users who make #Items, #Successors and Rating-sim large separately are different from each other.

In Figures 3(b) and 4(b), the results by Proposed achieve better scores than those by #Items and Rating-sim. Likewise, Proposed outperforms #Items and #Successors in terms of the rating similarity as in Figure 3(c) and 4(c). From these results, we can say that the conformity metrics learned by Proposed can be a good indicator to identify those who satisfy the above three necessary conditions simultaneously. In addition, from Figures 3(b) and 4(b), it is found that the number of successors of the users having the highest conformity metric is not necessarily large. This means a user with high conformity metric is not necessarily an early adopter.

On the other hand, it is suggested from Figures 3(a) and 4(a) that both PageRank and HITS can be better indicators than #Successors and Rating-sim to identify those users who rate many items. This is because these two ranking methods tend to give a higher score to a node that has a larger in-degree. Similarly, they outperform #Items and Rating-sim in terms of the number of successors in Figures 3(b) and 4(b). However, in Figures 3(c) and 4(c), their scores are comparable to or lower than those by #Items. This means that the scores of PageRank and HITS are no better indicators than the conformity metrics of Proposed to identify users who satisfy the three basic properties altogether.

5   Conclusion

In this paper, we addressed the problem of quantitatively assessing the conformity of a user in the context of rating items, and proposed an efficient algorithm that learns the conformity metric of each user from observed review scores. The idea behind is that a user often rates an item taking into account not only her own opinion but also scores already given to the item by other users, and the reliability of scores depends on who rated them. We modeled this rating process as a stochastic decision making process and used a modified Voter model. The proposed method can efficiently learn the conformity metrics based on an iterative algorithm within a few tens of iterations. Its generalization capability is insensitive to the value of the regularization factor. Empirical evaluation on the two real world review datasets uncovered some interesting findings about the conformity metrics learned by the proposed algorithm. The majority of people have an average conformity metric with adequate regularization factors, i.e., 1.0 and only a limited fraction of people have high or low conformity metrics, who are worth paying attention to. Conformity metric can be a good indicator to identify those who satisfy the following three basic properties simultaneously that are considered natural for a user to be of high conformity, i.e., 1) a multitude of rated items, 2) a multitude of followers, and 3) a high rating similarity between her own scores and her follower's. None of them can be a good indicator alone. We further found that users having a high PageRank score or a high HITS score tend to rate a large number of items and have a large number of followers, satisfying the above two properties, but their rating similarity is not as large as that of those who have high conformity metrics or those who rate a large number of items.

Acknowledgments. This work was partly supported by Asian Office of Aerospace Research and Development, Air Force Office of Scientific Research under Grant No. AOARD-13-4042, and JSPS Grant-in-Aid for Scientific Research (C) (No. 26330352).

References

1. Brin, S., Page, L.: The anatomy of a large-scale hypertextual web search engine. Computer Networks and ISDN Systems **30**, 107–117 (1998)
2. Even-Dar, E., Shapira, A.: A note on maximizing the spread of influence in social networks. In: Deng, X., Graham, F.C. (eds.) WINE 2007. LNCS, vol. 4858, pp. 281–286. Springer, Heidelberg (2007)
3. Jamali, M., Ester, A.: A matrix factorization technique with trust propagation for recommendation in social networks. In: Proceedings of the 4th ACM Conference on Recommender Systems (RecSys 2010), pp. 135–142 (2010)
4. Kleinberg, J.: Authoritative sources in a hyperlinked environment. Journal of the ACM **46**(5), 604–632 (1999)
5. Koren, Y.: Factorization meets the neighborhood: a mulifaceted collaborative filtering model. In: Proceedings of the 14th ACM SIGKDD International Conference on Knowledge Discovery and Data Mining (KDD 2008), pp. 426–434 (2008)
6. Ma, H., King, I., Lyu, M.R.: Learning to recommend with social trust ensemble. In: Proceedings of the 32nd International ACM SIGIR Conference on Research and Development (SIGIR 2009), pp. 203–210 (2009)
7. Ma, H., Yang, H., Lyu, M.R., King, I.: Sorec: social recommendation using probabilistic matrix factorization. In: Proceedings of the 17th ACM Conference on Information and Knowledge Management (CIKM 2008), pp. 931–940 (2008)
8. Ma, H., Zhou, D., Liu, C., Lyu, M.R., King, I.: Recommender systems with social regularization. In: Proceedings of the 4th ACM International Conference on Web Search and Data Mining (WSDM 2011), pp. 287–296 (2011)
9. Michael, L., Otterbacher, J.: Write like I write: herding in the language of online reviews. In: Proceedings of the Eighth International AAAI Conference on Weblogs and Social Media (ICWSM 2014), pp. 356–365 (2014)
10. Salakhutdinov, R., Mnih, A.: Probabilistic matrix factorization. In: Advances in Neural Information Processing Systems 20 (NIPS 2007), pp. 791–798 (2008)
11. Sood, V., Redner, S.: Voter model on heterogeneous graphs. Physical Review Letters **94**, 178701 (2005)
12. Tang, J., Hu, X., Gao, H., Liu, H.: Exploiting local and global social context for recommendation. In: Proceedings of the 23rd International Joint Conference on Artificial Intelligence (IJCAI 2013), pp. 2712–2718 (2013)

Are Tweets Biased by Audience? An Analysis from the View of Topic Diversity

Sandra Servia-Rodríguez[(✉)], Rebeca P. Díaz-Redondo, and
Ana Fernández-Vilas

I&C Lab, AtlantTIC Research Center, University of Vigo, Vigo, Spain
{sandra,rebeca,avilas}@det.uvigo.es

Abstract. The emergence of blogs, and especially microblogs, has granted users the possibility of publishing and sharing ideas, news, opinions and any other kind of content with their audience. But this has also brought them the arduous tasks of self-censorship and adaptation of the content to an audience previously envisioned in order to keep, and even increase, their social influence. Taking into account the impossibility of knowing this imagined audience and using Twitter as a case study, we analyse if the diversity of topics chosen by users in their tweets is biased by the size of their audience. Considering the number of followers as the users' audience and applying a methodology based on clustering the representative terms in tweets, we found that individuals with large audiences tend to deal with topics more diverse than those with small audiences. Understanding how audience size affects the range of topics chosen by a speaker have theoretical implications for sociological studies and even for the effective design of marketing campaigns.

Keywords: Topic diversity · Twitter · Users' behaviour · Audience

1 Introduction

The sociological theories of *self-presentation* and *impression management* [4] state that individuals attempt to influence the perception that others have of them by adapting their behaviour to their audience. As an example, the way in which we behave in presence of our boss is far from the one in presence of our relatives or close friends. In these face-to-face situations, the knowledge of our actual audience makes us adapt our behaviour, but in other situations as when we prepare a presentation or write a paper or book, our actual audience is unknown

Work funded by the Spanish Ministry of Economy and Competitiveness (EEBB-I-13-06425 and TEC2013-47665-C4-3-R); the European Regional Development Fund and the Galician Regional Government under agreement for funding the AtlantTIC Research Center; and the Spanish Government and the European Regional Development Fund under project TACTICA.

© Springer International Publishing Switzerland 2015
N. Agarwal et al. (Eds.): SBP 2015, LNCS 9021, pp. 193–202, 2015.
DOI: 10.1007/978-3-319-16268-3_20

and must be envisioned. This *imagined audience* [10], which is nothing more than a mental conceptualisation of the people with whom we are communicating, is conceived to help the speaker or the writer to compose the information.

The emergency of Social Web technologies have provided individuals with powerful tools to freely disseminate factual information, opinions, and, ultimately, any sort of content that they wish to share with others. These technologies have awarded people with the possibility of reaching many more individuals than with traditional face-to-face interactions, increasing considerably their potential audience and making even impossible to determine its real size. In this scenario, the *imagined audience* plays a key role in deciding what content (ideas, news, opinions,...) is going to be published. So, factors as self-censorship and adaptation of content to an audience previously envisioned [9] influence the writing and posting of messages (or any other kind of content) in this new conception of the Web. As an outstanding example of social technology, microblogging services as Twitter allow users to post short textual messages (tweets) in any time and from everywhere in an ubiquitous way. Since the vast majority of Twitter accounts are public and consequently the majority of tweets can be viewed by anyone, the potential audience of a tweet is unlimited. However, its actual audience is composed by only some of the *followers* of the publisher (followed user), being even different from the audience that he envisions. Moreover, Twitter users do not select their audience, but their audience is who selects them in some discovery process related with the content and topics that they address.

Putting aside the dynamic patterns and intrinsic opinions behind tweets, the subjects chosen by users in their tweets, both in the specific topics and, especially, in their diversity, are good indicators of their behaviour. Using Twitter as testbed, *we study if the size of the potential audience affects users' behaviour, and more specifically, the range of topics chosen by Twitter users.* That is, if users deal with few topics when their potential audience is small or, on the contrary, when they have a large audience, and vice versa.

A well-known algorithm for extracting topics from text is Latent Dirichlet Allocation (*LDA*) [1]. Although LDA-based solutions have been proposed to the problem of extracting topics from tweets [3,11], the short length of tweets, together with their frequent lack of correct textual structuring, led researchers to opt for alternatives based on clustering the representative terms in users' tweets according to their semantic relatedness [12,14,15]. This also allows a more accurate description of the topics chosen by users and a better appreciation of their diversity. Applying the methodology in [14], we conducted an experiment over a Twitter dataset [8], consisting of clustering the representative words used in users' tweets and comparing the number, size and distance of the resulting clusters with the size of their potential audience. For this study, the potential audience of a user u is made of his followers, i.e. those Twitter users that have established a unidirectional link with him, since u's tweets will be shown into his followers' respective Twitter homepages. Although the rest of Twitter users can also read u's tweets by accessing his profile page or by the Twitter searching tool, we assume that, if user v often visits user u's profile or reads his tweets, v will become u's follower before or after.

It could seem that users with large audiences (celebrities) tended to focus their tweets on those topics that have made them famous (sportsmen about their own sport, politicians about their own party, actors about their movies, etc.), minimising their content diversity. However, our analysis revealed that the diversity of topics of individuals with large potential audiences is, in general, higher than the one of users with a reduced number of followers ("ordinary" users). This suggests that large audiences are obtained by collating several smaller audiences and providing enough content to satisfy each of them. Next section explains the methodology to discover users' topics of interest from their tweets, whereas the description of the dataset together with the experiments conducted are provided in Section 3. Discussion and future work are outlined in Section 4.

2   Topics Extraction

The methodology, previously introduced and evaluated in [14], to analyse the relation between the size of the audience and the content diversity of users' tweets starts with the application of different NLP and data mining techniques to extract the topics treated by the user in his tweets. As a result, each topic is represented by means of a cluster of representative words for the given topic.

The mechanism of topics extraction begins obtaining relevant words from users' tweets by only considering lexical units that refer to fixed entities with meaning. We used *Stanford CoreNLP*[1], a tool that addresses all the basic levels of NLP. Then, the text is filtered by using POS tagging (which identifies each word part-of-speech category -Noun, Verb, etc.-) and lemmatisation (which identifies each word lemma), only keeping nouns in their citation form. The resulting words conform the set of input data points to the clustering algorithm. An explanation of the techniques applied is detailed below (see [14] for a complete description).

2.1   Semantic Relatedness Measure

Every clustering algorithm requires to know the similarity between data points which, in our case, is the semantic relatedness between the aforementioned relevant words. This *similarity is calculated as the weighted sum of two different measures*: the one based on an external source of background knowledge and the other on the personal knowledge of the user.

General Knowledge Based Measure: *Wikipedia Link-based Measure (WLM)* [16] is a semantic relatedness measure based on the hyperlink structure of Wikipedia. After identifying the Wikipedia articles that discuss the words of interests, the relatedness between the given articles is computed by means of two different measures: the one based on the links extending out of each article and the other on the links made to them. The first measure $sr_w(a, b)$ is defined by the angle between the vectors of the links found within the two articles of interest (a, b), where the weight w of the link $s \to t$ $(s \in \{a, b\})$ is:

---

[1] *http://nlp.stanford.edu/software/corenlp.shtml Last accessed on 12/11/14.*

$$w(s \rightarrow t) = log(\frac{|W|}{|T|}) \mid \text{if } s \in t, \text{ 0 otherwise} \tag{1}$$

where T is the set of articles that link to t and W the set of articles in Wikipedia. The latter is based on the *Normalized Google Distance* [2]:

$$sr_w(a, b) = \frac{\log(\max(|A|, |B|)) - \log(|A \cap B|)}{\log(|W|) - \log(\min(|A|, |B|))} \tag{2}$$

where A and B are the sets of articles that link to the articles of interest a and b respectively, and W is the entire Wikipedia. See [16] for a complete description.

Personal Knowledge Based Measure: Although a measure based on the general knowledge as *WLM* assesses with high accuracy the semantic relation between any two terms, it obviates the intentionality of the user when used the term. That is, the intrinsic relation that they acquire for being used together (in the same conversation, same tweet,...). With the aim of keeping the sense that the user gave to terms, our personal knowledge based measure $sr_u(a, b)$ states that two terms a and b are related for the user u if they appear together in, at least, one tweet t of u. Otherwise, there is no relation between them:

$$sr_u(a, b) = \begin{cases} 1 & \text{if } a, b \in t \\ 0 & \text{if } a, b \notin t \end{cases} \tag{3}$$

2.2   Clustering

Given the good results obtained with the hierarchical clustering in our closely related study [15], we opted again for the **Unweighted Pair Group Method with Arithmetic Mean (*UPGMA*)** [5] clustering algorithm. *UPGMA* is a hierarchical and agglomerative clustering algorithm that yields a dendrogram that can be cut at a chosen height to produce the desired number of clusters. The main steps of UPGMA algorithm are as follows:

1. Place each data point into its own singleton group.
2. Merge the two closest groups.
3. Update distances between the new cluster and each of the old clusters. Given a distance measure between points, UPGMA obtains the intergroup similarity between the clusters C and H as:

$$d(C, H) = \frac{1}{N_C N_H} \sum_{i \in C} \sum_{j \in H} d_{i,j} \tag{4}$$

where N_C (N_H) is the size of the cluster C (H) and $d_{i,j}$ is the distance between the data points i and j.
4. Repeat 2 and 3 until all the data are merged into a single cluster.

One of the key points of most of the clustering algorithms is the **selection of the number of resulting clusters**. In hierarchical clustering, this deals with identifying individual branches of the cluster tree: a process referred to as branch or tree cutting or dendrogram pruning. We used a tree cutting method that detects clusters in a dendrogram based on its shape: *Dynamic Tree Cut* [7]. This algorithm is based on an iterative process of cluster decomposition and combination that stops when the number of clusters becomes stable. After obtaining a few large clusters by the fixed height branch cut method, the joining heights of each cluster are analysed for a sub-cluster structure. Clusters with this sub-cluster structure are recursively split and, to avoid over-splitting, very small clusters are joined to their neighbouring major clusters. See [7] for a description.

3 Dataset and Results

Now, we detail the experiments to analyse the relation between users' topic diversity and their potential audience in Twitter and the results achieved. Specifically, we first obtained users' topics for latter comparing them with their followers.

3.1 Dataset

We used the Twitter dataset of Li et al. [8] obtained by crawling Twitter in May 2011. This dataset contains information about 139180 users including, for each one, at most 600 tweets and his social network (friends and followers). The distribution of potential audiences in this dataset, i.e. the number of users' followers, is similar to the one in Twitter in which up to 7 different orders of magnitude are present. Also, half of the users in the dataset have less than 10^3 followers and the 96% have less than 10^4, making that the huge amount of users be "ordinary" users and the presence of users with more than 10^6 followers (celebrities) be 34. We sampled this dataset to obtain a representative set of users in terms of number of followers. We define six different groups of users according to the order of magnitude of their audience: users with less than 100, users with more than 100 and less than 10^3 and so on until finally users with more than 10^6 followers, resulting in 2042 users in the first group, 76652 in the second, 54571 in the third, 4501 in the forth, 222 in the fifth and 34 in the group of users with more than 10^6 followers. Given that the minimum number of users in a group is 34, we randomly selected 34 users per group, ending with a total of 204 users. Figure 1a contains the CCDF of the number of followers/friends per user in the sampled dataset (note that, because of the sampling, the distribution of followers in our sample does not correspond with the one in the whole Twitter [6]). Figure 1b shows the distribution of the number of tweets per user in this final sample, where around 80% of the users have between 400 and 600 tweets.

3.2 Resulting Clusters (Topics)

We applied the topics extraction methodology explained in Section 2 over the tweets of each user in the sampled dataset. As a result, a set of representative clusters of tags emerged, each one representing one of his topics of interest. However, not all the clusters are representative of a topic, but we considered

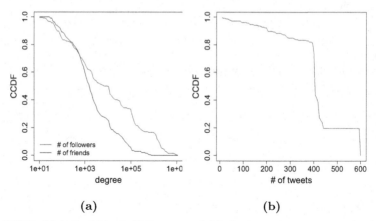

Fig. 1. CCDF of (a) # of friends/followers and (b) # of tweets per user in the dataset

that, for a cluster to be representative, its Silhouette width[2] [13] must be positive. That is, a cluster c represents a topic when the average dissimilarity (distance) from the point i (member of c) to all other points in c is lower than the lowest average dissimilarity (distance) from point i to all points in any other cluster different than c. The distribution of some parameters of the resulting clusters-topics are shown in Figure 2, whereas the average and standard deviation (std. dev.) of these parameters are provided in Table 1.

As viewed in Table 1, the average number of resulting clusters, 55.20, is higher than expected. However, the quality of many of these clusters is not good enough to be considered representative and, after keeping only clusters with positive Silhouette width, their number decreases drastically until 28.59. This methodology still produces a large number of topics per user which, together with the high standard deviation, lead us to clearly appreciate differences in the diversity of topics dealt by some users and others. With respect to the distribution of the number of representative clusters among users (Figure 2(a)), half of users have less than 30 clusters and users with more than 40 clusters are less than 15% of the total. In order to prove the significance of our findings, we calculated the Pearson correlation between the number of representative clusters per user and his number of tweets in the sample, obtaining that they are scarcely correlated (Pearson coefficient = 0.19).

Apart from the number of clusters (topics), the distance between clusters (both intra and inter cluster) is relevant to characterise users' topic diversity since the closer the clusters, the less diverse the topics are. At this respect, the average of the Silhouette width (in average for all the clusters of the user) is 0.148 with a standard deviation of 0.03. With respect to the distribution among users, the average Silhouette Width ranges from 0 (keep in mind that only clusters

---

[2] Sihouette width is a clustering validation measure which indicates the strength of a cluster or how well an element was clustered.

Table 1. Users' topics (clusters) parameters

	average	std. dev.
# of clusters	55.20	25.34
# of representative clusters	28.59	11.71
Silhouette width	0.148	0.03
# of tags per cluster (without r.)	7.07	1.37
# of tags per cluster (with r.)	22.21	27.91

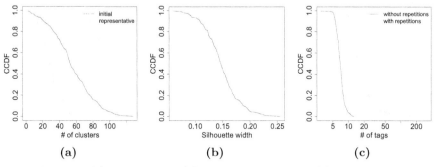

(a)   (b)   (c)

Fig. 2. CCDF of (a) # of clusters, (b) Silhouette width and (c) # of tags per cluster for all users in the dataset

with positive Silhouette width are considered) to 0.26. However, the majority of the users (around 80%) have an index between 0.08 and 0.18 (Figure 2(b)), which indicates that there are not huge differences between users according to the distance among their clusters.

The size of the clusters is also relevant when talking about diversity, since it indicates the relative importance of the topic for the user with respect to the rest of topics treated by him. Figure 2(c) shows that almost all users have clusters of, in average, less than 50 terms and the users with less tags per cluster have, in average, 7. But when no repetition of terms are taken into account (*without repetitions* in the figure), the differences among users with respect to the average of terms per cluster are drastically reduced, since almost all users have clusters with, in average, between 5 and 10 different terms. This is also observed in Table 1 and specifically in the difference between the standard deviations when taking into account tags repetition in clusters (*with r.*) and not (*without r.*).

3.3 Relating Content Diversity with Audience Size

We define the potential audience of a user as the set of Twitter users that follow him and his topics as the clusters of terms resulting from applying the methodology explained in Section 2. We calculated the Pearson correlation coefficient between the number of representative clusters (topics) and number of followers for all the users in our dataset. As the number of followers involves different orders of magnitude, we calculated this correlation between the number of clusters and the logarithm of the number of followers, obtaining a value of 0.218. This positive correlation means that users with many followers tend to have higher topic diversity than users with less followers. But, the increment of topic diversity is not fixed with the increment of the number of followers.

In order to put aside the influence of the accurate values of number of followers, we opted by an analysis per groups, grouping users' according to the order of magnitude of their number of followers (starting in 100 instead of 10). Figure 3 (a) shows the boxplots of the number of topics dealt by the users in each group. In view of the results, users with more than 10^6 followers (celebrities) clearly have a higher diversity of topics than the rest of users in the dataset (34.06 on average, versus the 31.44 in the case of users with between 10^4 and 10^5 followers or even less than 30 topics of users with less than 10^4 followers).

Although grouping users according to the order of magnitude of their number of followers seems a suitable classification, from Twitter's view classifying users into celebrities and "ordinary" users makes more sense. As the limit for being considered a celebrity according to followers is not clear, we did a new classification of users into three different groups: users with less than 10^4, users with between 10^4 and 10^6 and users with more than 10^6 followers. Results in Figure 3 (b) show that, in average, the number of clusters (topic diversity) is different for the different groups, being the lowest in the case of users with less than 10^4 followers (26.08) and the highest in the case of users with more than 10^6 followers (34.06). With all of this, what is clear is that the higher the audience, the higher the topic diversity.

Finally, the boxplots of the distance between clusters -Silhouette width- and the number of terms (tags) per cluster, are provided in Figures 3 (c) and 3(d) respectively. Although the average of Silhouette width is similar for the users in the different groups (around 0.15), the variance is higher in the case of "ordinary" users than in the case of celebrities, being 1.26×10^{-3} for "ordinary" users and 6.85×10^{-4} for celebrities. With respect to the number of tags per cluster, Figure 3 (d) shows that the average of tags per cluster is similar for all the users, but the variance is higher in the case of "ordinary" users than when the users are celebrities. This is in consonance with the results obtained in terms of number of clusters since, when considering approximately the same number of tweets per user, "ordinary" users tend to talk about less topics and much, since celebrities talk about more diverse topics but with lower intensity.

4   Discussion

This paper reports an experiment conducted over a Twitter dataset to analyse if the diversity of topics chosen by one user in his tweets is biased by his potential audience, understanding this as his number of followers. We found that "ordinary" users talk about less topics, but with more intensity, than celebrities. This confirms that users' behaviour is affected by their audience as expected from the theories of *self-presentation* and *impression management* [4]. It could seem that users with a huge amount of followers (celebrities) tended to minimise

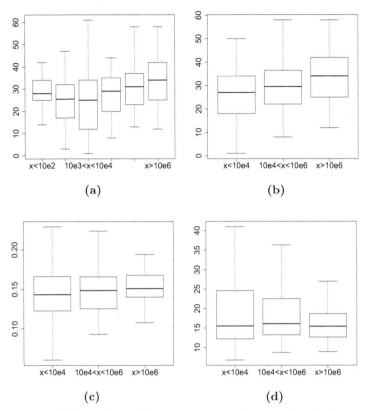

Fig. 3. CCDF of (a) # number of clusters per group for users classified into 6 groups; (b) # number of clusters per group, (c) Silhouette width and (d) # of tags per cluster for users classified into 3 groups.

their content diversity, dealing only with those topics that have made them famous (sportsmen about their own sport, politicians about their own party, actors about their movies, etc.). However, as Marwick and Boyd claimed [9] other factors come into play when celebrities post tweets. Apart from tweeting about their own interests and likes, celebrities make efforts to discover the interests of their followers to tweet accordingly and also to satisfy their sponsors promoting their products by sponsored tweets. They are unconsciously forced to keep the balance between keeping their authenticity, keeping their followers and keeping their sponsors' support, which would explain the inevitably increment of the content diversity of their tweets with respect to "ordinary" users.

However, having a large amount of followers does not guarantee that these followers pay attention to the user's tweets, which makes the real audience of a user is, in general, lower than his set of followers. As future work, we plan to enhance the analysis with a more accurate definition of audience, taking into account other Twitter signs that denote that a user u is really following the publications of other user v, as for example *retweets* and *mentions*. Once this is done, the next natural step should be analysing our data longitudinally instead of cross-sectionally, that is, studying whether or not people change the diversity of their topics as they get more followers. To this aim, the focus should be on a temporal network that records when contacts happen, in addition to whom has been in contact with whom. Therefore, we will need an extended dataset where contacts between every two users u and v were time stamped and tags (representing interests) were augmented with the duration of each interest for a specific user u.

References

1. Blei, D.M., Ng, A.Y., Jordan, M.I.: Latent dirichlet allocation. The Journal of Machine Learning Research **3**, 993–1022 (2003)
2. Cilibrasi, R.L., Vitanyi, P.M.B.: The google similarity distance. IEEE Transactions on Knowledge and Data Engineering **19**(3), 370–383 (2007). http://nlp.stanford.edu/software/corenlp.shtml (last accessed on December 11, 2014)
3. Dimitrov, A., Olteanu, A., Mcdowell, L., Aberer, K.: Topick: accurate topic distillation for user streams. In: IEEE 12th International Conference on Data Mining Workshops (ICDMW), pp. 882–885 (2012)
4. Goffman, E.: The presentation of self in everyday life (1959)
5. Jain, A.K., Dubes, R.C.: Algorithms for clustering data. Prentice-Hall, Inc. (1988)
6. Kwak, H., Lee, C., Park, H., Moon, S.: What is twitter, a social network or a news media? In: Proceedings of the 19th International Conference on World Wide Web, WWW 2010, pp. 591–600 (2010)
7. Langfelder, P., Zhang, B., Horvath, S.: Defining clusters from a hierarchical cluster tree: the Dynamic Tree Cut package for R. Bioinformatics **24**(5), 719–720 (2008)
8. Li, R., Wang, S., Deng, H., Wang, R., Chang, K.C.C.: Towards social user profiling: unified and discriminative influence model for inferring home locations. In: KDD, pp. 1023–1031 (2012)
9. Marwick, A.E., et al.: I tweet honestly, I tweet passionately: Twitter users, context collapse, and the imagined audience. New Media & Society **13**(1), 114–133 (2011)
10. Ong, W.J.: The writer's audience is always a fiction. Publications of the Modern Language Association of America, pp. 9–21 (1975)
11. Quercia, D., Askham, H., Crowcroft, J.: Tweetlda: supervised topic classification and link prediction in twitter. In: Proceedings of the 3rd Annual ACM Web Science Conference, WebSci 2012 (2012)
12. Rangrej, A., Kulkarni, S., Tendulkar, A.V.: Comparative study of clustering techniques for short text documents. In: Proceedings of the 20th International Conference Companion on World Wide Web, WWW 2011, pp. 111–112 (2011)
13. Rousseeuw, P.J.: Silhouettes: a graphical aid to the interpretation and validation of cluster analysis. Journal of Computational and Applied Mathematics **20**, 53–65 (1987)
14. Servia-Rodríguez, S., Fernández-Vilas, A., Díaz-Redondo, R., Pazos-Arias, J.: Inferring contexts from Facebook interactions: A social publicity scenario. IEEE Transactions on Multimedia **15**(6), 1296–1303 (2013)
15. Servia-Rodríguez, S., Fernández-Vilas, A., Díaz-Redondo, R.P., Pazos-Arias, J.J.: Comparing tag clustering algorithms for mining twitter users' interests. In: International Conference on Social Computing (SocialCom), pp. 679–684. IEEE (2013)
16. Witten, I., Milne, D.: An effective, low-cost measure of semantic relatedness obtained from wikipedia links. In: Proceeding of AAAI Workshop on Wikipedia and Artificial Intelligence: an Evolving Synergy, pp. 25–30 (2008)

Advantages of Cooperative Behavior During Tsunami Evacuation

Adam Slucki and Radosław Nielek$^{(\boxtimes)}$

Polish-Japanese Institute of Information Technology (PJIIT), Warsaw, Poland
{adam.slucki,nielek}@pjwstk.edu.pl

Abstract. Considering the case in which not every individual is able to efficiently evacuate, due to the lack of knowledge about safe spots and available routes to them, close cooperation between community members plays a critical role. Using agent-based simulation, we tested two hypothetical scenarios of human behavior during tsunami evacuation and their efficiency, considering time and number of rescued people. In the first scenario, individuals did not cooperate with each other. In the second one, community members were trying to organize in groups, even if they could evacuate by their own. The results showed that in the second scenario not only substantially higher percentage of citizens evacuated in a shorter time, but it was nearly as efficient as the evacuation of a community in which almost every individual know at least one safe spot.

1   Introduction

On the 11th March 2011 in Japan occurred an earthquake of magnitude 9 that was followed by a tsunami, of which the maximum noted height reached 40.5 m. Number of people reported as dead or missing exceeded 20 000 [1].

In the opinion of *Shiminkatsudou Information Center* number of victims would be even greater if not the close cooperation between members of the communities [2]. On the other hand, there are cases showing that a collective decision caused death of the whole or majority of a group. One of the examples can be the case of Okawa's primary school in Ishinomaki town. Despite the suggestion of one of the teachers, majority of citizens that had chosen Okawa's school as a shelter insisted on staying there, instead of evacuate to a spot situated on a higher altitude. Eventually, water level exceeded estimations and reached Okawa's school, causing death of 83 people. 25 citizens survived only because they decided to separate from the group and evacuate to another spot [3]. This example reveals potential risk of cooperation which is not only the possibility that a group will make a wrong decision. The process of organizing and sharing opinions may be time consuming and slow down evacuation of a whole group.

Using agent-based system we try to answer the following research questions: 1) *Is the cooperative behavior during evacuation more advantageous than selfish behavior?* 2) *How does the average knowledge about available evacuation points affect evacuation efficiency?* 3) *Does the information dissemination in cooperative commu-*

© Springer International Publishing Switzerland 2015
N. Agarwal et al. (Eds.): SBP 2015, LNCS 9021, pp. 203–212, 2015.
DOI: 10.1007/978-3-319-16268-3_21

nity significantly improves evacuation process considering that, in general, community has little knowledge available about the evacuation points?

Answering those questions may shed the light on important but often overlooked aspects of preparing emergency evacuation on community level, namely cultural differences regarding cooperation and selfishness. The most efficient strategy in Japan will not work properly in Europe, as people are in general more self-oriented.

The rest of this paper is organized as follows: Section 2 contains the summary of related work. Section 3 describes simulation model. Section 4 encloses the experimental setup. The results are presented in Section 5 followed by conclusions in Section 6.

2   Related Work

Although we are not aware of any other work that provides a direct, comparative analysis of evacuation effectiveness influenced by two different social behaviors of actors, number of researches have been made in the field of emergency evacuation.

For instance, there are psychological studies like Drury et al. [4] where authors analysed *post hoc* interviews with people who survived an event in which they shared the threat of death, to compare solidarity andoccurrence of supportive behavior among members in two groups of survivors with low and high identification with others. Psychological aspects of social behavior in response to emergency event were previously described in Mawson [5] who notes that references to mutual help and support should be found more often than acts driven by self-preservation. Some research focus on general reaction of community to a tsunami alert, without considering social aspects of its members' behavior. Kanai et.al [6], after Chilean tsunami in 2010, conducted a survey among citizens of Japanese coastal city, where evacuation order also was announced. The aim was to describe people's risk perception about tsunami, reaction to the official warning and evacuation scheme.

On the emergency management side, Lammel et al. [7] proposed traffic optimization for tsunami evacuation in the Indonesian city of Padang. Authors used agent based simulation to estimate the evacuation process, detect potential hindrances and locations where tsunami proof shelters would be absolutely necessary to allow a successful evacuation of citizens living in its area. In the other work, Zagorecki [8] focused on the aspect of information exchange between emergency response organizations. Zheng et al. [9] proposed a game-theoretical approach to simulate how evacuation urgency affects arising of cooperative behavior and what consequences it has for overall efficiency. Important issue mentioned also in [6] is that reaction to the emergency alert is often delayed, because citizens wait for additional confirmation. Tyshchuk et al. [10] used Natural Language Processing and Social Network Analysis to process Twitter messages published during tsunami in Japan in 2011 and automatically found leaders of communities, key players who were able to successfully urge others to take action. Arai [11] studied efficiency of information spreading from the linguistics perspective – how to compose the most persuasive warning.

Another important aspect in emergency management is an optimal utilization of rescue service. Gelenbe et al. [12] developed an algorithmic solution based on Random Neural Network that could solve optimization problem for simultaneously dispatching emergency teams to different locations. Presented approach allows us obtaining solutions for large-scale problems; hence, it could be used in real-time also during evacuation management process.

3   Model Description

3.1   World Representation

Model was implemented in NetLogo language and its source code is available on the Internet.[1] The terrain shape used in the model lowers the probability of reaching safe spot by moving in random direction and maintains realism what improves reliability of the results. It represents actual landscape of 4 km^2 of Kamaishi, Japanese coastal city, is based on the terrain map available in *Google Maps* service and divided into squares called patches. Every patch corresponds to 100 m^2 in reality. Altitude is marked with different colors; it helps to determine if given agent reached safe spot. Considering the fact that the region is surrounded by mountains seemingly any knowledge about available safe spots is necessary for an effective evacuation. However, photographs of the place reveal that there are many obstacles like high fences, concrete blocks preventing landslides on slopes and canals which significantly lower chances for successful evacuation of individuals without adequate knowledge what in this case is the knowledge about the location and route to at least one of spots specified by the Kamaishi officials and presented on the city homepage[2]. The model takes into account inaccessible terrain mentioned above.

Time in the model is counted in ticks. One tick represents 5 seconds in reality. Evacuation process begins immediately after the start of the simulation. Tsunami reaches the land after 180 ticks i.e. 15 minutes from the start of evacuation and moves with the speed 7 m/s. The maximum height, up to 40 m may be reached with 33% probability only after the whole terrain with latitude below 20 MASL has been flooded.

3.2   Agents Knowledge

In 2012 officials of Yuasa, another Japanese coastal city, organized evacuation training and conducted the survey with its 556 participants [13]. The results show that 86.9% of respondents knew at least one evacuation point. At the same time, 79% of them declared that they had already decided to which point they will evacuate in case of emergency. In the model, each agent knows given number which is a Poisson distributed random integer from 0 to 10 of available evacuation points. Agents who know more than one evacuation point always choose a spot, one of which they know, situated in the nearest location.

---

[1]   http://modelingcommons.org/browse/one_model/4115#model_tabs_browse_procedures

[2]   http://www.city.kamaishi.iwate.jp/index.cfm/6,18417,34,180,html

3.3   Agents Speed

The results presented in [13] show also that 81.8% of respondents plan to evacuate by walk (earthquakes usually cause infrastructure damages). Consequently, in the model every agent moves with walking speed which based on the research of Bohannon [14] is a double precision random number from normal distribution with mean of 1.4 m/s and standard deviation of 0.15 m/s (some agents may run instead of walk but it required a sufficient fitness). Seemingly, walking speed should not apply in case of emergency. However, the results of surveys conducted by Kanai [6] and Yuasa city [13] show that after official tsunami alert, without further confirmation, citizens often do not perceive the risk adequately and start evacuating with substantial delay. In 2010, when Kamaishi officials announced evacuation order at 9:34 am, predicting first wave coming at 1:30 pm, only 8% of 822 participants of the survey decided to evacuate immediately after announcement.

3.4   Behavior Types

In the model there are represented two opposite types of possible behaviors. In one setup, all the agents are eager to form a group and share the information about available evacuation points that they have, regardless being able to evacuate individually. In this case, after the start of the simulation every agent continuously tries to find the biggest group of other agents that are nearby. As the research of Helbing et al. [15] shows, in emergency situation people tend to show herding behavior. Joining the group with the greatest number of other participants is also the most advantageous in regard to information dissemination, because of the highest probability that at least one of its members knows location of one or more evacuation points.

In the other setup, all agents who know at least one evacuation point go there immediately after the start of the simulation. Agents without any knowledge about location of evacuation points try to follow random agent in a given radius, yet they do not exchange any information.

Agents may also try to help relatives. Every agent is assigned to a randomly selected group represented by a given number. If the agent notices relative behind him, in a given range, he moves toward the him and shares the information. Every agent may be helped only once during the simulation. In addition to information sharing, both the agents change their own speed to the speed of the faster one as we consider possibility of psychical help and greater motivation for moving faster.

3.5   Decision Process

The agent's internal decision process is presented in details in Appendix A. In general, it may be divided into three mains steps:

1. Agent chooses the nearest evacuation point that is known to him. If this process takes place after he shared information with another agent, he chooses given number of evacuation points, compare their distances with the location he has already chosen and heads to the closest one. If agent does not know any evacuation point, he approaches another, random agent in a given radius.

2. If the given agent do cooperate, he tries to locate another cooperative one in a defined radius, one who is surrounded by the greatest number of other agents (surrounding agent is defined as an agent located in 10 m radius). If such agent exists, he becomes a group center and has to change his speed to the mean speed of surrounding agents. All agents within a group share the knowledge about evacuation points. Additional parameter which determines size of a group that satisfy individuals and make them not to try to join bigger one, can be used to prevent forming too big groups.

3. An agent checks if behind, in distance of 6 to 15 m there is another agent whose group number difference is lower than given threshold and who has not been helped yet. If such agent, called relative exists, agent heads towards him, changes his own or relative's number of known evacuation points to the number known by better informed one and lastly changes speed of both to the speed of the faster one.

4   Experimental Setup

In order to test the influence of different behaviors on the efficiency of evacuation it is necessary to run multiple simulations with various setups where the key parameters are as follows.

- Number of evacuation points known by average citizen: [1; 4] - *this parameter is the λ coefficient for Poisson distribution and represents number of evacuation points locations known by average citizen. For the value 1 approximately 65% of all agents should know at least one evacuation point. For the value 4 this number grows to nearly 98%.*
- Percentage of cooperative agents: [0; 25; 50; 75; 100] – *it determines how many agents will try to form groups following the mechanism described in Section 3.4.*
- Group difference threshold: [0;1] – *determines maximum difference between numbers representing groups of two agents to consider them as relatives.*
- Radius: [50] – *determines radius in which agents can search for the other ones. Model does not directly take into account any buildings as a potential hindrance for the evacuation. However, considering existence of buildings and other objects we assume that mean range of vision could narrow to 50 meters on average.*

Every simulation for those sets of parameters was be repeated 30 times with initial number of agents equal to 300 and then averaged. Output value represents time in seconds from the start of each simulation to the state in which there is not any agent left in the model world as well as the percentage of agents who reached the evacuation point. Additionally, time was measured also in state where 10, 30, 50, 80 and 95% of agents evacuated.

5   Results

Simulations with setup in which agents do cooperate and exchange information show substantial difference in comparison with the opposite setup. In case of cooperative behavior, we may observe that agents quickly form well organized groups which head towards optimal, considering the distance, evacuation point.

In the opposite scenario agents also form groups. However, they are not organized and due to the lack of information dissemination, there are groups in which every agent knows at least one evacuation point's location. Such groups usually are not able to successfully evacuate. Fig. 1. presents the state in the 30th step of simulation for both scenarios.

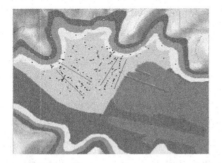 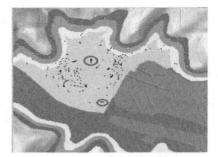

Fig. 1. Comparison simulation states for cooperative (left) and selfish community (right). Lines indicate well organized agents. Groups which do not face any evacuation point are encircled.

Output of simulations for different values of key parameters is presented in the tables below. First one shows the results for scenario of 0% of cooperative agents.

Table 1. Results of simulations where agents did not cooperate

No of evac. points known by agent	Group diff. thresh.	Mean of res-cued	Mean time
1	0	84.07%	1084 s
1	1	98.81%	942.16 s
4	0	99.82%	693.67 s
4	1	99.97%	631 s

The second table presents the results of simulations for the opposite scenario where all agents did cooperate.

Table 2. Results of simulations where agents did cooperate

No of evac. points known by agent	Group diff. thresh.	Mean of rescued	Mean time
1	0	98.39%	979.83 s
1	1	98.84%	930.52 s
4	0	99.94%	692.67 s
4	1	99.87%	721.79 s

Chart below presents time measured when given percentage of agents successfully reached one of evacuation points for communities with little knowledge about evacuation points and not eager to help their relatives. It is important to note that measures ends on the value of 95%. Final measure of time regards to the state in which all agents was rescued or died, hence cannot be included in the graph.

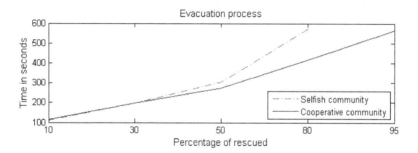

Fig. 2. Time measured when given percentage of agents with generally little knowledge evacuated

Similar chart was created to compare evacuation process of communities with good knowledge about available evacuation points.

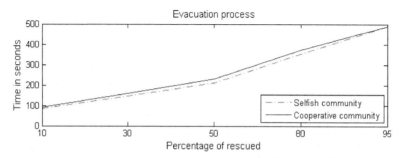

Fig. 3. Time measured when given percentage of agents with generally good knowledge evacuated

Means were compared with Student's t-test with 0.05 level of significance. Number of degrees of freedom was estimated with Welch–Satterthwaite equation.

Answering our research questions, tests showed that both mean of rescued and mean of time for scenario where members of a community had a little knowledge about evacuation points was significantly worse for the selfish community. As Fig. 2 shows, time when 50 or more percent of population reach safe spot is dramatically worsening for this community. What is more selfish community has not reached threshold of 95% rescued in any simulation. Tests show also that advantage of that 30% of selfish community who managed to reach safe spot on the beginning of simulation is not significant in comparison to results of cooperative community. Cooperative behavior that allows people better dissemination of information, substantially improves efficiency of evacuation process. Potential loss required for forming groups is compensated by its influence on community knowledge and better organization. Results for different number of cooperative agents show linear dependence of increase of that number and evacuation effectiveness improvement. Therefore awareness of importance of cooperation during emergency situation, which would lead to better information exchange, could help to increase evacuation efficiency to the level comparable with results of well informed community evacuation.

As expected, better knowledge significantly improved evacuation efficiency in comparison to any setup where agents had little knowledge about evacuation points. It emphasizes value of accurate information in the case of emergency.

Differences between results for two considered scenarios where average community member has good knowledge about available evacuation points are not significant when comparing overall results and time when community reached threshold of 95% rescued despite the fact that cooperative community performed little worse than selfish one until that threshold was reached as shows fig. 3. It is another confirmation that potential loss of time required for organization at the beginning of evacuation process is compensated even for communities with good knowledge in which most of the members could evacuate independently.

Surprisingly helping other individuals had not any noticeable influence on the evacuation efficiency, assuming that it would not be the only way of obtaining information.

6   Conclusion

We have proposed a multi-agent system approach to compare the influence of two different types of behavior, cooperative and selfish, on evacuation efficiency. Model used in the experiment allowed us to focus on knowledge about terrain and safe spots as key factor for successful evacuation.

The results confirmed that cooperative behavior and consequently sharing of information leads to a substantially higher rate of rescued people in less time, considering that given community, in general, has little knowledge about evacuation points. Therefore, experiment proved the importance of further research regarding improvement of information efficiency, including information provided by officials, dissemination between community members in case of emergency.

Finally, our model also highlighted problem of the lack of adequate signs warning that terrain ahead may be inaccessible for pedestrians or signs leading to the nearest evacuation point.

Appendix A

Diagram of agent's behavior in tsunami model

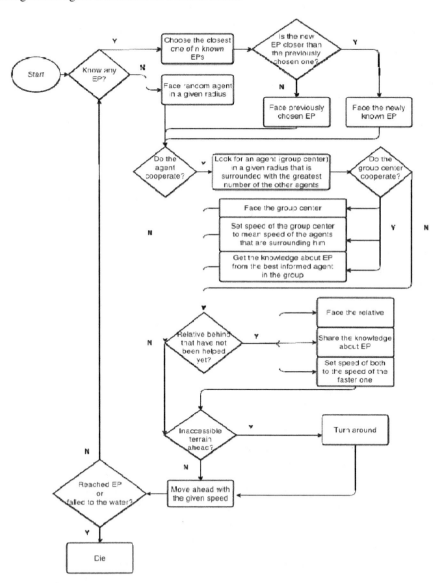

Acknowledgements. This work is supported by Polish National Science Centre grant 2012/05/B/ST6/03364.

References

1. Shaw, R., Takeuchi, Y.: East Japan Earthquake and Tsunami: Evacuation, Communication, Education and Volunteerism. Research Publishing Service (2012)
2. Report of Shiminkatsudou Information Center. The lesson we should learn from information dissemination and citizens' cooperation during 2011 earthquake off the Pacific coast of Tōhoku (2012). http://sicnpo.jp/saigai110311/imase-urado2012.pdf (retrieved)
3. Satoh, J.: Tragedy of Okawa's primary school (2012). http://www.geocities.jp/kxbrb316/report-20.htm (retrieved)
4. Drury, J., Cocking, C., Reicher, S.: Everyone for themselves? A comparative study of crowd soladarity among emergency survivors. British Journal of Social Psychology **48**(3), 487–506 (2009)
5. Mawson, A.: Understanding mass panic and other collective responses to threat and disaster. Psychiatry: Interpersonal and Biological Process **68**(2), 95–113 (2005)
6. Kanai, M., Katada, T.: Issues of Tsunami Evacuation Behavior in Japan: Residents Response in Case of Chilean Earthquake in 2010. In: Solutions to Coastal Disasters 2011, pp. 417–423. ASCE Publications (2011)
7. Lammel, G., Rieser, M., Nagel, K., Taubenbock, H., Trunz, G., Goseberg, N., Schlurmann, T., Klupfel, H., Setiadi, N., Birkmann, J.: Emergency preparedness in the case of tsunami evacuation analysis and traffic optimization for the Indonesian city of Padang. In: Pedestrian and Evacuation Dynamics 2008, pp. 171–183. Springer (2010)
8. Zagorecki, A., Ko, K., Comfort, L.K.: Interorganizational Information Exchange and Efficiency: Organizational Performance in Emergency Environments. Journal of Artificial Societies and Social Simulation **13**(3) 3 (2010)
9. Zheng, X., Cheng, Y.: Modeling cooperative and competitive behaviors in emergency evacuation: A game-theoretical approach. Journal Computers & Mathematics with Applications **62**(12), 4627–4634 (2011)
10. Tyshchuk, Y., Li, H., Ji, H., Wallace, W.A.: Evolution of communities on twitter and the role of their leaders during emergencies. In: Proceedings of the 2013 IEEE/ACM International Conference on Advances in Social Networks Analysis and Mining, pp. 727–133 (2013)
11. Arai, K.: How to Transmit Disaster Information Effectively: A Linguistic Perspective on Japan's Tsunami Warnings and Evacuation Instructions. International Journal of Disaster Risk Science **4**(3), 150–158 (2013)
12. Gelenbe, E., Timotheou, S., Nicholson, D.: Random neural network for emergency management. In: The Workshop on Grand Challenges in Modeling, Simulation and Analysis for Homeland Security (2010)
13. General Affairs Department of Yuasa City. Report of the results of evacuation training (2012). http://www.town.yuasa.wakayama.jp/bousai/hinankunren.pdf (retrieved)
14. Bohannon, R.W.: Comfortable and maximum walking speed of adults aged 20-79 years: reference values and determinants. Age and Ageing **26**, 15–19 (1997)
15. Helbing, D., Johansson, A.: Pedestrian, Crowd and Evacuation Dynamics. Encyclopedia of Complexity and Systems Science **16**, 6476–6495 (2010)

Learning Automated Agents from Historical Game Data via Tensor Decomposition

Peter Walker[1,2] and Ian Davidson[1,2(✉)]

[1] United States Navy, Bethesda, USA
peter.b.walker.mil@mail.mil
[2] Department of Computer Science, University of California, Davis,
Davis 95616, USA
davdison@cs.ucdavis.edu

Abstract. War games and military war games, in general, are extensively played throughout the world to help train people and see the effects of policies. Currently, these games are played by humans at great expense and logistically require many people to be physically present. In this work, we describe how to automatically create agents from historical data to replace some of the human players. We discuss why game-theoretic approaches are inappropriate for this task and the benefits of learning such agents. We formulate a tensor decomposition formulation to this problem that is efficiently solvable in polynomial time. We discuss preliminary results on real world data and future directions.

Keywords: War gaming · Modeling · Tensor decomposition

1 Introduction and Motivation

War games, and in particular military war games, at their core can be viewed as a group of individual players (red, green, yellow, blue etc.) each of which has a finite set of tasks (i.e. move troops, engage leaders, recruit locals) which can be scheduled to be played in a zone at a given time. Typical war games are simulated over hundreds of square miles with each zone being approximately one square mile. War games are quite complex in detail: population density, the local population attitudinal information and target information are all used in an adjudication system [1]. At the start of each turn, each player schedules their tasks (task type, location and time) and then an adjudication system determines the effect of their tasks that is fed back to each of the players at the start of the next turn. In this work and in games in general, the adjudication system is considered a black- box since it is realistic that each player will not know the effect of their behavior on each other.

On one hand, operating war games has many benefits. It allows easy training of individuals for one-off situations and, in many cases, is the only training these individuals can hope to receive. Due to this fact, war games are very popular and highly sought after.

© Springer International Publishing Switzerland 2015
N. Agarwal et al. (Eds.): SBP 2015, LNCS 9021, pp. 213–221, 2015.
DOI: 10.1007/978-3-319-16268-3_22

However, there are many challenges to running an efficient war game. Firstly, war games typically require a large number of people (upwards of twenty) to not only play the role of the players, but also the local population (to gauge their opinion on the core players). In addition, typical war games require these players to be physically co-located. Perhaps the greatest challenge in developing war games with high fidelity is that if a slight variation in the game is required, such as a more aggressive player, the entire game must be rerun.

Though the focus of this work is on war games, it should be noted that our work is generalizable to any large spatial multi-player game where one player is to be replaced by an agent. Physical world examples include "games" where players are police groups and drug smugglers as well as cyber examples where the players are network administrators and various hacking groups intent on deploying denial of service attacks.

A key benefit of our tensor decomposition formulation is that we can easily represent the underlying model as a stimuli-response network and that it allows analysis of the underlying model used by the player. In this way, we can see our work as an example of a **Reflex Agent** using the nomenclature of Russell and Norvig [9]. In the next section, we outline the core problems centered around extracting this stimuli-response network and several innovative uses of the network. An obvious question is, since we are building agents to play a game, why not use the extensive literature on game-theory. We answer this question in section 3. Section 4 outlines our tensor formulation and the underlying algorithms used in the decomposition. We show results in Section 6 and draw conclusion on our work and list future work in section 7.

2   Three Problem Statements

The data available for each player is for a past war game: the task types the players performed along with location and time. This can easily be modeled for a single player as a **binary-valued** order four tensor with the dimensions being longitude, latitude, time and event type. We will use the notation χ_i to represent the ith players behavior at a space time with $\chi^{long,lat,time,task}$ indicating if the player performed *task* at *long, lat* at time *time*. We can view such a tensor for a single player as shown in Figure 1. To lower the dimensionality of the data to three, each shade represents a different type of task. Such tensors will exist for all players and will represent the training data we shall build our agents from. We can now sketch the problems we wish to investigate as described below and shown diagrammatically in Figure

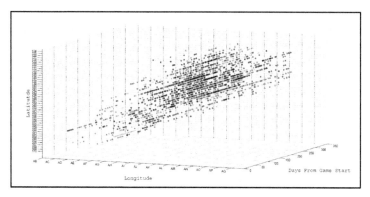

Fig. 1. The tensor representing a single player's behavior in a 22 week game. Each color represents a different type of task. Such a tensor (χ_i) are the input in the problem definitions.

Problem 1: Learning Stimuli-Response relationships

In this problem we aim to discover what stimuli-response relationships exist between the various players. This knowledge can then be summarized into networks to simplify the model so it can be easily verified.

Definition Problem 1:

Input: $\chi_i \cdots \chi_m$

Output: $\forall i, j \quad (long, lat, time, event) \in Stimuli_{i \rightarrow j} \rightarrow$
$\qquad\qquad (long', lat', time', event') \in Response_{j \rightarrow i}$

Here, we see the input into the problem is the various players' behavior tensor and the output is two sets: stimuli and response for each two-player combination. Each player can be viewed as a set of stimuli and a set of responses for each and every other player.

We can summarize (by aggregating over time and space) a collection of stimuli-response relationships into a graph with the nodes being all possible tasks for all players and the edges being the particular stimuli-response relationship. We refer to such a graph as the stimuli-response graph and note that we can simplify this graph by limiting the nodes and edges to be for a particular player. Therefore, we have a stimuli-response graph for each of the players in the war game. An example of such a graph is shown in Figure 2.

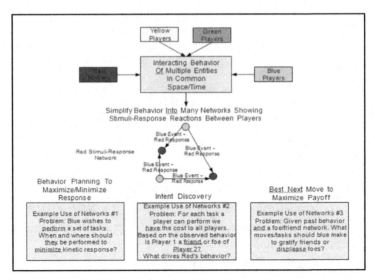

Fig. 2. Core Problem of Inducing a Stimuli-Response Graph with Several Uses

Problem 2: What is the best set of Moves to Minimize/Maximize Response

This problem addresses the important application of given a set of actions/tasks a player wishes to make, where and when should he play them to minimize or maximize the response from the other players. We note that we do not consider a player's response to his own actions as part of the optimization problem.

Definition Problem 2:

Input: $\forall i, j \quad Stimuli_{i \to j}, Response_{j \to i}$: the output of Problem 1

$e_1 \ldots e_n$: A series of events that player p wishes to play

Output: $(long_1, lat_1, time_1) \ldots (long_n, lat_n, time_n)$ such that:

$argmin_{(long_1, lat_1, time_1) \ldots (long_n, lat_n, time_n)} \quad i \quad k, k =_p |Response_{p \to k}(long_i, lat_i, time_i, event_i)|$

where $Response_{j,k}()$ is the function that takes in the events of player j and returns a set of responses for player k.

Problem 3: Alliance/Intent Discovery

Given a player's historical tasks/moves in a game and a cost function for each player, our aim here is to determine what their intent was with respect to each other. We can quantify the relationship between player i to player j (which is asymmetric) by determining the cost of player i's actions less the cost of those same actions to player j. A positive value indicates a friendly relationship since the cost to i is greater than the cost to j.

Definition Problem 3:

Input: $\forall i, j$ $Stimuli_{i \to j}, Response_{j \to i}$: the output of Problem 1

$\forall i \; f_i(long, lat, time, event)$: a cost function that for a given event, location and time combination

returns the cost to player i

$(long_1, lat_1, time_1, event_1) \ldots (long_n, lat_n, time_n, event_n)$ are all the events that player i played dur- ing the game.

Output: $Intent_{i \to j} = \;_a\{f_i(Response_{i \to i}(long_a, lat_a, time_a, event_a)) - f_j(Response_{i \to j}(long_a, lat_a, time_a, event_a))\}$

where $Response_{j \to k}()$ is the function that takes in the events of player j and returns a set of responses for player k.

For a given set of values $j = 1 \ldots m$ we can then examine $Intent_{i \to j}$ to determine a "social" network amongst the players. Such friend/foe networks (positive edge weights indicate friends, negative edge weights foes) have been studied before with our own work [3] looking at how to segment such networks.

3   Why Game Theory is Not Applicable

Game theory has been used extensively to created automated players in two player and multi- player games: the world checkers champion is a machine [2], chess players have beaten world champions and competent automated players exist for Go, bridge and back-gammon [9] [8].

All these automated players use variations of the standard mini-max search and pruning algorithms such as alpha beta to efficiently calculate the Nash equilibrium. However, calculating the Nash or other equilibrium for war game style games is not applicable for a number of reasons:

- *No agreed upon terminal state.* Most games with equilibriums require agreement on the end state (i.e. Check mate with chess). However no such agreement exists in our situation and most likely evolves as the game progresses.

- *Not zero-sum.* Most (but not all) algorithms to determine an optimum series of moves re- quires the game to be zero-sum. In our context it would be mean that the cost of the red player for an action is the benefit to say the blue player. This is clearly not the case.

- *Not complete information.* The requirement of complete or perfect information means that each player can see each other players actions. This is not the case since only the **successful** tasks that impact the other players is shown to the players on the subsequent turns.

- *Non stationary payoff function.* Typical games are not only not zero-sum games but the payoff functions evolve over time. Consider the benefit to the Red player of recruiting insur- gents. The benefit of doing this early in the game is much more (they can be used for longer periods of time) than later in the game.

- *Non stationary alliances.* As is often the case, the alliances that form can change and even reform over time. This makes placing a constant set of payoffs at leaf nodes in the game tree not possible.

For all of the above reasons and more we cannot use traditional game playing algorithms and instead turn to a tensor decomposition formulation since it allows us to efficiently extract stimuli- response relationships (problem 1) and also address problems 2 and 3.

4   A Tensor Decomposition Formulation

Here we sketch the tensor decomposition formulations we have used. As a primer, we briefly overview tensors and, in doing so, introduce our notation. For a complete overview of tensors and decompositions, the reader is referred to the excellent survey of Kolda et. al [7]. For clarity, a tensor can be considered as a high order matrix.

In this formulation, all tasks played by all players are in one tensor. This means that if we have m players there will be an $m+3$ mode tensor since for each player we have their tasks/events recorded at longitude, latitude and time which we shall refer to as χ. We can loosely view χ as a concatenation of the player tensors mentioned earlier. For example, in a two player game, our aim is to decompose this tensor into f factors denoted by the matrices $\mathbf{A}$, $\mathbf{B}$, $\mathbf{C}$, $\mathbf{D}$ and $\mathbf{E}$ where the factors are stacked column-wise. These represent player1-tasks, player2-tasks, longitude, latitude, and time respectively. Then the ith factor is simply $\hat{\mathbf{X}}_{\mathbf{i}} = \mathbf{A_i} \circ \mathbf{B_i} \circ \mathbf{C_i} \circ \mathbf{D_i} \circ \mathbf{E_i}$ and can be considered as one part of the simplification of the behavior. Our standard least squares objective function for the canonical decomposition is shown in equation 1.

$$\arg\min_{A,B,C,D,E} \|\chi - A \circ B \circ C \circ D \circ E\|F \tag{1}$$

Throughout the rest of the paper $\circ$ is the tensor outer product, formally defined as $A \circ B \circ C \circ D \circ E = \sum_i A_i \circ B_i \circ C_i \circ D_i \circ E_i$ and $\circ$ is the Kronecker product [6]. The solution to this problem can be cast as an alternating least squares problem solvable with the algorithm shown in Figure 3. For simplicity, the algorithm is described solving an order three tensor. Extension to high order tensors is trivial and involves solving for the additional modes in D and E. Though the objective function is not convex, others have shown that alternative least squares solution for such problems have many desirable properties [5]. Other methods to improve the efficiency of decomposing large tensors are described in our earlier work [4].

Alternating Least Squares for Canonical Order Three Tensor Decomposition
Input: $\mathcal{X}$: The tensor to decompose, ε: The minimum error.
Output: $\mathbf{A}, \mathbf{B}, \mathbf{C}$

1. Calculate matricizations X_A, X_B, X_C corresponding to where $X_A = A(B \circ C)$
2. Solve and set $A = \arg\min_{A} \; \|X_A - A(C \circ B)^T\|_F$
3. Solve and set $B = \arg\min_{B} \; \|X_B - B(C \circ A)^T\|_F$
4. Solve and set $C = \arg\min_{C} \; \|X_C - C(B \circ A)^T\|_F$
5. $\|\mathcal{X} - A \circ B \circ C\| < \varepsilon$
   - Not Satisfied: Goto step 2.
   - Satisfied: **return** A, B, C.

Fig. 3. The ALS algorithm for solving the canonical decomposition. The algorithm is shown for an order three tensor but is easily changed to an arbitrary order tensor

5   Using Tensor Decompositions to Address Our Core Problems

We now show how tensor decompositions can be used to address the problems addressed in section 2. It is straight forward to see that we need only show how to extract out the stimuli and response sets from the tensor decomposition (Problem 1) and that our Problem Definitions for 2 and 3 can then easily be solved.

Recall we wish to determine the stimuli-response relationship between the players. This can be achieved by examining the factors from the decomposition. We can visually interpret the stimuli-response by considering the example factor in the tensor decomposition of the data shown in figure 4. The top plot shows those entries in the outer product of $long \times lat \times events$ that are greater than some minimum threshold with the different players tasks/events color coded. The way to interpret this "thresholded" factor is that when the red player performs the task (shown as a red square) such as Kidnapping at location (AP, 169) the blue player responds with all the other tasks shown at various locations in the figure. However, this relationship is not consistent over time, instead it varies as is shown in the lower plot. There will be such stimuli responses for each factor in the tensor decomposition.

We can formalize the process of extracting repetitive stimuli responses as follows. We examine each factor (denoted by χ_i) and for each **pair** of entries in the factor greater than we create a stimuli response entry.

Creating Stimuli-Response Entries from Tensor Decomposition

$Stimuli_{i \to j} = \emptyset, Response_{j \to i} = \emptyset$

$\forall a: \chi_a (player_i, long, lat, event) > \alpha$ and $\chi_a (player_j, long', lat', event') > \alpha$ then

$Stimuli_{i \to j} = Stimuli_{i \to j} \cup (long, lat, time, event)$

$Response_{j \to i} = Response_{j \to i} \cup (long', lat', time', event')$

---

6   Experimental Results

Here we provide some of our evaluation results using leave one out cross fold validation. We take all the turns in the game and break them in two parts. The first part consisting of all turns except one and is the training set while the second is the test set consisting of just one turn. This is replicated so that each and every turn is a test set. We then perform a tensor decomposition on the training set and using the stimuli response entries created from the decomposition (see previous section) as the input into problem 2 with the aim to maximize the response from all other players. We can then feed in the test turn of the game as stimuli and compare the responses against the actual responses of each player in the turn. Figure 5 shows the errors by task and week for the red player. Given that the red player performed over 1400 tasks in the game, we see that the number of errors is relatively small being less than 15%.

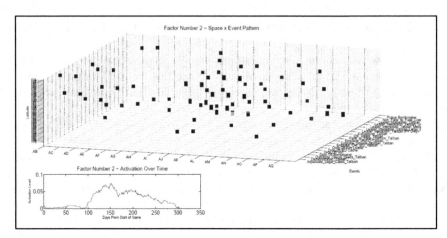

Fig. 4. Factor 1 of the PARAFAC decomposition of the game data. Note the upper panel is the spatial event patterns and the temporal activation (**e**) is shown in the lower panel.

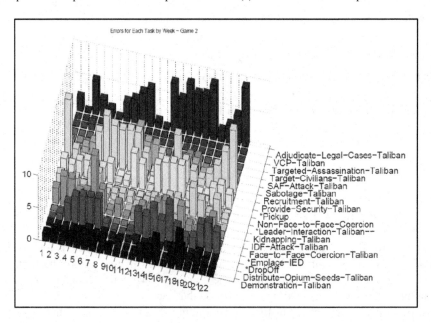

Fig. 5. The error of the red agent created from the stimuli-response relationships discovered from the tensor decomposition compared to the ground truth of what the red player actually performed

7   Conclusion and Future Work

Multi-player geographically spatial games such as war games are popular. They consist of multiple players that interact with one another with the core aim to understand the stimuli-response relationships between each player which we described as Problem 1. These stimuli-response relationships can be considered as the basis of reflex agent. Once we have these relationships we can tackle interesting problems such as determining the location of moves to minimize responses, understand the intent between players and even determining the next best move. We discussed how traditional game playing approaches are not appropriate to building such agents and how tensor decomposition can be used instead. Preliminary results are promising with the agent play- ing the red side being more than 85% accurate. In future work we plan to move beyond a stimuli response model and incorporate agent utility/intentions.

Acknowledgements. The research reported in this paper was supported by Office of Naval Research Grant (NAVY 00014-11-1-0108). The opinions of the authors do not necessarily reflect those of the United States Navy or the University of California - Davis.

LCDR Peter B Walker is a military service member. This work was prepared as part of his official duties. Title 17 U.S.C. 101 defines U.S. Government work as a work prepared by a military service member or employee of the U.S. Government as part of that person's official duties.

References

1. Brennan, R.: Protecting the Homeland: Insights from Army Wargames. No. 1490. Rand Corporation (2002)
2. http://webdocs.cs.ualberta.ca/~chinook/
3. Davidson, I.: Knowledge Driven Clustering. In: International Joint Conference on Artificial Intelligence (IJCAI) (2009)
4. Davidson, I., Gilpin, S., Walker, P.: Adversarial Event Behavior and its Analysis. Journal of Knowledge Discovery and Data Mining (DMKD), November 2012
5. Lawson, C.L., Hanson, R.J.: Solving least squares problems. Classics in Appl. Math., No.15. SIAM, Philadelphia (1995)
6. Carroll, J.D., Chang, J.: Analysis of individual differences in multidimensional scaling via an N-way generalization of "Eckart-Young" decomposition. Psychometrika 35 (1970)
7. Kolda, T., Bader, B.: Tensor Decompositions and Applications. SIAM Review **51**(3), 455–500 (2009)
8. Mitchell, T.: Machine Learning. McGraw Hill (1997)
9. Norvig, R., Russel, S.: Artificial Intelligence: A Modern Approach. Pearson (2002)

A Comparative Study of Symptom Clustering
On Clinical and Social Media Data

Christopher C. Yang[1($\boxtimes$)], Edward Ip[2], Nancy Avis[2], Qing Ping[1], and Ling Jiang[1]

[1] College of Computing and Informatics, Drexel University, Philadelphia, USA
chris.yang@drexel.edu
[2] Wake Forest School of Medicine, Winston-Salem, USA

Abstract. While the so called Big Social Data – a generic term referring to the massive amount of data automatically collected from various sources such as search queries and text generated from social media – have emerged as an important theme in computational sciences for purposes such as predicting the movement of stock prices, trends in diseases, and rates of adverse drug effects, studies that directly compares results from such analyses and those derived from formal studies such as surveillance and clinical trials are rare and far-between. Some studies have seemed to suggest that the results from the analysis of Big Social Data may not entirely agree with observed results or results from structured surveys. For example, the rate of influenza as predicted from Google-based search data was found to overestimate the actual rate. However, the number of such important comparative studies is still very limited. In this paper, we compare findings derived from data collected from a medical forum on the Internet and those derived from a formal clinical study. Specifically, we examine the co-occurrence of keywords in symptoms from participants on a breast cancer online forum and directly compare the symptom cluster patterns to data obtained from a clinical study of breast cancer survivors. The clinical study used data from a highly structured symptom checklist collected from N = 653 breast cancer-survivors. Our findings suggest that the symptom clusters obtained from the two studies have substantial overlap, but inconsistencies do remain, especially for context-sensitive symptom items. In summary, the study demonstrates the potential of mining unstructured, text-based, online forum data for supplementing and validating structured quantitative data collected from clinical studies.

Keywords: Symptom clustering · Social media · Health informatics · Medhelp · Breast cancer

1 Introduction

Since about 2000, the rapid emergence of social media such as FaceBook, Twitter, and Flickr has created new opportunities for researchers in the social and behavioral sciences. The promises and challenges of the so-called Big Social Data have been delineated by many researchers [1]. Academically, and perhaps even more so commercially, Big Social Data is attractive to social science researchers working in

© Springer International Publishing Switzerland 2015
N. Agarwal et al. (Eds.): SBP 2015, LNCS 9021, pp. 222–231, 2015.
DOI: 10.1007/978-3-319-16268-3_23

various fields. For the purpose of this paper we focus on online social-media data. There are many obvious advantages of using this type of data in social science research. First, social-media data, in digitized form, reflect the behavior of users, and so such data ensure a certain degree of ecological validity [2]. Specifically, social-media data allow the study of human interaction in a natural setting, which otherwise could be distorted by more intrusive methods or artificial settings such as clinical experiments, in which observer effects, or other motivations could obscure actual behavior. The sheer amount of information offers rich opportunities for new research questions that might not otherwise be available [3]. As pointed out by Mahrt and Scharkow [4], the large amount of data can also serve as a first step in a research study – e.g., hard-to-reach populations or rare events could be "discovered" as a proverbial needle in the digital haystack, allowing follow-up study of samples on a smaller scale.

Albeit slowly, we are learning from different applications of Big Social Data regarding both its power and limitation. For example, Google Flu Trends (GFT) reported using search queries to identify trends in seasonal influenza and to predict the rate of the spread of the disease [5], thus demonstrating the power of Big Social Data. However, the use of Big Social Data for predicting trends and patterns is not without its critics. Boyd and Crawford [6] have argued that the quantity of data does not mean that one can ignore foundational issues of measurement, construct validity and reliability, and dependencies among data sources. In 2013, GFT made unfortunate headlines because it was reported that it was predicting more than double the proportion of doctor visits for influenza-like illness than were reported by the U.S. Centers for Disease Control (CDC). As Lazer et al. [7] pointed out in a postmortem analysis of the GFT mistake, the most important challenge in using Big Data is that "most big data that have received popular attention are not the output of instruments designed to produce valid and reliable data amenable for scientific analysis." The authors also reported that by combining GFT and lagged CDC data, as well as dynamically recalibrating GFT, one can substantially improve on the performance of either the GFT or the CDC alone for predicting influenza-like illness. This is an important observation: while online data are massive in quantity, they generally lack the high level of validity, reliability, and accuracy of data collected from small-scale, structured, formal scientific studies. A rewarding approach for making the best use of Big Social Data is to first compare and validate results derived from online data against those obtained through rigorous scientific methods such as surveillance monitoring or clinical studies. If prediction is the purpose, then a second step could be the development of combined models using data from both sources [7].

In this paper, we investigate a concrete example of comparing results derived from online social-media data and data collected from a clinical study. Specifically, we compare and contrast results of symptom cluster patterns derived from data crawled from an online medical forum and data collected from a clinical study, and then we discuss the implications of our findings.

The delineation of symptom clusters in breast cancer patients is an important medical research topic that is central to the management and control of the disease in survivors. Symptoms in cancer survivors rarely occur alone. For example, fatigue, weight gain, and altered sexuality could be side effects of breast cancer treatment. The presence of multiple symptoms may have an additive or even multiplicative negative effect on patients' functioning, state of mind, and quality of life. The range of symptoms could be rather broad, ranging from general ones such as fatigue to cancer-specific ones such as tenderness in breast. For example, the instrument Symptom Checklist [8], which was used in the current study, includes a total of 39 symptoms.

While the majority of clinical studies on cancer-related symptoms have focused on one symptom, there is growing interest in evaluating multiple symptoms – i.e., treating symptoms as clusters [9]. A symptom cluster was defined as three or more concurrent and related symptoms frequently found in cancer patients [10]. There are several reasons for this growing trend. First, coexisting symptoms may share a common underlying etiology, and so trying to deal with symptoms one by one is both inefficient and likely to be less effective than by examining coexisting symptoms as clusters. It has been shown that particular biomarkers such as serum cortisol, melatonin, and serotonin are all related to symptoms such as fatigue, sleep, and depressive moods during chemotherapy [11]. Understanding covariation in symptoms helps in the discovery of physiological mechanisms that lead to the manifestation of both the disease and the side-effects of treatment. A second and perhaps more important observation, at least from a public-health perspective, is that previous studies of the treatment of symptoms have suggested that intervention concurrently improves multiple symptoms. In other words, there are both physiological and clinical bases for studying symptoms in cancer patients not as isolated outcomes but as clusters.

2   Methods

2.1   Data Sources

The data for the present study come from two distinctive sources: a clinical study and a social media website. The clinical study was conducted among women aged 25 years and older who were newly diagnosed with stage I, II, or III breast cancer. Recruitment was conducted at two sites, both of which are academic cancer centers in the U.S., from 2002 to 2006. Only baseline data were used, and a total of N = 653 women completed the baseline survey, which included a symptom checklist. All of the baseline questionnaires were completed within 8 months of diagnosis. Five age categories were used: 25–44 (N = 132), 45 – 54 (N = 209), 55 – 64 (N = 167), 65 – 74 (N = 102), and 75+ (N = 43). In terms of cancer stages, 34.1% were in Stage 1, 55.3% were in Stage 2, and 10.6% were in Stage 3. Of the women who participated, 81% received chemotherapy and 86% received radiation. A Symptom Checklist [8] was used to assess symptoms, and the instrument contained 39 different symptoms. The symptoms were presented to patients as a checklist with a scale of 1–4, with 1 standing for "Symptom did not occur," 2 standing for "Symptom occurred but was mild," 3 standing for "Symptom occurred but was moderate," and 4 standing for "Symptom occurred and was severe." Details about the design of the clinical study are reported in [12].

We transformed the 39 × 653 data table from a Likert scale of 1–4 to 0/1: categories 1–3 were mapped to 0, and category 4 was mapped to 1.The similarity between the two symptoms i and j was defined as:

$$Similarity[symptom(i), symptom(j)] = (Co\_occurrence[symptom(i), symptom(j)])/$$
$$(Occurrence[symptom(i)] \times Occurrence[symptom(j)]), \tag{1}$$

where co-occurrence[symptom(i), symptom(j)] is the number of occurrences in which both symptoms i and j co-exist, and occurrence[symptom(i)] and occurrence [symptom(j)] are the number of occurrences of symptom i and j respectively.

The social-media data set was collected from the social media Website Medhelp.com. The Website has over 12 million users each month, making it one of the most used social media for Internet users interested in health issues. Participants in forums hosted by Medhelp.com shared and discussed health information and related life experiences. We specifically used data from one of the many Health Community Forums, which ranged from different types of cancer to other specific disorders such as diabetes and addiction. Given proper permission from Medhelp.com, we extracted 50,426 publicly available messages (including both the posts and the comments) from the Breast Cancer forum spanning the period October 1, 2006 to September 21, 2014. We parsed the raw text into structured information that included user IDs and time stamps on the content of each message, then we selected 100 users with at least 15 messages within a period of 6 months.

To identify the symptoms mentioned by the users, we used a vocabulary list constructed in the following way. First, we compiled from the Consumer Health Vocabulary (CHV) [13,14] a list of breast cancer symptoms that were included in the Symptom Checklist. Some of the symptoms from the Symptom Checklist were not mentioned at all in the posts, and they were eventually deleted from the analysis, resulting in a total of 34 symptoms. Next, for each breast cancer symptom, we constructed a consumer health vocabulary set that social media users may use to highlight the symptom [15,16]. For example, the symptom "diarrhea" is linked to a consumer health vocabulary set of words that include "water stools," "loose stools," "bowel movement," etc. Whenever a phrase in the health vocabulary set of a particular symptom appeared in a message, we considered that to be an occurrence of the corresponding symptom.

2.2   Data Analysis

The two matrices – the 39 ×39 similarity matrix derived from the clinical data and the 34 × 34 similarity matrix derived from social-media data – formed the basis of the cluster analysis. The K-medoid Clustering method [17] was applied to each matrix, which is a partition algorithm somewhat similar to K-means. However, unlike K-means, which calculates the centroid of a cluster as the average of the members of that cluster at each new iteration, K-medoid Clustering assigns the members of the cluster with the lowest overall cost as the new centroid, or medoid, at each iteration. For the current study, an advantage of using the K-medoid method is that it improves the interpretability of the cluster solution. A medoid can be viewed as forming an anchoring point for interpreting the entire cluster.

To conduct K-medoid Clustering for each matrix, the number of clusters needed to be determined first. We used the Average Silhouette Width (ASW) [18] to determine the number of clusters. When the clustering is working properly, the dissimilarity of a given point from other points within the same cluster will be smaller than its dissimilarity from other points in the next closest cluster, and in such cases the ASW will be closer to 1. On the other hand, if the clustering is not fitting the data well, the ASW would be closer to -1. An ASW value of 0 indicates that the points are sitting on the border between two hard-to-distinguish clusters.

3   Results

3.1   Symptom Clustering Results of Social Media Data

We computed the ASWs for different numbers of clusters (k = 1, 2, 3, ..., 15). Figure 1 shows the ASW values plotted against the number of clusters for the social-media data. We selected the smallest number of clusters K = 8, which attains the maximum value of ASW.

Figure 2 depicts the corresponding cluster solution, with symptoms, as represented by nodes, that belong to the same cluster being fenced with rectangles. The size of the node is in proportion to the prevalence of the symptom, and the thickness of the link between nodes is in proportion to the degree of similarity between the two symptoms. The medoid, or "anchor," of each cluster is denoted by a solid dark circle.

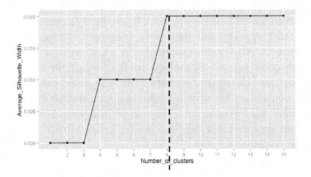

Fig. 1. Average silhouette width for social-media data

The 8-cluster solution result appears to achieve a rather coherent structure. For example, thicker edges (more similar symptoms) mostly reside in the same cluster. Each cluster contains two to five symptoms, and the symptoms within a cluster are generally closely linked. The clusters also provide meaningful clinical interpretations. For example, Cluster 1 (Fig. 1), which contains hot flashes (medoid), vaginal dryness, night sweats, and mood changes, can be described as menopausal symptoms. Cluster 7 contains nausea and diarrhea and can be described as a gastrointestinal cluster. On the other hand, difficulty concentrating, forgetfulness, and nervousness could all be side-effects due to cancer treatment such as radiation and chemotherapy.

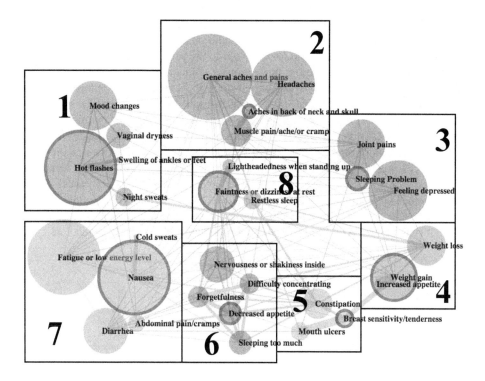

Fig. 2. Clustering results of social-media data

3.2 Symptom Clustering Results of Clinical Data

For the clinical data, the ASW reached its maximum at 7 clusters (Fig. 3). Figure 4 depicts the 7-cluster solution for the clinical data.

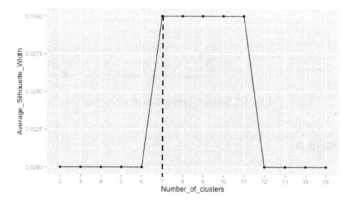

Fig. 3. Average silhouette width for clinical data

The sizes of clusters in the 7-cluster solution for clinical data are relatively even, and like the cluster solution for the social-media data, symptoms that have a higher degree of similarity (thicker links) are assigned to the same cluster. Each cluster has 3 to 13 symptoms, and compared with the cluster solution derived from social-media data, there are a number of overlapping clusters. For example, symptoms related to pain (Muscle Pain and General Ache, Headache and Ache in the Back), symptoms related to cognition (Forgetful and Difficulty Concentrating), and symptoms related to weight (Weight Gain and Increased Appetite) overlap substantially. Clusters 2,4,5,6, and 7 approximately map to Clusters 5,1,3,6, and 7 derived from social-media data. However, Cluster 1 in the clinical data appears to encompass three clusters identified in social media – fatigue, aches in the back and neck, headaches, and dizziness. There are also other incongruences across the two clustering solutions. For example, feeling depressed is associated with fatigue, mood, and emotional functioning in the clinical data but is associated with joint pain and sleep problems in the social-media data.

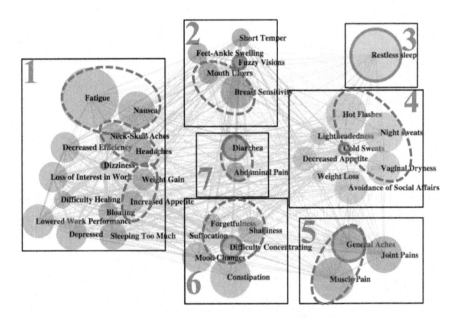

Fig. 4. Clustering results of clinical data

4   Discussion

By comparing the symptom clustering result of social-media data with medical record data, we made several observations.

First, there are substantial overlaps between the two sets of symptom clusters. Many symptoms are logically and semantically related, so it is not surprising that they appear together. For example, General Ache, Headache, Pain in the Back, and Muscle

Pain appear together, the weight-related symptoms Weight Gain and Increased Appetite also appear together. Besides the obvious semantically related symptom clusters, which could be a consequence of how the health vocabulary set was established for the social-media data, the other clusters derived from social media are also clinically meaningful and offer useful interpretations. The menopausal symptom cluster, for example, agrees quite well with that derived from the clinical data. Given the nature of the unstructured text in the social-media data, the high level of consistency between the two sets of cluster solutions is rather surprising.

At the same time, there are also important discrepancies between the two sets of clusters. Weight Loss and Weight Gain, for example, both appear in the same cluster in the social-media solution but in distinct clusters in the clinical data. As a matter of fact, weight loss and weight gain are clinically distinct issues among cancer patients – the former is related to loss of appetite and sometimes menopausal symptoms, whereas the latter is associated with increased appetite and often general fatigue. However, because of the way health vocabularies were extracted from text messages, mentions of "weight loss" and "weight gain" that co-occur in the same message – regardless of context – would be treated as evidence of association. Thus, in examining the results from the social-media data, it is important to know the context in which the symptoms appear.

We also observed that the similarity matrix derived from the social-media data is relatively sparse compared with that derived from clinical data. Indeed, this is partly our reason for using the Severe Symptom category vs. other categories for dichotomizing the clinical data. Had we used Presence of Symptom vs. any level of severity, the co-occurrences would have been much higher in the clinical data. In the clinical study, all 39 symptoms were presented to the patients so that they had to go through them one by one. This could have led participants to check more symptoms than they would have reported in an open-ended questionnaire. Discussion of symptoms in an online forum is more likely to be similar to responding to an open-ended questionnaire. Furthermore, online participants are more likely to mention only the most bothersome symptoms in posts and comments instead of offering a laundry list of symptoms they experienced. Apparently, such a dynamic in online interaction is quite different from the dynamic of responding to a questionnaire in a clinical study when the participants could be interviewed by a clinical associate in a clinic.

The different dynamics are reflected in the denseness, or the lack of it, of the respective similarity matrices.

5   Conclusion

There is an increasing tendency to use Big Social Data to automate learning, compute trends, and make predictions. While some have advocated the use of such data to study many different kinds of phenomena without even requiring significant knowledge of data-analysis techniques, others have argued for more balanced approaches such as combining Big Data with ethnographic methods and small-scale, structured studies. In this study, we demonstrated that Big Social Data can be used in conjunction with clinical data to inform an important clinical question. Although there is substantial agreement between the results derived from online forum data and from clinical study data, there are also significant discrepancies. Understanding the context of a problem is important in explaining such discrepancies and for integrating knowledge derived from both types of analyses.

Acknowledgment. This work was supported by the grant award NIH 1R21AG042761-01(PI: Edward Ip) and NSF DIBB (PI: Chris Yang).

References

1. Manovich, L.: Trending: the promises and challenges of big social data. In: Gold, M.K. (ed.) Debates in the Digital Humanities, pp. 460–475. University of Minnoestoa Press, Minneapolis (2012)
2. Mehl, M.R., Gill, A.J.: Automatic text analysis. In: Gosling, S.D., Johnson, J.A. (eds.) Advanced Methods for Conducting Online Behavioral Research, pp. 109–127. American Psychological Association, Washington (2010)
3. Vogt, W.P., Garnder, D.C., Haeffele, L.M.: When To Use What Research Design. Guilford, New York (2012)
4. Mahrt, M., Scharkow, M.: The Value of Big Data in Digital Media Research. J. Broadcast & Elect. Media **57**, 20–33 (2013)
5. Ginsberg, J., Mohebbi, M.H., Patel, R.S., Brammer, L., Smolinski, M.S., Brilliant, L.: Detecting Influenza Epidemics Using Search Engine Query Data. Nature **457**, 1012–1014 (2009)
6. Boyd, D., Crawford, K.: Critical Questions for Big Data. Informat., Commun., & Soc. **15**, 662–679 (2012)
7. Lazer, D., Kennedy, R., King, G., Vespignani, A.: The Parable of Google Flu: Traps in Big Data Analysis. Science **343**, 1203–1205 (2014)
8. De Haes, J.C., Van Knippenberg, F.C., Neijt, J.P.: Measuring Psychological and Physical Distress in Cancer Patients: Structure and Application of The Rotterdam Symptom Checklist. Br. J. Cancer **62**, 1034–1038 (1990)
9. Miaskowski, C.: Symptom Clusters: Establishing the Link Between Clinical Practice and Symptom Management Research. Supp. Care Canc. **14**, 792–794 (2006)
10. Dodd, M., Janson, S., Facione, N., et al.: Advancing the Science of Symptom Management. J. Adv. Irs **33**, 668–676 (2001)
11. Payne, J.K., Piper, B.F., Rabinowitz, I., Zimmerman, M.B.: Biomarkers, Fatigue, Sleep, and Depressive Symptoms in Women With Breast Cancer: a Pilot Study. Oncology Nursing Forum **33**, 775–783 (2006)
12. Avis, N., Levine, B., Naughton, M., Case, L.D., Naftalis, E., Van Zee, K.J.: Age Related Longitudinal Changes in Depressive Symptoms Following Breast Cancer Diagnosis and Treatment. Breast Cancer Res. Treat **139**(10), 199–206 (2013)
13. Zeng, Q.T., Tse, T.: Exploring and Developing Consumer Health Vocabularies. Journal of the American Medical Informatics Association **13**, 24–29 (2006)
14. Tse, T., Soergel, D.: Exploring medical expressions used by consumers and the media: an emerging view of consumer health vocabularies. In: AMIA Annual Symposium Proceedings, pp. 674–678 (2003)

15. Jiang, L., Yang, C.C., Li, J.: Discovering consumer health expressions from consumer-contributed content. In: Greenberg, A.M., Kennedy, W.G., Bos, N.D. (eds.) SBP 2013. LNCS, vol. 7812, pp. 164–174. Springer, Heidelberg (2013)
16. Jiang, L., Yang C.C.: Using Co-occurrence analysis to expand consumer health vocabularies from social media data. In: Proceedings of IEEE International Conference on Healthcare Informatics 2013, pp. 74–81 (2013)
17. Kaufman, L., Rousseeuw, P.J.: Clustering by means of medoids. In: Dodge, Y. (ed.) Statistical Data Analysis Based on the L1–Norm and Related Methods, pp. 405–416. North-Holland, Birkhäuser (1987)
18. Rousseeuw, P.J.: Silhouettes: a Graphical Aid to the Interpretation and Validation of Cluster Analysis. Journal of Computational and Applied Mathematics **20**, 53–65 (1987). doi:10.1016/0377-0427(87)90125-7

Women's Right to Drive: Spillover of Brokers, Mobilization, and Cyberactivism

Serpil Yuce, Nitin Agarwal[(✉)], and Rolf T. Wigand

Department of Information Science,
University of Arkansas at Little Rock, Little Rock, USA
{sxtokdemir,nxagarwal,rtwigand}@ualr.edu

Abstract. The advent of modern forms of information and communication technologies (ICTs), such as social media, have modified the ways people communicate, connect, and diffuse information. Social media has played an unprecedented role in coordinating and mobilization of social movements. Further, cross-influence among social movements has been observed during the 2011 Arab Spring, ongoing Saudi Women's movements against inequitable gender laws, and other social movements. Moreover, this has been a topic of research in the social movement spillover literature, where resource sharing, spillover of activists, supporters, and coalitions within social movements have been the focus of these studies. However, much of the work published in this area dates back to the 1960s. It is crucial to re-evaluate traditional theories of spillover in the modern ICT landscape. The evolution of Internet-driven collective actions triggered the examination and discovery of some essential factors of spillover remaining theoretically underdeveloped and call for innovative fundamental research that can provide observations in reconceptualizing spillover effects in online environments. In this research, we examine three campaigns during 2013 of the ongoing women's right to drive movement, namely the 'October 26 Driving Campaign', the 'November 31 Driving Campaign', and the 'December 28 Driving Campaign'. By analyzing the Twitter data on the networks of the three campaigns, we identify the factors that help us study spillover between the campaigns. We aim to assess social movement spillover effects by identifying – (1) common activists and supporters, (2) the inter-campaign brokers/coalitions, and (3) the diffusion of hashtags and other resources. We envision the findings of this study to shed insights on information diffusion and mutual influence across movements and provide a deeper understanding of interconnected social movements.

Keywords: Collective action · Social movement · Spillover · Social media · Hashtag diffusion · Women's right to drive · Twitter · Methodology · Resource sharing · Coalition

1   Introduction

The prevalence of contemporary forms of information and communications technologies (ICTs), such as social media, have transformed the ways people interact, commu-

© Springer International Publishing Switzerland 2015
N. Agarwal et al. (Eds.): SBP 2015, LNCS 9021, pp. 232–242, 2015.
DOI: 10.1007/978-3-319-16268-3_24

nicate, and share information. As evident in the mass protests during the Arab Spring, the Occupy and other recent movements, social media platforms helped the protesters to instantly spread messages, organize, and mobilize support for their campaigns. This has been a topic of research in the social movement spillover literature, where resource sharing, spillover of activists, supporters, and coalitions within social movements have been the focus of these studies. Contemporary movements are especially affected by "spillover" because the Internet increases the speed and effectiveness with which allied social movements reach out to one another's activists. Social movement spillover is the most basic type of inter-movement interdependency. Movements may be sequenced such that one movement is clearly the "initiator" and the other a "spinoff". Social movements must be considered in terms of their connections to past, contemporaneous, and upcoming social movements. However, much of the work published in this area dates back to the 1960s. It is crucial to re-evaluate traditional theories of spillover within the modern ICT landscape. The evolution of Internet-driven collective actions triggered the examination and discovery of some essential factors of spillover remaining theoretically underdeveloped and call for innovative fundamental research that can provide observations in reconceptualizing spillover effects in online environments. In this research, we examine three campaigns during 2013 of the ongoing Saudi women's right to drive movement, namely the 'October 26 Driving Campaign', the 'November 31 Driving Campaign', and the 'December 28 Driving Campaign'. By analyzing the Twitter data on the networks of three campaigns, we identify the factors that help us study spillover between the campaigns. We are looking for answers for the following questions including: "Who are the bridging nodes in the networks and how does their brokerage give direction to the diffusion of ideas through interrelated social movements?" and "How does the interaction of networks and coalitions trigger the spillover effect and helps to shape the evolution of future social movements?". We aim to assess social movement spillover effects by identifying – (1) common activists and supporters, (2) the inter-campaign brokers/coalitions, and (3) the diffusion of hashtags and other resources. We envision the findings of this study to shed insights on information diffusion and mutual influence across movements and provide a deeper understanding of interconnected social movements.

2   Women's Right to Drive Movements

Saudi women face some of the most inequitable laws and practices when compared to international standards, including the prohibition of driving motorized vehicles. Saudi Arabia is the only country worldwide with this type of ban in effect. Although there is no official ban, due to social and cultural restrictions there are no women drivers in the Kingdom (United Nations 2007). In order to create awareness about these inequitable laws and practices, Saudi women have organized several campaigns as part of a bigger movement. The origin of 'Women's Right to Drive', a movement that emerged to address this issue, can be traced to the November 6, 1990 event in which 47 Riyadh women staged a protest and drove in a Riyadh parking lot in resistance to this ban. More than a decade later, on International Women's Day in 2008, the YouTube video of Wajeha al-Huwaider driving had garnered the interest of social media sites and women across the world making her protest international news [1]. Following the al-Huwaider protest, in 2011, a group of women led by Manal al-Sharif, started the Facebook campaign supporting women's driving rights in Saudi Arabia, 'Women to Drive.' As the movement evolved over time, naturally there are various online platforms used for the same cause. Until recently, the most active participation in this movement has been predominantly from female Muslim bloggers who use blogs as primary means to converse. However since 2013, with the Oct26Driving campaign, Twitter became one of the most widely used platforms to coordinate and mobilize protests around the right to drive movement. The platform has been used to extend the campaign network and gain additional traction. The Oct26Driving campaign is considered as one of the most successful campaigns in the history of Saudi Arabia led entirely via social media, especially Twitter. The Oct26Drving campaign is a part of the Women's Right to Drive movement in order to create awareness about inequitable laws and practices, especially the ban on driving. Our earlier studies [2, 3, 4, and 5] have analyzed these campaigns to understand various aspects of online collective action. Known as "The 26th October Campaign", it quickly gained momentum with its online petition garnering more than 16,000 signatures (according to the official campaign website) despite the Kingdom's restrictions on protests. The campaign website was hacked on October 9, 2013 leading to a surge in Twitter activity. The October 26th campaign is a grassroots campaign with the participation of the women and men of Saudi Arabia and aims to revive the demand to lift the ban on women driving. Although King Abdullah bin Abdulaziz, now deceased, left this matter to society, the government's reaction makes it very clear that this is not a societal decision but a political decision. The next day of "mass" driving is November 31, 2013. The ban on women driving in Saudi Arabia has become a symbol of a far wider lack of gender equality in the conservative Islamic Kingdom, where women must have permission from a male guardian to travel. Eventually, a new hashtag emerged on Twitter: #Nov31Driving. Close to October 26, 2013, activists and supporters started swapping and/or combining the original hashtag dedicated to Oct26Driving campaign with hashtags dedicated to a newly forming campaign, named as Nov31Driving campaign, with a date that does not exist, i.e. the date is only symbolic, and it is intended to continue indefinitely. Although it could cause confusion since November only has 30 days, various tweets explain it as a way to make the campaign continuing and open-ended. However, the campaign was extended to increasingly challenge the law ahead of a new nationwide day of defiance on December 28, 2013. A Saudi female activist, Nasima al-Sada, has called on other women to get behind the wheel again on December 28, where the call for action is a "reminder of the right so it is not forgotten". Women activists are now driving weekly and documenting their confrontations with law enforcement on social media to increase pressure on the conservative country and keep the issue in the public eye. The campaign aims to revive the demand to lift the ban on women driving while stressing that the initiative has no anti-Islamic or political agenda. As mentioned earlier, the Oct26Driving campaign is considered as one of the most successful campaigns in the history of Saudi Arabia when compared to previous and subsequent Women's Right to Drive campaigns.

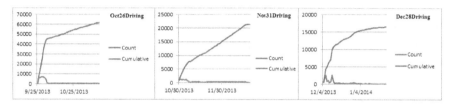

Fig. 1. Tweet activity for Oct26Driving, Nov31Driving, and Dec28Driving campaigns

As shown in Figure 1 above, we observe a decline in participation of members of Nov31Driving and Dec28Driving campaigns, when comparing the tweet volume of Nov31Driving and Dec28Driving campaigns to that of Oct26Driving campaign. Though activists were encouraging women, the Saudi government toughened the ban on female drivers by increasing the arrests and flogging punishments after the success of the Oct26Driving campaign. It is an inevitable fact that daunts the Saudi Women's courage and voluntarism in participation.

3   Literature Review

There has been a recent increase in the literature on the influence of Twitter in political contestations in the Middle East and North Africa (MENA) region, starting with #iranelection in 2009 that attempted to position popular demonstrations as the "Twitter Revolution" [6]. More recent literature has turned to the "Arab Spring" uprisings. According to Skinner, "Hashtags are used both to coordinate planning for certain events or to tie one's tweets with a larger discussion on the subject" [7]. The Saudi women's driving campaign used the Twitter network in both capacities. Hashtag diffusion in different languages has been linked to drawing local collective action into the transnational Twittersphere networks through bridging mechanisms [6], which contribute to transnational support for localized movements. The shifts in popularity of hashtags, in the case of the Egyptian revolution, was related to the growth of an elite group of Twitter users, elsewhere referred to as "key actor types" [8], who were most influential in orienting discourses on the uprisings [6]. These shifts are important in gauging not only key actors but also how movements spread on Twitter. Lerman and Ghosh (2010) [9] outline the importance of users in information cascades, which rely on a triad of network-retweets-followers in the Twittersphere. We are interested in identifying key actors, information cascades, and transnational bridges within Twitter supporting the Saudi women's driving campaign.

While the "Arab Spring" on-the-ground tweets were largely unstructured real-time reporting of events, leading to the dissemination of information and mobilization efforts under various hashtags, our paper investigates a loosely organized grassroots example of collective action, which manifested itself on Twitter as @Oct26driving, so there were fewer possibilities of sub-clusters formulating under divergent hashtags. In other words, since @Oct26driving was established as a Twitter profile, Twits interested in finding out about the movement and sharing information had a central platform to engage with the campaign. However, since the Saudi women's driving movement was disseminated by two hashtags in Arabic and English, the development of these hashtags were locally and transnationally positioned to mobilize linked clusters within the larger movement.

3.1   Social Movement Spillover

The concept of spillover is observationally built on a case study of the interaction between the women's and the peace movement in the United States and theoretically refers to political process approaches and to the New Social Movement Theory [10, p. 278]. Contemporary movements are specifically influenced by spillover because the Internet raises the speed and efficiency with which ideologically connected social movements reach out to one another's activists [11, 12, and 13]. Meyer and Whittier's analysis of "social movement spillover" is the process of mutual influence and support between two or more allied movements. The Social Movement Spillover Theory is mostly focused around the perception that people frequently exchange their support and participation from group to group and points to two ways a movement can impact subsequent movements. Movements can affect subsequent movements, first, by passing information about public actions and their effects on organizational responses to collective action, and identifying potential strengths and vulnerabilities in the political structure or affecting changes in the external environment that restructure political opportunities, and, second, by altering the people, groups, and standards within the movement itself [10, p. 279]. Different movements and organizations within movements share individuals, and these shared individuals can move from one group to another or cooperate across groups (interorganizational support). In the context of these women's rights movements our focus is on the spillover of supporters, activists, and resources within the movements. Movements embrace and diffuse between one another by adopting a range of movement characteristics including frames (organizational structures), collective identities, tactics, and movement culture.

3.2   Computational Analysis of Events

Most of the existing research in computational analysis of events examines events in isolation. These analysis consider events as having discrete boundaries, with almost zero interaction between events. Our work differs in the sense that we consider interaction, exchange, and spillover between events in terms of actors, ties, resources, and networks. Next, we briefly review the computational studies literature on event analysis. User-generated data from various social media platforms, related to real-life events, have been studied to perform wide range of analysis. Socio-political inferences are drawn by studying sentiments and opinions of people towards public and political events from Twitter [14], as well as blogs [15]. Twitter has been extensively used as a source for analyzing information circulated during natural disasters and crisis situations [16], [17]. Tweets related to events have been extracted, summarized and visualized, in order to have a deeper understanding of the events [18], [19]. Event related contents have been found leveraging the tagging and location information associated with the photos shared in Flickr [20]. Becker et al. [21], studied how to identify events and high quality sources related to them from Twitter. User-generated data from various social media platforms, related to real-life events, have been studied to perform a wide range of analyses. Platforms like TwitterStand [22], Twitris [23], TwitInfo [18] and TweetXplorer [24] have developed techniques to provide analytics, and visualizations related to different real-life events. Similar tools have been used for tracking earthquakes, providing humanitarian aid during the time of crisis, and analyzing political campaigns. Most of the event analysis frameworks rely on finding relevant keywords and networks between the content producers in order to analyze events.

4   Methodology

Across the board, social movement researchers have addressed the study of social networks in ways that highlight only particular forms of connections which limits the exploration of particular challenges that have concerned scholars of social movements for decades. These challenges include questions about: diffusion of social movement performances or activities, identities, and organizational forms; certain activists and organizations' positions in movements both as 'brokers' of information, ideas, or resources, and as leaders or focal points of power and prestige; the role of social networks in engaging a movement's members; the framework of movement organizations and coalitions. In this research, we examine three campaigns during 2013 of the ongoing women's right to drive movement, namely the 'October 26 Driving Campaign', the 'November 31 Driving Campaign', and the 'December 28 Driving Campaign'. The main focus of this research is to analyze data to demonstrate and reveal information about how ideas diffused and generated a mutual influence and support between several congruent social movements. By analyzing the Twitter data on the networks of three campaigns, we identify the factors that help us study spillover between the campaigns. We aim to assess social movement spillover effects by identifying – (1) common activists and supporters, (2) the inter-campaign brokers/coalitions, and (3) the diffusion of hashtags and other resources. For this purpose, we are seeking answers for the following questions: (1) "Who are the bridging nodes in the networks and how does their brokerage give direction to the diffusion of ideas through interrelated social movements?" – (2) "How does the interaction of networks and coalitions trigger the spillover effect and help to shape the evolution of future social movements?"

4.1   Data Collection for the Campaigns

The content from 116,565 tweets was collected (Oct26Driving: 83,433 tweets, Nov31Driving: 22,466 tweets, and Dec28Driving:10,666 tweets) using the Scraper-Wiki (www.scraperwiki.com) program. Tweets are collected based on if they include dominant hashtags dedicated to the three campaigns, viz. '#oct26driving', '#قيادة_26أكتوبر', '#نوفمبر31قيادة', 'Nov31Driving', '#ديسمبر28قيادة' and 'Dec28Driving'. Other available information, such as language, name, retweeted user name (RT), and other included hashtags, is also collected. Since these tweets are updated with frequencies varying between hundred to thousands of tweets per day, a crawler (viz., ScraperWiki), was configured with the above mentioned dominant hashtags running constantly to systematically collect, parse, and index the data.

4.2   Tweet Classification and Overlap Detection

The indexed tweets were filtered based on their hashtag usage, particularly
'#oct26driving' and '#قيادة_26اكتوبر' hashtags for Oct26Driving campaign,
'#نوفمبر31قيادة' and 'Nov31Driving' hashtags for Nov31Driving campaign, and
'#ديسمبر28قيادة' and 'Dec28Driving' hashtags for Dec28Driving campaign. The tweets
were further grouped based on if they used either or all the hashtags. Our first aim
was to focus on the cumulative traffic of the tweets and track the flow of common
activists and supporters through these three campaigns to identify the social movement spillover. Tweet-retweet networks are created during the time periods corresponding to the formation of campaigns. Tweet-retweet networks helped us track the
inter-campaign brokers and coalitions to able to identify the spillover effect across the
campaigns. To study the social movement spillover effect, as a second filtering
process, we created hashtag and resource networks for each campaign and studied the
role of hashtag and resource sharing in social movement spillover. The networks of
each campaign were compared to identify the overlap(s) between these networks.

5   Analysis and Results

5.1   Diffusion of Activists and Supporters

As discussed earlier, Meyer and Whittier's analysis of "social movement spillover" is
the process of mutual influence and support between two or more allied movements.
In the content of the Women's Right to Drive movements the focus is on the spillover
of supporters, activists, and resources within the movements. In order to be able to
identify and track the formation of the spillover of activists and supporters among
three campaigns, we compared the user and re-tweet networks of each campaign and
detected the overlap between them as shown in Figure 2.

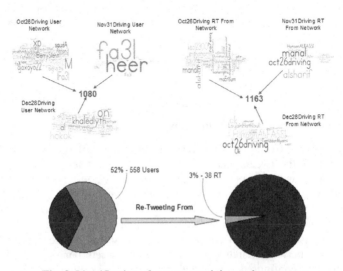

Fig. 2. Identification of common activists and supporters

Out of 45,276 distinct members of the three campaigns, we identified 1,080 users, who participated in all three campaigns to support the Women's Right to Drive movement. The result indicates that there is a strong presence of inter-campaign collaboration among the three campaigns. In addition to this collaboration, to emphasize the spillover effect, we further analyzed our data to identify the common activists who act as brokers in bridging three overlapping networks. From the retweet (RT) network, our results indicate that 1,163 activists out of 9,319 activists are common to the three campaigns. When we compare these two separate overlapping networks, we discover the important fact that 52% of the shared users are retweeting from 38 activists common to all three campaigns. As hypothesized earlier, there is an obvious interaction and cooperation between three campaigns that help us to identify the social movement spillover among them.

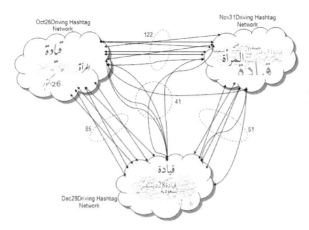

Fig. 3. Hashtag diffusion of Oct26Driving, Nov31Driving, and Dec28Driving campaigns

5.2 Diffusion of Hashtags and Other Resources

Given the definitive nature of hashtags, we investigated the diffusion of hashtag usage among three campaigns to identify and track the formation of the social movement spillover. For this purpose, we created a hashtag network for each campaign and compared each individual network with each other to identify overlap and diffusion. Figure 3 shows the number of hashtags that diffused from one campaign to the subsequent. As hypothesized earlier, increased co-occurrence of the hashtags between campaign networks result in formation of the social movement spillover among three campaigns.

As mentioned above, to identify the social movement spillover we analyze the resource sharing in terms of spillover of media platforms including Twitter, YouTube, and other media channels. We analyze our data to track the diffusion of YouTube videos (www.youtube.com/watch?v=xiuOY5xM_So&feature=youtube_gdata_player) and audio (cdn.top4top.net/d_4628b8faa42.mp3) recordings. We found shared audio and video recordings common to the three campaigns. This illustrates the spillover of resources shared on social media platforms to help campaign organization efforts and raising awareness. Figure 4 illustrates the relation of the user interaction network and the social media interaction network. The user interaction network depicts those users who supported the movement through Twitter by tweeting about the campaigns.

As evident in the Figure 4, there are users who have supported multiple campaigns depicted by the overlapping set of nodes in the network. For instance, users who support the Oct26Driving campaign and the Nov 31Driving campaign have an orange and a blue edge. Users who are supporting all the three campaigns are connected by orange, blue, and green edges. The social media interaction network depicts the social media platforms used by the supporters of the campaigns. As evident in the figure, there are media platforms such as YouTube and Top4toP that are common to all the campaigns, meaning the supporters have used these platforms to share resources across the different campaigns.

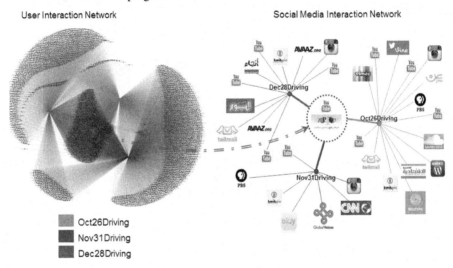

Fig. 4. Interaction between user network and YouTube video network

6   Conclusion

As evident in the mass protests during the Arab Spring, the Occupy and other recent movements, social media platforms helped the protesters to instantly spread messages, organize, and mobilize support for their campaigns. This has been a topic of research in the social movement spillover literature, where resource sharing, spillover of activists, supporters, and coalitions within social movements have been the focus of these studies. Contemporary movements are especially affected by "spillover" because the Internet increases the speed and effectiveness with which allied social movements reach out to one another's activists. The Social Movement Spillover Theory is mostly focused around the perception that people frequently exchange support and participation across movements. It further points to ways a movement can impact consequent or subsequent movements. In the context of the Women's Right to Drive movements the focus is on the spillover of supporters, activists, and resources within the movements. In this study, we examined the networks of three campaigns related to Women's Right to Drive movement viz., 'Oct26Driving', 'Nov31Driving', and 'Dec28Driving', to study the spillover effect among them. By extracting the common supporters' and activists' networks, we discovered brokers who bridge different campaigns' networks. Further, we studied the sharing of hashtags and other resources (e.g., YouTube videos and audio recordings) among campaign networks to identify the spillover among three campaigns. Such resource sharing manifests itself into deeper and faster penetration, wider spread, and mobilization and communication costs. We envision the findings of this study to shed insights on information diffusion and mutual influence across movements and provide a deeper understanding of interconnected social movements.

Acknowledgments. This research is funded in part by the U.S. National Science Foundation's Social Computational Systems (SoCS) research program (award numbers: IIS-1110868 and IIS-1110649) and the U.S. Office of Naval Research (award numbers: N000141010091 and N000141410489). The authors are grateful to the support.

References

1. Young, J.: Wajeha Al-Huwaider: A Brave Heart! Al Waref Institute. http://www.alwaref.org/fr/component/content/article/37-figure-of-the-month/191-wajeha-al-huwaider-a-brave-heart (last accessed February 6, 2014)
2. Agarwal, N., Lim, M., Wigand, R.T.: Finding her master's voice: The power of collective action among female muslim bloggers. In: Proceedings of the 19th European Conference on Information Systems (ECIS), Helsinki, Finland, June 9-11, Paper 74 (2011a)
3. Agarwal, N., Lim, M., Wigand, R.T.: Collective action theory meets the blogosphere: A new methodology. In: Fong, S. (ed.) NDT 2011. CCIS, vol. 136, pp. 224–239. Springer, Heidelberg (2011)
4. Yuce, S.T., Agarwal, N., Wigand, R.T.: Mapping cyber-collective action among female muslim bloggers for the women to drive campaign. In: Greenberg, A.M., Kennedy, W.G., Bos, N.D. (eds.) SBP 2013. LNCS, vol. 7812, pp. 331–340. Springer, Heidelberg (2013)
5. Yuce, S., Agarwal, N., Wigand, R.T., Lim, M., Robinson, R.S.: Studying the evolution of online collective action: saudi arabian women's 'Oct26Driving' twitter campaign. In: Kennedy, W.G., Agarwal, N., Yang, S.J. (eds.) SBP 2014. LNCS, vol. 8393, pp. 413–420. Springer, Heidelberg (2014)
6. Brun, A., Highfiel, T., Burgess, J.: The Arab Spring and Social Media Audiences: English and Arabic Twitter Users and Their Networks. American Behavioral Scientist **57**(7), 871–898 (2013)
7. Skinner, J.: Social Media and Revolution: The Arab Spring and the Occupy Movement as Seen through Three Information Studies Paradigms. Sprouts: Working Papers on Information Systems **11**(169) (2011). http://sprouts.aisnet.org/11-169
8. Lotan, G., Graeff, E., Ananny, M., Gaffney, D., Pearce, I., Boyd, D.: The revolutions were tweeted: Information flows during the 2011 Tunisian and Egyptian Revolutions. International Journal of Communication **5**, 1375–1405 (2011)
9. Lerman, K., Ghosh, R.: Information contagion: An empirical study of the spread of news on digg and twitter social networks. In: ICWSM, vol. 10, pp. 90–97 (2010)
10. Meyer, D.S., Whittier, N.: Social Movement Spillover. Social Problems **41**, 277–297 (1994)
11. Carty, V., Onyett, J.: Protest, Cyberactivism and New Social Movements: The Reemergence of the Peace Movement Post 9/11. Social Movement Studies **5**(3), 229–249 (2006)

12. Chadwick, A.: Digital network repertoires and organizational hybridity. Political Communication **24**(3), 283–301 (2007)
13. Nah, S., Veenstra, A.S., Shah, D.V.: The Internet and anti-war activism: A case study of information, expression, and action. Journal of Computer-Mediated Communication **12**(1), 230–247 (2006)
14. Younus, A., et al.: What do the average twitterers say: A twitter model for public opinion analysis in the face of major political events. In: 2011 International Conference on Advances in Social Networks Analysis and Mining (ASONAM), pp. 618–623. IEEE (2011)
15. Agarwal, N., Liu, H.: Blogosphere: Research Issues, Tools, and Applications. SIGKDD Explorations **10**(1), 18–31 (2008)
16. Vieweg, S., et al.: Microblogging during two natural hazards events: what twitter may contribute to situational awareness. In: Proceedings of the 28th International Conference on Human Factors in Computing Systems, pp. 1079–1088. ACM (2010)
17. Cheong, F., et al.: Social media data mining: A social network analysis of tweets during the 2010–2011 Australian floods. In: PACIS 2011 Proceedings (2011)
18. Marcus, A., et al.: Twitinfo: aggregating and visualizing microblogs for event exploration. In: Proceedings of the 2011 Annual Conference on Human Factors in Computing Systems, pp. 227–236. ACM (2011)
19. Popescu, A., et al.: Extracting events and event descriptions from twitter. In: Proceedings of the 20th International Conference Companion on World Wide Web, pp. 105–106. ACM (2011)
20. Rattenbury, T., et al.: Towards automatic extraction of event and place semantics from Flickr tags. In: Proceedings of the 30th Annual International ACM SIGIR Conference on Research and Development in Information Retrieval, pp. 103–110. ACM (2007)
21. Becker, H., et al.: Selecting quality twitter content for events. In: Proceedings of the Fifth International AAAI Conference on Weblogs and Social Media (ICWSM 2011) (2011)
22. Sankaranarayanan, J., Samet, H., Teitler, B.E., Lieberman, M.D., Sperling, J.: Twitterstand: News in tweets. In: Proceedings of the 17th ACM SIGSPATIAL International Conference on Advances in Geographic Information Systems, pp. 42–51. ACM (2009)
23. Jadhav, A., Purohit, H., Kapanipathi, P., Ananthram, P., Ranabahu, A., Nguyen, V., Mendes, P.N., Smith, A.G., Cooney, M., Sheth, A.: Twitris 2.0: Semantically empowered system for understanding perceptions from social data. In: Semantic Web Challenge (2010)
24. Morstatter, F., Kumar, S., Liu, H., Maciejewski, R.: Understanding twitter data with tweetxplorer. In: Proceedings of the 19th ACM SIGKDD International Conference on Knowledge Discovery and Data Mining, pp. 1482–1485. ACM (2013)

Poster Presentations

A Mathematical Epidemiology Approach for Identifying Critical Issues in Social Media

Segun M. Akinwumi$^{(\boxtimes)}$

University of Alberta, Edmonton, AB T6G 2G1, Canada
segunmic@ualberta.ca

Abstract. This work innovated mathematical epidemiology concepts to create a model used to identify critical social issues. A compartmental SEI (acronym for Susceptible Exposed Infected) model was developed to investigate the flow of issues in social media. The basic reproduction number R_0, the number of secondary cases resulting from the introduction of an index case into an otherwise uninfected population, was derived for the model. Any social issue with R_0 greater than one was defined to be a critical issue. The model was evaluated under about 4.7 million tweets from Nigeria streamed over a six-month period and R_0 was estimated for each of the top twenty social phenomena in the dataset. Security ($R_0 = 30.784$), Ebola ($R_0 = 51.949$), and Transport System ($R_0 = 4.166$) were found to be critical issues in Nigeria using this methodology.

Keywords: Social media · Social epidemiology · Methodology · Twitter · Compartmental model · Mathematical epidemiology

1 Introduction

The Arab Spring impacts can be attributed to social media [4], [5]. There has been a restricted access to prominent social media such as Facebook, Twitter, and YouTube in countries such as Saudi Arabia, China, Syria, Egypt and Turkey [7], [8]. Rather than being perceived as a threat, many key government agencies have started using social media as platforms for social good. For instance, social media played an important role in identifying the masterminds of the tragic explosions at the 2013 Boston Marathon in Massachusetts, United States [2].

Social media posts of criminals can be analysed to understand the motives behind their crimes [10]. Social media posts can be used to identify individuals at risks of committing suicide [9]. There are various active online groups that daily share information about current affairs in their countries [2]. Mathematical modelling can be applied to formulate policies that will optimally address impending crises in a society based on these online activities.

The purpose of this work is to describe how mathematical epidemiology principles can be deployed to identify critical issues in social media data. Specifically, in this paper a compartmental model was developed to investigate the flow of

© Springer International Publishing Switzerland 2015
N. Agarwal et al. (Eds.): SBP 2015, LNCS 9021, pp. 245–250, 2015.
DOI: 10.1007/978-3-319-16268-3_25

information in social media. The basic reproduction number R_0, the number of secondary cases resulting from the introduction of an index case into an otherwise uninfected population, was derived for the model. The model was fitted to about 4.7 million tweets from Nigeria and R_0 for social issues that might be important to Nigerians was estimated.

Secion 2 of this article contains a review of some recent studies related to this work. The dataset used for this study is described and summarised in Secion 3. A mathematical epidemiology model of Twitter activities and the basic reproduction number for this model are presented in Secion 4. Model fittings and R_0 estimates are discussed in Secion 5. This work is concluded in Secion 6.

2   Related Work

The propagation of disease in human population has been extensively studied using mathematical modeling [3] and these various studies have provided control strategies for disease outbreaks [3], [1]. These models essentially partition the total population into compartments that correspond to possible health states of individuals in the population. This modeling paradigm was first applied to news and rumors on Twitter by the authors of [4] using a SEIZ (Susceptible Exposed Infected Skeptics) model. In the context of news and rumour spread on Twitter, SEIZ model assumes: a Twitter user who has not seen a tweet about the news is susceptible (S); a user who has tweeted about the news is infected (I); a user who has seen a tweet about the news but has not tweeted about it is exposed (E); and a user who has seen a tweet about the news but has chosen not to tweet about it is considered to be a skeptic (Z). Individual Twitter user is allowed to transfer between compartments at specified constant rates but the model assumes vital dynamics, which include the influx and outflux rates into the system, is negligible.

The SEIZ model can be simplified by combining the E and Z compartments. This makes me propose an SEI (Susceptible Exposed Infected) model that includes the vital dynamics, allowing for situations where a user can follow or unfollow a tweet, or subscribe or unsubscribe from Twitter over the study period. The model was then used to find leading thoughts in tweets by deriving the popular basic reproduction number R_0 for the model using the next generation matrix approach [6].

3   Datasets

This work used Twitter datasets without loss of generality since tweets are publicly available for research purposes. I streamed 4, 744, 929 tweets from Nigeria between Apr. 1, 2014 to Sep. 30, 2014, 6:00 am to 11:00 pm local time daily. This period was assumed to be the time when most Nigerian Twitter users are active. The top 25% of the active users contributed 2, 160, 562 tweets over the study period.

Hashtags (words with #, e.g. #qna) identify messages on specific topics. The top 20 hashtags (Fig. 1) were used to summarise the 4.7 million tweets and the SEI model developed in the next section was applied to the dynamics of tweets containing each of these hashtags. For the purpose of model evaluation, an infected tweet (from an infected Twitter user) was defined as a tweet that contained a hashtag of interest. The R_0 for each hashtag was estimated to determine if the hashtag connotes a critical message ($R_0 > 1$) or otherwise.

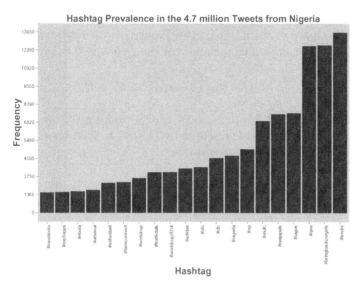

Fig. 1. Frequency Distribution of the Top 20 Hashtags in the 4.7 million Tweets from Nigeria Streamed from Apr. 1, 2014 to Sep. 30, 2014, 6:00 am to 11:00 pm Local Time Daily.

4   The Mathematical Model

It was assumed that the total population of Twitter users could be partitioned into three groups of individuals: the susceptible group S of Twitter users who have not seen tweets about the news of interest; the infectious group I of users who have tweeted about the news of interest; and the exposed group E of users who have seen tweets about the news of interest but have not tweeted about it.

It was further assumed that users subscribe to Twitter and consequently become susceptible at a rate given by dN, where $N = S + E + I$, that susceptible individuals can see an infected tweet and become infected at a rate given by $\beta \frac{SI}{N}$, and that susceptible individuals leave the class at a rate given by dS. The dynamics of the susceptible class is then given by

$$\frac{dS}{dt} = dN - \beta \frac{SI}{N} - dS. \tag{1}$$

In addition, a proportion p of the infected susceptible $\beta\frac{SI}{N}$ can move to the exposed class E. Exposed users can become infected by re-exposure to tweets from infected individuals at a rate given by $\alpha\frac{EI}{N}$ or these users can simply move to the infected class at a rate given by εE. The dynamics of the exposed class is as follows

$$\frac{dE}{dt} = p\beta\frac{SI}{N} - \alpha\frac{EI}{N} - \varepsilon E - dE, \tag{2}$$

where dE is the rate at which exposed individuals leave the class.

As a consequence, the contribution of the exposed users to the infected class is $\alpha\frac{EI}{N} + \varepsilon E$. The remaining proportion $(1-p)$ of the infected susceptible $\beta\frac{SI}{N}$ was assumed to move to the infected class I and dI is the rate at which infected individuals leave the class. The dynamics of the I class is then given by

$$\frac{dI}{dt} = (1-p)\beta\frac{SI}{N} + \alpha\frac{EI}{N} + \varepsilon E - dI. \tag{3}$$

The transfer diagram for this model can be represented by Fig. 2e with the corresponding dynamics represented by (1) to (3). The basic reproduction number R_0 of (1) to (3) was derived to be

$$R_0 = \frac{(1-p)\beta}{d} + \frac{p\beta}{d}\frac{\varepsilon}{\varepsilon+d},$$

using the next generation matrix approach [6].

5   Model Evaluation

The model in Section 4 was fitted to the tweets for each of the hashtags in Fig. 1 and the parameter set that minimised the objective function

$$\sum_{i=1}^{n}\left(I(t_i) - \text{number of infected tweets}(t_i)\right)^2$$

for the data of each hashtag was used to estimate the corresponding R_0. The **lsqnonlin** function in [11] was used for the least squares fitting. Only the fitting results for #bringbackourgirls, #ebola, #traffictalk, and #nigeria are reported as these are more likely to contain interesting social information in contrast to the other hashtags. The fittings are shown in Fig. 2 with the respective basic reproduction number R_0.

Insecurity, typified by the #bringbackourgirls hashtag which emanated from the struggle of activists to intensify efforts on the release of the Chibok girls adopted by Boko Haram, a terrorist group, is clearly critical to Nigerians. This was found in the model fitting with $R_0 = 30.784$ for #bringbackourgirls. Ebola Virus Disease (EVD) is also of concern to Nigerians as revealed by the $R_0 = 51.949$ for #ebola. Similarly, transport system, typified by #traffictalk with $R_0 = 4.166$, is critical to Nigerians.

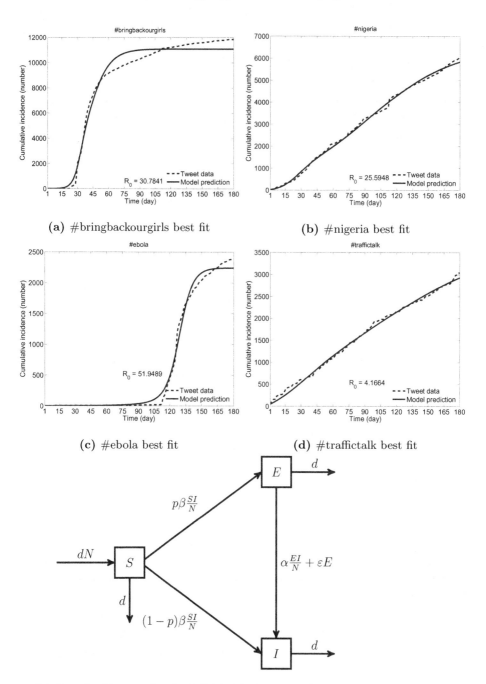

(a) #bringbackourgirls best fit

(b) #nigeria best fit

(c) #ebola best fit

(d) #traffictalk best fit

(e) Transfer Diagram for the SEI Model of Twitter Activities.

Fig. 2. Best Model Fits for Hashtags with Social Interests, Fig. 2a to Fig. 2d, and Transfer Diagram for the SEI Model, Fig. 2e.

6   Conclusion

Remarkably, both Security and Ebola have been suggested by opinion leaders to be critical factors that might influence the outcome of 2015 General Elections in Nigeria. This work identified Security ($R_0 = 30.784$), Ebola ($R_0 = 51.949$), and Transport System ($R_0 = 4.166$) as critical issues in this country. It is worthy to note that further research is needed to identify the topics in these hashtags, especially the #nigeria tag, as it is unclear what messages are being discussed in the contexts of these tags.

Acknowledgments. I acknowledge the Pacific Institute for the Mathematical Sciences-International Graduate Training Center Fellowship for funding.

References

1. Hethcote, H.W.: The Mathematics of Infectious Disease. SIAM Rev. **42**(4), 599–653 (2006)
2. Mosley, R.C.: Social Media Analytics: Data Mining Applied to Insurance Twitter Posts. Casualty Actuarial Society E-Forum **2** (2012)
3. Brauer, F.: Mathematical Epidemiology is not an Oxymoron. BMC Public Health **9**(Suppl 1), S2 (2009)
4. Jin, F., Dougherty, E., Saraf, P., Cao, Y., Ramakrishnan, N.: Epidemiological modeling of news and rumors on twitter. In: SNAKDD 2013 Proceedings of the 7th Workshop on Social Network Mining and Analysis, Article No. 8. ACM, New York (2013)
5. Sarihan, A.: Is the Arab Spring in the Third Wave of Democratization? The Case of Syria and Egypt. TJP **3**(1) (2012)
6. Diekmann, O., Heesterbeek, J.A.P., Roberts, M.G.: The construction of next-generation matrices for compartmental epidemic models. J. R. Soc. Interface **7**(47), 873–885 (2010)
7. Human Rights Watch: World Report 2013. http://www.hrw.org/sites/default/files/wr2013_web.pdf (retrieved November 13, 2014)
8. Safranek, R.: The Emerging Role of Social Media in Political and Regime Change. http://md1.csa.com/discoveryguides/social_media/review.pdf (retrieved November 13, 2014)
9. Robinson, J., Rodrigues, M., Fisher, S., Herrman, H.: Suicide and Social Media: Findings from the Literature Review. Young and Well Cooperative Research Centre, Melbourne (2014)
10. Frank, R., Cheng, C., Pun, V.: Social Media Sites: New Fora for Criminal, Communication, and Investigation Opportunities. Ottawa, ON: Public Safety Canada (2011)
11. MATLAB version 8.1.0.604. Natick, Massachusetts: The MathWorks Inc. (2013)

Analyzing Deviant Cyber Flash Mobs
of ISIL on Twitter

Samer Al-khateeb$^{(\boxtimes)}$ and Nitin Agarwal

Department of Information Science,
University of Arkansas at Little Rock, Little Rock, AR 72204, USA
{sxalkhateeb,nxagarwal}@ualr.edu

Abstract. Transnational crime organizations and terrorist groups such as, ISIL use social media in a highly sophisticated and coordinated manner to disseminate their propaganda. Various social media platforms, such as Twitter, YouTube, Facebook, ask.fm, Tumblr, etc. are used by ISIL to recruit, radicalize, and raise funds. By doing so, they have more control on their messages and more assurance that their messages (text, photo, or video) reach many people instantly. Here, we study ISIL's social media activity (propaganda messages, recruitment messages, etc.) and model their sociotechnical behavior as a manifestation of a deviant cyber flash mob (DCFM) behavior. We operationalize our conceptual model for analyzing DCFM, demonstrate the model's applicability to real-world setting, evaluate the research using Twitter data, and in doing so, highlight the challenges, contributions, and broader implications of the work. The findings of the research reveal a strategic choice of platforms used by ISIL's top disseminators for propaganda dissemination. Further, the network activity of ISIL's top disseminators reflect a well-coordinated organization with designated roles for outreach and intra-organization communications.

Keywords: Flash mob · ISIL · Social media · Twitter · Radicalize · Propaganda · Campaigns · Collective action · Social network · Cyber warfare

1 Introduction

Social media is largely considered as a positive and powerful vehicle of change enabling people to connect with friends, family, or professionals. Unfortunately, that power has been harnessed by extremists and terrorist groups to spread propaganda and influence mass thinking posing security and safety risks significant enough to merit its study. Social media sources could be highly biased, distorted, and may be part of deliberate deception campaigns, attempts to incite crowd violence and provoke hysterical reactions. Several examples indicate the effectiveness of using social media in cyber warfare tactics (e.g., #AMessage-FromISIStoUS [1]) or cyber campaigns (e.g., #NotInMyName campaigns [2]). The sophisticated and highly coordinated use of social media to disseminate

© Springer International Publishing Switzerland 2015
N. Agarwal et al. (Eds.): SBP 2015, LNCS 9021, pp. 251–257, 2015.
DOI: 10.1007/978-3-319-16268-3_26

propaganda and effect cyber campaigns assisting information and tactical operations is alarmingly understudied. Through this research, we develop a socio-computational model to study the deviant sociotechnical behaviors of ISIL on Twitter, which resembles that of a DCFM [3]. Specifically, we seek answers to the following questions – what strategies do transnational crime organizations and terrorist groups, such as ISIL, use to disseminate propaganda through social media?, how can we determine the social media strategy of a specific terrorist group?, who are the powerful actors in their network?, and can we model and operationalize the DCFM behavior of such groups? We make the following contributions in this paper – (i) we identify a strategy that terrorist groups use to disseminate propaganda through the social media, (ii) we operationalize the conceptual framework to model the DCFM behavior on real world social media data, (iii) the research reveals a strategic choice of platforms used by ISIL's top powerful actors (or, disseminators) for propaganda dissemination, and (iv) we identify a methodology to discover powerful actors/disseminators on Twitter.

2   Literature Review

Many recent studies investigate ISIL focusing on who they are, what they want, what is their state structure, and where they get their money from [4]. A study conducted by the International Centre for the Study of Radicalization and Political Violence (ICSR) shows a group of individuals who help ISIL disseminate their propaganda on Twitter and other social media platforms [5]. These individuals either self-identified or provided their *affiliation* information through other means, e.g., Tweets or Facebook updates. There is a strong sense of *motivation* among these individuals to disseminate ISIL propaganda. One of the interviewee replied, "it was because of a strong sense of Muslim identity and the fact that there is lots of propaganda against the Muslim side". He also said "I don't intend to contribute to the conflict, [but] I intend to contribute to the toppling of these regimes." Moreover, he valued the anonymity of Twitter, saying that if it was compromised, "I won't hesitate to quit." [5].

The highly sophisticated and well coordinated use of social media by terrorist groups (especially, ISIL) motivated this study to investigate the deviant sociotechnical behavior. We develop a conceptual model that focus on identifying the factors that lead to these DCFMs with underpinnings in sociological theories of collective action and collective identity formation [3]. We systematically examine the choices an actor has in case of success or failure of a DCFMs [6]. In this study, we examine ISIL's Twitter activity from a DCFM perspective. We analyze their information dissemination strategies that reach many people instantly including texts, videos, and pictures.

3   Conceptual Framework of DCFM

We developed a conceptual framework that identifies the factors governing success or failure of a DCFM [3], leveraging social science theories such as collective action, collective identity formation, collective decision, social capital, and

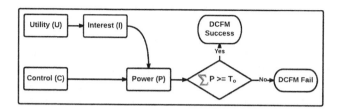

Fig. 1. Identified factors that govern the DCFM's success or failure

Table 1. Top 20 hashtags and keywords used in social media cluster sorted by their usage from the most used to the least used

1. #abu bakr al-baghdadi	11. # ابو بكر البغدادي
2. #caliphate instagram photos	12. ISIL
3. #daash	13. # islamicstate
4. #isil	14. daash
5. #isil instagram photos	15. islamic state of iraq and the levant
6. #isis	16. الدولة الإسلامية في العراق
7. #isis instagram photos	17. الدولة الإسلامية في العراق والشام
8. #islamic state in iraq and sham	18. # الدولة الإسلامية في العراق والشام
9. #islamic state in iraq and syria	19. داعش
10. #khalifa instagram photos	20. # داعش

network modularity. These theories have often been used to explain the group dynamics underlying collective behavior such as the parkours, flash mobs, campaigns, and social movements [7]. The identified factors are – *Utility (U)* (the benefits an individual gain if the DCFM success or fail), *Interest (I)* (how much interest this node has based on the utility gained), *Power (P)* (how powerful a node is in the group), and *Control (C)* (how much control the node has on the outcome of the DCFM). These factors are used to build a computational model to predict a DCFM's success or failure (Figure 1). Here, we operationalize the conceptual model for the DCFM behavior. We estimate the aforementioned factors on the Twitter network of powerful actors of the group spreading ISIL propaganda, whose behavior resembles that of a DCFM.

4   Methodology

The ICSR study reported that powerful actors or, "top disseminators" [5] act as important brokers of news to the online community because their political, moral, and spiritual messages are considered attractive by a number of foreign fighters. These *powerful* actors have much *interest* in disseminating messages and exert much *control* on the dissemination process, since they are followed by many users on Twitter. The powerful actors might not be foreign fighters but they are ideologically supporting the terrorist groups and helping them in the process of propaganda dissemination. To study their social media strategy and examine the DCFM behavior, we operationalize the conceptual model described in [3] (see Figure 1).

The Group Usage of Social Media. Here, we investigate whether there is a natural preference towards a particular social media platform by a terrorist group, and if there exists such a preference then what it is. To answer this question, we study ISIL's behavior on social media and specifically their DCFM. We identified a list of keywords and hashtags in Arabic and English most relevant for ISIL activities (see Table 1). The keywords are then used to monitor the usage of various social media platforms by ISIL. The results are shown in Figure 2. From figure 2 we can see that the keywords and hashtags can be grouped into three clusters. Cluster 1 consist of keywords and hashtags used by the *International news media*, cluster 2 consists of keywords and hashtags used by *regional/local news media*, and cluster 3 consists of keywords and hashtags used by *social media*. Upon zooming in on the social media cluster (Figure 3), we found that ISIL is mainly using 4 social media platforms to disseminate their propaganda ranked in decreasing order of usage as follows, 1. Twitter, 2. YouTube, 3. Facebook, and 4. Tumblr. Since Twitter is ranked as the number one source, we operationalize our conceptual framework on Twitter.

Data Set and Software Used. To operationalize our conceptual framework we collected Twitter data of the top 10 ISIL disseminators (i.e., most followed individuals disseminating ISIL propaganda) identified in the ICSR report [5] (see Figure 4). The dataset contains 13,789 unique vertices, 19,155 unique edges, 40 unique web links, 1,313 re-tweets and mentions, and 54 unique hashtags (see Figure 5). Figure 4 shows ISIL's top 10 disseminators' network (friends and followers) that was collected. To collect data, analyze, and generate figure 4, NodeXl was used. TouchGraph SEO browser was used to determine the dominant keywords as they are used on various websites and most widely used social media platforms for ISIL's propaganda dissemination.

Parameter Estimation. DCFM can be considered as a form of a cyber-collective action that is defined as an action aiming to improve a group's conditions (such as, status or power) [3]. For the purpose of this study we define DCFM as the cyber manifestation of disseminating a propaganda message (text, photo, or video) using Twitter, e.g. the spread of James Foley beheading video [8]. Here, we define a way to estimate the factors viz., Utility (U), Interest (I), Control (C), and Power (P) in Twitter network that govern the success or failure of DCFM [3]. As mentioned earlier, the extremist groups such as ISIL use powerful actors/disseminators to exert direct influence and control over their messages. These actors are *powerful* because they have *interest* in disseminating the propaganda and *control* over the dissemination of the message. The power of those nodes (powerful actors) can be estimated by finding both the amount of interest and control they have. To operationalize the parameters, we consider *the number of mentions and re-tweets* a node makes (about a specific propaganda or message) as a measure of their **Interest (I)** in the diffusion of the message. To calculate **Control (C)**, which is defined as the control of individual/actor/node

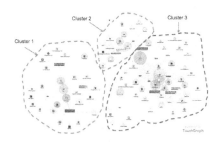

Fig. 2. Keywords and hashtags usage clustered into three groups: 1. International News Media, 2. Regional/Local News Media, and 3. Social Media.

Fig. 3. A closer view on the social media cluster (i.e. cluster 3). Four main social media platforms are used to disseminate ISIL propaganda, viz., 1. Twitter, 2. YouTube, 3. Facebook, and 4. Tumblr.

Fig. 4. Friends and followers network of ISIL's top 10 disseminators with **green** edges indicating their **outgoing** edges and **red** edges indicating their **incoming** edges.

Fig. 5. The usage of the 54 hashtags in the collected dataset.

on spreading the message or propaganda, we use the *in-degree centrality* an actor has. The more in-degree centrality an actor has the more control this node has in spreading a message in the group. **Power (P)** of an actor is a function of her control (C) and interest (I). Table 2 summarizes the parameter estimation.

According to the conceptual framework presented in Figure 1, summation of the power of all the powerful actors (disseminators) in the network helps determining the success or failure of the DCFM. If the sum is greater than or equal to a pre-determined threshold (T_o) then the DCFM is likely to succeed, i.e., $\sum_1^j P_j >= T_o$. T_o could be learned based on historic data from successful and failed DCFMs. We calculated the power (P) of the top 10 disseminators of

Table 2. Parameters used in the conceptual framework and their estimation using Twitter data

Parameter	Method of Estimation
Control (C)	In-degree Centrality
Interest (I)	*(Number of retweets + mentions)/(total number of tweets)*
Power (P)	Control * Interest

Table 3. The Top 10 Powerful Actors of ISIL on Twitter

Powerful Actors	# of Tweets	# of Mentions and Retweets	In-Degree Centrality	Power
shamiwitness	114907	82	2002	1.428
abusiqr	19706	32	2000	3.247
saqransaar	812	69	1526	129.67
ash_shawqi	5990	15	2003	5.01
troublejee	20110	74	2001	7.36
khalid_maqdisi	588	32	1043	56.76
nasserjan2	2257	72	1998	63.73
jabhtanNusrah	245	19	2003	155.33
ahmadMusaJibri1	0	0	58	0
musaCerantonio	698	82	2000	234.95

ISIL on Twitter as well as the summation of their power, i.e., $\sum_1^{10} P_j = 657.485$. The results are shown in table 3.

5   Looking Ahead

In this paper, we advance our study of deviant sociotechnical behaviors on social media platforms. Numerous accounts have reported the sophisticated usage of social media platforms by ISIL for information dissemination, recruitment, radicalization, and raising funds. This study in specific, looks at the social media activity of ISIL. We identified the top social media platforms used by the ISIL for communication, of which Twitter clearly stood out. By analyzing the Twitter network of ISIL, we identified the powerful actors and further modeled their deviant sociotechnical behaviors by operationalizing the DCFM model. The research presents a framework that currently considers basic network structural measures to model the behavior with theoretical underpinning in collective action theories. However, to model the complexities of the behaviors exhibited by transnational crime organizations such as ISIL, more research needs to be done. Our future directions include, evaluating the DCFM model on different social media platforms, test the necessary and sufficiency conditions for the parameters of the model, and further enhance the model by exploring the possibility to include content based features.

Acknowledgements. This research was made possible by a grant from the U.S. Office of Naval Research under Grant Number N000141410489.

References

1. Bhatia, M.S.: World War III: The Cyber War. International Journal of Cyber Warfare and Terrorism (IJCWT) **1**(3), 59–69 (2011)
2. Franceschi-Bicchierai, L.: Muslims Launch Powerful Social Media Campaign Against ISIS With #NotInMyName (September 22, 2014). http://mashable.com/2014/09/22/notinmyname-muslims-anti-isis-social-media-campaign/
3. Al-khateeb, S., Agarwal, N.: Developing a Conceptual Framework For Modeling Deviant Cyber Flash Mob: A Socio-Computational Approach Leveraging Hypergraph Constructs. JDFSL **9**(2), 113–128 (2014)
4. Arango, T., et al.: How ISIS Works (2014). http://www.nytimes.com/interactive/2014/09/16/world/middleeast/how-isis-works.html
5. Carter, J.A., Maher, S., Neumann, P.R.: #Greenbirds: Measuring Importance and Influence in Syrian Foreign Fighter Networks (2014)
6. Al-khateeb, S., Agarwal, N.: Modeling Flash Mobs in Cybernetic Space. In: Proceedings of the JISIC. IEEE, September 2014
7. Coleman, J.S.: Collective actions. In: The Mathematics of Collective Action, vol. (3), pp. 61–90. Transaction Publishers (1973)
8. Lee, D.: James Foley: Extremists battle with social media (August 20, 2014). http://www.bbc.com/news/technology-28870777

Social Behavior Bias and Knowledge Management Optimization

Yaniv Altshuler[1(✉)], Alex (Sandy) Pentland[2], and Goren Gordon[2]

[1] Athena Wisdom, Cambridge, USA
yaniv@athenawisdom.com
[2] MIT Media Lab, Cambridge, USA
{pentland,ggordon}@media.mit.edu

Abstract. Individuals can manage and process novel information only to some degree. Hence, when performing a perceptual novel task there is a balance between too little information (i.e. not getting enough to finish the task), and too much information (i.e. a processing constraint). Combining these new findings to a formal mathematical description of efficiency of novel information processing results in an inverted U-shape, wherein too little information is not effective to solving a problem, yet too much information is also detrimental as it requires more processing power than available. However, in an information flooded economic environment, it has been shown that humans are rather poor at managing information overload, which results in far from optimal performance. In this work we speculate that this is due to the fact that they are actually trying to maximize the wrong thing, e.g. maximizing monetary gains, while completely disregarding information management principles that underlie their decision-making. Thus, in a social decision-making environment, when information flows from one individual to another, people may "misuse" the abundance of information they receive. Using the model of individual novelty management, and the empirical statistical nature of investors' inclination to information, we have derived the social network information flow dynamics and have shown that the "spread" of people's position along the inverted U-shape of efficient information management leads to an unstable and inefficient macro-scale dynamics of the network's performance. This was in turn validated through a global inverted U-shape, observed in the macro-scale network performance.

Keywords: Behavioral modeling · Social networks · Social physics · Information flow · Information management

1 Introduction

In the Information Age, getting the correct amount and type of information is essential. In many scenarios, the decision-making process is driven by information, even though the goal is not information-specific, e.g. investment gains. Moreover, recent platforms enable the social connectivity among participants to be the source of information flow,

© Springer International Publishing Switzerland 2015
N. Agarwal et al. (Eds.): SBP 2015, LNCS 9021, pp. 258–263, 2015.
DOI: 10.1007/978-3-319-16268-3_27

i.e. information flows from one agent to another, thus influencing not only the individual performance but also the global performance [17-19]. This work follows a statistical physics approach such that the individual decision making agents are akin to atoms and their interaction, which results in the network, is akin to an ensemble. The first step in our approach is presented in this paper, wherein we projected recent results and models on how biological agents, humans [22] and animals [20], process information in an information-oriented task, onto a social trading platform (called "eToro", serving over 3 million financial investors worldwide [16]). The models enable the unique possibility to have a specific quantifiable prediction on performance of individuals in a social trading platform, where information is key to efficiency. We show that a basic underlying principle of biological decision-making agents is novelty management, i.e. the trade-off between too little and too much information. Two complementary studies have presented mathematical models, validated in experiments in humans [22] and rodents [21] that stress the importance of a processing cost to information.

These models predict that in an information-rich environment there will be an inverted-U relation between the amount of information an individual agent receives and its efficiency in that system. We have analyzed data from a social trading platform, where investors participate in a 'collaborative decision making process', manifested through the fact that each individual agent makes a decision based on information sources available in the system – the previous investment decisions of the other participants. We show that our prediction holds, i.e. the revenues of an individual agent is related to the number of information sources in an inverted-U manner. We thus establish a unique relation between neuroscience findings and individual agents' performance in a social interaction scenario.

2 Related Work

We live in the "big data" era. Many of our daily activities, and specifically those that has to do with our social or financial behaviors, are heavily influenced by information we obtain from our surroundings through various media channels, and through direct or indirect interactions with our peers [17,18]. The sudden influx of data is transforming social sciences at an unprecedented pace [14,15]. Various recent works have shown how large communities can be modeled as "swarms", allowing interactions meta-data to be used as the sole (or major) data source for the generation of accurate predictions regarding dominating crowd behaviors and trends [6-12]. In many cases these predictions are dominated by the topological properties of the network formed by these interactions [13]. Recent studies have already illustrated the potential that extensive behavioral data sets (e.g., Google trends, Wikipedia usage patterns, and financial news developments) could offer us a better understanding of collective human behavior in financial markets [1-4]. In this work we focus on economic decision under risk, a key subject of behavior economics [5]. Successful behavior economic theories acknowledge the complexity of human economic behavior and introduce models that are well grounded in psychological research.

There have been several psychological studies showing the merit and sources of an inverted-U shape of virtues and attributes. These claimed to be sources of wellbeing in the sense that "too much of a good thing" is not necessarily good [25,27]. However, these studies have not presented a neuroscience-based mathematical model that can serve as the basis of a comprehensive framework for individual and social interactions.

Recent studies in neuroscience have enabled the development of rich models that attempt to explain complex behaviors of biological agents in information-oriented tasks. The first study was done with rodents who explored new arenas for the first time [26]. The rodents explored the new environment in a growing dimensional complexity, first exploring the zero-dimension entrance to the arena, then the one-dimension wall and only then the two-dimensional open-space. These exploration excursions were intermittent with retreats, in which the rodents went back to their home cage. A new model was presented that accounted for this complex behavior from a single principle of novelty management [20]. It was shown that information-seeking animals maximized their novelty signal-to-noise ratio, i.e. they aimed for a stable flow of information.

The rodents in the study all exhibited optimality, in terms of maximizing the novelty signal-to-noise ratio. Humans, however, have been shown to be sub-optimal in purely information tasks, compared to rodents [23]. In other words, while rodents have evolved to optimize this SNR and thus do not exhibit in their natural behavior the full spectrum of information flows, and hence sub-optimality, we predict that humans will do so. Namely, we speculate that through their constant attempts to increase revenues, human participants will advocate various strategies for the rate and diversity of their information acquisition, thus resulting in an explicit observable inverted-U shape of information-efficiency curve.

In a different study with human participants, a perceptual task was designed to test the behavior of people with new senses, i.e. artificial whiskers on their fingers [22]. The behavior they exhibited spontaneously in this information-seeking task showed a convergent exploration behavior, similar to those exhibited in rodents. A Bayesian perception optimal-control model suggested that there is a processing cost to too much information, which then leads to a stable information flow. This study suggests that humans follow the same principles as the exploring mice.

3   Data Collection and Results

The financial transaction data used in this work was received from an online social financial trading platform for foreign exchanges, equity indices and commodities, called *eToro* [16]. This trading platform allows traders to take both long and short positions, with a minimal bid of a few dollars as well as leverage up to 400 times. Every investor has complete knowledge of the trading decisions of all the other investors, and may use this information as an input in their own decision making process. There are approximately 3 million registered accounts in this online social trading platform. Our data are composed of over 40 million trades during years 2011 to 2014.

As mentioned in previous sections, whereas mice were evolutionary "calibrated" to behave according to this optimal SNR, in an information flooded economic environment, human investors have been shown to perform poorly at managing information overload, which results in far from optimal performance ([23] and many others). We believe that this is due to the fact that human financial investors are constantly trying to maximize the wrong element of their decision making process, e.g. maximizing monetary gains (instead of information sources), and disregarding the basic information management principles that govern their decision-making. In other words, people do not internalize the fact that too much information is sometimes detrimental, since it confounds the decision-making process. While rodents appear to be (almost-) always on the top of the inverted U-shape, we have discovered that humans appear to be distributed along it, with some people getting too little information, while other too much. This is shown in Figure 1, illustrating the dependency of the average return of investment on the number of information resources, demonstrating an inverted-U pattern.

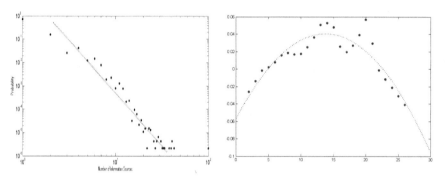

Fig. 1. The Figure depicts the average financial transaction gains as a function of the amount of information sources used.   The data was provided by a leading social-trading platform, representing over 3 million investors, for a period of 2 years. Power low of the information sources distribution (left), and the mean gain as a function of the number of information sources (right).

Thus, in a social decision-making environment, when information flows from one individual to another, people may "misuse" the abundance of information they receive. Using the model of individual novelty management [20], and the empirical statistical nature of investors' inclination to information, we have derived the social network information flow dynamics and have shown that the "spread" of people's position along the inverted U-shape of efficient information management leads to an unstable and inefficient macro-scale dynamics of the network's performance. This was in turn validated through a global inverted U-shape, observed in the macro-scale network performance [25].

4   Conclusions

In this paper we have conjectured that human financial investors behave similarly to rodents, with respect to their inability to process unlimited amounts of data sources, and subsequently – the existence of an optimal signal-to-noise ratio with respect to the number of different information sources they handle during their decision making process. We have further predicted that while rodents were evolved to maintain an optimal number of information sources, human, while trying to maximize financial gains, will stray from this optimal configuration (usually by acquiring and handling a larger number of information sources). We have shown that if such behavior indeed exists, it will be manifested through an observable inverted-U ratio between the average financial gain and the number of information sources used in order to obtain it.

We have tested our hypothesis using a highly detailed large financial trading database, containing the complete financial trades of 3 million financial investors. We have demonstrated the existence of the predicted inverted-U ratio, validating our assumption regarding the strong dependency of decision making (in finance at least) by the number of information sources used, the existence of an optimum for it, and the inability of human investors to maintain it.

We suggest that changing the distribution of people's position along the information management axis can have drastic effects on the network performance. Two basic manipulations can be considered from a physical system analogy: (i) changing the "temperature" of the system, i.e. either raising it to create a more diverse spread or lowering it to make a more homogenous network; (ii) by lowering the system's temperature one can then tune the distribution center to be more in the optimal efficient information management regime.

References

1. Moat, H.S., et al.: Quantifying Wikipedia usage patterns before stock market moves. Sci. Rep. **3**, 1801 (2013)
2. Alanyali, M., Moat, H.S., Preis, T.: Quantifying the relationship between financial news and the stock market. Sci. Rep. **3**, 3578 (2013)
3. Preis, T., Moat, H.S., Stanley, H.E.: Quantifying trading behavior in financial markets using google trends. Sci. Rep. **3**, 1684 (2013)
4. Krasny, Y.: The effect of social status on decision-making and prices in financial networks. In: Altshuler, Y., Elovici, Y., Cremers, A.B., Aharony, N., Pentland, A. (eds.) Security and Privacy in Social Networks, pp. 85–131. Springer, New York (2013)
5. Diamond, P., Vartiainen, H.: Behavioral Economics and Its Applications Hardcover. Princeton University Press (2007)
6. Altshuler, Y., Elovici, Y., Cremers, A.B., Aharony, N., Pentland, A.: Security and Privacy in Social Networks. Recherche **67** (2012)
7. Altshuler, Y., Fire, M., Shmueli, E., Elovici, Y., Bruckstein, A., Pentland, A., Lazer, D.: Detecting anomalous behaviors using structural properties of social networks. In: Greenberg, A.M., Kennedy, W.G., Bos, N.D. (eds.) SBP 2013. LNCS, vol. 7812, pp. 433–440. Springer, Heidelberg (2013)

8. Altshuler, Y., Fire, M., Shmueli, E., Elovici, Y., Bruckstein, A., Pentland, A.: The Social Amplifier—Reaction of Human Communities to Emergencies. Journal of Statistical Physics **152**(3), 399–418 (2013)
9. Altshuler, Y., Aharony, N., Elovici, Y., Pentland, A., Cebrian, M.: Stealing Reality: When Criminals Become Data Scientists (or Vice Versa). IEEE Intelligent Systems **26**(6), 22–30 (2011)
10. Altshuler, Y., Fire, M., Aharony, N., Elovici, Y., Pentland, A.: How many makes a crowd? on the correlation between groups' size and the accuracy of modeling. In: Social Computing, Behavioral-Cultural Modeling and Prediction (SBP) (2012)
11. Altshuler, Y., Fire, M., Aharony, N., Elovici, Y., Pentland, A.: Incremental Learning with Accuracy Prediction of Social and Individual Properties from Mobile-Phone Data. Arxiv preprint arXiv:1111.4645 (2011)
12. Altshuler, Y., Yanovsky, V., Wagner, I., Bruckstein, A.: Swarm intelligence—searchers, cleaners and hunters. In: Nedjah, N., de Macedo Mourelle, L. (eds.) Swarm Intelligence Systems. SCI, vol. 26, pp. 93–132. Springer, Heidelberg (2006)
13. Altshuler, Y., Wagner, I.A., Bruckstein, A.M.: On Swarm Optimality In Dynamic And Symmetric Environments. Economics **7**, 11–18 (2008)
14. Lazer, D., et al.: Computational social science. Science **323**, 721–723 (2009)
15. Battiston, S., Puliga, M., Kaushik, R., Tasca, P., Caldarelli, G.: Debtrank: Too central to fail? Financial networks, the fed and systemic risk. Sci. Rep. **2** (2012)
16. www.etoro.com
17. Pan, W., Altshuler, Y., Pentland, A.: Decoding Social Influence and the Wisdom of the Crowd in Financial Trading Network. IEEE Social Computing (2012)
18. Altshuler, Y., Pan, W., Pentland, A.: Trends prediction using social diffusion models. In: Yang, S.J., Greenberg, A.M., Endsley, M. (eds.) SBP 2012. LNCS, vol. 7227, pp. 97–104. Springer, Heidelberg (2012)
19. Liu, Y.Y, Nacher, J.C., Ochiai, T., Martino, M., Altshuler, Y.: Prospect Theory for Online Financial Trading. PLOS ONE (2014)
20. Gordon, G., Fonio, E., Ahissar, E.: Emergent Exploration via Novelty Management. Journal of Neuroscience **34**(38), 12646–12661 (2014)
21. Gordon, G., Fonio, E., Ahissar, E.: Learning and control of exploration primitives. Journal of Computational Neuroscience, 1–22 (2014)
22. Saig, A., Gordon, G., Assa, E., Arieli, A., Ahissar, E.: Motor-sensory confluence in tactile perception. Journal of Neuroscience **32**(40), 14022–14032 (2012)
23. Zweig, J.: The Trouble With Humans. Money Magazine **11**, 67–71 (2000)
24. Altshuler, Y., Pentland, A.: Accurate Decision Making in Social Networks. NetSci (2013)
25. Grant, A.M., Schwartz, B.: Too Much of a Good Thing. The Challenge and Opportunity of the Inverted U. Perspectives on Psychological Science **6**(1), 61–76 (2011)
26. Fonio, E., Benjamini, Y., Golani, I.: Freedom of movement and the stability of its unfolding in free exploration of mice. Proceedings of the National Academy of Science (PNAS) **106**(50), 21335–21340 (2009)
27. Lazer, D., Friedman, A.: The network structure of exploration and exploitation. Administrative Science Quarterly **52**(4), 667–694 (2007)

Passive Crowd Sourcing for Technology Prediction

Erica J. Briscoe[✉], Scott Appling[✉], and Joel Schlosser[✉]

Georgia Tech Research Institute, Georgia Institute of Technology,
Atlanta, GA 30332, USA
{erica.briscoe,scott.appling,joel.schlosser}@gtri.gatech.edu

Abstract. Technology prediction methods use various types of information to make systematic forecasts about technological innovations. Forecasting approaches vary, including quantitative (such as patent analysis) and qualitative methods (such as expert elicitation using the Delphi method). We discuss a new method and system for predicting technology futures by harnessing the predictive information made available by society in open sources. Our approach automatically discovers future-looking temporal phrases associated with technology topics and presents predictions deemed significant using the G^2 statistic. Here, we evaluate the phrase discovery component using a dataset of 782 technology forecast statements. We hope to demonstrate that 'passive crowd-sourcing' may be a meaningful source of technology-related predictive intelligence.

1 Introduction

Though bibliometric methods have had some success in predicting technology futures, there are limitations to these approaches that stem from their concentrated use of data sources with inherent limitations and biases. To supplement the use of such traditional sources (e.g. publication and patent data), informal media (e.g. online articles or blog entries) can be employed as a rich source of technology information. [19] wrote that the "communication of science and technology is assisted by the media, which acts as a source of information as well as the mode by which it is transferred." The media represents a pervasive and influential means of educating the public, making it likely that societal participation in the discussion of technology is a driving force to innovation.

In this paper, we discuss the utilization of human input for producing technology forecasts. Rather than using a solicitation-based approach, we mine open text for passively made predictions, thereby casting a wider net and leveraging the knowledge of the 'crowd'. To accomplish this, we take advantage of natural language processing methods that focus on temporal recognition and event detection. By combining crowd intelligence and temporal technology event recognition, we are able to characterize a technological 'vision' of the future.

© Springer International Publishing Switzerland 2015
N. Agarwal et al. (Eds.): SBP 2015, LNCS 9021, pp. 264–269, 2015.
DOI: 10.1007/978-3-319-16268-3_28

2   Related Work

2.1   Deriving Intelligence from the 'Crowd'

In the past few years, the concept of acquiring information from large networks of people in online spaces, often known as the "crowd", has been the subject of much research publication and media attention [22]. Methods for processing the unstructured information that people openly provide have grown considerably [21] and are increasingly being refined and combined so as to allow practitioners to make more complex and accurate assessments of specific areas of interest [1], including those of a predictive nature that effectively mine 'collective intelligence' [13]. The practice of institutions openly outsourcing specific needs to external groups is often referred to as crowdsourcing [10]. This practice has been deployed in a number of different domains, with varying success ([16], [12], [9]). One type of crowdsourcing application aims at harnessing external expertise to produce intelligence of a greater quality than that derived from sole individuals [3]. The strengths in doing so lie in the diversity of opinion that is acquired through these methods, which serve to remove biases commonly manifest in individually-derived intelligence. Similar to our approach is the 'SciCast' system, a research project led by George Mason University [24], that increases the accuracy of crowd-sourced forecasting of science and technology (S&T) events using mathematical methods to aggregate individual forecasts into an overall forecast for each question. In contrast to this approach, we use passive crowd-sourcing, which is characterized through the utilization of open content that has not been specifically solicited [5].

2.2   Technology Forecasting Methods

Technology forecasting, like many types of forecasting, is an imperfect, but necessary practice for any organization that can be affected by technological change. Much previous work has been focused on the development of automated methods for mining information sources, specifically those relevant to science and technology, with the goal of producing insight into future technological trends (e.g. [26]). Many of these efforts rely on computationally detecting the relationships between science and technology domains [25], the researchers within those domains [6], or other contextual factors [23]. Although lacking examples relying on mass communication artifacts such as news articles or text collected from online social networks, [8] provides a robust overview of methods for technology forecasting grouped according to their central approaches.

2.3   Event Detection

Event detection is a method by which natural language text is processed in order to determine specific phenomena (e.g. concerts, sporting events, protests). [18] describes methods for automatically detecting events about pre-specified entities by considering temporally located sets of social media artifacts. [17] presents an approach that operates over large streams of social media artifacts to extract event-relevant keywords from temporally coherent sets; while the approach is promising, the authors were limited by their lack of access to datasets with ground truth. [20] presented a domain-free approach to event recognition over large corpora of social media artifacts through a process involving the shallow parsing of artifacts followed by temporal extraction, event tagging, and then noise reduction through use of the G^2 statistic [7]. Our approach is inspired by [20]; we create a method by which to extract both technology *events* and their temporal information, indicating when the writer anticipated the technology innovation would happen in the future.

3   Technology Emergence Prediction via Open Media

To gather crowd intelligence, we build models to automatically recognize technology emergence based on the detection of technology and temporal phrases from single sentence forecast statements.[1] We gather ground truth information using a majority vote of human annotators. Below, we describe the technology forecasts corpus, temporal phrase detection, and then technology phrase detection. Figure 1 depicts our overall approach that starts with a large collection of technology articles and leads to technology predictions from forecast statements. In this work, we focus on evaluating the "Technology Phrase Detection" component.

Fig. 1. System process diagram describing the path from technology articles to significant technology predictions. The current research focuses on "Technology Phrase Detection" in the system pipeline.

3.1   Technology Prediction Corpus

Using Amazon Mechanical Turk[2] [4], human annotators were tasked to select parts of 782 forecast statements which they thought most likely represented the technology phrase[3]. The forecast statements were collected from various technology articles over multiple data sources (articles, books, primarily) over the past 60 years. These forecast were also labeled by their broad technology

---

[1] Here we focus on forecast statements where both the event emergence phrase and the temporal expression are components of a single sentence. More complex text structure, such as when the temporal and the event phrase are located separately in a paragraph, will be addressed in future work.

[2] http://www.mturk.com

[3] Use phrases were also collected but not used in this work.

areas. For example, if the forecast statement was, "1-Gb DRAMs will not appear on the market until 2000." then the technology phrase would be "1-Gb DRAMs." Below is a list of example forecast statements and the annotations provided by Turkers. Technology topic phrases are bolded.

1. 17,500 **new commercial planes** will be delivered by 2024.
2. In 1993, **space technology** will be able to track military targets from space.
3. **Carbon dioxide emissions** in the Republic of Korea will be 141 MtC in 2000.

3.2 Future-Leaning Temporal Expressions

We use a version of TempEx from NLTK [2] that we have extended to recognize temporal expressions in future technology forecast statements. For example, we look for prediction oriented phrasing such as "by 2020" or "In 5 years". In cases where relative references are used, we supply the article creation date to infer the absolute date.

3.3 Technology Phrase Detection

We train a linear chain condition random field [15] on a corpus of annotated forecast statements where the three possible labels correspond to: 1) the beginning; 2) inside; or, 3) outside position of a technology phrase. Following [14,20], we use similar features to inform our model about how to best recognize technology phrases within a single sentence forecast statement; we also include a new feature related to the token distance before and after the the temporal phrase. We hypothesize that encoding information about the direction and distance from the temporal phrase to the technology phrase better informs our model, helping to avoid tagging use phrases as technology phrases.

4   Technology Topic Phrase Tagging Results

We evaluated our technology topic phrase model using two versions of precision and recall. The first is traditional precision and recall where a true positive is a complete match of the technology phrase and no other tokens are added or removed in the tagging. The second metric used we refer to as partial precision and recall, where we count a tagging as a true positive if it contains at least one token in the ground truth technology phrase. Since in each statement there is only one technology phrase (true positive) to label, precision and recall are the same in calculation. 10-fold cross validation resulted in an average precision score of 54.7% and an average partial precision score of 63.2% where the best score fold (using partial precision) was 72%. We surmise the relatively small size of our dataset accounted for some amount of error and are in the process of developing a larger technology prediction dataset.

5   Discussion

While our work is motivated by [17,20], we see the supervised learning shared tasks like that of [11] useful for comparing the quality of our technology tagging performance. There, the authors were focused on detecting events of interest in bioinformatics (e.g. predicting protein phrases in biology related natural language texts). After utilizing 7,449 sentences for training, 1,450 for development, and 2,447 for testing, the average precision score for all 24 teams was approximately 65% while the Top 5 teams was 77.6%. In light of these results we see our initial performance as comparable and believe with larger training and testing corpora our scores would likely see, at minimum, moderate improvement.

6   Conclusions and Future Work

In this initial work, we briefly described the social nature of technology communications, the use of passive crowdsourcing as a source of intelligence, and developed a new approach to recognizing significant technology predictions using open source data, mined mainly from technology articles. Areas for future work include the following: 1) analysis of the distribution of parse structures across the corpus; 2) the detection of technology forecasts from multi-sentential statements; 3) the identification and assignment of sub-topic labels to specific groups of technology forecasts; and finally 4) using informal discussions like those found in online social networks, along with blog posts, as sources for technology forecasting knowledge.

References

1. Appling, S., Briscoe, E., Ediger, D., Poovey, J., McColl, R.: Deriving disaster-related information from social media. In: KDD-LESI 2014: Proceedings of the 1st KDD Workshop on Learning about Emergencies from Social Information at KDD14, pp. 16–22 (2014)
2. Bird, S., Klein, E., Loper, E.: Natural language processing with Python. O'Reilly Media, Inc. (2009)
3. Brabham, D.C.: Moving the crowd at istockphoto: The composition of the crowd and motivations for participation in a crowdsourcing application. First Monday 13(6) (2008)
4. Buhrmester, M., Kwang, T., Gosling, S.D.: Amazon's mechanical turk a new source of inexpensive, yet high-quality, data? Perspectives on Psychological Science 6(1), 3–5 (2011)
5. Charalabidis, Y., Loukis, E.N., Androutsopoulou, A., Karkaletsis, V., Triantafillou, A.: Passive crowdsourcing in government using social media. Transforming Government: People, Process and Policy 8(2), 7 (2014)
6. Daim, T.U., Rueda, G., Martin, H., Gerdsri, P.: Forecasting emerging technologies: Use of bibliometrics and patent analysis. Technological Forecasting and Social Change 73(8), 981–1012 (2006)
7. Dunning, T.: Accurate methods for the statistics of surprise and coincidence. Computational Linguistics 19(1), 61–74 (1993)

8. Firat, A.K., Woon, W.L., Madnick, S.: Technological forecasting-a review. Composite Information Systems Laboratory (CISL). Massachusetts Institute of Technology (2008)
9. Franklin, M.J., Kossmann, D., Kraska, T., Ramesh, S., Xin, R.: Crowddb: answering queries with crowdsourcing. In: Proceedings of the 2011 ACM SIGMOD International Conference on Management of Data, pp. 61–72. ACM (2011)
10. Howe, J.: The rise of crowdsourcing. Wired Magazine **14**(6), 1–4 (2006)
11. Kim, J.D., Ohta, T., Pyysalo, S., Kano, Y., Tsujii, J.: Overview of bionlp'09 shared task on event extraction. In: Proceedings of the Workshop on Current Trends in Biomedical Natural Language Processing: Shared Task, pp. 1–9. Association for Computational Linguistics (2009)
12. Kittur, A., Chi, E.H., Suh, B.: Crowdsourcing user studies with mechanical turk. In: Proceedings of the SIGCHI Conference on Human Factors in Computing Systems, pp. 453–456. ACM (2008)
13. Lévy, P., Bonomo, R.: Collective intelligence: Mankind's emerging world in cyberspace. Perseus Publishing (1999)
14. Marrero, M., Sanchez-Cuadrado, S., Lara, J.M., Andreadakis, G.: Evaluation of named entity extraction systems. Advances in Computational Linguistics, Research in Computing Science **41**, 47–58 (2009)
15. McCallum, A.K.: Mallet: A machine learning for language toolkit (2002)
16. Nam, T.: Suggesting frameworks of citizen-sourcing via government 2.0. Government Information Quarterly **29**(1), 12–20 (2012)
17. Parikh, R., Karlapalem, K.: Et: events from tweets. In: Proceedings of the 22nd International Conference on World Wide Web Companion, International World Wide Web Conferences Steering Committee, pp. 613–620 (2013)
18. Popescu, A.M., Pennacchiotti, M., Paranjpe, D.: Extracting events and event descriptions from twitter. In: Proceedings of the 20th International Conference Companion on World Wide Web, pp. 105–106. ACM (2011)
19. Rennie, L., Stocklmayer, S.M.: The communication of science and technology: Past, present and future agendas. International Journal of Science Education **25**(6), 759–773 (2003)
20. Ritter, A., Etzioni, O., Clark, S., et al.: Open domain event extraction from twitter. In: Proceedings of the 18th ACM SIGKDD International Conference on Knowledge Discovery and Data Mining, pp. 1104–1112. ACM (2012)
21. Segaran, T.: Programming collective intelligence: building smart web 2.0 applications. O'Reilly Media, Inc. (2007)
22. Shirky, C.: Here comes everybody: The power of organizing without organizations. Penguin (2008)
23. Turró, A., Urbano, D., Peris-Ortiz, M.: Culture and innovation: the moderating effect of cultural values on corporate entrepreneurship. Technological Forecasting and Social Change (2013)
24. George Mason University: Scicast (2014). https://scicast.org/#!/
25. Watts, R.J., Porter, A.L., Newman, N.C.: Innovation forecasting using bibliometrics. Competitive Intelligence Review **9**(4), 11–19 (1998)
26. Zhu, D., Porter, A.L.: Automated extraction and visualization of information for technological intelligence and forecasting. Technological Forecasting and Social Change **69**(5), 495–506 (2002)

How to Predict Social Trends by Mining User Sentiments

Iuliia Chepurna$^{(\boxtimes)}$, Somayyeh Aghababaei, and Masoud Makrehchi

University of Ontario Institute of Technology,
2000 Simcoe Street North, Oshawa, ON, Canada
{iuliia.chepurna,somayyeh.aghababaei,masoud.makrehchi}@uoit.ca

Abstract. The majority of techniques in socio-behavioral modelling tend to consider user-generated content in a bulk, which may ignore personal contributions of specific users to predictability of the system. We propose a novel user-based approach designed specifically to capture most predictive hidden variables which can be discovered in a context of specific individual only. User content is assessed to determine both the subset of best, "expert", users able to reflect particular social trend of interest, and their transformation into feature space used for modelling. The technique is tested on a case study of Chicago crime rate trend prediction using historical tweets of selected citizens. We also propose a new user ranking approach which exploits the concept of user credibility.

Keywords: User-based prediction · Socio-economic trends forecasting

1 Introduction

Social networks are proliferating with personal content which was successfully utilized in various applications, such as user profiling, personalized targeting, opinion mining, content and user recommendation. Professionals are tracked online and their insights are extensively exploited by expert-based systems to predict stock market movements, share revenues, box office and other trends. Is it possible to apply similar approach to a broader spectrum of social prediction problems where user-generated data was only considered as a whole? In this paper we propose user-based approach for social trends forecasting which, unlike expert-based techniques, is not restricted to specific area of competence and does not require explicit recommendations. The method was examined on a case study of crime trend prediction for Chicago, IL. There is a number of underlying factors affecting future crimes, such as, unemployment and divorce rates, education level and average income, community welfare, general happiness of the population, and many others. We hypothesize that having "perfect sample" of "expert users" from a pool of citizens active on Twitter, we would be able to extract those hidden variables and translate them into predictive signals.

Various approaches were undertaken to deal with crime prediction problem. Conventional techniques, used by law enforcement agencies are mostly concerned

© Springer International Publishing Switzerland 2015
N. Agarwal et al. (Eds.): SBP 2015, LNCS 9021, pp. 270–275, 2015.
DOI: 10.1007/978-3-319-16268-3_29

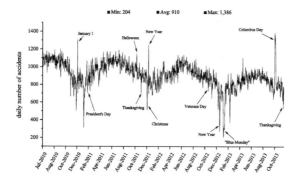

Fig. 1. Daily aggregated incidents. Spikes were observed during statutory holidays.

with generating hotspot maps [1,2] which are peculiar to specific location and thus cannot be generalized. To overcome this issue further approaches incorporated background knowledge about spatial features, such as distance to intersections and highways, schools and businesses, and additional information about the neighborhood [7,10]. Another line of research considered the neighborhood social fabric as a key factor influencing criminal activities [4,6].

Wang *et al.* [8] were the first to blend social media data into traditional models. The authors predicted city-wide hit-and-run incidents based on topics extracted from tweets of manually selected news agencies. Another closest work to our research was presented by Gerber [3] and suggested to extend Kernel Density Estimation (KDE) with topics extracted from Twitter. Although usage of tweets introduced some additional context, the model built by KDE lacks portability. In other words it can not be used for predicting crime in other cities. This approach also does not take into account temporal changes in criminal activities.

2 Dataset

Chicago police reports: Our dataset was collected from official Data Portal of Chicago[1] and covers time range between July 1, 2010 and November 30, 2013 with a total of 1.1M incidents (see Fig. 1). Each incident is reported together with its exact location (both longitude-latitude pair and full address), timestamp and type of a crime.

Twitter data set: Corresponding set of Twitter users from Chicago was collected utilizing Online Coupling from the Past [9] which guarantees the convergence on a "perfect sample" of the whole user network, yet being unbiased towards individuals with extreme number of connections. Historical timelines of selected people were retrieved and restricted to the same timeframe – between July 1, 2010 and November 30, 2013. Per-user metrics are reported in Table 1.

---

[1] https://data.cityofchicago.org/Public-Safety/Crimes-Map/dfnk-7re6

Table 1. Statistics on user activity during period of observation. Total number of users: 2753. Total number of days: 1249. Users active more than average number of days: 1035 (83%); users with more then average number of total posts: 892 (32%).

statistic	days with posts	overall tweets
min	1	1
max	1237	3246
mean	187	675
std	195	838

3   Data Representation

Timelines of selected users exhibit certain behavior which is examined to derive latent factors that correlate with future crimes. In this work we assume that sentiment can be treated as a sufficient proxy to individual's feelings and intentions. For this task, users' daily tweets are first fed to lexicon-based sentiment analyzer LIWC [5] and then discretized using the following procedure:

$$
\begin{cases}
\max\{positive, negative\}, \text{ if } \frac{|positive-negative|}{\max\{positive,negative\}} \geq 0.1, \\
neutral, \text{ otherwise.}
\end{cases}
$$

where *positive* and *negative* were calculated by LIWC and could take values between 0 and 100. For days with no posts neutral sentiment is assigned. Additional feature is kept to indicate whether a user was active on that specific day or not. We also examined other sentiment representations, such as normalized positive and negative, negative-to-positive ratio, and positive minus negative, but none offered better results than the original representation. Therefore feature matrix has the following form: $\{U_{ij}\} = \{(s_{ij}, a_{ij})\}$, where U_{ij} is the user i on day j, s_{ij} is user discrete sentiment on day j, and a_{ij} is binary activity indicator. Associated labels are assigned as next day crime trend either increasing or decreasing.

4   Experiments

4.1   User Daily Sentiment

This experiment considered all users for the task of the prediction. Here and after we only mention linear SVM as selected binary classifier, since it offered best results. Set of different features was examined (*content-based*: agreed daily sentiment for all users, and LIWC-derived features capturing fraction of words which can be described as related to "death" and "swear"; *temporal*: month), but the representation described in the previous section was eventually kept as providing better results. The achieved F-measure was equal to 0.5. Taking into consideration extreme sparsity of the feature matrix ($\sim$84%) we decided to introduce user ranking discussed in the following section.

4.2   User Ranking

As can be seen, even "perfect sampling" does not guarantee uniform distribution of the tweets. To mitigate this issue we would like to select a subset of users active enough to provide sufficient amount of content to our model. Besides, we are actually not interested in all active users, only in "experts" reflecting changes in crime trend best among all initially sampled individuals.

Feature Selection Based User Filtering: To detect most relevant users we employed traditional feature selection techniques, such as χ^2, Pearson's r, and normalized information gain, with the following modification: only sentiment for days with activity was considered. All approaches offered approximately same results, with the best one (F-measure=0.55) yielded by χ^2.

Credibility We also proposed a new ranking technique which exploits the notion of user's credibility. Let $\{U_{ij}\}$ be user matrix that consists of $\{(s_{ij}, a_{ij})\}$, where s_{ij} is discrete sentiment of user i on day j and is defined as

$$s_{ij} = \begin{cases} -1, \text{ negative,} \\ 0, \text{ neutral or absent, ,} \\ +1, \text{ positive.} \end{cases}$$

a_{ij} is a binary activity indicator, l_j is crime trend on day j, and $(s'_i, l') = (s_i, l)$ such that $a_{ij} \neq 0$. Then $credibility_i = \text{f-measure}(s'_i, l')$ the way s'_i is treated as predicted labels.

Since individuals who correlate negatively with the trend also have significant predictive power, we decided to introduce *inverse credibility* to detect those users. $inv\_credibility_i = credibility_i$ with $s'_{ij} = -s_{ij}$.

Clearly, scheme that ignores days with no tweets, brings the least active users to the top. To account for that we also considered *weighted credibility* which penalizes such candidates:

$$credibility\_w_i = \begin{cases} credibility_i \cdot ad\_norm_i, \text{ if } ad\_norm_i \neq 0, \\ credibility_i \cdot tp\_norm_i, \text{ otherwise.} \end{cases}$$

where ad_norm_i is normalized number of days when user i was active, and tp_norm_i is normalized number of posts of the user. The best result offered by this ranker was F-measure=0.58.

Activity-Based Filtering As we mentioned earlier, the best subset of users would include those who are both active and capable of following crime trend. Proposed ranking is two-step: first, the most active individuals (according to metrics in the Table 1) are selected, then they are ranked based on their importance. We tested several activity filters: thresholding based on each metric, clustering and lexsort. The best result was achieved for lexsort and Pearson's r with F-measure=0.59.

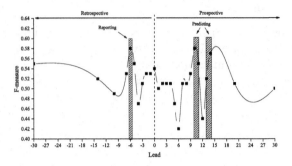

Fig. 2. Predictability observed for different lead/lags

4.3 Long-Term Predictability

It would be useful to have a model able to infer crime index behavior for a long time span. We undertook a series of experiments to assess predictive power of user-generated content. Results are depicted on the Fig. 2. As it can be seen, the best forecasting occurs within a two-week delay. Interestingly, tweets are also highly correlated with the last week events, which also confirm the hypothesis of "reporting" behavior of users.

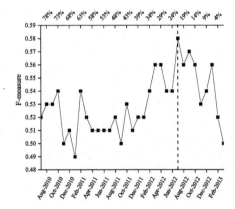

Fig. 3. Test set is equal to 20% of the dataset and is fixed between April 1, 2013 and November 30, 2013. Beginning of the test set is used as a starting point of training set which increases towards July 1, 2010.

4.4 Dependency on Historical Data

Retraining the classifier with up-to-date observations, in general, improves the results. But it is not clear whether the model should be fed all historical data available or the most recent only. Experimental results show that actually only 20% of historical set is needed to reach optimal performance (see Fig. 3), in fact, expanding training set after that point drastically decreases F-measure.

5 Conclusions and Future Work

We proposed a novel user-based framework for predicting social signals of various nature. It was tested on crime trend prediction for Chicago, which, in contrast to previous work, can be easily applied to any other location. We also introduced a new user ranking technique capable of selecting relevant active users based on their credibility. Directions of future work may include exploring different user representations and examining predictability for particular crime types.

References

1. Chainey, S., Tompson, L., Uhlig, S.: The utility of hotspot mapping for predicting spatial patterns of crime. Security Journal **21**(1), 4–28 (2008)
2. Eck, J., Chainey, S., Cameron, J., Wilson, R.: Mapping crime: Understanding hotspots (2005)
3. Gerber, M.S.: Predicting crime using twitter and kernel density estimation. Decision Support Systems **61**, 115–125 (2014)
4. Hipp, J.R., Butts, C.T., Acton, R., Nagle, N.N., Boessen, A.: Extrapolative simulation of neighborhood networks based on population spatial distribution: Do they predict crime? Social Networks **35**(4), 614–625 (2013)
5. Pennebaker, J.W., Francis, M.E., Booth, R.J.: Linguistic inquiry and word count: Liwc 2001. Lawrence Erlbaum Associates, Mahway (2001)
6. Tita, G.E., Boessen, A.: Social networks and the ecology of crime: using social network data to understand the spatial distribution of crime. The SAGE Handbook of Criminological Research Methods, p. 128 (2011)
7. Wang, X., Brown, D.E.: The spatio-temporal modeling for criminal incidents. Security Informatics **1**(1), 1–17 (2012)
8. Wang, X., Gerber, M.S., Brown, D.E.: Automatic crime prediction using events extracted from twitter posts. In: Yang, S.J., Greenberg, A.M., Endsley, M. (eds.) SBP 2012. LNCS, vol. 7227, pp. 231–238. Springer, Heidelberg (2012)
9. White, K., Li, G., Japkowicz, N.: Sampling online social networks using coupling from the past. In: 2012 IEEE 12th International Conference on Data Mining Workshops (ICDMW), pp. 266–272. IEEE (2012)
10. Xue, Y., Brown, D.E.: Spatial analysis with preference specification of latent decision makers for criminal event prediction. Decision Support Systems **41**(3), 560–573 (2006)

'Linguistics-Lite' Topic Extraction from Multilingual Social Media Data

Peter A. Chew[✉]

Galisteo Consulting Group, Inc., 4004 Carlisle Blvd NE, Suite H, Albuquerque,
NM 87107, USA
PAChew@galisteoconsulting.com

Abstract. To achieve accurate situation assessments and information domin-
ance the commander needs accurate and rapid insight into the socio-cognitive
landscape of his communities of interest. This requires insight into the key actors, groups, and their issues and concerns, and to have early indicators of
changes. Social media (which by its nature is noisy and multilingual) is increasing the amount and type of data available for early assessment of rapidly
emerging and changing situations such as disasters or crises. In this paper, we
present a way of extracting topics from this kind of data in a principled and
scalable fashion – regardless of the mix of languages, subject matter, or prove-
nance of data (e.g. Twitter, VKontakte). Using a non-trivial validation task, we
demonstrate that the technique is highly accurate (around 92%). We then show
the results of applying the technique to a sample of around 100,000 Twitter
posts generally relating to the early-2014 conflict in Ukraine, and explain how
these results – or comparable results of applying the technique to other datasets
– would enable a busy analyst quickly to gain a top-down understanding of a
large set of data and help him or her to decide where to focus more detailed attention.

Keywords: Text analysis · Methodological innovation · Multilingual · Validation

1 Introduction

According to a recent Gartner Group report [1], the scale of data everywhere is growing at a rate of 40% to 60% a year. This is a phenomenal rate and reflects a trend
which is observable within the wider world – that data is being generated faster amd
faster, and is becoming more and more widely available. For example, the internet has
grown from around 25 million webpages in 1996 to over a trillion today, in a multitude of languages.

Within government circles, there is considerable interest in using all this openly
available data to help inform smarter decisions broadly relating to national security.
For example, when terrorist attacks or disasters like hurricanes strike around the
globe, one can often learn about actual events on the ground more quickly through
social media than via first responders [2]. Or, if terrorist 'chatter' is predictive of an
impending attack [3], and at least some of this chatter is evident (or has surrogates) in

© Springer International Publishing Switzerland 2015
N. Agarwal et al. (Eds.): SBP 2015, LNCS 9021, pp. 276–282, 2015.
DOI: 10.1007/978-3-319-16268-3_30

open-source data, could real-time monitoring of social media help in interdiction of such attacks? Could a 'social radar' [4] allow this kind of monitoring? What would computational tools to allow this kind of real-time social media monitoring look like, how do we know we can rely on them, and how exactly could they present information to the human user? In this paper, we suggest a possible answer.

In section 2, we outline key general criteria to which we think any 'social radar' system dealing with text in particular ought to conform. In section 3, we give a brief introduction to how Singular Value Decomposition (SVD) has previously been applied to text analysis, and contrast this with the novel element of our current approach. Section 4 shows through non-trivial validation that our approach significantly outperforms the previous state-of-the-art, while actually being easier to implement. In Section 5 we showcase the application of this approach to Twitter data from early 2014 relating to the crisis between Ukraine and Russia, and conclude with suggestions of how the results might be useful to the intelligence analysis community.

2   System Criteria

We previously argued [5] that given the inherent size, noisiness, multilinguality, and heterogeneity of the open-source data that we would like to analyze, it is absolutely essential for any analytics algorithms or techniques that we use to have at least the following interrelated characteristics:

- efficiency and scalability
- generality
- not rule-based
- not 'supervised'
- language-independent

These criteria are discussed more detail in our previous paper, and here we shall focus on the one which is most important to our results here, **language-independence**. Language-independence in an analytics system is related to generality. Just as we want not to have to rely on a domain expert in developing the system, to the extent possible we also want to avoid reliance on linguistic experts. This is why we call our approach 'linguistics-lite'. For example, if we are analyzing Twitter data relating to Ukraine, a large portion of that data may be in Russian. One could envisage a traditional (also rules-based) approach to intelligence analysis where a Russian linguist is brought in, both to read and sort through the Twitter posts and perhaps to develop lists of Russian keywords to flag the posts which merit closer attention. But linguists are expensive and not always available (especially for minority languages), and reading through posts is time-consuming. Perhaps machine translation could instead be used? But what could get lost in translation? Or is there instead a way of using statistical principles that might apply language-universally to sort through text data and help an analyst who might not have linguistic expertise understand the recurring patterns that emerge?

3   Application of Singular Value Decomposition to Text

Singular Value Decomposition (SVD) [6] is a technique that achieves precisely the goals we have just outlined. Neither SVD nor its application to text is new (the history of SVD in text analysis goes back more than 20 years [7]); instead, the specific adaptation of SVD to multilingual content is the key contribution of this paper. Here, we solve the following two problems not addressed by previous approaches:

1. If the documents the analyst wants to analyze are in a language that he does not have the ability to read, can we provide the top n topics in a way which enables him to avoid having to rely on the services of a linguist?
2. If the input documents are in *multiple* languages, and if some of the same topics (e.g. the Russian annexation of Crimea) are discussed in more than one language, can we structure our output to take this into account, so that each row in the table corresponds to a single cross-language topical pattern?

These problems are solved using multilingual LSA. Multilingual LSA, like LSA itself, is not new (e.g. [9]). However, existing multilingual LSA approaches will tell us about the topical patterns of the parallel corpus only, and very few corpora that we want to analyze in real life (such as social media data) are parallel. Thus, we cannot use an approach such as [9], or subsequent approaches such as [8], to produce an analysis of the most prevalent topical patterns in *any* multilingual (but not necessarily parallel) set of text. This is a serious limitation.

We overcome this limitation in a novel way by formulating the input to SVD differently, but in a way which still adheres to the five criteria of section 2. Previous multilingual LSA approaches compute SVD of the term-by-document X constructed from the multilingual social media data. However, this produces output where each pattern (principal component) is monolingual, failing to address either 1 or 2 above.

Thus, we need to reformulate X to make it 'multilingual', but while preserving the inherent structure of X so that we eventually obtain the topics of X, not the topics of some other corpus. This can be done without resorting to machine translation as follows. A multilingual 'translation' matrix M is constructed from a parallel corpus. In this matrix, the probability that each term i_1 in language L1 maps to one or more terms i_2, i_3, ... in language L2 is entered in the appropriate cell. The number of rows and the number of columns in M are both equal to the number of terms or words that occur in *any* language in X. For example, French 'maison' might map in English to 'house' with probability 0.7 and to 'home' with probability 0.3. The means by which these probabilities are computed from a parallel corpus, and the means by which specific terms are statistically aligned across languages even when word alignments are not specifically given in the parallel corpus, are both component parts of Statistical Machine Translation [10], and used in previous approaches to multilingual LSA [8]. This sub-component of our approach adheres to the five principles listed in section 2.

To make X 'multilingual', in linear algebra terms, we now simply compute the matrix multiplication $MX = Y$. This allows SVD (when applied to Y) to correlate topics across languages, but while preserving the topical structure of X.

Doc #	Source text
10433	This should calm things right down. RT @AFP: UPDATE: 2,000 Russian soldiers land in \
35286	They call themselves \
570	Russia seizes control of Crimea! I really should have been keeping up with this. #Ukrain
8082	Casi 2.000 soldados rusos han aterrizado en las últimas horas en Crimea: Casi 2.000 soldados rusos
8085	Casi 2.000 soldados rusos han aterrizado en las últimas horas en Crimea: Casi 2.000 soldados rusos
8059	Casi 2.000 soldados rusos han aterrizado en las últimas horas en Crimea: Casi 2.000 soldados rusos
8048	Casi 2.000 soldados rusos han aterrizado en las últimas horas en Crimea: Casi 2.000 soldados rusos
64251	Now they are using armoured vehicles against protesters. This is madness. #Ukraine #e
8081	Casi 2.000 soldados rusos han aterrizado en las últimas horas en Crimea http://t.co/V1Ivdl4YXW
8093	Casi 2.000 soldados rusos han aterrizado en las últimas horas en Crimea http://t.co/KAFSJJJTQb

Fig. 1. A topical pattern from multilingual Twitter data

Applying this approach (an SVD of Y rather than X) to a dataset of 69,238 Twitter posts relating to the Ukraine crisis, over 98% of which were in one of 7 languages (English – 45%, Ukrainian – 22%, Russian – 18%, Italian – 6%, French – 4%, German – 2%, and Spanish – 2%), for example, a prominent PC that emerges clearly relates to the landing of 2,000 Russian soldiers in Crimea. Despite the language barrier, this component appropriately groups English document 10433 with Spanish documents 8082, 8085, 8059, and 8048 (see Figure 1).

On the face of it, therefore, the approach seems to produce plausible cross-linguistic results. An English-speaking analyst reviewing these results would be able to make sense of what the topical pattern is 'about' even without the ability to read Spanish, simply because English-language documents are also represented within the output. The approach of computing an SVD of Y rather than X appears to solve the problems we posed at the beginning of section 5. We now consider the validity of these results.

4   Validation

While, as stated above, we have constructed an approach which does not rely on any inherent linguistic parallelism in the input corpus, this of course does not exclude using our approach with a parallel corpus. Conceptually, our approach to validation takes the position that documents which are translations of one another ought to receive very similar weightings in the different PCs. The interested reader can refer to [8] for more details, since the validation task we use is exactly the same as outlined there. Overall, the results we obtained from this experiment, with like-for-like comparisons to previously published work, are shown below.

These results show clearly that our current approach outperforms the previous state-of-the-art in accuracy by around 3.5 percentage points. This outperformance is highly statistically significant ($p < 10^{-9}$) given the number of documents in the parallel corpus used for testing (570). So, our proposed approach is demonstrably more accurate, but more importantly, it solves the problem that the previous state-of-the-art completely failed to solve: how to extract the PCs of a multilingual but non-parallel corpus in a way which preserves that corpus's topical structure, but still correlates documents in different languages on the same topic. This is most analogous to the problem the intelligence analyst wants to solve in real life: quickly gaining a top-down understanding of a large multilingual corpus about which we know very little *a priori*.

Approach	% of cases where translations are nearest neighbors
Our approach: SVD of (DX) = Y)	**91.7%**
Best of [8] (Note 1)	88.2%
Best of [8] without morphology (Note 2)	87.8%

Note 1: Use unsupervised morphological analysis to compute a mapping of terms in M to morphemes (sub-word components). Use this to transform M into M1 and X into X1 (where M1 is a morpheme-by-morpheme multilingual mapping matrix, and X1 is a morpheme-by-document matrix). Eigenvalue decomposition of M1 to produce U. Compute the matrix multiplication U.X1 to obtain document vectors for X (and X1) according to the PCs of M1. Note that this approach (1) has considerably more 'moving parts' than our approach which does not use morphological structure, and (2) has the drawback, as already discussed, that the PCs are those of the parallel corpus, not the corpus from which we derive X that we are really interested in analyzing.

Note 2: Eigenvalue decomposition of M to produce U. Compute the matrix multiplication U.X to obtain document vectors for X according to the PCs of M. This approach has the second, but not the first drawback, stated in Note 1.

5   Applications of the Approach and Conclusions

We have now demonstrated the validity of our approach and, in passing, shown some of the output that it can produce from raw social media data. Social media data, of course, contains attributes other than just text, and this can also be used in concert with the approach presented in this paper to produce visualizations of other kinds, which perhaps also contribute to the approach's credibility. In particular, with Twitter data, each post is also timestamped, and in addition we often have geospatial coordinates associated with each tweet. Relating this back to our linear algebraic conceptualization of the problem, each column of X (the term-by-document matrix used as input to SVD), and therefore by extension V (the document-by-concept matrix output by SVD), is associated with a latitude, longitude, and date and time.

For any PC, therefore, we can visualize *where* and *when* the associated topical pattern was most prominent. An example is shown in Figure 2, which represents the third PC obtained from our Ukraine Twitter data. This corresponds to the third most prevalent topical pattern. First, it is plausible that the third most prevalent topical pattern should be 'Euromaidan' (Russian 'Евромайдан'). The topic appears to have been tweeted from around the world. It 'bursts' in late January 2014, precisely when the movement erupted (a dead man was found hanging in Kiev's Maidan Nezalezhnosti [Independence Square] on January 27, 2014). For an analyst reviewing a continuous stream of Twitter data, without prior knowledge of the Maidan revolution, this could perhaps have provided an early warning.

To conclude: in this paper we have shown a novel application of Singular Value Decomposition, specifically adapted to deal with multilingual input, whereby an analyst can be informed of the key topical patterns discussed in large numbers of documents (e.g. social media posts). The approach is scalable, general, adaptable, resilient to noise, efficient, language-independent, has a minimum of 'moving parts', and produces the same results from the same input every time. In particular, the approach is a 'linguistics-lite' one, relying solely on patterns of word occurrence in whatever data is used as a starting-point, and completely avoiding the use of components that require linguistics expertise to build or compile (such as stemmers, stoplists, parsers, part-of-speech taggers, or named-entity recognition). Not only are the results credible and plausible, we also demonstrate that they are also extremely accurate, based on our empirical and non-trivial validation. We hope our approach will point the way, technically, to greater efficiencies in making sense of the ever-increasing and ever more heterogeneous repositories of openly available information on the World Wide Web.

PC#	Weight	Top 10 keywords in any language for this principal component
3	15.6182	euromaidan, Евромайдан, євромайдан, RT, Ukraine, http, Майдан, co, t, Dbnmjr,

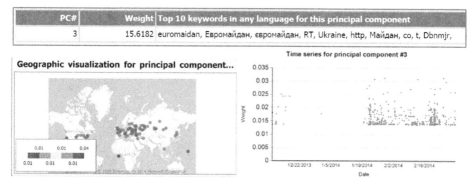

Fig. 2. PC #3 in the Ukraine Twitter data

References

1. Gartner Group: User Survey Analysis: Key Trends Shaping the Future of Data Center Infrastructure Through 2011. Gartner Report ID G00208112 (2011)
2. Lindsay, B.: Social Media and Disasters: Current Uses, Future Options, and Policy Considerations. Congressional Research Service 7-5700, R41987 (2011)
3. Drozdova, K., Samoilov, M.: Predictive analysis of concealed social network activities based on communication technology choices: early-warning detection of attack signals from terrorist organizations. Comp. and Math. Org. Theory 16(1), 61–88 (2010)
4. Costa, B., Boiney, J.: Social Radar. MITRE Technical Report #120088 (2012)
5. Chew, P.: Critiquing Text Analysis in Social Modeling: Best Practices, Limitations, and New Frontiers. Soc. Computing, Behavioral-Cultural Modeling & Prediction, pp. 350-358 (2013)
6. Golub, G.H., Van Loan, C.F.: Matrix Computations, 3rd edn. Johns Hopkins University Press, Baltimore (1996)

7. Deerwester, S., Dumais, S.T., Furnas, G.W., Landauer, T.K., Harshman, R.: Indexing by Latent Semantic Analysis. Journal of the American Soc. for Inf. Science **41**(6), 391–407 (1990)
8. Chew, P.A., Bader, B.W., Helmreich, S., Abdelali, A., Verzi, S.J.: An Information-Theoretic, Vector-Space Model Approach to Cross-Language Information Retrieval. Journal of Natural Language Engineering **17**(1), 37–70 (2011)
9. Young, P.: Cross Language Information Retrieval Using Latent Semantic Indexing. Master's thesis, University of Knoxville, Tennessee: Knoxville, TN (1994)
10. Brown, P.F., Della Pietra, V.J., Della Pietra, S.A., Mercer, R.L.: The Mathematics of Statistical Machine Translation: Parameter Estimation. Comp. Ling. **19**(2), 263–311 (1993)

Mining Business Competitiveness from User Visitation Data

Thanh-Nam Doan, Freddy Chong Tat Chua, and Ee-Peng Lim[✉]

Singapore Management University, Singapore, Singapore
{tndoan.2012,freddychua,eplim}@smu.edu.sg

Abstract. Ranking businesses by competitiveness is useful in many applications including business (e.g., restaurant) recommendation, and estimation of intrinsic value of businesses for mergers and acquisitions. Our literature reveals that previous methods of business ranking have ignored the competing relationship among businesses within their geographical areas. To account for competition, we propose the use of PageRank model and its variant to derive the *Competitive Rank* of businesses. We use the check-ins of users from Foursquare, a location-based social network, to model the winners of competitions among stores. The results of our experiments show that Competitive Rank works well when evaluated against ground truth business ranking.

1 Introduction

In consumer and retail business, determining the value of a store has always been an important task. It concerns the assessment of revenue that can be generated by the store, its rental worth as well as other measures that affect investment-related decisions. Traditionally, one could judge a store's value by referring to business reports published by some authority sources. This kind of reports while giving expert-level assessment suffers from a few major shortcomings such as high cost in engaging experts and obsolete measures used in assessment.

With the advent of Web 2.0 and social media, we can now crowdsource user ratings and reviews of stores to distinguish the good stores from the bad ones. These self-reporting user-generated content may however be biased according to user preferences [4] and may not be trustworthy [5]. This approach also does not consider users who actually visit the stores and choose to be the silent supporters of good stores.

This research therefore introduces a novel way to determine the value of business stores using user visitation data which are now easily available from location social media such as Foursquare, Twitter, etc. While these data may not cover the entire user population, they still represent a significant user population sample which is still growing at a fast rate fueled by the pervasive use of mobile phones and location aware applications.

© Springer International Publishing Switzerland 2015
N. Agarwal et al. (Eds.): SBP 2015, LNCS 9021, pp. 283–289, 2015.
DOI: 10.1007/978-3-319-16268-3_31

This paper performs an exploratory research into the use of visitation data available from Foursquare to determine store competitiveness. Our objective is to derive competition from visits performed by large number of users and to develop models to ranks the stores based on this data. Our proposed approach is to first construct a neighborhood graph of stores, and convert user visits to competitive probabilities among stores. We then define PageRank style store ranking models using the competitive probabilities.

We summarize our research contributions of this work as follows:

- This research is the first that quantifies store competitiveness using user visitation data. Several *PageRank* style ranking models have been defined.
- We evaluate our proposed six models on a real dataset and present case examples to illustrate the characteristics of these models.
- We further conduct experiments to evaluate the proposed models against the external ground truth information.

Among the previous works, the Huff model [2] is relevant to our problem. The model derives the probability that a user visits a store based on the distance between them. Huff model assumes that user location are known, an assumption that does not hold in practice due to privacy concerns. Our proposed models on the other hand only requires the locations of stores but not the users' home locations. Shenghua Bao et al. [1] proposed a method to rank competitors based on text crawled from website. Theodoros Lappas et al. [3] constructed the competitiveness among venues or products by coverage. Given users' preference or item, they could find the top-k competitors but it is hard to know the ranking among them.

2   Proposed Store Ranking Models

We now present two assumptions of our proposed framework to rank stores using check-in data extracted for a set of stores. The first assumption is that the stores to be ranked are of the same type. The second assumption is that there are competitions between stores that are near each other.

Our proposed store ranking framework consists of the following major steps. Firstly, we construct an undirected graph G consisting of stores as vertices. Two stores i and j are connected by an edge (i,j) if i and j are not more than λ apart, a distance threshold. Secondly, we provide different store competitive probability definitions p_{ji}'s for the edges in G in Section 2.1. Thirdly, we apply the PageRank-style models to compute the store competitive probability values. The end results are stores' ranks. In the following, we shall elaborate the details of Steps 2 and 3.

2.1   Modeling Store Competitive Probability

Given a store adjacency graph with stores as nodes, we derive the competitive probability p_{ji} from one node j to another node i based on how much the store value of j could be "distributed" (or lost) to i. Suppose i and j are in competition of some candidate users, the more users visiting i the more j is losing the competition. Ideally, we would like to know: (a) the set of users considering to visit store j, and (b) the subset of them actually visiting store i instead. In most practical settings, we can observe (b) but not (a) unless the users' store preferences are provided. Without infringing the user privacy preferences, we present three different approaches to infer (a) using already observed visit data with different assumptions.

Equal probability (EPR) assumption. Suppose store j has $deg(j)$ neighboring stores. Without referring to any observed visit data, we assume that every neighboring store of j will get equal share of visits. We define $p_{EPR}(j,i)$ under the equal probability assumption as: $p_{EPR}(j,i) = \frac{1}{deg(j)}$

Neighborhood check-in ratio (NCR) assumption. Suppose n_i denote the number of check-ins for any store i. The neighborhood check-in ratio assumption states that the set of potential visits to a store j is the sum of observed visits to j and its neighboring stores. Hence, under the NCR assumption, the competitive probability from store j to store i is defined as: $p_{NCR}(j,i) = \frac{n_i}{\sum_{j \leftrightarrow k} n_k + n_j}$ where $j \leftrightarrow k$ denotes that j is a neighbor of k.

Neighborhood user ratio (NUR) assumption. Suppose m_i denote the number of users performing check-ins on any store i. The neighborhood user ratio assumption states that the set of potential users to a store j is the sum of observed users to j and its neighboring stores. Hence, under the NUR assumption, the competitive probability from store j to store i is defined as: $p_{NUR}(j,i) = \frac{m_i}{\sum_{j \leftrightarrow k} m_k + m_j}$

Next, we will apply the above competitive probability definitions to a few PageRank-style models that compute store values.

2.2 PageRank Model

PageRank [6] was originally designed to compute the importance of web pages based on the directed links among the pages. The key idea of PageRank is that an important page should be linked from other important pages. In our context, we define the first PageRank-style model with the competitive probabilities derived by the equal probability assumption. Let $PR_{EPR}(i)$ denote the PageRank of store i and is defined as: $PR_{EPR}(i) = (1-\alpha) \cdot \frac{1}{N} + \alpha \cdot \sum_{j \leftrightarrow i} PR_{EPR}(j) \cdot p_{EPR}(j,i)$ where N denotes the total number of stores, and α is called *damping factor* to control the weight given to random walk in the PageRank computation. In our experiments, we set $\alpha = 0.85$ by default. Replacing p_{EPR} by p_{NCR} or p_{NUR} gives us the definition of PR_{NCR} or PR_{NUR}.

2.3 CompetitiveRank Model

Other than the definition of competitive probability, we also explore other variants of PageRank style models by changing the random visits to any stores in the

Table 1. Jaccard Coefficient@top k. All models have $\alpha = 0.85$. All Jaccard coefficient scores greater than 75% are in bold text. The unit in table is percentage.

Top K	PR_{NCR}			PR_{NUR}			CR_{EPR}			CR_{NCR}			CR_{NUR}		
	100	200	300	100	200	300	100	200	300	100	200	300	100	200	300
PR_{EPR}	0.5	5	6.2	1	4.4	5.8	0.0	3.1	4.3	0.0	1.5	3.3	0.0	1.8	2.9
PR_{NCR}	-	-	-	70.9	**79.4**	**79.6**	10.5	12.4	15.6	30.7	36	41.2	27.4	33	39
PR_{NUR}	-	-	-	-	-	-	8.7	13	14.9	23.5	35.1	39	21.2	35.6	39.5
CR_{EPR}	-	-	-	-	-	-	-	-	-	39	37.9	39.8	39.8	37.9	40.2
CR_{NCR}	-	-	-	-	-	-	-	-	-	-	-	-	**81.8**	**85.2**	**89.9**

adjacency graph. In the PR_X models, every store is visited with an equal probably $\frac{1}{N}$. This random visit scheme can be modified to create a hybrid PageRank-style model incorporating the observed visit data.

The new PageRank style model, known as **CompetitiveRank (CR)**, aims to combine the earlier PageRank models with the observed check-in data. We define the CompetitiveRank model. $CR_X(i) = (1-\alpha) \cdot \frac{n_i}{\sum_k n_k} + \alpha \cdot \sum_{j \leftrightarrow i} CR_X(j) \cdot p_X(j,i)$ where X denotes one of EPR, NCR and NUR. By varying the α parameter, we can moderate the effect of *check-in ratio* $\frac{n_i}{\sum_k n_k}$ of store i, relative to the random walk effect. When $\alpha = 0$, CR_X reduces to check-in ratio.

3   Experiments on Real Datasets

3.1   Dataset

Our data consists of public check-ins of 55,891 Singapore users in their Twitter timelines from 15 Aug 2012 to 3 June 2013. In our experiments, we only extract venues that are restaurants and their check-ins. There are 121,439 check-ins at 7,290 restaurants in Singapore.

To determine a suitable distance threshold λ for defining the neighborhood of a restaurant, we observe that less than 12% of the restaurants have their nearest neighbors more than 100 meters away. This is not a surprise given that the city of Singapore is densely populated with food-related stores. We therefore set λ to be 100 meters to construct the network of restaurants. Large number of restaurants have a few neighbors and a few restaurants have more than 50 neighbors. Besides, there are 835 restaurants which do not have any neighbors. The restaurants with the largest number of check-ins received 1,373 check-ins while 2,078 restaurants have only one check-in each.

3.2   Evaluation

Correlation Analysis: By considering $k = 100, 200, 300$, we derive the Jaccard Coefficient of the top k ranked stores returned by each model in Table 1. Generally, PR_{EPR} model is most different from the other models. CR_{EPR} is also different from other models but is more similar to other CR models than PR_{EPR} and other PR models. The most similar model pairs however go to the (PR_{NCR}, PR_{NUR}) and (CR_{NCR}, CR_{NUR}) pairs with more than 70% overlaps

Table 2. Spearman correlation coefficient. Coefficients greater than 0.70 are boldfaced.

	PR_{NCR}	PR_{NUR}	CR_{EPR}	CR_{NCR}	CR_{NUR}
PR_{EPR}	0.15	0.16	0.29	0.228	0.23
PR_{NCR}	-	**0.96**	-0.0069	**0.73**	0.667
PR_{NUR}	-	-	-0.0096	0.692	0.68
CR_{EPR}	-	-	-	0.581	0.62
CR_{NCR}	-	-	-	-	**0.974**

between their top k ranked stores. The difference between PR and CR models can be explained by the damping factor. In CR model, it is usually larger than PR's one because the number of venues is smaller than the number of check-ins.

Now, we evaluate the Spearman rank correlation of the full rank lists returned by each pair of models as shown in Table 2 to verify the similarity among them. Table 2 essentially confirms that (PR_{NCR},PR_{NUR}) and (CR_{NCR},CR_{NUR}) model pairs are most similar. In fact, both model pairs enjoy > 0.9 correlation coefficient values. The result is consistent with that of Table 1.

Case Examples: In this section, we show two case examples to illustrate how CR_{NUR} model differs from check-in count when ranking the stores.

Case Study 1. The first part of Table 3 shows the *The Manhattan Fish Market* restaurant. The restaurant has about 139 check-ins ranked 136th according to check-in count. By CR_{NUR} model, however, it has higher rank(39th). The result can be explained by the CR_{NUR} values of its neighbors. According to the Table 3, the average CR_{NUR} of its neighbors is high given the average rank 2682.31 is higher than the middle rank of $\frac{7890}{2} = 3945$.

Table 3. Case Studies of Our Model

Store Name	# Check-in's (Rank)	CR_{NUR} (Rank)	# Neighbors	Avg CR_{NUR} of neighbors	Avg CR_{NUR} Rank of neighbors
Case study 1					
The Manhattan Fish Market	139 (136th)	0.0019(39th)	42	0.00025	2682.31 th
Case study 2					
BALIthai	59 (494th)	0.00071 (298th)	55	-	2433.82 th
Xin Wang Hong Kong Cafe	130 (158th)	0.00068 (312th)	10	-	3149.6^{th}

Case Study 2. The second part of Table 3 shows two stores *BALIthai* and *Xin Wang Hong Kong Cafe(XWHKC)* that are ranked in different order by check-in count and by CR_{NUR}. By check-in count, *BALIthai* is ranked lower than *XWHKC*. By CR_{NUR}, however, we have the reverse rank order due to the higher average CR_{NUR} rank of *BALIthai*'s neighbors. The better ranked neighbors suggest that *BALIthai* must be quite good so as to win visits from these neighboring competing stores. Moreover, the Foursquare score of *BALIthai* is 6.9 with 6 likes from users while *XWHKC*'s score is 5.71 with 4 likes. This fact gives us more confident about the superior of CR_{NUR}. Although the two empirical examples are based on CR_{NUR}, there are many other case examples we can cite for other PageRank style models.

Evaluation with Foursquare score and number of likes data: Foursquare scores ranging from 0 to 10 reflect users' opinions about venues by combining user's response such as tips, check-ins, likes. Thus, we could use Foursquare score to evaluate our models.

Table 4 shows the average Foursquare score of top k restaurants returned by each model. PR_{NUR} and PR_{NCR} are the winners as they have higher scores in three out of four cases. CR_{NUR} performs worse than PR_{NUR} and PR_{NCR} but its result is similar to the Check-in count.

Table 4. Performance - top k restaurants.

Top k	Check-in count	PR_{EPR}	PR_{NUR}	PR_{NCR}	CR_{EPR}	CR_{NUR}	CR_{NCR}
10	7.737	2.52	8.071	8.081	6.027	7.61	7.61
20	6.749	2.405	7.9325	7.9825	5.942	6.8895	7.1865
50	7.002	2.532	7.11	7.0862	6.2682	6.936	6.952
100	6.9491	3.0628	6.8307	6.8331	5.4108	7.01	6.93

Table 5 shows the Spearman correlation between the Foursquare scores and ranking scores of restaurants returned by the proposed models. CR_{EPR} and PR_{EPR} have negative correlation while PR_{NUR} and PR_{NCR} have strong positive correlation with Foursquare scores. CR_{NUR} and CR_{NCR} have positive correlation with Foursquare scores but the correlation is weak, in fact weaker than *Check-in count*. The result between Table 4 and Table 5 are consistent because both tables show the superior performance of PR_{NCR} and PR_{NUR} over the other models.

Table 5. Spearman correlation of Foursquare score and all models.

Check-ins count	PR_{EPR}	PR_{NUR}	PR_{NCR}	CR_{EPR}	CR_{NUR}	CR_{NCR}
0.0476	-0.0488	0.1148	0.1358	-0.07	0.027	0.0417

4   Conclusion

We have proposed ranking methods using data from location-based social media by turning check-ins into competitions between restaurants and their neighbors. We have evaluated our models on real dataset from Foursquare and found probability options p_{NCR} and p_{NUR} behave similarly. We have also quanlitatively analyzed the results through cases studies and verify the correctness of our models via the "ground truth". In our future work, we plan to incorporate features like the distance between user and store; comments and reviews from users; social relationships.

References

1. Bao, S., Li, R., Yu, Y., Cao, Y.: Competitor mining with the web. IEEE TKDE **20**(10), 1297–1310 (2008)
2. Huff, D.L.: A probabilistic analysis of shopping center trade areas. Land Economics (1963)

3. Lappas, T., Valkanas, G., Gunopulos, D.: Efficient and domain-invariant competitor mining. In: KDD (2012)
4. Lauw, H.W., Lim, E.-P., Wang, K.: Bias and controversy in evaluation systems. IEEE TKDE **20**(11) (2008)
5. Lim, E.-P., Nguyen, V.-A., Jindal, N., Liu, B., Lauw, H.W.: Detecting product review spammers using rating behaviors. In: CIKM (2010)
6. Page, L., Brin, S., Motwani, R., Winograd, T.: The pagerank citation ranking: bringing order to the web. In: WWW (1998)

Automatic Tonal Music Composition
Using Functional Harmony

Michele Della Ventura[✉]

Department of Technology, Music Academy "Studio Musica", Treviso, Italy
michele.dellaventura@tin.it

Abstract. The application of Artificial Intelligence technology to the field of music composition has always been fascinating. Different algorithms are created for automatic music composition and in all cases it could be possible that the sequence of the notes of the melody doesn't permit to obtain a correct sequence of the chords (building on the base of the notes of the melody) on the base of the musical grammar. This paper, which outlines key ideas of our research in this field, provides a step to pass this gap: it proposes a method based on a self-learning model that combines De La Motte's theory of Functional Harmony in a Markov process.

Keywords: Artificial intelligence · Automatic composition · Functional harmony

1 Introduction

Automatic music composition is an interesting but challenging task, because computers do not possess any form of creativity, which is necessary to create music. A musical piece is a multi-dimensional space with different interdependent levels: pitch and duration of sounds, vertical and horizontal sonorities, musical phrases and so on. An automatic music composition system could be useful in various ways: for example, human composers could seek inspiration for their own compositions.

This article is going to present an algorithm able to generate a musical idea, of assistance to the composer, on the basis of a self-learning system, focused on the concept of "Functional Harmony". This paper is structured as follows. Section 2 method and related works. Section 3 theory of the Functional Harmony. Section 4 the Process of Markov. Section 5 method and obtained results. Section 6 conclusions.

2 Method and Related Works

This study provides an automatic music composition system inspired by existing works where a musical melody is created randomly, or on the base of specific music rules to determine the sequence of the notes, or by means of generation techniques based on Markov Models [1] [2] [3] [4] [5] [6]. In all these cases it could be possible that the sequence of the notes of the melody doesn't permit to obtain a correct

© Springer International Publishing Switzerland 2015
N. Agarwal et al. (Eds.): SBP 2015, LNCS 9021, pp. 290–295, 2015.
DOI: 10.1007/978-3-319-16268-3_32

sequence of the chords (building on the base of the notes of the melody): for example, the transition between two chords doesn't respect the rules of the tonal music (such as the VII grade goes to the II grade) or the motif doesn't ends with a "Cadence".

The method proposed tries to pass this problem by means of the use of the theory of Functional Harmony of De La Motte in a Markov process.

3   Functional Harmony

In the functional theory [7], the goal is to identify in a sound, a chord or a chord succession, the "intrinsic sonorous value" assumed, compared to a specific reference system polarized in a center, or the capacity to establish organic relations with other sounds, chords or chord successions of the same system.

The chord is concerned, the functional theory tends to research, beyond what it represents by itself in comparison to a certain reference system [8] (for instance, the chord G-B-D, compared to the tonal system and the tonality of C Major, is the dominant chord), the harmonic function performed, the organic relation established with the one that comes before and the one that comes after it.

The pillars of the functional theory are the harmonic functions of tonic (T), subdominant (S) and dominant (D).

It follows that all the chords will have a harmonic function of relaxation or of tonal center T, or of tension towards such center D, or of breakaway from it S.

The three harmonic functions of I, IV and V degree are called main functions and the chords relating to the rest of degrees on the scale (II, III, VI and VII) are considered "representatives" of the I, IV and V degree (see fig 1).

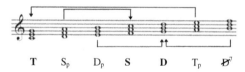

$$\text{T} \quad \text{S}_p \quad \text{D}_p \quad \text{S} \quad \text{D} \quad \text{T}_p \quad \cancel{D}^7$$

Fig. 1. Sequence of the degrees of the C major scale with their related harmonic function

Based on the above considerations, it is important to point out that generally:

1. the function of Tonic (T) goes towards a function of Subdominant (S) that can be represented by the IV degree (S) or by the II degree (S_p) of the scale;
2. the function of Subdominant (S) goes towards a function of Dominant (D) that can be represented by the V degree (D), by the III degree (D_p) or by the VII degree ($\cancel{D}^7$);
3. the function of Dominant (D) goes towards a function of Tonic (T) that can be represented by the I degree (T) or by the VI degree (T_p).

Musical grammar provides the composer with a series of tools allowing him to vary, within the same musical piece, an already presented melodic line, by inserting notes which are extraneous to harmony. The sounds of a melodic line, in fact, may belong to the harmonic construction or may be extraneous to it. The former sounds, which fall in the chordal components, are called real, while the latter sounds, which belong to the horizontal dimension, take the name of melodic figurations (passing tones, turns or escape tones).

They are complementary additional elements of the basic melodic material that lean directly or indirectly on real notes and also resolve on them.

4   Hidden Markov Model (HMM)

The Markov chains are a stochastic process, characterized by Markov properties.
It is a mathematical tool according to which the probability of a certain future event to occur depends uniquely on the current state [9] [6].

This method is based on a designed algorithm, that uses a matrix of the transitions to construct a compositional logic able to create a musical idea [10]: the matrix represents the probabilities for a type of harmonic function to resolve to another type of harmonic function (Fig. 2) [11] [12].

Fig. 2. Matrix of the transitions of the harmonic functions

A first and main task of the algorithm is to read music compositions in MIDI format, recognize the harmonic functions of the different musical degrees [13] and update the matrix of transitions. By reading an ever bigger number of music compositions, the algorithm will be able to propose ever more pleasant musical ideas: and this is because, by reading the music compositions, the probabilities of transition, but also the individual *"state-transitions"* (T, S, D) change, i.e. if a new harmonic function is identified (for instance the function S_p) this function will automatically be inserted in the matrix as a new *"state"* generating new transition probabilities.

5   Method and Obtained Results

The algorithm realized has the objective of proposing a new tonal musical idea as a source of inspiration for a new composition that has the typical characteristic of a musical phrase, i.e. it does not contain modulations (passage from one tonality to another) and the first and last harmonic functions are the tonic ones. Later the composer can use the melodic figuration to refine them. The only parameters required as input for the elaboration are the musical tempo and the number of beats. According to these parameters, it is possible to define the total number of movements that will compose the new idea and we will determine for every single function the number of times to be repeated within the idea on the basis of the transition percentages derived from the transition matrix.

An example of a musical idea in a 6/8 and four beats, generated after the reading of only three music compositions by different authors and different ages is illustrated below (Fig. 3): Flute Sonata in A major by Bach, KV 265, Symphony in G major by Stamitz and Song without words No. 9 by Mendelssohn.

	T	Tp	S	Sp	D
T		1	5		2
Tp			1		1
S			1	1	4
Sp		1			1
D	7				1

Fig. 3. Example of functional analysis and final transition matrix

By means of the last transition matrix the algorithm determines the transition percentage from one state to the other (Fig. 4a) and therefore, on the basis of the total number of movements required for the new musical idea (in this example there are 24 because 4 beats of 6 subdivisions each are required), the number of times every function may occur (Fig. 4b).

	T	Tp	S	Sp	D
T		4	18		8
Tp			4		4
S			4	4	15
Sp		4			4
D	27				4

a)

	T	Tp	S	Sp	D
T		1	4		2
Tp			1		1
S			1	1	3
Sp		1			1
D	7				1

b)

Fig. 4. Representation of the transition percentages from one state to the other (a) and of the number of every function within the musical idea

The results of figure 4b represent the basis for the random generation of the harmonic functions of the new idea. It is immediately deduced that there won't be a unique possible combination of harmonic functions, but, on the contrary, many different combinations may be obtained. The only common element of all these combinations is that the first and the last harmonic function will always be the Tonic one inasmuch as all the tonal music compositions always begin on the Tonic chord because it is representative of the main tonality. In figure 5 below there is a representation of one of the possible combinations of harmonic functions and some possible examples of melodies, defined according to the rules of traditional harmony: every harmonic function is determined by the structure of the chord from which it derives and the chord is formed (fundamentally) by three sounds at a third distance one from the other. In the example in figure 5 the main tonality is C major and therefore the harmonic functions will be represented by the following sounds:

1. T (representative of the first degree): C, E, G;
2. Sp (representative of the second degree): D, F, A;
3. Dp (representative of the third degree): E, G, B
4. S (representative of the fourth degree): F, A, C;
5. D (representative of the fifth degree): G, B, D.
6. Tp (representative of the sixth degree): A, C, E;
7. $\mathcal{D}^7$ (representative of the seventh degree): B, D, F;

In this case, too, it is easy to understand how the presence in the transition matrix of the secondary harmonic functions may generate more appreciable melodies thanks to the presence of different combinations of sounds.

Fig. 5. Example of functional structure and related possible melodies

An example of how the third melody of figure 5 may be modified by the composer by using the melodic figurations is given in figure 6.

Fig. 6. Example of a melody

6   Conclusions

We have designed, implemented and evaluated an approach utilizing machine learning technique, based on Markov's process, for algorithmic generation of monophonic musical melody.

The method described in this article permits to obtain melodies in quick time (typically, few seconds to a few minutes): the melodies are simple, yet pleasant and they represent a musical idea of assistance to the composer as a source of inspiration for a new composition.

The results highlighted that the use of the theory of Functional Harmony permits to obtain a melody that include the concept of "Cadence" which is very important on the compositional level for the definition of the musical phrase.

As future work, this idea can be extended to generating monophonic melodies which contain the concept of modulation and sound like a more complete musical piece.

The tools presented in this article, represent a means of support to the didactic activities: a useful tool to allow specific in-depth analysis, to stimulate the recovery of abilities that are not entirely acquired or as a simple tool of consultation and support to the explanation of the reader.

References

1. Cambouropoulos, E.: Markov Chains as an Aid to Computer-Assisted Composition. Musical Praxis **1**(1), 41–52 (1994)
2. Wadi, A.: Analysis of Music Note Patterns Via Markov Chains, Senior Honors Projects. Paper 2 (2012)
3. Lichtenwalter, R.N., Lichtenwalter, K., Chawla, N.V.: Applying learning algorithms to music generation. In: Proceedings of the 4th Indian International Conference on Artificial Intelligence, IICAI (2009)
4. Cope, D.: Experiments in Musical Intelligence. A-R Editions Inc, Madison (1996)
5. Chan, M., Potter, J., Schubert, E.: Improving algorithmic music composition with machine learning. In: Proceedings of the 9th International Conference on Music Perception and Cognition, ICMPC (2006)
6. Moroni, A., Manzolli, J., Van Zuben, F., Gudwin, R.: Vox Populi: An interactive Evolutionary System for Algorithmic Music Composition. Leonardo Music Journal **10**, 49–54 (2000)
7. de la Motte, D.: Manuale di armonia. Bärenreiter (1976)
8. Coltro, B.: Lezioni di armonia complementare. Zanibon (1979)
9. Bengio, Y.: Markovian Models for Sequential Data. Neural Computing Surveys **2**, 129–162 (1999)
10. Bini, D.A., Latouche, G., Meini, B.: Numerical Methods for Structured Markov Chains. Oxford University Press (2005)
11. Rabiner, L.R.: A Tutorial on Hidden Markov Models and Selected Applications in Speech Recognition. Proceedings of the IEEE **77**(2), 257–286 (1989)
12. Kazi, N., Bhatia, S.: Various Artificial Intelligence Techniques for automated melody generation. International Journal of Engineering Research & Technology **2**, 1646–1652 (2013)
13. Della Ventura, M.: Influence of the harmonic/functional analysis on the musical execution: representation and algorithm. In: Proceedings of the International Conference on Applied Informatics and Computing Theory, Barcellona (Spain) (2012)

Gaming the Social System: A Game Theoretic Examination of Social Influence in Risk Behaviour

Bahareh Esfahbod$^{(\boxtimes)}$, Kurt Kreuger, and Nathaniel Osgood

Department of Computer Science, University of Saskatchewan,
Saskatoon, SK, Canada
bae480@mail.usask.ca, {kurt.kreuger,nathaniel.osgood}@usask.ca

Abstract. In this research we study the effect of social network in risk behaviour. Using an agent-based model based on very simple game-theoretic assumptions, we build a toy model involving donation games over a population. We considered two different variations of a strategy (individually focused and social group focused) and observed drastic differences at the collective level between each. Stable trust patterns were not evolvable in our model with completely social agents. Individually-oriented agents were required. But when trust patterns were able to form among the group-focused agents, large cliques tended to form, contrary to the individually-oriented agents.

Keywords: Social network · Game theory · Liability · Transitivity of trust

1 Introduction

The *Prisoner's Dilemma* is a classic game-theory thought experiment. The general idea is that a situation is created in which two completely rational, self-interested agents would choose to go against a strategy that would lead to a globally ideal solution for the pair, resulting in a less-ideal global solution.

The original game is relatively straight-forward: Two criminals are caught for a crime and are separated with no communication between them. They are interviewed by police, who have enough evidence to convict them of a lesser crime, but not enough for the more serious crime. If prisoner A betrays prisoner B but B stays silent, then A gets off with no jail time while B serves 3 years in prison, and vice versa. If neither prisoner betrays, they each spend 1 year in prison. If both betray, they each serve 2 years in prison.

For each individual prisoner, because betraying the partner gives them less years in prison, then the rational solution is for each one of them to betray their partner and serve 2 years in jail, each. But the best general solution (fewer global years spent in jail) is for both to stay silent, thereby serving 1 year each. One way to see this is to look at your individual possibilities if you betray or stay silent, and compare them. If you stay silent, you either serve 3 years (when the

© Springer International Publishing Switzerland 2015
N. Agarwal et al. (Eds.): SBP 2015, LNCS 9021, pp. 296–301, 2015.
DOI: 10.1007/978-3-319-16268-3_33

partner betrays), or 1 year (when the partner stays silent as well). If you betray, you either serve 2 years (when the partner betrays), or are set free (when the partner stays silent). For any decision your partner makes, choosing to betray always gives you a lower jail time than to stay silent. Thus, a rational solution would be to always betray. But in that case, two rational prisoners would always spend 2 years each in prison, rather than the more ideal 1 year each.

This has been extended to repeated dilemmas involving the trading of goods rather than prison time. These are called *donation games*[3]. It is found that, when the number of trades is not fixed, certain properties emerge about ideal trading strategies. In such case, the rational solution is not as clear as the single dilemma case above. In fact, many strategies have been placed in competition with others to attempt to find an ideal strategy, depending on other parameters of the game.

In this study, we developed an agent based model which uses this repeated donation game and includes components of agent trust, history, and, more importantly, trading networks. The goal of the experiment is to add the transitivity factor of trust as a new strategy to the population and observe the behaviour that such strategy emerges.

2   Model Description

The model developed for this research is a modification of the closed-bag exchange variant of the Prisoner's Dilemma, first described by Douglas Hofstadter[4]. It was developed as an agent-based model in *AnyLogic* simulation software[1]. An environment is created in which a population of N ($= 100$) agents make repeated and randomly assigned trades with one another according to one of two versions of a common strategy. Agents retain a trade history for each other agent, which is a score determined by the past history of that agent. Each agent (A) initiates a trade with a random agent (B), and based on A's strategy and the trade history score of B in A's record, A calculates the chance of honesty, and vice versa. After the trade is done, the trade history matrix is updated based on whether the other agent chose honesty or dishonesty. The agent who initiated the trade waits on average 1 time unit until initiating the next trade.

The basic trade strategy is a type of stochastic tit-for-tat. Rather than a definitive choice before each trade, as in the classic tit-for-tat, here the choice is probabilistic. The trade history in any given trade is recorded as a number between -100 and +100. The function, C_H shown in Figure 1, called *chance of honesty*, maps the value of the trade history to a percentage. It starts at 0, neutral, but weighted slightly towards the honest. If two agents with no trade history trading naively, they trade honestly 60% of the time.

The $N \times N$ trade history matrix is called S and the value S_{AB} is the trade history value that A stores for B. After a trade between two agents A and B, they update their trade history values for each other. For example, A increases S_{AB} by $trustBuild$ ($= 10$) if B was honest for that last trade, and decreases S_{AB} by $trustBreak$ ($= 15$) if B wasn't honest. The fact that $trustBreak$ is greater than *trustBuild* is an attempt to include the assumption that in the real world, building trust is harder than breaking it. This means that 10 honest trades in a row ensures 100% honesty in the future and 7 dishonest trades ensures 0% honesty. Therefore, as expected, an environment exclusively of agents that trade with this strategy will lead to a trust network whose trust pattern is random.

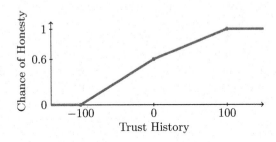

Fig. 1. Chance of honesty function

The modification to this strategy is having *grouper* agents. Agents are called *grouper* if they only consider the trade history values from their *trust network* (the collection of agents whom they trust). Rather than using their own trade history, like *selfers*, *grouper* agents use the collective trust history value from their trust network to calculate their chance of honesty in a trade (Note that these agents still record their own trade history vector, which can be used by agents that trust them). In this case, $C_H(G_{AB})$ calculates the chance of honesty of agent A for agent B. G_{AB} is called agent A's *group value* and is calculated by the following formula, where T_A with m members is the trust network of agent A:

$$G_{AB} = \frac{1}{m} \sum_{I \in T_A} \left(\frac{S_{AI}}{100} \right)^2 S_{IB}.$$

In general, we allow for a spectrum between *selfers* and *groupers*. In this case, the trust value of agent A for agent B is a convex combination between A's own trust history of B and the collective trust history value of B from A's trust network. This is seen by the following formula where the *self-ratio* value, $0 \leq \alpha \leq 1$, is set individually and determines how the self- and group-factors are effective.

$$T_{AB} = \alpha S_{AB} + (1 - \alpha) G_{AB}.$$

3   Experimental Runs

Here, we briefly show the main behaviour of the model under 4 main conditions: a population entirely composed of *selfers*, entirely of *groupers*, a hybrid population, and a simple possible intervention condition. Initially, the trust values are set to 0 for each person and undergo a period where they gradually become polarized. The figures below are when the model reaches a stable state.

3.1 *Selfer* Population

In this example, we show the simple situation of having only *selfers*. The outcomes are very predictable. Each agent chooses their trust history with every other agent based solely on their individual interactions with that other agent, and so we see Figure 2a. In this figure, we represent the population trust state with a 100 × 100 matrix for a population of 100. The cell in row i column j visualizes the trust value that agent i has for agent j. But rather than cells presenting numbers, we encoded them to present the color corresponding to the value of the cell, from black (totally distrustful - always be dishonest) to white (perfectly trusting - always be honest). This matrix is symmetric, which means every trusting relationship is symmetric, i.e., if A trusts B, then B trusts A.

This example shows that roughly half of the possible relationships are white, and their distribution is random. This is what we expect as each trust relationship is decided separately from the rest.

3.2 *Grouper* Population

Figure 2b shows the outcome of a typical simulation with only *groupers*. There is no strict steady state in this case since no stable trust relationships are formed. But if we run it for an indeterminately long time, we see that the pattern as a whole is stable. This pattern shows a largely distrustful population, with no consistent relationships formed. This is expected since everyone's decision about their honesty in a given trade is based only on the state of their other temporarily-trusted neighbours. Each trust decision is so coupled with every other decision that each will consistently pull and push every other and no stable relationships form.

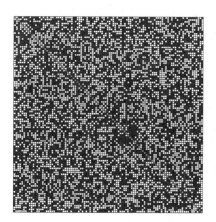 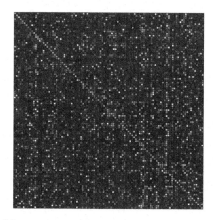

(a) Trust matrix of the *selfer*-only population

(b) Trust matrix of the *grouper*-only population

Fig. 2. Trust matrix outputs for example extreme experimental runs

3.3   Hybrid Population

In this section, we begin to mix populations. Figure 3a shows the same trust matrix for a population of 50 *selfers* and 50 *groupers*. We can see 4 distinct quadrants in the matrix. The top left quadrant is the trust within the *grouper*-only population, and the bottom right is the *selfer*-only. The bottom left is the trust that members of the *selfers* have for members of the *groupers*, and vice versa in the top right. What we see is that the *selfer* population is unchanged in appearance from the example above; a random distribution of cells with roughly half white and apparently symmetric. The *grouper* population, however, is quite different. It is a completely trusting sub-population; each member trusts each other member. Moreover, the off-axis quadrants indicate that all the *groupers* have the same trust pattern for individuals from the *selfer* population. Their individual trust history vectors are all identical. This was never achieved without a *selfer* population to effectively seed the system with a stable trust network around which the *groupers* could form their pattern. This is especially clear when we reduce the size of the *selfer* population.

Figure 3b shows a simulation where the *selfer* population has been reduced to 5, with 95 being *groupers*. Again, we see the 4 sections, and again, we see that the *selfer* network is as expected, random and symmetric, though harder to see because it is small. The larger top left section has now a more complex pattern. Upon examination, what is seen is that there are 2 distinct trust patterns with no overlap. This means that the *groupers* have divided into 2 completely separate cliques. We observed anywhere from 1 to 4 groups form when the *grouper* ratio is very large, depending on the stochastics.

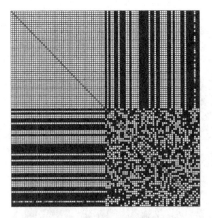 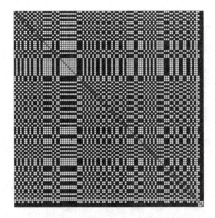

(a) Trust matrix of the hybrid population with 50% *selfers* and 50% *groupers*

(b) Trust matrix of the hybrid population with 5% *selfers* and 95% *groupers*

Fig. 3. Trust matrix outputs for example hybrid experimental runs

It is interesting to note that it does not take much of a *selfer* population to expect a dramatic change in the pattern of the groups. With no *selfers*, no stable state solution has ever been observed. With only 5% as *selfers*, a stable state solution always occurs in a short enough time to examine, though the specific structure of *groupers* can drastically depend on the stochastics.

3.4 Intervention Condition

The final example is when we have a population entirely of *groupers*, with 1 agent that is always honest. If this agent is always honest from the start, then the system as a whole often, if not always, moves relatively quickly towards total trust. Every agent trusts every other agent perfectly. Without this agent, the population of *groupers* never manages to get into a stable trust pattern.

4 Conclusion and Future Work

Using very simple assumptions of agent trading and social behaviour, we model a system that displays some complex structures. For extremely social individuals who use only their social networks to make trust decisions, they are unable to form a stable trust relationship pattern without the presence of more individually-minded agents. Furthermore, these highly social agents tend to make mutually exclusive cliques. There are some intriguing, if stylistic, similarities with the formation of inter-competitive groups of humans.

In this model, the networks arise from the underlying trust behaviour of participants. But some studies [2] indicate how the reverse occurs as well, how networks influence trust relationships. These dynamics could be explored in more depth by adding an initial network for agents and investigating how it might affect trust development.

References

1. AnyLogic Software, version 7.0.1. http://www.anylogic.com
2. Barr, A., Ensminger, J., Johnson, J.C.: Social networks and trust in cross-cultural economic experiments. In: Whom Can We Trust? : How Groups, Networks, and Institutions Make Trust Possible, pp. 65–90. Russell Sage Foundation (2009)
3. Hilbe, C., Nowak, M., Sigmund, K.: Evolution of extortion in Iterated Prisoner's Dilemma games. Proc. Natl. Acad. Sci. U. S. A. **110**, 6913–6918 (2013)
4. Hofstadter, D.R.: Computer tournaments of the Prisoner's Dilemma suggest how cooperation evolves. Sci. Am. **248**(5), 14–20 (1983)

MECH: A Model for Predictive Analysis of Human Choices in Asymmetric Conflicts

Stephen George[1], Xing Wang[2], and Jyh-Charn Liu[2(✉)]

[1] U.S. Department of Defense, New York, USA
ticom.dev@gmail.com
[2] Texas A&M University, College Station, TX 77840, USA
{xingwang,liu}@cse.tamu.edu

Abstract. This paper presents a novel behavior model to characterize functional requirements and environmental constraints for attackers planning attacks in asymmetric conflicts (AC). The *Monitor, Emplacement,* and *Command/Control in a Halo* (MECH) model offers a flexible representation of the physical relationships between belligerents, based on behaviors and risk preferences. MECH supports automated reasoning of locational utilities of AC based on risk-informed human behaviors. We populate the model with a set of 77 features derived from visibility analysis, terrain and local population and show that it is effective for statistics-based classification of roadside attack locations and ranking of tactical values for situational awareness applications.

Keywords: Asymmetric conflict · Behavior modeling · Decision making · Risk aversion · Situational awareness · Command and control · C2

1 Introduction

Asymmetric conflict (AC) is warfare between unequal opponents. Their engagements are often shaped by the weaker party in the conflict by clever selection of attack locations, and ambush its stronger adversary. Engagement control is most evident in the site selection and employment of resources in the attacks, where small decisions can alter the outcome of the AC event significantly. While available resources like the number of attackers physically constrain some choices, specific attack details like overwatch placement and explosive charge emplacement are strongly influenced by subtle criteria like past successes and failures, attacker training, attacker experience, adherence to doctrine, organizational structure, and desired outcome. All attack-related choices are constrained by local terrain, the sophistication of the target, support from the local populace, and available of critical support like long-distance communications. In spite of unknown site selection criteria and attacker constraints, execution of an AC event exposes the outcome of attack-related decisions. Observable outcomes include the attack location and type, estimated number and placement of

This work was supported in part by an ONR grant N00014-12-1-0531 and a National Defense Science and Engineering Graduate (NDSEG) fellowship. Opinions, findings and conclusions or recommendations expressed in this material are the author(s) and do not necessarily reflect those of the sponsors.

© Springer International Publishing Switzerland 2015
N. Agarwal et al. (Eds.): SBP 2015, LNCS 9021, pp. 302–307, 2015.
DOI: 10.1007/978-3-319-16268-3_34

attackers, likely overwatch sites, and the advantage offered by terrain selection. Collection and analysis of features associated with these outcomes can improve predictability and recognition of high-threat locations.

This research develops an AC event model that incorporates select factors important for preparation and execution of an attack or event. This human behavior model constrained by terrain and tactics is based on two key observations: (1) Humans instinctively avoid risk and seek benefit with their choices. Risk aversion influences choices related to siting, timing and use of terrain. (2) Humans tend to make consistent decisions.

Over time, when faced with similar inputs, these (consistent) choices may form patterns detectable by algorithms. Based on these two observations, historical AC events and environs are modeled in this work to capture the site-related choices made by the humans involved, as well as common factors studied in conventional terrain analysis. The study leads to the MECH model, the features that populate it and statistical learning techniques that use MECH for pattern detection and predictive analysis.

Regarding some literatures related to our work, predictive analysis of AC events incorporates elements from diverse including general conflict knowledge, feature extraction, and predictive analytics. Perry and Gordon provide an intelligence-oriented view of asymmetric warfare in [1]. The U.S. Army Ranger Handbook [2] offers a concise compendium of small unit tactics suitable for both attack and defense roles in AC. In this work, the intervisibility between attackers and targets [3] is used to extract candidates of machine learning features like viewshed and cumulative viewshed [4], which was used to derive other additional features (like 'openness'). Geomorphometric measures and terrain classification are addressed in [5][6]. Statistical learning and its feature selection were discussed in [7] [8].

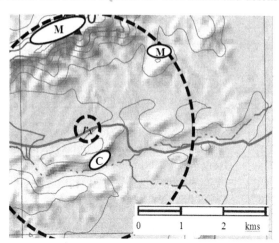

Fig. 1. The MECH Model

2   MECH: A Model for Asymmetric Conflict

The MECH model (Monitor, Emplacement and Control in Halo, illustrated in Fig. 1) describes the locational relationship between belligerents in the planning and execution of an AC event. It aims to model combat actions, particularly the movement and arrangement of the personnel and material with respect to terrain and opposing forces. In Figure 1, r_x is a possible AC site that lies on a road in Afghanistan near 32.1N 66.9E. The area between the inner and outer dashed circles is the Halo where *Monitor* and *Control* actors are found. White ellipses, marked with 'M' and 'C', represent possible *Monitor* and *Control* sites appropriate for the local terrain. Topographic contour lines at 100 meters vertical intervals provide an indication of terrain configuration.

The planning and execution of an AC event is accomplished in steps: (1) The attacker selects the combat operation, like IED or direct fire, and a general area for the event. (2) The attacker narrows down to a specific attack site, called *Emplacement* (E), based on the analysis of the aforementioned factors. (3) The environs of the E are inspected to set the *Control* (C) and *Monitor* (M) sites with intervisibility to E. (4) The location E is prepared and human actors are situated in M and C to provide overwatch and early warning, and initiates the AC event when a suitable target reaches E. Conceivably, a rationale attacker maximizes the chance of success by optimal utilization of the local terrain, target vulnerabilities, and its capabilities. If when optimality is difficult to achieve, the attack configuration for an AC event must be good enough for the attacker launch the action.

A list of 77 features developed for MECH is listed in Table 1. In most cases, features are collected at multiple resolutions or across windows of different sizes. To reflect tactical constraints and variations on the ground, we use window radius w to be short range (50, 100), medium range (350, 500), and far range (1000, 3000 (w_{max})).

Table 1. Feature group list

G_1	Visibility Index (4 features)	Window radii 100-350, 350, 500, 1K meters
G_2	Shape Complexity (4 features)	
G_3	Distance to invisible region (min/mean/max) (15 features)	From M/C, based on sparse viewshed (# of radials 4, 8, 16, 32 and 64)
G_4	Local openness (5 features)	
G_5	Planimetric area (5 features)	
G_6	Rugosity (5 features)	
G_7	Shape Complexity 3D (5 features)	
G_8	Optimal CEA (min/median/max) (3 features)	From M/C (route range 100, 250, 500 and 1K meter)
G_9	Route visibility near *emplacement* (min/median/max) (12 features)	
G_{10}	Elevation, Slope, Convexity and Texture (1 feature each)	Window radii 50, 100, 350, 500, 1K meters
G_{11}	Elevation range (5) , Elevation roughness (5 features)	
G_{12}	Minimum distance to city of at least certain scale (5 features)	Population threshold (1, 1K, 10K, 50K, 100K people)

The ASTER Global Digital Elevation Map[1] (DEM) (~30 meters resolution) is used for geomorphometric and visibility analyses. Fig 2.a shows a DEM with the emplacement site at the center of the map. The elevation of a point p is represented as $elev(p)$. Let the line between two points p_i and p_j be denoted by a set of points $L(p_i, p_j)$ where inter-visibility is defined as

---

[1] https://lpdaac.usgs.gov/data_access

$$v(p_i, p_j) = \begin{cases} 1, \forall p' \in L(p_i, p_j), slope(p_i, p_j) \geq \max(\{slope(p_i, p')\}) \\ 0, otherwise \end{cases}$$

, and $slope(p_i, p_j) = (elev(p_j) - elev(p_i))/dist(p_i, p_j)$.Extending this concept to all locations within range w_{max} of r_x produces a viewshed (Fig. 2.b). Escape adjacency (EA) locations at the boundary of viewshed are shown in blue in Fig. 2.b.

Based on viewshed, we identify two groups of features, each defined over different study regions. $P_w(r)$ denotes the set of discrete locations within distance w from r. Feature G_1 ,visibility index, is the count of pixels in $P_w(r)$ visible from *Emplacement* r. Feature G_2, shape complexity index, assesses the degree of pixel dispersion in the viewshed. It is derived as perimeter-to-circumference ratio, with perimeter being the count of EA pixels in $P_w(r)$ and circumference of the circle whose area equals the visibility index.

Sparse Viewshed. Sparse viewshed aims to capture the major terrain structures in $N = \{4, 8, 16, 32, 64\}$ directions from the view of potential *Emplacement* r. Denote $SV(r)$ as the set of farthest visible points viewed from r along N directions. Then the sparse viewshed is the polygon constructed from $SV(r)$. By measuring the distance from r to points in $SV(r)$, we derive feature G_3 as the minimum, mean, and maximum values. Feature G_4, openness, is defined as the mean of absolute value of the slope from r to each point in $SV(r)$. Feature G_5, planimetric area, is the area of the sparse viewshed polygon. Feature G_6, rugosity, is the ratio between the three-dimensional surface area of the polygon and its planimetric area. Feature G_7 extends G_2 into three dimensions.

EA, CEA, Optimal CEA and Route Visibility. Both viewshed and sparse viewshed are derived from the perspective of the *Emplacement*. However, proximity to concealment is important for *Monitor* and *Control*. Thus, the concept of escape adjacency may capture actor risk tolerance. A point p is EA for *Emplacement* r when

$$EA(p, r) = \begin{cases} 1, & if\ v(p, r) = 1, \sum_{p' \in n_8(p)} v(p', p) < 8 \\ 0, otherwise \end{cases}$$

, where $n_8(p)$ is an 8-connected neighbor as commonly used in image processing.

In some cases, visibility of the approaches to r is also important. We propose $CEA(p) = \sum_{r \in R} EA(p, r)$ to evaluate the utility of locations within range w_{max} of p. The CEA map is illustrated in Fig. 2.c. *CEA*-based features are extracted from locally optimal locations that are selected by

$$CEA_{opt}(p) = \begin{cases} CEA(p), if\ CEA(p) = \max\{CEA(p')|dist(p', p) < w'\} \\ 0, otherwise \end{cases}$$

, and w' is the radius of the local window used for screening. Feature G_8 for r is the minimum, median and maximum $CEA_{opt}(p)$ of locations selected from a set of *Monitor/Control* candidate sites $MC(r) = \{p|CEA_{opt}(p) > 0\}$. These candidates are shown as points in Fig. 2.d. Feature G_9, describes the visibility of locations R' along route R near r and defined as the ratio of locations in R' that share intervisibility to $MC(r)$. Feature G_9 is based on route visibility for $p \in MC(r)$.

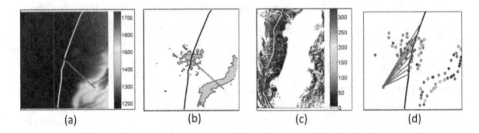

Fig. 2. Visibility structures. (a) DEM; (b) viewshed and EA; (c) EA with weight from CEA; (d) optimal weighted EA, Monitor and Control candidates; (f) route visibility.

Other Features. Feature group G_{10} contains elevation and the geomorphometric features slope gradient, texture and convexity [9]. Elevation range and roughness in G_{11} are the change in and standard derivation of elevation within a study region. Features in G_{12} are based on population estimates[2] and consist of the minimum distance from *Emplacement* r to the nearest inhabited site with a population of η.

3   MECH in the Real World: Afghanistan

Two stages of experiments have been conducted to study the effectiveness of statistical learning algorithms for roadside attacks. A dataset spanning 19 months (Feb. 2011- Aug. 2012) and containing the dates and locations of AC events in Afghanistan was used for the study. In the first stage of study, a total of 250 distinct seeds were randomly selected from event data set. For each seed, geographically and temporally constrained training and test sets were assembled from the set of all IED events near roads and a set of randomly selected non-event sites near roads. An ensemble of classifiers is used for the final classification. After normalization and feature selection, the overall accuracy is in the range of 76% to 81%, as given in Table 2, albeit the accuracy of IED events ranges from 93% to 96%. Due to space constraints, additional predictive analysis results are omitted that address temporal, spatial and combined constraints; normalization; feature selection and dimensionality reduction; and training and test set sizes.   The second stage of experiments is summarized in [9].

Table 2. Preliminary Analysis of MECH Model, Predictive Accuracy for IED/DF Attacks

IED		DF	
kNN	SVM	kNN	SVM
0.7676	0.8119	0.7970	0.8387

---

[2]   http://www.fallingrain.com/world/AF/

4   Conclusion

This paper presents a tactical behavior model for selection of AC attack locations. Visibility analysis, local population and geomorphometry transform measurable environs to a set of 77 tactical behavior features to support situational awareness studies. As of writing of this paper, MECH model has been further explored for its utility in statistical pattern classification [9] and behavior based simulation [10]. Results from both studies suggest that the selection of road side locations are not random but follow detectable but shifting patterns.   This work and related results [9], [10] represent an early look at ongoing research into the complexity and behaviors of AC tactics.

The applications of this research span situational awareness at both strategic and tactical levels, battlefield reconnaissance, sensor placement and cuing, and battlefield resource allocation. Research topics worthy of investigation include the impact of territorial claims in tribal regions of Afghanistan, seasonal pattern shifts related to weather and cultivation, and the optimality of planning with respect to external factors like road conditions and population support.

References

1. Perry, W.L., Gordon, J.: Analytic Support to Intelligence in Counterinsurgencies, RAND National Defense Research Institute (2008)
2. Ranger Handbook. Fort Benning, GA: United States Army Infantry School, Ranger Department (2013)
3. Richbourg, R., Olson, W.K.: A Hybrid Expert System that Combines Technologies to Address the Problem of Military Terrain Analysis. Expert Systems with Applications **11**(2), 207 (1996)
4. Lake, M., Woodman, P., Mithen, S.: Tailoring GIS software for archaeological applications: an example concerning viewshed analysis. Journal of Archaeological Science **25**(1), 27–38 (1998)
5. Hengl, T., Reuter, H.I.: Geomorphometry: Concepts, Software, Applications. Elsevier (2009)
6. Iwahashi, J., Pike, R.J.: Automated classifications of topography from DEMs by an unsupervised nested-means algorithm and a three-part geometric signature. Geomorphology **86**(3) (2007)
7. Guyon, I., Elisseeff, A.: An introduction to variable and feature selection. Journal of Machine Learning Research **3**, 1157–1182 (2003)
8. Hastie, T., Tibshirani, R., Friedman, J.H.: The elements of statistical learning: data mining, inference, and prediction. Springer (2001)
9. Wang, X., George, S., Lin, J., Liu, J.-C.: Quantifying tactical risk: a framework for statistical classification using MECH. In: SBP 2015, Washington, D.C. (2015)
10. Lin, J., Qu, B., Wang, X., George, S., Liu, J.-C.: Risk management in asymmetric conflict: using predictive route reconnaissance to assess and mitigate threats. In: SBP 2015, Washington, D.C. (2015)

Using Topic Models to Measure Social Psychological Characteristics in Online Social Media

Ian Wood[✉]

Australian National University, Canberra, ACT 0200, Australia
ian.wood@anu.edu.au

Abstract. Despite a growing body of research into computational models of social psychological processes, direct empirical grounding for these models remains an elusive goal. This is largely due to the difficulty of measuring modelled characteristics of social groups. This paper presents a methodology combining supervised topic models with traditional psycho-linguistic research as a first step towards such a goal. The method is applied to a collection of over a million tweets from the Twitter 'pro-anorexia' community.

Keywords: Psycho linguistics · Text mining · Social psychology · Empirical grounding · Methodology

1 Introduction

Detailed computational models of socio-psychological processes often suffer from a lack of direct empirical grounding. Traditional techniques for measuring relevant characteristics and relationships require controlled interventions and/or intensive expert annotations, restricting the number of individuals that can be assessed, and any interventions may perturb the very processes under study.

In cases where a substantial portion of group interaction can be captured as text, notably communities that operate over online social media, analysis of those texts and associated metadata promises an avenue for direct and unobtrusive observation of relevant traits of individuals and their communications.

Recent advances in data mining have produced new methods for extracting information from large collections of text data. However applying those techniques to reveal social psychological features, and in particular as an empirical grounding for computational models, remains largely unexplored. The methodology presented here attempts to make a modest start to this endeavour by combining topic models with frequency based psycholinguistic methods.

As an example, the technique is applied to a collection of over a million tweets from the Twitter "pro-anorexia" and eating disorder community. LIWC word categories (Linguistic Enquiry with Word Count [6]) known to be linked to the salience of personal or gender identity [2] are used as a proxy for salience.

© Springer International Publishing Switzerland 2015
N. Agarwal et al. (Eds.): SBP 2015, LNCS 9021, pp. 308–313, 2015.
DOI: 10.1007/978-3-319-16268-3_35

In Section 2 the methodology and the techniques used therein are presented. In Section 3 the collection and preparation of the example data set is described and the reference study linking to identity salience is outlined. In Section 4 the outputs of application to the example data are analysed. Finally, Section 5 summarises the results and proposes further avenues of study.

2   Methodology

Our methodology utilises lists of words whose frequency is linked to characteristics of interest. These lists could be generated by expert analysis of document samples or chosen from word frequency based tools such as LIWC [6] combined with research that links LIWC word classes with characteristics of interest.

Topic models are then applied to identify patterns of word co-occurrence (topics) across a collection of group interactions in text form. The discovered topics contextualise the frequency based tools, identifying other words that occur alongside those indicated by the tool. A novel modification to a topic regularisation technique [4] with the aim to focus topics on words from the lists. Posterior predictive checks [3] are used to ensure the identified topics accurately represent true structures in the data.

2.1   Probabilistic Topic Models

Probabilistic topic models are Bayesian techniques for discovering latent themes in a corpus of (typically text) documents. Themes or 'topics' in a corpus are represented as distributions over words and each document is then considered as distribution over topics. A generative process is defined using these distributions, and its most likely parameter settings given the corpus (the posterior distribution) is estimated. In standard Latent Dirichlet Allocation [1,5](LDA), Dirichlet priors are given for topic-document ($\theta_{t|d}$) and word-topic ($\phi_{w|t}$) multinomial distributions. For the ith token in document d, a topic assignment z_{id} is drawn from $\mathtt{Mult}[\theta_{t|d}]$ and the word x_{id} is drawn from $\mathtt{Mult}[\phi_{x|z_{id}}]$. Given hyperparameters α and β, the generative process is then:

$$\theta_{t|d} \sim \mathtt{Dirichlet}[\alpha] \qquad \phi_{w|t} \sim \mathtt{Dirichlet}[\beta]$$
$$z_{id} \sim \mathtt{Mult}[\theta_{t|d}] \qquad x_{id} \sim \mathtt{Mult}[\phi_{w|t}]$$

The posterior distribution can be estimated using the standard Gibbs sampling update:

$$p(z_{id} = t | x_{id} = w, \mathbf{z}^{\neg id}) \propto \frac{N_{wt}^{\neg id} + \beta}{N_t^{\neg id} + W\beta}(N_{td}^{\neg id} + \alpha)$$

The $N^{\neg id}$ terms represent counts excluding token id (e.g.: $N_{wt}^{\neg id}$ is the number of times topic t is assigned to word w) and $\mathbf{z}^{\neg id}$ is the list of topic assignments (excluding token id).

2.2 Posterior Predictive Checks

Though topic models are a powerful tool, they provide no guarantees on the accuracy of the inferred model. Posterior predictive checks [3] are a mechanism for testing Bayesian models. A discriminant function of input data is chosen to capture some quantity of interest. The inferred model is then used to generate many artificial data sets, and the values of the discriminant function on these data sets provide an estimate of probable function values. If the value of the discriminant function applied to the real input data is improbable, the generative model and/or prior has failed to capture relevant structure in the data. Mimno et.al [3] proposed the mutual information between word allocations to a topic and the allocations of those words to documents as a discriminant function.

$$MI(W, D|k) = \sum_{w,d} P(w, d|k) \log \frac{P(w, d|k)}{P(w|k)P(d|k)}$$

The intuition is that the probability of a word given it's topic allocation should be independent of the document in which it falls. If this is the case, the stated mutual information will be zero. Due to sampling errors this can only be expected for corpora of near infinite size. In practical settings some small positive value will result from random fluctuations.

Important here is the interpretation of failure of this discriminant function. Topic-document mutual information that is higher than expected indicates that words assigned to a particular topic are not evenly distributed among documents. Conversely, values within the expected range indicate that words are as evenly distributed as can be expected, and so topic probabilities are a reasonable estimate for word frequencies among words assigned to the topic. If we restrict the discriminant function in turn to words from each word list of interest, topics with acceptable values make reasonable proxies for words in that list.

2.3 Topic Model Regularisation

In an attempt to focus topics on words from the word lists of interest, a novel adaptation to a topic model regularisation technique [4] was used. The regularisation technique essentially replaces the usual prior for topic-word probabilities by a structured prior that favours known word associations given in a matrix $\mathbf{C}$.

$$p(\phi_t|\mathbf{C}) \propto \left(\phi_t^T \mathbf{C} \phi_t\right)^\nu$$

Optimising the log posterior with respect to ϕ and adapting the LDA Gibbs update results in the following update equations:

$$\phi_{w|t} \leftarrow \frac{1}{N_t + 2\nu} \left(N_{wt} + 2\nu \frac{\phi_{w|t} \sum_{i=1}^{W} \mathbf{C}_{iw}\phi_{i|t}}{\phi_t^T \mathbf{C} \phi_t}\right)$$

$$p(z_{id} = t|x_{id} = w, \mathbf{z}^{\neg id}, \phi_{w|t}) \propto \phi_{w|t}(N_{td}^{\neg id} + \alpha)$$

In the original approach, word associations are drawn from a large, relevant reference corpus. Here we have no reference corpus, and instead use words from the corpus under investigation and choose artificially strong associations between words from the same list, leaving other word pairs 'unassociated'. The word associations in [4] are given in a matrix of word dependencies represented by the pairwise mutual information (PMI) between words in a reference corpus.

$$\text{PMI}_{ref}(w_i, w_j) = \log \frac{p_{ref}(w_i, w_j)}{P_{ref}(w_i)P_{ref}(w_j)}$$

In our case, we calculate PMI values for word pairs in our corpus with the (false) assumption that words from the same list of interest always co-occur. In this case $P(w_i, w_j) = \min(p(w_i), p(w_j))$ and

$$\text{PMI}_{corp}(w_i, w_j) = -\log\left(\max(p_{corp}(w_i), p_{corp}(w_j))\right)$$

3 Identity Salience and the "pro-ana" Twitter Community

3.1 Identity Salience

Research in psychology has found that there are typically many facets to a persons sense of identity. The social and cognitive context in which a person acts determines which of these facets are active or 'salient' at any given time. In terms of group dynamics, the identity salience of group members during interaction is an important factor in the determination, propagation and reinforcement of the groups social identities. These, in turn, can have a significant impact on the decisions of and opinions formed within the group.

A recent study [2] investigated identity salience in 142 young women, mostly undergraduates at the Australian National University. Respondents were first given a priming task designed to make either the their gender or personal identity salient. Several self-report psycho-metric tasks and a writing task on dieting and weight loss were then performed. The psychometric measures confirmed that the priming task had succeeded. In-sample logistic regression on LIWC scores from the writing task was able to predict the prime condition in 73.9% of cases.

3.2 Data Set

Data was collected on a selection of Twitter tags such as *#proana*, *#edproblems* and *#thinspiration* found to be used by the Twitter "pro-anorexia" and eating disorder community between December 2012 and September 2014. The identity salience study (Section 3.1) has some relevance here as this community consists predominantly of young women and diet and especially weight loss are significant topics of discussion. For this analysis, a "document" was taken to be a single tweet. Tweets were tokenised by standardising numerous text 'smiley' forms, isolating punctuation as individual word tokens (these are a LIWC variable of some interest) and converting mixed case words to lower case (all caps words were retained). Url's, #tags, @mentions and apostrophised words (eg: "didn't") were left unchanged. Further pre-processing included removal of word tokens appearing less than 5 times[1] and removal of tweets with less than 3 word tokens. This resulted in a corpus of 262736 documents and a vocabulary of 18713 words.

4   Measurements

The eight most significant LIWC variables in the logistic regression mentioned in Section 3.1 were chosen as a proxy for differentiating between personal and gender identity salience. Thirteen 50 and thirteen 20 topic models were estimated, five each using standard LDA ($\alpha = 0.05N/DT$, $\beta = 0.01$, N words and D documents in corpus, T topics) and eight each using regularised LDA (α as for standard LDA, $\nu = 0.01V/2$, V words in corpus vocabulary). These values of α allocate 5% of the probability mass for smoothing. The choice of ν reflects some equivalence to the choice of β in that the denominator of the update equation for ϕ is the same. This value encourages tweets to exhibit few topics.

All models performed acceptably in the posterior predictive checks, with topic-document mutual information falling within the span of 100 simulated corpora in all but 0.3% (50 topics) and 1.2% (20 topics) of topic/LIWC/model combinations. The ordinary LDA models performed somewhat worse, with an average 47.6 (out of 400) and 12.8 (out of 160) topic/LIWC combinations falling outside the middle 80% of simulated values (20 and 50 topic models respectively). The regularised model averaged 29.6 and 8 outside 80% of simulated values.

To assess the difference between the models with respect to identity salience, we used the conditional entropy of the predicted probability of a salient personal identity in each tweet, given word-topic allocations.

$$H(\texttt{salience}|\texttt{model}) = - \sum_{\text{topic } t} P(t) \sum_{\text{salience } s} P(s|t) \log_2 P(s|t)$$

This quantity measures the amount of extra information (in binary bits) needed to obtain the predicted probabilities, given word-topic allocations. The difference in encoded information about salience between the regularised and standard LDA models was not significant ($p = .37$ and $p = .21$ for 20 and 50 topics respectively in 2-tailed Kolmogorov-Smirnov tests). As expected, the 50 topic models encoded more information (median 0.73) than the 20 topic models (median 0.77, K-S test $p < 10^{-6}$). Other choices of ν with both stronger and weaker regularisation were also tried, however all models with lower entropy than standard LDA also had unacceptable posterior predictive checks. The total entropy of personal vs. gender salience is very close to one (> 0.99). Thus the topic assignments alone account for about a fifth of that information.

Figure 1 indicates that in these models, about a third of the topics exhibit very high personal salience probability, much higher than the average 0.51 for

---

[1] Word tokens from LIWC word classes under study were retained.

documents in the corpus. Those topics represent a coherent context with high personal salience, demonstrating the utility of this approach. Again, the regularised models are more or less equivalent to the standard LDA models.

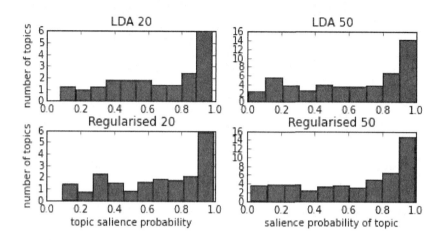

Fig. 1. Average number of topics with high personal salience.

5   Conclusion and Further Work

Topic models combined with word frequency based psychometric tools are shown to provide useful contextualisation and a measure of the features those tools detect. Results such as this can provide insights into the psychological processes active within a group as well as provide some measure of their activity. Topic regularisation was found not to improve the method.

References

1. Blei, D.M., Ng, A.Y., Jordan, M.I.: Latent dirichlet allocation. The Journal of Machine Learning Research **3**, 993–1022 (2003)
2. Dann, E.: The Thin Ideal, Female Identity and Self-Worth: An Exploration of Language Use. Honours thesis, Department of Psychology, The ANU, Australia (2011)
3. Mimno, D., Blei, D.: Bayesian checking for topic models. In: Conference on Empirical Methods in Natural Language Processing, EMNLP 2011 (2011)
4. Newman, D., Bonilla, E., Buntine, W.: Improving topic coherence with regularized topic models. In: Advances in Neural Information Processing Systems (2011)
5. Pritchard, J.K., Stephens, M., Donnelly, P.: Inference of population structure using multilocus genotype data. Genetics **155**(2), 945–959 (2000)
6. Tausczik, Y.R., Pennebaker, J.W.: The psychological meaning of words: Liwc and computerized text analysis methods. Journal of Language and Social Psychology **29**(1), 24–54 (2010)

Expanding Consumer Health Vocabularies by Learning Consumer Health Expressions from Online Health Social Media

Ling Jiang [(✉)] and Christopher C. Yang

College of Computing and Informatics, Drexel University,
Philadelphia, USA
L.Jenny.Jiang@gmail.com

Abstract. The language gap existing between health consumers and health professionals hinders effective healthcare information retrieval and communication. To bridge this language gap, many efforts have been taken to develop the Consumer Health Vocabularies (CHVs). One crucial task in developing CHVs is extracting consumer health expressions. However, most of existing studies of consumer health expressions extraction involve heavy human efforts. In this work, we presented automatic methods based on co-occurrence analysis for extracting consumer health expressions from consumer-contributed content in social media data. The experiment results showed that our proposed methods are effective.

Keywords: Consumer Health Vocabulary (CHV) · Social media · Co-occurrence analysis

1 Introduction

Even though more health related information is available online than before, consumers' attempts to obtain demanded healthcare information are still hampered by their lack of professional knowledge and, more importantly, by the language gap existing between health consumers and health professionals. The healthcare domain is filled with medical jargons, intimidating health consumers who are unfamiliar with those highly professional terminologies. Consumers often find it difficult to understand medical terminologies. This language gap between health consumers and professionals hinders effective healthcare information retrieval and communication.

In order to bridge this language gap, many researchers have been working on developing Consumer Health Vocabularies (CHVs). Most existing studies focus on identifying consumer medical vocabularies and mapping them to professional vocabularies [1-4]. An important subtask in developing CHV is extracting consumer health expressions. The most direct way is asking a representative sample of consumers to review lists of expressions[2]. However, this approach is not feasible given the enormous number and range of health related concepts. Another practical approach to identifying consumer health expressions is to mine existing text [5-8]. Currently, very

© Springer International Publishing Switzerland 2015
N. Agarwal et al. (Eds.): SBP 2015, LNCS 9021, pp. 314–320, 2015.
DOI: 10.1007/978-3-319-16268-3_36

few studies have explored the value of social media data for consumer expressions extraction. In addition, most current methods involve a large part of human efforts. Given the enormous amount of healthcare social media data, identifying consumer health expressions manually is very time-consuming and inefficient.

In this work, we introduce an automatic method to extract consumer health expressions from consumer-contributed content in social media data. Co-occurrence analysis is used to extract candidate terms that are then ranked by a ranking score function. Finally, experiment was conducted to evaluate the performances of proposed methods. We also discussed the effectiveness of the expanded CHV in extracting health related threads from social media data.

2   Automatic Extraction of CHV from Social Media

2.1   Co-occurrence Analysis

We propose to extract consumer health expressions using co-occurrence analysis. The rationale behind this method is that, "Two (or more) words that tend to occur in similar linguistic context (i.e. to have similar co-occurrence patterns) tend to resemble each other in meaning" [9]. For one specific health concept, there would be a great variety of expressions. We can consider one common expression as the starting point and use co-occurrence analysis to identify the other expressions. Based on this theory, we started with a seed term list to find those terms that resemble the seed terms in meaning by using co-occurrence analysis. One thing we need to keep in mind is that the relationship between co-occurring terms could be synonym or relatedness. So, we expected to extract two types of terms: alternative terms that are synonyms for seed terms, and related terms that are highly related to seed terms.

We are particularly interested in extracting consumer health expressions for Adverse Drug Reactions (ADRs). In our previous study [10-12], we studied how to detect ADR signals from social media data, and the ADRs lexicon is crucial for effective detection of ADRs signals. To find the expressions that are used by most consumers, the seed terms should be similar with consumer language. Therefore, we used the first generation CHV [3] to generate the seed terms. For each ADR, we collected a set of seed terms by searching CHV wiki[1] for a list of related expressions.

2.2   Candidate Terms

Given the dataset and the seed term lists, we are able to extract candidate terms for each ADR. First of all, we need to determine the granularity of co-occurrence analysis. In this study, we used online discussion threads as the dataset, so the granularity could be either a whole thread or a message. In this work, we chose message as the granularity considering topic digression could happen in a thread especially when the thread contains a large number of comments. Then we could extract candidate terms by analyzing co-occurrence of all terms with the seed terms in the collection of

---

[1] http://consumerhealthvocab.chpc.utah.edu/CHVwiki/

messages of each ADR. The form of candidate terms could be single words and/or multi-word phrases. In our preliminary study [13], we only extracted bigrams as candidate terms from the dataset. However, only considering bigrams will leave out other valid expressions in unigrams, trigrams or quandrigrams. Therefore, we expanded our study and extract n-grams (n=1, 2, 3, 4) ADR expressions from online healthcare social media data.

In order to extract all n-gram candidate terms from the dataset, all messages have to be Part-Of-Speech tagged [14] beforehand. After we tagged all the messages, the following pre-defined patterns were used to match the n-grams in the dataset:

1) unigram: "Noun";   2) Bigram: "Noun+Noun", "Noun+Verb", "Adjective+Noun";   3) Trigram: "Noun/Adjective/Verb+Noun/Adjective/Verb/Preposition/Conjunction+Noun/Adjective/Verb"; 4) Quandrigram: "Noun/Adjective/Verb+Noun/Verb/Adjective/Preposition/Conjunction+Noun/Verb/Adjective/Preposition/Conjunction+Noun/Adjective/Verb"

2.3   Ranking Score

Candidate terms are not necessarily valid consumer expressions for certain ADR. On one hand, some general terms may be identified if we simply depended on the co-occurrence frequency. On the other hand, terms that are considered to be valid consumer expressions for one ADR would also co-occur with seed terms of another ADR. To remove these noisy terms while keeping relevant terms, we propose a ranking score function s_{ij} to measure the strength of association between seed terms and candidate terms. If one term occurs in several ADRs, the term will be penalized according to how many unrelated ADRs it occurs in, no matter how frequently it appears in each ADR.

$$s_{ij} = \frac{cf_{ij}}{M_j} \times \log \frac{N}{df} \ (i = 1,2,3 \dots ; j = 1,2,3 \dots) \tag{1}$$

where
s_{ij} is the ranking score of phrase i in ADR j;
cf_{ij} is the sum of co-occurrence frequency of phrase i with all seed terms of ADR j;
M_j is the total number of messages in which seed terms of ADR j appear;
N is the total number of ADRs;
df is the number of ADRs that phrase i appears in across all the N ADRs.

3   Experiment

In order to evaluate the effectiveness of proposed method, experiments were conducted. The data were collected from a popular online health community – MedHelp. We collected all the original discussion threads of 45 drugs from MedHelp using an automatic web crawler. There are at least 500 threads for each drugs, and in each thread, the original post and following comments were considered as separate messages, which are the granularity of the co-occurrence analysis. 173,187 messages

were
collected in total. We then built up seed term lists for each ADR by searching the
CHV wiki for ADR terms. The proposed algorithms aim to penalize those general
terms that appear in almost every ADR while increase the weight for important terms
that only occur frequently in one ADR. Under this circumstance, the larger the number of ADRs, the more effective the algorithms would be. Hence, we built up seed
terms list for 100 ADRs, which are randomly selected.

3.1   Extracting Consumer ADR Expressions

After co-occurrence analysis was conducted on each ADR, n-gram candidate terms
were extracted and ranked in a descending order of their ranking score. For each
ADR, we obtained 400 terms in four lists, each of which contains 100 top ranking
unigrams, bigrams, trigrams and quandrigrams respectively. To set up the gold standard, we consulted a health professional to identify alternative and related terms from
the results. In Table 1, terms in bold type are alternative terms, and terms in italic type
are related terms.

As shown in Table 1, many valid expressions were detected for most ADRs.
However, only two valid expressions were identified for diarrhea, and this may be
because diarrhea itself is the most common term used by consumers. Nevertheless,
diarrhea was not included in the top 10 because there are so many noisy unigrams that
diarrhea ranked only in top 30. This indicates that unigram expressions are hard to be
identified due to the impact of a great number of noisy. The results showed that the
proposed method is effective in extracting valid ADR expressions for most ADRs.
Meanwhile, the ADR itself could impact the performance depending on the diversity
of expressions that could convey the same meaning with the original expression.

Table 1. Alternative & Related Terms for the 5 ADRs

Heart Disease	Depression	Diarrhea	Suicide	Kidney Disease
psvt	*paxil*	stones	suicide	dc
heart disease	partial seizure	pain attacks	suicidal ideations	*kidney function*
cardiac disorders	mental depression	pain attacks returned	mood stabilizer	chronic kidney
high cholesterol	depressive disorder	**urine and watery stools**	mood stabilizers	kidney disease
heart diseases	depression disorders	cancers and immune diseases	suicidal thoughts	kidney diseases
cholesterol levels	depression and anxiety	weeks her pain attacks	complex ptsd	current soc
heart disorders	anxiety and depression	doctor did a sonogram	suicide thoughts	kidney disorders
heart conditions	major depressive disorder	**pain and loose stool**	*borderline personality disorder*	heart center website
risk of vascular	*paxil for two years*	mild pain and loose	order to be diagnosed	difference to entirely close
high risk of vascular	example the great day	old daughter has suffered	mind as you read	function to some degree

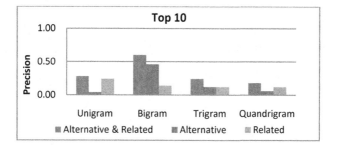

Fig. 1. Comparison among n-grams (Top10)

Another noticeable pattern in Table 1 is that bigrams usually occupy a higher percentage in the resultant lists, which confirms our previous observation that bigram is the commonest form of ADR expressions. To show the difference of performance among n-grams, we used each n-gram individually to extract ADR expressions for the five ADRs. Fig. 1 illustrates the average performance of the five ADRs for Top 10 terms. As we can see, only considering bigrams achieves the best performance expect for related terms. The results indicate that unigrams performed better than bigrams in finding related terms. This can be explained by the fact that unigrams usually tend to be determined as related terms rather than alternative terms, and that is also why unigrams performed the worst in identifying alternative terms. Unigrams usually could not provide enough semantic information to be identified as resembling other terms in meaning. For example, *stool* was marked as related term to the ADR diarrhea by the health professional while *watery stool* was identified as alternative term. Obviously, *stool* is what consumers would mention when discussing diarrhea and it is highly related to diarrhea; however, it does not really resemble the meaning of *diarrhea* like *watery stool* does. Overall, the results showed that only using bigrams could cover most ADR expressions in the dataset. However, we could still miss out some important ADR expressions if we only considering bigrams.

3.2   Extracting ADR Related Threads Using Expanded CHV

To further evaluate the expanded CHV using the proposed methods, we used the original CHV and the expanded CHV to detect ADR related threads from online healthcare social media and compared the two sets of results. With the gold standard, we added all the identified alternative terms into the original CHV and obtain the expanded CHV. Then we used the original and expanded CHV to extract ADR related threads separately. We added all the alternative terms in Top 30 into the original CHV for expanding the CHV.

Table 2. Extraction of ADR Related Threads

Drug Name	# threads		# threads of heart disease	# threads of depression	# threads of suicidal	# threads of kidney disease
Lansoprazole	591	Original	10	77	3	6
		Expanded	112	81	9	10
		Increased by	1020.0%	5.2%	200.0%	66.7%
Luvox	566	Original	17	324	71	6
		Expanded	65	355	110	7
		Increased by	282.4%	9.6%	54.9%	16.7%
Prozac	718	Original	7	387	46	2
		Expanded	45	415	80	3
		Increased by	542.9%	7.24%	73.9%	50%
Simvastain	782	Original	109	82	2	15
		Expanded	269	87	3	23
		Increased by	146.8%	6.1%	50%	53.3%
Zocor	813	Original	150	96	8	16

		Expanded	454	109	11	22
		Increased by	202.7%	13.5%	37.5%	37.5%

We used the original and expanded CHV to identify threads related to four ADRs, including heart disease, depression, suicidal and kidney disease, from five drugs. As we can see in Table 2, the results indicate that the proposed methods have effectively identified valid consumer expressions for the ADRs. Using the expanded CHV can extract substantially more threads related to each ADR. The results imply that consumers not only use the expressions from the original CHV like *heart disease*, but also use other expressions like *heart problem, heart condition*. Extracting a larger number of ADR related threads could contribute to better qualitative study and content analysis.

4   Conclusion

In this work, we proposed automatic algorithms for identifying consumer health expressions from consumer-contributed content in social media. The experiment results showed that our methods are effective in discovering valid consumer expressions. For future study, we should focus on expanding the CHVs by adding new identified terms into current CHVs iteratively and automatically. Consumer health vocabulary problem has been considered to be a fundamental issue in health information study. Our study could not only contribute to the development of CHVs, but also be beneficial for other healthcare studies, such as ADRs signals detection.

References

1. Zeng, Q.T., Tse, T.: Exploring and developing consumer health vocabularies. Journal of the American Medical Informatics Association **13**(1), 24–29 (2006)
2. Zeng, Q.T., et al.: Identifying consumer-friendly display (CFD) names for health concepts. In: AMIA Annual Symposium Proceedings, pp. 859–863 (2005)
3. Zeng, Q.T., et al.: Exploring lexical forms: first-generation consumer health vocabularies. In: AMIA Annual Symposium Proceedings. American Medical Informatics Association (2006)
4. Zeng, Q.T., et al.: Term identification methods for consumer health vocabulary development. J. Med. Internet Res. **9**(1), e4 (2007)
5. Patrick, T.B., et al.: Evaluation of controlled vocabulary resources for development of a consumer entry vocabulary for diabetes. Journal of Medical Internet Research **3**(3), E24 (2001)
6. Tse, T., Soergel, D.: Exploring medical expressions used by consumers and the media: an emerging view of consumer health vocabularies. In: AMIA Annual Symposium Proceedings, pp. 674–678 (2003)
7. Doing-Harris, K.M., Zeng-Treitler, Q.: Computer-assisted update of a consumer health vocabulary through mining of social network data. Journal of Medical Internet Research **13**(2) (2011)

8. Jiang, L., Yang, C.C., Li, J.: Discovering consumer health expressions from consumer-contributed content. In: Greenberg, A.M., Kennedy, W.G., Bos, N.D. (eds.) SBP 2013. LNCS, vol. 7812, pp. 164–174. Springer, Heidelberg (2013)
9. Lancia, F.: Word Co-occurrence and Theory of Meaning (2005)
10. Yang, C.C., et al.: Detecting signals of adverse drug reactions from health consumer contributed content in social media. In: Proceedings of ACM SIGKDD Workshop on Health Informatics, August 12, 2012 (2012)
11. Yang, C.C., et al.: Social media mining for drug safety signal detection. In: Proceedings of the 2012 International Workshop on Smart Health and Wellbeing. ACM (2012)
12. Yang, H., Yang, C.C.: Harnessing social media for drug-drug interactions detection. In: Proceedings of IEEE International Conference on Healthcare Informatics 2013, Philadelphia, PA (2013)
13. Jiang, L., Yang, C.C.: Using co-occurrence analysis to expand consumer health vocabularies from social media data. In: Proceedings of IEEE International Conference on Healthcare Informatics 2013, Philadelphia, PA (2013)
14. Toutanova, K., et al.: Feature-rich part-of-speech tagging with a cyclic dependency network. In: Proceedings of the 2003 Conference of the North American Chapter of the Association for Computational Linguistics on Human Language Technology, vol. 1, Association for Computational Linguistics (2003)

Identifying Correlates of Homicide Rates in Michoacán, Mexico

Carl M. Kruger[✉] and Matthew S. Gerber

Department of Systems and Information Engineering, University of Virginia,
Charlottesville, VA, USA
{cmk3pt,msg8u}@virginia.edu

Abstract. Violent crime rates in Mexico have reached epidemic levels recently due to a corrupt and ineffective government and violent drug-trafficking organizations (DTOs). This research analyzes social, economic, and demographic data, to identify factors that correlate with homicide rates. Previous research efforts have focused on trends of homicide rates and provided subjective recommendations without supporting empirical evidence. This research provides an objective analysis of homicide rates, using a large database consisting of census data provided by the Instituto Nacional De Estradstica Y Geografa (INEGI) website. Through formal experimentation, we found that the most significant indicators were economic and education variables (e.g., Occupants in Refuge and Population over 5 Years Without Schooling), supporting recommendations on policy changes from other research efforts. This paper presents our experiments, results, and insights regarding future work.

Keywords: Homicide · Mexico · Linear regression

1 Background

The United Nations Office on Drugs and Crime ranks Mexico as one of the most dangerous countries in the world, as the twenty-second country with the highest homicide rate [4]. A major source of violence in Mexico stems from the drug war. Confluence of rival Drug Trafficking Organizations (DTOs) and government corruption has increased violent crime rates to epidemic proportions in Mexico. In December 2006, President Felipe Calderon mobilized the Mexican military to engage in a major offensive with DTOs in the state of Michoacán. The drug war continues today and has claimed well over 50,000 lives [1].

Previous research efforts offer expert analysis of the history of the Mexican Drug War as well as recommended solutions to reverse the trends of violence in Mexico. The Justice in Mexico Project (JMP) is an organization that addresses security and human rights issues related to drug violence.[1] They draw few conclusions and recommendations based on data analysis but rather make a subjective analysis of the increase in drug-related homicides as a result of corrupt

---

[1] https://justiceinmexico.org/

© Springer International Publishing Switzerland 2015
N. Agarwal et al. (Eds.): SBP 2015, LNCS 9021, pp. 321–326, 2015.
DOI: 10.1007/978-3-319-16268-3_37

Mexican politics [1]. Likewise, Sibel McGee, et al., discuss the Mexican Cartel problem from a systemic perspective [3], breaking down the problem into several domains containing dynamic relationships and critical variables involved in DTO operations. They provide a holistic assessment of the cartel problem with recommendations on long-term solutions to combat cartel aggression, to include education and economic reform.

This paper presents research aimed at identifying correlates of homicide rates in Michoacán, Mexico. Using linear regression methods, we analyzed indicator variables that correlate with homicide rates. Our results show there are strong correlations between education, economics, and homicide rates. Our results substantiate other research efforts and current plans by the Mexican Government to reduce homicide rates as well as highlight other areas not previously explored. To date, such plans have not been backed by substantial amounts of empirical research.

2   Problem and Objectives

We have focused our research on identifying and understanding correlates of homicide rates in Mexico, focusing on the state of Michoacán. Michoacán continues to be relevant in the drug war, as the Mexican military as well as emerging vigilante groups continue to engage DTOs [5] [6]. Our primary objective was to empirically identify correlates of homicide rates within Michoacán, and highlight areas not previously addressed. By identifying such correlates and quantifying their strengths, we believe decision makers will be in a better position to understand the recent increase in homicides within Michoacán and develop effective solutions. Although the drug war is one of the primary motivating factors, this research considers homicides of all types.

3   Data and Methodology

We obtained data from the Instituto Nacional De Estradstica Y Geografa (INEGI) website.[2] INEGI publishes data, consisting of 965 indicators in four categories: (1) Economy, (2) Environment, (3) Population, Households, and Housing, and (4) Society and Government. We aggregated the data at the municipality level and used homicide rate per 100,000 inhabitants as our response variable. All other variables were also normalized to reflect a rate per 100,000 inhabitants. INEGI measures most indicators annually, but some are measured in increments of five years. For the latter, years with no measurement assume the value of the most recent measurement. We re-normalized the data on a scale of 0 to 1, where 0 corresponds to the minimum value for each indicator and 1 corresponds to the maximum value for each indicator. Other crime rates strongly correlated with homicide rates; however, since our intent is to identify social correlates of homicide, we removed these variables. Additionally, the economic variables oriented

---

[2] http://www.inegi.org.mx

around publishing statistics of the various economic industries in Michoacán, Mexico, and did not provide statistics of individual economic strength or poverty levels, so we removed these variables as well. Any indicators with missing values were removed from the data set, yielding 149 indicators for analysis.

Focusing on 2000-2010, there are only 11 observations per municipality. We pooled observations from all municipalities, resulting in a data set with 1,243 observations. We fit a linear regression model to the response and indicator variables using the generic form of linear regression and performed two analytic tasks based on the resulting model. First, we assessed the indicator variables for significance and evaluated their weights in the context of prior research findings. Second, we evaluated the linear regression model in a 10-fold cross-validation experimental setup designed to assess the model's ability to explain previously unseen data.

4   Results and Discussion

We developed a main effects model using all 149 variables, and then we conducted stepwise regression. The model showed evidence of multicollinearity, so we removed them in a stepwise manner with a variance inflation factor cutoff of 5 [10]. The resulting model consisted of 38 variables and was statistically significant at a level of 0.05. Additionally, we calculated both the R^2 and adjusted R^2 values to determine each model's accuracy. The R^2 was 0.458 and Adjusted R^2 was 0.441. The diagnostic plots in Figure 1 show some evidence of heteroscedasticity and non-normality in our data. This is also evident in the Residuals vs. Leverage Plot as the influential points in the data consist of extremely high homicide rate observations.

We calculated the p-value for each indicator variable using the same statistical significance threshold, and 15 of the 38 variables were statistically significant. Table 1 shows a summary of the most interesting, statistically significant variables, their coefficients, and statistical significance. The population of men and general deaths of women being positively correlated with homicide rates could suggest that a larger male population accounts for higher homicide rates. Economically, with occupants in refuge positively correlated with homicide rates, it might suggest increased economic hardships force more families into refuge, increasing vulnerability to criminal activity. Both education variables show how a less educated population positively correlates with homicide rates, while a more educated population negatively correlates with homicide rates. Finally, Private Homes with 5-8 Inhabitants could suggest that these are extended family homes. Being negatively correlated with homicide rates, extended family homes could be less likely to fall prey to criminal activities due to strong family structure.

We used 10-fold cross-validation to evaluate how well the model performed on different subsets of data, using Mean Squared Error (MSE) to evaluate our model's performance. The MSE was 452. In order to investigate what might account for such high MSEs, we broke down the results by municipality. The five municipalities with the highest MSEs recorded some of the highest homicide rates. The 95% confidence interval of all homicide rate observations falls

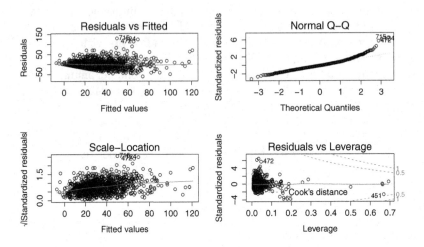

Fig. 1. Diagnostic Plots for Linear Regression Model

Table 1. Variable Statistics

Variable	Coefficient	Stat. Sig.
Total Population, Men	61.470	< 0.001
General Deaths, Women	15.574	0.0159
Occupants in Refuge	10.690	< 0.0182
Private Homes With 5-8 Inhabitants	-30.525	< 0.001
Population Over 5 Years Without Schooling	22.061	< 0.001
Students in Basic and Higher Secondary Education	-19.208	0.0131

between 0 and 100. Over 30% of all observations of the five municipalities with the highest MSE results recorded homicide rates above 100. Meanwhile, the five municipalities with the lowest MSE results remained well within this 95% confidence interval, (Table 2). This suggests that the model does not perform well in cases of extremely high homicide rates. In cases with more moderate homicide rates the model performs better.

Our research corroborates research done by McGee, et al., which posits that economic and education reform are paramount in developing a population that will resist the allure of criminal business [3]. However, neither McGee, et al., nor JMP address the male population as a primary concern for criminal behavior. Based on prison statistics for Michoacán, Mexico, over 85% of inmates are male.[3] This suggests criminal activity is predominantly executed by male citizens. Additionally, another aspect not previously addressed is the impact of

---

[3] These statistics provided by INEGI Bank of Information for 2009-2011.

Table 2. MSE and Homicide Rate Statistics

Municipality	MSE	Min H.R.	Mean H.R.	Max H.R.	H.R. > 100
La Piedad	54.5	10.6	31.2	49.4	0
Contepec	70.3	10.0	28.4	52.1	0
Tzintzuntzan	77.1	0.0	20.4	40.8	0
Ocampo	81.0	10.6	23.8	37.2	0
Zamora	82.7	3.7	28.6	49.2	0
Tumbiscato	1279.6	0.0	36.3	131.5	1
Aquila	1624.7	13.5	71.4	153.0	2
Taretan	1705.1	22.6	67.8	179.0	2
Nocupetaro	1795.4	13.1	85.0	128.0	5
Coahuayana	2860.6	21.5	112.5	193.2	7

extended family homes on criminal activity. This could suggest that areas with more extended family households develop stronger family networks that are more immune to the allure of DTOs.

5 Conclusions and Future Work

Violence in Mexico continues to be a primary concern for the Mexican and United States governments. Policy makers and humanitarian efforts have called for reform in various areas (e.g., education), but there has been a lack of empirical research supporting such reforms. We have developed a linear model of homicides within Michoacán based on data collected by INEGI. Our model provides empirical support for reform in areas like education and economics, corroborating past research [1] [3]. Our model also highlights other areas of reform that have not received as much attention (e.g. Total Male Population and Extended Family Homes). Efforts to address some of these factors are already under way. In February 2014, the Peña Nieto administration launched a social development program called "Plan Michoacán," focused on key areas of economic development, education, infrastructure and housing, public health, and social development and sustainability [2]. The results of this study substantiate their areas of focus and possibly identify other areas that were not included in Plan Michoacán.

Future work will focus on three areas. First, we will investigate interactions between independent variables. Second, we will explore alternative modeling techniques including Random Forest Regression and its associated variable importance measures [9]. While linear regression offered interesting results, MSE was high for some municipalities, which may be a result of non-linearities. Random Forest Regression may overcome this shortcoming and offer greater insight as to correlating factors. Third, we hope to focus more specifically on cartel-related homicides. The present investigation focused on homicides generally because such data are readily available. However, a large component of the homicide process is driven by cartels. By focusing specifically on cartel-related homicides, we hope to uncover key contributory factors that are not currently being considered by anti-violence efforts in Mexico.

References

1. Shirk, D.A., et al.: Drug Violence in Mexico: Data and Analysis from 2001–2009. Technical Report, Justice in Mexico Project, Trans-Border Institute, Joan B. Kroc School of Peace Studies (2010)
2. Heinle, K., et al.: Drug Violence in Mexico: Data and Analysis through 2013. Technical Report, Justice in Mexico Project, Trans-Border Institute, Joan B. Kroc School of Peace Studies (2014)
3. McGee, S., et al.: Mexico's Cartel Problem: A Systems Thinking Perspective. Technical Report, Applied Systems Thinking Institute, Analytic Services, Inc. (2014)
4. Global Study on Homicide, United Nations Office on Drugs and Crime. http://www.unodc.org.gsh
5. The War on Michoacán: A Brief Chronology. http://www.borderlandbeat.com/2014/01/the-war-in-michoacan-brief-chronology.html
6. Flannery, N.P.: Mexico Media Roundup: Another Month of Violence and Vigilantes in Michoacán. http://www.forbes.com/sites/nathanielparishflannery/2014/03/31/
7. Maindonald, J.H., et al.: DAAG: Data Analysis and Graphics data and functions. R package version 1.20 (2014). http://CRAN.R-project.org/package==DAAG
8. Alfons, A.: cvTools: Cross-validation tools for regression model. R package version 0.3.2 (2012). http://CRAN.R-project.org/package==cvTools
9. Liaw, A., et al.: Classification and Regression by randomForest. R News **2**(3), 18–22 (2002)
10. Fox, J., Weisberg, S.: An R Companion to Applied Regression, 2nd edn. Sage, Thousand Oaks. http://socserv.socsci.mcmaster.ca/jfox/Books/Companion

A Preliminary Model of Media Influence on Attitude Diffusion

Kiran Lakkaraju[✉]

Sandia National Labs, Albuquerque, USA
klakkar@sandia.gov

Abstract. We (Sandia National Laboratories is a multi-program laboratory managed and operated by Sandia Corporation, a wholly owned subsidiary of Lockheed Martin Corporation, for the U.S. Department of Energys National Nuclear Security Administration under contract DE-AC04-94AL85000. This document has technical report number: SAND2013-9802C.) study the effects of agenda setting on attitude diffusion using the "Multi-Agent, Multi-Attitude Framework" (MAMA). MAMA captures the interaction between attitudes (through cognitive consistency effects) and interpersonal interaction. Agenda setting is when media's focus on certain stories increases their importance in the minds of the viewers. Using the MAMA model, we show that agenda setting plus strategic choice of seed nodes minimizes diffusion time.

1 Introduction

Attitudes are "general and relatively enduring evaluative responses to objects" where objects can be "a person, a group, an issue or a concept" [1, Page 1]. Attitudes form a cornerstone of social psychology and are incredibly important to the decisions and behaviors of humans. In the past 50 years, numerous studies and meta-reviews have shown that attitudes impact behaviors in a variety of domains (see [2,3] for reviews), such as product purchasing, or health related behaviors (such as choosing to vaccinate children,[4]). Attitudes can change from a variety of factors, including persuasive messages from friends, family and peers, and media influences [2].

Given that attitudes inform important behavior and that attitude can change due to social influence, we feel it is critically important to study "attitude diffusion" – how attitudes spread throughout a population. However, current diffusion models do not capture the key characteristics of attitude change and attitude diffusion.

While there are numerous models of diffusion (e.g., "voter models" [5], thershold models [6]), most of these models do not take into account the underlying complexity and differentiating features of attitude diffusion described above.

Our objective is to use a simple computational model that captures two important concepts in attitude diffusion: (1) the interaction of attitudes through cognitive consistency; and (2) media influence on attitudes. The model is called the "Multi-Agent, Multi-Attitude" (MAMA) model.

© Springer International Publishing Switzerland 2015
N. Agarwal et al. (Eds.): SBP 2015, LNCS 9021, pp. 327–332, 2015.
DOI: 10.1007/978-3-319-16268-3_38

2   MAMA Model Description

The MAMA model contains two levels, a *social* level – which captures inter-personal interaction between agents – and a *cognitive* level that captures the interactions of attitudes *within* an agent.

Let $G_s = <V_s, E_s>$ be an undirected graph that represents the social level of the model. Let $a_i \in V_s$ be the set of agents, and $(a_i, a_j) \in E_s$ represent a bidirectional influencing relationship between agents i and j. Each agent has a cognitive network associated with it. A cognitive network is a weighted undirected graph, $G_c = <V_c, E_c>$ that represents cognitions and the interactions between them. We use the term cognitions to refer to any entity towards which an individual can have an attitude, such as people, places and things; but also to more abstruse entities like values.

Let $c_k \in V_c$ be the set of cognitions, and $(c_k, c_q) \in E_c$ represents a bidirectional influencing relationship between cognitions k and q. $w(k, q)$ is the weight of edge (c_k, c_q); the weight can either be $+1$, or -1: $w(k, q) \in \{1, -1\}$. The weight represents the relationship between cognitions, as we describe later on. For convenience, we let $n_c = |V_c|$.

An attitude towards a cognition is represented as either -1 (a negative attitude) or $+1$ (a positive attitude). We call this the *value* of the cognition. Let $v(i, k)$ be the value of cognition k of agent i.

The theory of cognitive consistency says that individuals have a drive toward maintaining a consistent set of attitudes. For instance, it is generally inconsistent for a person who has a strong positive attitude towards recycling to also have a strong positive attitude towards littering. Cognitive consistency is an important part of attitude dynamics.

Let $\chi_i(k, q)$ be the *consistency* of an edge (c_k, c_q) in the cognitive network of agent a_i. The value of $\chi_i(k, q)$ is:

$$\chi_i(k, q) = \begin{cases} 1 & \text{if } w(k, q)v(i, k)v(i, q) > 0, \\ 0 & \text{Otherwise} \end{cases} \tag{1}$$

This equation merely says that if an edge has a negative weight, the edge is consistent if the two cognitions have values with differing signs. If an edge has a positive weight, the edge is consistent if the two cognitions have values with the same sign.

Let the *state* of a cognitive network be an assignment of values to its cognitions. There are $m = 2^{n_c}$ states for a cognitive network, labelled: $s_1 \ldots s_m$. $s_p(k)$ is the value of cognition k in state p.

The consistency of a cognition k for agent i is:

$$\phi_i(k) = \frac{\sum_{(c_k, c_q) \in I_{i,k}} \chi_i(k, q)}{|I_{i,k}|} \tag{2}$$

where $I_{i,k}$ is the set of edges incident to cognition k in agent i. Intuitively, consistency increases as a cognition has more edges that are consistent.

2.1   Attitude Change

Attitude change is initiated by interpersonal interaction, but mediated by the state of the cognitive network. P_{base} represents the baseline probability for an agent to change their attitude, which in this work is fixed and the same for all agents.

The drive for cognitive consistency is modeled as a multiplicative weight on P_{base}. The more consistent a cognition is with its neighboring cognitions, the less likely it is to change. Let $f_{con}(k, i)$ represent this weight on changing cognition k of agent i. $f_{con}(k, i)$ is a sigmoid curve:

$$f_{con}(k, i) = \epsilon + \frac{2}{1 + e^{-10(\phi_i(k) - .5)}} \qquad (3)$$

For cognitions that have more than 50% of their neighbors in an inconsistent state, $f_{con}(k) > 1.0$, thus increasing the probability they will change (with a maximum multiplicative increase of 2). For those with less than 50% of their neighbors in an inconsistent state, $f_{con}(k) < 1.0$, decreasing the probability to change (with a minimum of ϵ).

The *embeddedness* of a cognition refers to how well it is connected to other cognitions in the cognitive network. Embeddedness is related to a resistance to change (see [2, Chap. 12] for a review).

We represent this resistance to change as a multiplicative weight on P_{base}.

Let $f_{deg}(k, i)$ be the *resistance* to change cognition k of agent i based on the cognition's embeddedness, which we measure through its degree ($deg(k)$). Intuitively, we want $f_{deg}(k, i)$ to decrease as we increase the degree of the concept.

$$f_{deg}(k, i) = \begin{cases} 1.0 & \text{if } deg(k) < deg_{max,i}/2, \\ 0.5 & \text{else} \end{cases} \qquad (4)$$

where $deg_{max,i}$ is the highest degree in the cognitive network of agent i.

Probability of Change. Bringing everything together, let $P_{change}(k, i)$ be the probability of cognition k of agent i changing value, given that i is interacting with another agent with the opposite value for cognition k. Then:

$$P_{change}(k, i) = P_{baseline} \cdot f_{degree}(k, i) \cdot f_{con}(k, i) \qquad (5)$$

2.2   Model Dynamics

We run the model for a set of timesteps. At each timestp t, two random, neighboring, agents (a_i and a_j) are chosen, and a *topic cognition* is chosen based on the current agenda. With probability P_{change}, the attitude of a_j for the topic is changed to the attitude of a_i.

Since we have multiple cognition in our model, we designate a single state s^* as the *goal* state. Once a cognitions switches to the value in the progressive state, it cannot switch back.

2.3   Agenda Setting

We assume that issues are represented by cognitions, and that increasing issue importance leads to more discussion. Thus we define agenda setting in terms of the choice of topic cognition. We define an *agenda* $\pi = [P(c_1), \ldots, P(c_{n_c})]$ as a probability distribution over cognitions. The agenda defines which cognition is chosen.

A *Time-Independent Agenda* (TIA) is a fixed probability distribution over the cognitions. A special case is the uniform distribution, where each cognition has a probability of $1/n_c$ of being chosen.

A *Time Varying Agenda* (TVA) is an agenda that changes over time. Essentially, it is some number of agendas which are active at certain times. From timestep 0 to timestep 1000, the agenda may be $\pi_{1000} = [1/3, 1/3, 1/3]$, but from 1000 onwards, the agenda may be: $\pi_\infty = [1/9, 1/9, 7/9]$. The "boundary value" is the time at which the agenda changes.

3   Experiments

We will use the metric of *mean diffusion time* (or convergence time) – the number of timesteps it takes for 90% of the population to reach the goal state – to compare the impact of agenda setting and node choice.

All agents will have the same type of cognitive network with 3 cognitions (c_2, c_1, c_3), connected in a line in that order. c_1 is called the central cognition (since it has two neighboring cognitions) and the others are called "ancillary cognitions".

10% of the population is initially set to the the goal state of $s^* = <+1, +1, -1>$ and the rest of the agents are assigned to $s' = <-1, -1, +1>$.

Two social networks were studied: ; **Scale Free Graph** [7]: $N = 2000$ and $m = 3$; **Facebook Circles** [8] with $N = 4039$ and $88,234$ edges.

We are interested in two conditions, whether agenda setting is on or off, and how the seeds are chosen. When agenda setting was on, we used a time varying agenda. Let π_b^k be an agenda that was used from timestep 0 to timestep b which sets cognition k to $p = .9$ and the other two cognitions to $(1 - p)/2$. After timestep b (the boundary value), we set the agenda to $\pi = [1/3, 1/3, 1/3]$.

Without agenda setting (the "Agenda off" condition), we assume that each cognition can be a topic with equal probability. This corresponds to an agenda of $[1/3, 1/3, 1/3]$.

The seed nodes were set according to four different methods: **Random**: Uniformly random selection of vertices; **Degree**: Top 10% of vertices according to decreasing degree were chosen; **Betweenness**: Top 10% of vertices based on decreasing betweenness; **Eigenvector Centrality**: Top 10% of vertices based on decreasing eigenvector centrality.

Figure 1 shows the mean convergence time (over 100 runs) for different choice of seeds and for varying boundary values.

In the "Agenda off" situation, as expected, we see that the random setting performs the worst by more than 200 timesteps. This matches expectations – existing work in the influence maximization setting indicates that strategically choosing the seed nodes can increase the influence spread under a linear cascade model [6].

As we increase the boundary value, the influence of the non-uniform strategy become more prevalent. The minimum mean time of convergence occurs when the agenda setting is on, and the seed nodes are set (the results for Between and Degree almost entirely overlap).

These results indicate the the best performing option is to include agenda setting and strategic node choice. For the scale free network, choosing the degree or betweenness measure provides the best results.

Figure1(bottom) show the mean convergence time for the Facebook network and the degree, random, and betweenness node choice metrics. Surprisingly, the eigenvector centrality measures performs very poorly (approximately 3500 timesteps at it's minimum). It is not included in this Figure 1 for space reasons.

Interestingly, the random node choice performs better than the degree based node choice. This is probably due to the topology of the Facebook graph.

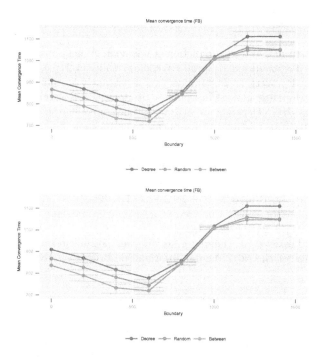

Fig. 1. Top: Mean time to convergence +/- 1 standard deviation. Betweenness and Degree are overlapping. For the Scale Free network. **Bottom:** Mean time to convergence +/- 1 standard deviation for the Facebook network. Excluding the Eigenvectgor Centrality measure.

4   Related Work and Conclusions

Agent-based simulation is an important tool that allows empirical study of complex interactions, in our case between interpersonal influence and attitudes. In this work, we developed a novel agent based model that captures social and cognitive factors that affect decision making (the MAMA model). We used agent-based simulation to study the impact of agenda setting on diffusion time within the MAMA model.

We found that:

1. Agenda setting paired with strategic seed choice results in the quickest diffusion time.

References

1. Visser, P.S., Clark, L.M.: Attitudes. In: Kuper, A., Kuper, J., (eds.): Social Science Encyclopedia. Routledge (2003)
2. Eagly, A.H., Chaiken, S.: The Psychology of Attitudes. Wadsworth Cengage Learning (1993)
3. Kraus, S.J.: Attitudes and the prediction of behavior: A meta-analysis of the empirical literature. Personality and Social Psychology Bulletin **21**(1), 58–75 (1995)
4. Peretti-Watel, P., Verger, P., Raude, J., Constant, A., Gautier, A., Jestin, C., Beck, F.: Dramatic change in public attitudes towards vaccination during the 2009 influenza a(H1N1) pandemic in france. Euro surveillance: bulletin Européen sur les maladies transmissibles = European communicable disease bulletin **18**(44) (2013)
5. Masuda, N., Gibert, N., Redner, S.: Heterogeneous voter models. Physical Review E **82**(1), 010103 (2010)
6. Kempe, D., Kleinberg, J., Tardos, É.: Maximizing the spread of influence through a social network. In: Proceedings of 2003 SIGKDD (2003)
7. Barabási, A.L., Albert, R.: Emergence of scaling in random networks. Science **286**(5439), 509–512 PMID:10521342 (1999)
8. McAuley, J., Leskovec, J.: Learning to discover social circles in ego networks. In: Proceedings of the 2012 Neural Information Processing Systems Conference (2012)

A Study of Daily Sample Composition on Amazon Mechanical Turk

Kiran Lakkaraju[✉]

Sandia National Labs, Albuquerque, USA
klakkar@sandia.gov

Abstract. Amazon Mechanical Turk (AMT) has become a powerful tool for social scientists due to its inexpensiveness, ease of use, and ability to attract large numbers of workers. While the subject pool is diverse, there are numerous questions regarding the composition of the workers as a function of when the "Human Intelligence Task" (HIT) is posted. Given the "queue" nature of HITs and the disparity in geography of participants, it is natural to wonder whether HIT posting time/day can have an impact on the population that is sampled. We address this question using a panel survey on AMT and show (surprisingly) that except for gender, there is no statistically significant difference in terms of demographics characteristics as a function of HIT posting time.

1 Introduction

Amazon Mechanical Turk (AMT) is an online crowdsourcing platform that allows "requestors" to post tasks (called "Human Intelligence Tasks" or HITs), such as image classification and text categorization, for "workers" (or Turkers as they are called) to complete. Increasingly, AMT is being used by the academic community to conduct experiments that would normally have taken place in the lab, for instance in cognitive behavioral experiments [3], identifying subjects with particular psychiatric symptoms [14], and behavioral economics [6].

Numerous studies have shown the significant diversity of subject pools recruited from AMT [1,3,12], thus addressing significant concerns of participant homogeniety in lab based experiments (i.e., the WEIRDness issue [4]. However, the labor market structure of AMT, and the diversity of its subject pool, may in fact cause a problem.

Ten's of thousands of tasks are available at any time on AMT. To manage this abundance of tasks, workers often sort the tasks by their posting date [2]; this leads to newly posted tasks showing up in the first pages of the results and older tasks moving down in the queue. The probability of a worker seeing a task far down in the search results is low (perhaps following the "law of surfing" which suggests an inverse Gaussian distribution for the number of items an Internet user views [5,7]).

Since tasks can quickly be pushed several pages down, it's clear that the sample of workers a task draws can be influenced by the time the task is posted. In addition, several surveys have shown that workers come from two main countries, the US and India [1,3,12]. Thus, due to time zone differences, demographic characteristics can vary based on HIT posting time.

© Springer International Publishing Switzerland 2015
N. Agarwal et al. (Eds.): SBP 2015, LNCS 9021, pp. 333–338, 2015.
DOI: 10.1007/978-3-319-16268-3_39

Some anecdotal evidence suggests this may be the case. For instance, in [8] it was found that there was a not-inconsequential difference in the gender of workers when posting a task in the morning vs. evening.

Given the importance of understanding the demographics of the AMT sample, in this work our objective is to evaluate the potential impact of time of HIT posting on the demographics of the responders. This will allow us to understand potential disparities in sample composition that may influence the outcome of a task.

1.1 Data Collection

We took a "panel survey" approach. We conducted 3 waves in which we posted a survey instrument (described in more detail below) as a HIT on AMT. This analysis will focus on data from Wave 1.

In the first wave we posted the survey instrument twice a day (at 8:00am MT) every day for a calendar week (7 days), starting on a Monday. We allowed 50 responses per posting of the HIT. During this week we made no limitation on the number of times participants could complete the survey (at different times). For instance, a respondent could complete the survey on Monday 8/18 at 8:00am, AND again on Monday, 8/18 in the evening. However, each respondent could only complete the survey once per batch.

Payment was set to $1.00 per survey response. The HIT stayed open for 3 hours, although we usually had 50 responses well before then.

Wave 2 took place 1 week and 2 days after the start of Wave 1. We invited all participants from Wave 1 to retake the survey for the same payment. We kept Wave 2 open for 6 days. Only one HIT was provided, and each participant could only take the survey once.

Wave 3 took place two weeks after Wave 2 and 3 weeks and 2 days after Wave 1. We followed the same procedure as Wave 2.

The survey instrument consisted of approximately 120 questions that ranged from basic demographics to attitudinal, personality and psychological measures. Of pertinence to this work will be the 5 demographic questions:

- In which country do you currently live? (choose from a list of countries).
- What age are you? (answer options will be age ranges).
- What is your household income? (answer options will be in ranges).
- What is your gender?
- What is the highest level of education completed in your household?

While HITS were used to recruit participants from AMT, the actual survey was hosted on SurveyMonkey. A "double validation code" approach was used to link AMT users to the survey response. When accepting the HIT, Turkers were given a validation code which they were to enter into the survey instrument on SurveyMonkey. At the end of the survey they were given a "survey code" to enter back into the HIT.

The main purpose of the codes were to link the respondents between AMT and SurveyMonkey without having to use respondents AMT ID (given the PII nature of the AMT ID [9]). In addition, this verified that the Turker completed the HIT.

For this study we did not penalize the Turker if they did not correctly put in the validation or survey code.

In the following we will use this terminology:

Survey Instance. A survey instance is one batch of surveys released. It is uniquely determined by the day of the week and the time at which it was released.

Survey response (or just response). A survey response is one response by a subject to a survey. A survey response is uniquely determined by the subject id, creation date and time of the survey instances release.

Starting with 1463 responses, we removed responses where:

- The subject did not put in the correct validation code (we kept responses from individuals who put in their AMT ID for their validation code): 150
- The subjects did not correctly answer all three instruction manipulation check questions [11]: 123
- The subjects took less than 1 minutes to do the survey: 6 responses.
- The subjects did not consent to the experiment: 80 responses
- The subjects answered less than 90% of the questions: 67 responses

This left us with *1056* survey responses.

2   Results

Table 1 shows distribution of demographic characteristics as a function of day of the week and time of the day that the HIT was posted. A χ^2-test was performed on all demographic categories over the two categories (day of week and time of day). In all cases except gender, no statistically significant difference was found – indicating that neither the day of the week nor time-of-the day was associated with a change in the demographics of Turkers who responded to the HIT.

Gender did have a statistically significant difference.

3   Discussion

The results are somewhat surprising, especially for the country measure. Given the time zone difference, and the task-queue nature (as described in Section 1 we expected a significant difference in the demographics of users.

One possible explanation is that the time of acceptance of the HIT was quite different from the time the HIT was posted – thus masking the impact of time. We set the HITs to have an expiration date of 12 hours, however all of the Wave

Table 1. Demographic characteristics of participants divided by day of week and by time of day. Only Gender has a statistically significant difference over days of the week ($\chi^2 = 13.96, p = .03$ using simulated p-value with 5000 replicates).

| | | W1 Days | | | | | | | W1 Time | |
| | Mon | Tues | Wed | Thurs | Fri | Sat | Sun | | morning | evening |
N =	(64)	(66)	(70)	(69)	(72)	(73)	(80)		(247)	(280)
Gender**										
Male	0.516	0.364	0.420	0.565	0.380	0.361	0.551		0.455	0.449
Female	0.484	0.636	0.580	0.435	0.620	0.639	0.449		0.545	0.551
Income										
<$25K	0.281	0.273	0.243	0.232	0.250	0.205	0.200		0.223	0.255
$25k-$49K	0.312	0.364	0.414	0.449	0.333	0.384	0.438		0.417	0.356
$50k-$74k	0.172	0.152	0.143	0.130	0.264	0.192	0.212		0.170	0.194
$75k-$99k	0.062	0.106	0.129	0.130	0.042	0.123	0.062		0.089	0.097
$100k-$124k	0.125	0.061	0	0.014	0.083	0.055	0.038		0.049	0.057
$125k-$149k	0.016	0.015	0.057	0.029	0	0.014	0.050		0.036	0.016
>$150k	0.031	0.030	0.014	0.014	0.028	0.027	0		0.016	0.024
Education										
Some H.S.	0	0.015	0.014	0	0.028	0.027	0		0.016	0.008
H.S. Grad.	0.062	0.045	0.057	0.072	0.125	0.096	0.112		0.089	0.077
Some College	0.234	0.242	0.243	0.188	0.181	0.178	0.212		0.190	0.231
Assoc. Degree	0.141	0.030	0.129	0.159	0.042	0.096	0.062		0.113	0.073
Bachelor's	0.344	0.409	0.371	0.348	0.389	0.356	0.375		0.360	0.381
Some Grad.	0.016	0.015	0.014	0.087	0.042	0.082	0.012		0.036	0.040
Master's	0.172	0.212	0.114	0.116	0.181	0.151	0.200		0.162	0.166
Doctoral	0.031	0.030	0.057	0.029	0.014	0.014	0.025		0.032	0.024
Country										
USA	0.797	0.788	0.829	0.783	0.875	0.877	0.838		0.846	0.810
India	0.188	0.197	0.129	0.217	0.111	0.082	0.150		0.130	0.174
Other	0.016	0.015	0.043	0	0.014	0.041	0.012		0.024	0.016
Age										
0-17	0	0	0	0	0	0	0		0	0
18-25	0.141	0.121	0.114	0.188	0.181	0.205	0.175		0.178	0.146
26-35	0.391	0.364	0.400	0.377	0.403	0.411	0.350		0.324	0.445
36-45	0.203	0.197	0.171	0.174	0.222	0.205	0.275		0.227	0.190
46-55	0.078	0.061	0.143	0.130	0.125	0.096	0.088		0.101	0.105
56-65	0.172	0.182	0.143	0.116	0.069	0.082	0.088		0.146	0.093
66+	0.016	0.076	0.029	0.014	0	0	0.025		0.024	0.020

1 HITs were completed within 6 hours, and in fact most HITs were submitted within 30 minutes of posting the HIT (median of 26.3 minutes).

Another reason may be that users are explicitly searching for certain HITs to perform – thus the impact of queue placement is mitigated. This is unlikely, as previous work ([2]) indicated that only 20% of subjects use a keyword search.

Another reason for these results may be that subjects are lying about their demographic information. This is also an unlikely possibility, as [10,13] indicate workers self-reported demographic characteristics are stable over time. In fact,

[13] conducted an independent verification of country of residence (via IP address analysis) showing that, at least for country of residence, there is a high (97.2%) match rate.

One final reason may be a bias in which subject responses are kept. That is, certain demographics will be more likely to satisfy criteria to have their surveys kept in the analysis. This could mask demographic differences.Even if this were true, post-hoc data cleaning is common among AMT based studies. Thus, most studies will be analyzing data from cleaned responses anyway, and these sample composition results would hold.

A more detailed analysis of these issues is planned for future work.

4   Conclusion

The composition of the samples from AMT is an important issue that is, as of yet, understudied [12]. Through a panel survey design, we show, surprisingly, that there is no statistically significant difference in sample composition on the country of residence, income, age, and education measures as a function of when a HIT was posted (morning/evening) and day that HIT was posted (Mon-Sun).

Gender, however, does show a statistically significant difference ($\chi^2, p < .05$) as a function of day of the week. This matches results from [8] which also showed a difference in gender. However, their difference appeared as a function of the day and time of posting, whereas we were able to identify the difference as stemming from the day.

Acknowledgments. Sandia National Laboratories is a multi-program laboratory managed and operated by Sandia Corporation, a wholly owned subsidiary of Lockheed Martin Corporation, for the U.S. Department of Energys National Nuclear Security Administration under contract DE-AC04-94AL85000.

The author thanks Alisa Rogers for help in deploying the survey.

This document has technical report number: SAND2014-19979 C

References

1. Berinsky, A.J., Huber, G.A., Lenz, G.S.: Evaluating online labor markets for experimental research: Amazon.com's mechanical turk. Political Analysis **20**(3), 351–368 (2012). http://pan.oxfordjournals.org/content/20/3/351
2. Chilton, L.B., Horton, J.J., Miller, R.C., Azenkot, S.: Task search in a human computation market. In: Proceedings of the ACM SIGKDD Workshop on Human Computation, HCOMP 2010, pp. 1–9. ACM, New York (2010). http://doi.acm.org/10.1145/1837885.1837889
3. Crump, M.J.C., McDonnell, J.V., Gureckis, T.M.: Evaluating amazon's mechanical turk as a tool for experimental behavioral research. PLoS ONE **8**(3), e57410 (2013). http://dx.doi.org/10.1371/journal.pone.0057410
4. Henrich, J., Heine, S.J., Norenzayan, A.: The weirdest people in the world? The Behavioral and Brain Sciences **33**(2–3), 61–83; discussion 83–135 (2010)

5. Hogg, T., Lerman, K., Smith, L.M.: Stochastic models predict user behavior in social media. HUMAN **2**(1), 25–39 (2013). http://ojs.scienceengineering.org/index. php/human/article/view/72
6. Horton, J.J., Rand, D.G., Zeckhauser, R.J.: The online laboratory: conducting experiments in a real labor market. Experimental Economics **14**, 399–425 (2011)
7. Huberman, B.A., Pirolli, P.L.T., Pitkow, J.E., Lukose, R.M.: Strong regularities in world wide web surfing. Science **280**(5360), 95–97 (1998). http://www.sciencemag. org/content/280/5360/95
8. Komarov, S., Reinecke, K., Gajos, K.Z.: Crowdsourcing performance evaluations of user interfaces. In: Proceedings of the SIGCHI Conference on Human Factors in Computing Systems, CHI 2013, pp. 207–216. ACM, New York (2013). http://doi. acm.org/10.1145/2470654.2470684
9. Lease, M., Hullman, J., Bigham, J.P., Bernstein, M.S., Kim, J., Lasecki, W., Bakhshi, S., Mitra, T., Miller, R.C.: Mechanical turk is not anonymous. SSRN Scholarly Paper ID 2228728, Social Science Research Network, Rochester, NY, March 2013. http://papers.ssrn.com/abstract=2228728
10. Mason, W., Suri, S.: Conducting behavioral research on amazon's mechanical turk. Behavior Research Methods **44**(1), 1–23 (2012). http://link.springer.com/article/ 10.3758/s13428-011-0124-6
11. Oppenheimer, D.M., Meyvis, T., Davidenko, N.: Instructional manipulation checks: Detecting satisficing to increase statistical power. Journal of Experimental Social Psychology **45**(4), 867–872 (2009). http://www.sciencedirect.com/science/ article/pii/S0022103109000766
12. Paolacci, G., Chandler, J.: Inside the turk understanding mechanical turk as a participant pool. Current Directions in Psychological Science **23**(3), 184–188 (2014). http://cdp.sagepub.com/content/23/3/184
13. Rand, D.G.: The promise of mechanical turk: How online labor markets can help theorists run behavioral experiments. Journal of Theoretical Biology **299**, 172–179 (2012). http://www.sciencedirect.com/science/article/pii/S0022519311001330
14. Shapiro, D.N., Chandler, J., Mueller, P.A.: Using mechanical turk to study clinical populations. Clinical Psychological Science **1**(2), 213–220 (2013). http://cpx.sagepub.com/content/1/2/213

The Controlled, Large Online Social Experimentation Platform (CLOSE)

Kiran Lakkaraju$^{(\boxtimes)}$, Brenda Medina, Alisa N. Rogers, Derek M. Trumbo,
Ann Speed, and Jonathan T. McClain

Sandia National Labs, Albuquerque, USA
klakkar@sandia.gov

Abstract. We present a new platform to do online, social influence experiments – the Controlled, Online Social Experimentation (CLOSE) system. We describe it's development, potential uses and justification for use. The CLOSE platform can be used to do long term (weeks to months) experiments in which we can manipulate the interaction networks (within the experiment) of a diverse (drawn from Amazon Mechanical Turk (AMT)) subject pool.

1 Introduction

The increasing use of social media has provided social scientists with a variety of extremely useful data sets to study social influence. However, the types of influence seen in these domains are often restricted to "lightweight" activities such as clicking a link, or retweeting. These behaviors do not have much long-term impact, nor are economically costly to the subject. In contrast, we are interested in "heavyweight" decisions of the sort characterized by economic impact and long-term effects. For example, purchasing residential solar panels is capital intensive, and has a long-term impact on your mortgage and energy consumption.

With these "heavyweight" decisions, we believe *attitudes* play a much larger role. Attitudes, which are "general and relatively enduring evaluative responses to objects" where objects can be "a person, a group, an issue or a concept" [10, Page 1]. Attitudes have a long history in social psychology, and have been shown to have an impact on the behaviors of individuals (e.g., voting behavior [8], consumer purchases [6]).

Studies of online behavior/attitudes can shed light on how attitudes shift and diffuse in populations as a function of, for instance, topological properties of social networks.

However, most existing studies are observational. While providing immense data sets (thousands of subjects) it is difficult to separate correlation from causality. Observational studies have limited ability to manipulate the interaction network (who subjects interact with, often the same as their "social network") of individuals or account for hidden channels of communication between subjects. Experiments in lab settings can address some of these issues (for instance,

© Springer International Publishing Switzerland 2015
N. Agarwal et al. (Eds.): SBP 2015, LNCS 9021, pp. 339–344, 2015.
DOI: 10.1007/978-3-319-16268-3_40

through physical separation of individuals) but are significantly limited in scale, length of study, and subject diversity [7].

We propose a new, online, digital experimentation software platform, the Controlled, Large, Online Social Experiment (CLOSE) system, that allows researchers to conduct online, randomized controlled trials in which the interaction network of subjects can be controlled. Through this platform, we can conduct longitudinal experiments with hundreds of participants to test the effects of interaction network topology, group dynamics, and peer effects. The CLOSE system can create large scale social experiments, manage the large number of subjects and collect data from the subjects. Unique aspects of the CLOSE system are:

1. Inherently social: The platform provides the means to specify the "social" networks of subjects – allowing the study of network topology on influence.
2. Scalable: CLOSE uses a database (H2) backend, allowing us to scale to thousands of subjects.
3. Flexible: CLOSE is domain-agnostic.
4. Longitudinal: CLOSE can be used to create longitudinal experiments, requesting subjects to return to the experiment over weeks or months.

2 Overview of the CLOSE System

Figure 1 shows a screenshot of the current version of the CLOSE system. In this experiment, subjects log in and are asked to read articles and answer questions regarding the articles. Some questions are factual while others are about attitudes/opinions.

Subjects are assigned "friends" by the experimenter. The "friends" may be other subjects, or experimenter controlled agents. The "friends" answers can be relayed to the subject. Through this we can impose an interaction network topology on the subject pool. Because of the online nature, we can scale the interaction network to hundreds or thousands of subjects with arbitrary complexity.

To prevent collusion among subjects, each subject will be assigned a user name by the experimenter. This name will not be known to the subject. From the subjects perspective they have chosen a particular user name which appears whenever they are using the CLOSE system.

Similarly, each subject will choose an avatar to represent themselves. However the experimenter can decide how a subjects avatar looks to others within the CLOSE system.

For instance in Figure 1 the picture of the dog and the name "G4T5BY" were chosen by the subject. However, when the subject appears to others (in a "friend list"), their avatar and username will be the ones chosen by the experimenter.

This mechanism has two benefits:

1. It prevents collusion: Even if subjects discuss the task on online forums (such as the popular www.mturkforum or www.turkernation.com when subjects

are recruited from Amazon Mechanical Turk), it will be highly unlikely that they will discover who their friends are. This allows us to conduct experiments free from the worry of collusion.

2. It allows us to induce group similarity effects by assigning avatars and usernames to a subjects friends that will be similar to the user.

Subjects will see the responses of their "friends" to questions they are being asked. This is the primary method by which we can influence subjects. Since friends can be other subjects, or experimenter controlled agents, we have fine grained control over what information a subject sees.

2.1 Architecture

The CLOSE system is built on the GRAILS web framework [2] to support the display of webpages for subjects. A configuration/administration page is available that lets experimenters define:

- Articles and information for subjects to read.
- Questions for subjects to answer.
- Public avatar/username's for subjects.

A CLOSE experiment takes place in a series of *rounds*. In each round:

1. The interaction network for subjects is specified.
2. The article and questions are chosen.

Having the interaction network defined per round allows us to address subjects who fail to complete the task (in the previous round). We can also arbitrarily change the time between rounds.

2.2 Subject Pool

Subjects can be recruited through any means. Currently we are focusing on Amazon Mechanical Turk (AMT) [1] for recruitment. AMT has been used extensively for psychological and economic research and several studies on the representativeness of the subject pools have been conducted [4,5,9]. Subjects will be recruited through posting a task on AMT which will require the subjects to log into the LIFE platform and complete a round (once registered).

3 Related Work

Several other experimental platforms exist that provide varying capabilities, such as "PlanOut" [3], and Volunteer Science (http://volunteerscience.com/). The CLOSE platform is focused on social experiments, explicitly allowing for interaction between subjects, which is not a component that is easily added to the above platforms.

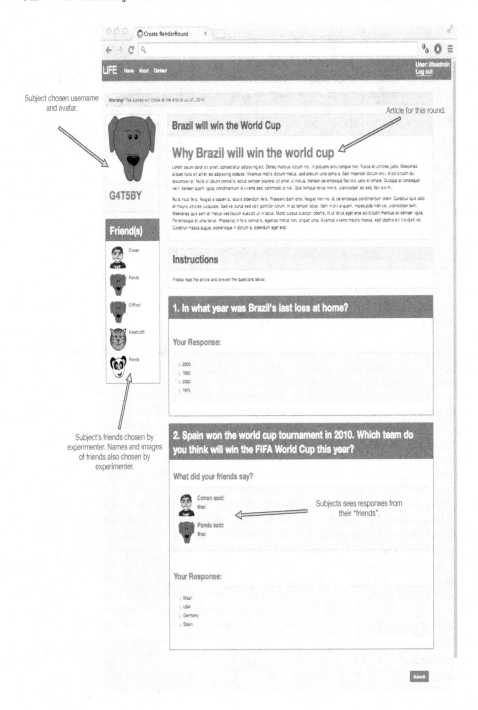

Fig. 1. Screenshot from the current version of the CLOSE platform. The article is randomly generated text.

4   Conclusion

The CLOSE system provides a way for researchers to create scalable experiments that can test for the impact of interaction network topology, among other aspects, on social contagion. Through online experiments we can better understand the causal underpinning of attitude change through social influence. We have described the platform and shown that (1) there is evidence that social influence can occur in an experimental setting like the CLOSE platform; (2) Subjects drawn from AMT will return to take the same experiment; (3) Subjects are relatively stable over time.

5   Conclusion

The CLOSE system provides a way for researchers to create scalable experiments that can test for the impact of interaction network topology, among other aspects, on social contagion. Through online experiments we can better understand the causal underpinning of attitude change through social influence. In this paper we have described the platform its capabilities. Work is currently being done to develop experiments within this platform.

Acknowledgements. Sandia National Laboratories is a multi-program laboratory managed and operated by Sandia Corporation, a wholly owned subsidiary of Lockheed Martin Corporation, for the U.S. Department of Energys National Nuclear Security Administration under contract DE-AC04-94AL85000.

This document has technical report number: SAND2014-16512A

References

1. Amazon mechanical turk (www.mturk.com) (August 2014). www.mturk.com
2. Grails (www.grails.org) (August 2014). www.grails.org
3. Bakshy, E., Eckles, D., Bernstein, M.S.: Designing and deploying online field experiments. In: Proceedings of the 23rd International Conference on World Wide Web, WWW 2014, International World Wide Web Conferences Steering Committee, Republic and Canton of Geneva, Switzerland, pp. 283–292 (2014). http://dx.doi.org/10.1145/2566486.2567967
4. Berinsky, A.J., Huber, G.A., Lenz, G.S.: Evaluating online labor markets for experimental research: Amazon.com's mechanical turk. Political Analysis 20(3), 351–368 (2012). http://pan.oxfordjournals.org/content/20/3/351
5. Crump, M.J.C., McDonnell, J.V., Gureckis, T.M.: Evaluating amazon's mechanical turk as a tool for experimental behavioral research. PLoS One 8(3), e57410 (2013). http://dx.doi.org/10.1371/journal.pone.0057410
6. Fazio, R.H., Olson, M.A.: Attitudes: Foundations, functions, and consequences. In: Hogg, M.A., Cooper, J. (eds.) The Sage Handbook of Social Psychology. Sage (2003)
7. Henrich, J., Heine, S.J., Norenzayan, A.: The weirdest people in the world The Behavioral and Brain Sciences 33(2–3), 61–83; discussion 83–135 (2010)

8. Krosnick, J.A.: The role of attitude importance in social evaluation: A study of policy preferences, presidential candidate evaluation, and voting behavior. Journal of Personality and Social Psychology **55**(2), 196–210 (1988)
9. Paolacci, G., Chandler, J.: Inside the turk understanding mechanical turk as a participant pool. Current Directions in Psychological Science 23(3), 184–188 (2014). http://cdp.sagepub.com/content/23/3/184
10. Visser, P.S., Clark, L.M.: Attitudes. In: Kuper, A., Kuper, J. (eds.) Social Science Encyclopedia. Routledge (2003)

Early Prediction of Movie Success —
What, Who, and When

Michael Lash[1(✉)], Sunyang Fu[2], Shiyao Wang[1], and Kang Zhao[2(✉)]

[1] Department of Computer Science, The University of Iowa, Iowa City, USA
{Michael-Lash,Shiyao-Wang}@uiowa.edu
[2] Tippie College of Business, The University of Iowa, Iowa City, USA
{Sunyang-Fu,Kang-Zhao}@uiowa.edu

Abstract. Leveraging historical data from the movie industry, this study built a predictive model for movie success, deviating from past studies by predicting profit (as opposed to revenue) at early stages of production (as opposed to just prior to release) to increase investor certainty. Our work derived several groups of novel features for each movie, based on the cast and collaboration network (who'), content ('what'), and time of release ('when').

Keywords: Predictive model · Data analytics · Social network · Text mining · Decision support system

1   Introduction

In the U.S., the motion picture industry produces approximately 500 movies in a year [10] garnering, on average, $60 million of investment capital per film [12]. Despite the large capital investment that must be made prior to movie production, the success, or profitability, of a movie is largely uncertain. Past studies have two limitations: First, the focus is almost solely on total box office revenue [2], [9], [11] or theater admissions [8]; Second, many of these studies employed features that are only available just prior to [1,2], or even after [7] [12] [14], the official release of a movie.

For investors, though, the profit of a movie is a better indicator of success than the total revenue. Also, it is worth noting that the prediction of a movie's success should leverage features that are available only during the time in which investments are being garnished.

To the best of our knowledge, this research represents the first attempt to predict the profitability of movies at early stages of production, which we achieve by examining 'who' factors-- its actors, directors, and social networks based on previous collaboration; the 'what' factors—a movie's genre, rating, and plot synopsis; as well as the 'when' factor—when a movie will be released and the temporal ebbs and flows of the movie industry.

© Springer International Publishing Switzerland 2015
N. Agarwal et al. (Eds.): SBP 2015, LNCS 9021, pp. 345–349, 2015.
DOI: 10.1007/978-3-319-16268-3_41

2   Feature Engineering

2.1   Who are Involved

The movie industry is characterized by movie stars, which function as a name brand, drawing crowds, and thus increasing sales [2], [6] . Thus we included the following features to measure the 'star power' of a movie.

   a. Total and average tenure: Total/average between first and most recent appearance.

   b. Total and average actor gross are the sum/average of all actors' gross revenues across all movies they have appeared in.

   d. Total and average director gross are the total/average revenues of all movies directed by the director of target movie m.

Another important factor for movie success is team cohesion [8]. Thus we took a social network approach, which provide a wealth of valuable information about various types of inter-personal relationships, such as dating [15], collaboration [16], and communication [5]. In this research, we built a collaboration network among actors based on their co-star relationship; we aggregated this undirected, weighted (collaborations) network to 1999 and created 11 separate, yearly networks (through 2010); each year is used in deriving features to predict a subsequent years profit.

   Network metrics that fall into the former category can be summarized as follows:

   a. Team heterogeneity: It is believed that successful movies have a certain degree of originality to them [8], which is accomplished by a hetergenous team of actors, and which we capture via cosine similarity using cast members' neighborhood vector, denoted in Equation 1.

$$\frac{1}{(n(n-1)/2)}\sum_{i=0}^{n-2}\sum_{j=i+1}^{n-1}\frac{Act_i \bullet Act_j}{\| Act_i \| \| Act_j \|} \tag{1}$$

   b. Average degree: overall centrality among team members.

   c. Total and average betweenness centrality: experienced actors who are not well known may bring an 'unseen' benefit to a production.

Network metrics that measure a movie's effect on the structure of an existing social network are summarized as follows:

   d. Decrease in average shortest path: Structural holes are an important concept in networks pertaining to movies [13], as individuals who span these structural holes are said to have high social capital [4].

   e. Change in clustering coefficient: A feature that captures the diffusion of information across the network [16].

We capture content-based features, such as those found in a script or, in our case, a plot synopsis, by using Latent Dirichlet Allocation (LDA) [3], which his ultimately expressed as a topic distribution vector for each movie in our dataset.

2.2   When a Movie will be Released

Temporal features are an important component to consider in the ever-evolving multi-billion dollar motion picture industry. As such, we incorporated the following features:

   a. Average annual profit in the year prior to release.

   b. Annual profitability percentage by genre is the percentage of profitable movies in the previous year, that fall into the same genre as m.

   c. Annual weighted profitability by genre (AWPG): Is the sum of cosine similarity, genre vector-wise, between a given movie and each movie of the previous year, weighted by the profitability of that genre.

$$AWPG_m = \sum_{m'_i \in y-1} sim\big(g(m), g(m'_i)\big) * p(m'_i) \qquad (2)$$

   d. Release dates: *Season* (spring, summer, etc) and *Holiday Release* (the week leading up to, and including, Christmas).

3   Experiments and Results

Data was collected from a movie archive website, BoxOfficeMojo, including data of 5,440 actors as well as 14,097 movies from year 1921 to 2014.

3.1   Experiment Setup and Results

The goal of our research is to predict whether a movie will be profitable, which we define as a profit (revenue minus budget) at .25 of 1 standard deviation above mean, in order to insure a reasonable ROI.

Our dataset includes 1353 movies (revenue and budget info available), 384 of which were profitable. We excluded sequels (confounding) and features pertaining to cast members were derived from the full set of 14097 movies.

Table 1. Classification outcome from the logistic regression classifier

	All features	without When	without What	without Who
AUC	0.801	0.736	0.803	0.72
Accuracy	0.771	0.738	0.763	0.734
F1 score	0.757	0.714	0.743	0.708

Various classification algorithms were used for the prediction (with 10-fold cross validation) with logistic regression yielding the best results, (please refer to the 'All features' column in Table 1).

3.2   Discussions

Results in Table 1 also show those obtained from our logistic regression classifier upon omitting each class of features. As the reader will notice, the classifier deteriorates when

'When' and 'Who' features are removed, indicating the contribution of such features; 'What' features, then, likely overlap those of 'When' and 'Who'.

Table 2 shows the top contributing features by coefficient value of the "non-profitable" class (negative coefficient are indicative of profit), which admittedly only indicate their role within our predictive model. However, there are still some very interesting findings. For example, the Horror Genre contributes positively to profitability, perhaps due to the relatively low budget needed for success (ie. Paranormal Activity); "Documentaries" may be in a similar situation. It was also found that annual profit percentage by genre contributed to movie profitability, indicating the importance of current trends. In terms of "who" is involved in a movie, both average director gross and average actor gross contribute positively to movie profitability (name brand and skill).

Table 2. Top features in each feature group for non-profitable movies

Group	Feature	Coefficient
What--Rating	"NC-17" rating	12.13
	"G" rating	-1.047
What--Genre	"Documentary" genre	-1.609
	"Horror" genre	-1.297
What--Plot topic	Topic #13	0.343
	Topic #7	0.183
Who--Individual	Avg. director gross	-0.963
	Avg. actor gross	-0.906
Who--Network	Avg. degree	0.362
	Total betweenness centrality	0.159
When	Annual prof. pctg. by genre	-1.007
	Winter release	-0.297

On the other hand, some features were noteworthy with regard to movie losses; topic 13, represented by familial-type words (eg. "wife"), indicating that perhaps movies that focus their content on relationships are less often profitable. While the NC-17 rating is of no surprise, average degree and total betweenness centrality are; we believe this may be a product of collinearity, better address by an explanatory model in future work.

4   Conclusions and Limitations

In this paper, we predicted the profitability of movies at early stages of movie production, leveraging "What", "Who" and "When" features. Applying this predictive model to empirical data, we showed that our model is able to achieve decent performance, using a wide variety of features, which could be employed in decision support to aid in investment decisions

The authors recognize that there are a few limitations, including collinearity and possible sampling bias.

References

1. Apala, K.R., Jose, M., Motnam, S., Chan, C.-C., Liszka, K. J., de Gregorio, F.: Prediction of movies box office performance using social media. In: Proceedings of the 2013 IEEE/ACM International Conference on Advances in Social Networks Analysis and Mining, ASONAM 2013, pp. 1209–1214. ACM, New York (2013). doi:10.1145/2492517.2500232
2. Asur, S., Huberman, B.A.: Predicting the future with social media. In: Proceedings of the 2010 IEEE/WIC/ACM International Conference on Web Intelligence and Intelligent Agent Technology, Washington, DC, USA, pp. 492–499 (2010). doi:10.1109/WI-IAT.2010.63
3. Blei, D.M., Ng, A.Y., Jordan, M.I.: Latent dirichlet allocation. J. Mach. Learn. Res. **3**, 993–1022 (2003). doi:10.1162/jmlr.2003.3.4-5.993
4. Burt, R.: Structural holes: The social structure of competition. Harvard Univ Press (1995)
5. Diesner, J., Frantz, T., Carley, K.: Communication Networks from the Enron Email Corpus 'It's Always About the People. Enron is no Different'. Computational & Mathematical Organization Theory **11**, 201–228 (2005). doi:10.1007/s10588-005-5377-0
6. Elberse, A.: The Power of Stars: Do Star Actors Drive the Success of Movies? Journal of Marketing **71**(4), 102–120 (2007). doi:10.2307/30164000
7. Gopinath, S., Chintagunta, P.K., Venkataraman, S.: Blogs, Advertising, and Local-Market Movie Box Office Performance. Management Science (2013). doi:10.1287/mnsc.2013.1732
8. Meiseberg, B., Ehrmann, T.: Diversity in teams and the success of cultural products. Journal of Cultural Economics **37**(1), 61–86 (2013). doi:10.1007/s10824-012-9173-7
9. Meiseberg, B., Ehrmann, T., Dormann, J.: We don't need another hero–implications from network structure and resource commitment for movie performance. Schmalenbach Business Review (sbr) **60**(1), 74–98 (2008)
10. Mestyán, M., Yasseri, T., Kertész, J.: Early prediction of movie box office success based on Wikipedia activity big data. PloS One **8**(8), e71226 (2013)
11. Sharda, R., Delen, D.: Predicting box-office success of motion pictures with neural networks. Expert Systems with Applications **30**(2), 243–254 (2006). doi:10.1016/j.eswa.2005.07.018
12. Simonoff, J.S., Sparrow, I.R.: Predicting Movie Grosses: Winners and Losers, Blockbusters and Sleepers. Chance **13**(3), 15–24 (2000). doi:10.1080/09332480.2000.10542216
13. Zaheer, A., Soda, G.: Network Evolution: The Origins of Structural Holes. Administrative Science Quarterly **54**(1), 1–31 (2009). doi:10.2189/asqu.2009.54.1.1
14. Zhang, W., Skiena, S.: Improving movie gross prediction through news analysis. In: Proceedings of the 2009 IEEE/WIC/ACM International Joint Conference on Web Intelligence and Intelligent Agent Technology, Washington, DC, USA, pp. 301–304 (2009). doi:10.1109/WI-IAT.2009.53
15. Zhao, K., Wang, X., Yu, M., Gao, B.: User recommendation in reciprocal and bipartite social networks–an online dating case study. IEEE Intelligent Systems **29**(2), 27–35 (2013). doi:10.1109/MIS.2013.104
16. Zhao, K., Yen, J., Ngamassi, L.-M., Maitland, C., Tapia, A.: Simulating Inter-organizational Collaboration Network: a Multi-relational and Event-based Approach. Simulation **88**, 617–631 (2012). doi:10.1177/0037549711421942

Risk Management in Asymmetric Conflict: Using Predictive Route Reconnaissance to Assess and Mitigate Threats

Jason Lin[1], Benke Qu[1], Xing Wang[1], Stephen M. George[2], and Jyh-Charn Liu[1]([✉])

[1] Department of Computer Science and Engineering, Texas A&M University,
College Station, TX 77840, USA
{senyalin,qubenke,xingwang,liu}@cse.tamu.edu
[2] U.S. Department of Defense, College Station, USA
ticom.dev@gmail.com

Abstract. This paper presents novel computing algorithms to generate *tactical risk maps* (TRM) based on the MECH (*Monitor, Emplacement*, and *Command/Control* in a *Halo*) model to evaluate locational values for attackers to launch improvised explosive device (IED) vs. direct fire (DF) attacks. Given a *study area* R, its proximity P can be mapped to explore noticeable characteristics associated with the attack locations. Within the distance constraints of the Halo, a simple optimization formula is proposed to support flexible representations of risk preferences of the attackers in ranking of locations for the M, C and E functions across R. Several case studies on major corridors find a significant number of attack locations were near or at local maxima of the measurement *route exposure*. It was found that IED sites tend to have good visibility and more uniform line-of-sight (LOS) distances. On the other hand, most DF locations are near the boundary of the viewshed suggesting careful selection of the sites to provide cover in the attack.

Keywords: Asymmetric warfare · Optimization · Simulation · Risk aversion

1 Introduction

Roadside attack is a highly effective tactic in asymmetric conflicts (AC), in which a red team emplaced an IED at a select location, observes the movements of the blue team, and detonates the planted explosive devices when the blue team reaches the locations. Within control distance of the devices, the red team can select the best attack position that provides good visibility to monitor the movement of the blue team, and good cover against their return fire. To model the roadside attack, several general warfare models have been developed in the literature. Lanchester [1] proposed that

---

This work was supported in part by an ONR grant N00014-12-1-0531 and a National Defense Science and Engineering Graduate (NDSEG) fellowship. Any opinions, findings and conclusions or recommendations expressed in this material are the author(s) and do not necessarily reflect those of the sponsors.

N. Agarwal et al. (Eds.): SBP 2015, LNCS 9021, pp. 350–355, 2015.
DOI: 10.1007/978-3-319-16268-3_42

the combat power of a military force equals to the unit personnel size squared. Later Deitchman [2] modified the model in [1] to accommodate guerrilla warfare. In the geo-spatial field, some works [9,10] focus on terrain analysis to determine the optimal conditions of certain operations. Some economic models [4-6] suggest that human decisions are often made based on risk analysis. Shakarian et al. [3] formalized the optimal adversary strategy and the maximal counter-adversary strategy based on geospatial abduction problems.

This paper presents an interactive computing framework for the MECH model [11] to support tactical risk analysis based on *visibility* and *cover*. Given a study area R (i.e., an isolated location, an area, or a route.) and its proximity P, the analysis may be initiated from the perspective of R towards P, or vice versa. The system also supports visual inference of viewsheds at and around known attack locations. Fusion of different reasoning approaches contributes to higher level reasoning about attack decisions.

2   MECH Behavior Model and Tactical Risks

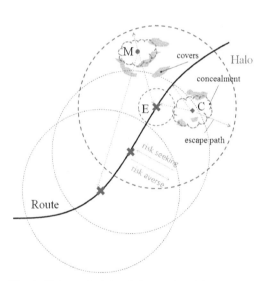

Fig. 1. Halo tactical level factors for risk preference modeling

The tactical risk analysis of a study area like a route R needs to consider both R and its proximity P. The baseline MECH model in Figure 1 illustrates the locational relationship between the red team at some location(s) of the proximity P of R, along which the blue team needs to travel through. In this picture an IED is located at the center of the Halo. Considering that the red team has the freedom to place its monitoring and control positions anywhere within the Halo, the tactical analysis is based on locations that are observable to the red team. Both offensive and defensive tactics are considered.

The *visibility* of a location p_y in P is defined as the sum of the LOS between p_y and each point in R. It essentially gauges how easy it is for p_y to monitor or control R. A *cover* for p_y is a patch of area at its nearby location out of the LOS of some points in R. Firing ranges of the red and blue teams, device control range and blast range together define the main distance constraints of red team attacks. Concealment is defined as the degree of blockage between p_y and R.

A sound modeling approach should also accommodate different perceptions about risks vs. rewards [7,8] of individuals. To support different risk preferences, we propose a composite utility function U for the red team, based on the weighted sum (+) or product (×) of the offense utility U_O and defense utility U_D to reflect the additive and multiplicative effects of offensive and defensive functions, respectively. The risk preference of the actor is represented by the weight coefficients of the two types of functions ω_O and ω_D, $\omega_O + \omega_D = 1$, respectively. As such, we get $U = S_O \cdot S_D [\omega_O \cdot U_O(F_M) \Delta \omega_D \cdot U_D(F_C)]$ as an initial format of the tactical utility function, where $\Delta \in \{+, \times\}$, F_M is the score function for observability and F_C for concealment. The two score functions are normalized from their physical measurements to a percentile scale $[0,100]$. As a threshold-based qualifier, a location is not considered if either of U_O or U_D are below some threshold (i.e. $S_O \rightarrow \{1 \ (if \ U_O > \tau_O), \ 0 \ (otherwise)\}$, and $S_D \rightarrow \{1 \ (if \ U_D > \tau_D), \ 0 \ (otherwise)\}$.

To reduce the computing cost of observability, we compute the viewshed $VS(r_x)$ for discrete points $r_x \in R$ at fixed intervals with the observability of point p_y in P calculated as $F_M(p_y) = \sum_x \text{LOS}(r_x, p_y)$, where LOS: $R \times P \rightarrow \{1 \ (visible), \ 0 \ (invisible)\}$. In contrast to the observability score, the concealment score of p_y is defined as the sum of points within the radius γ of circular range O that are invisible to a specified E location r_x. $F_C(p_y) = \sum_i {}^\wedge\text{LOS}(r_x, p_i) , p_i \in O(p_y, \gamma)$, and ${}^\wedge$LOS: $R \times P \rightarrow \{1 \ (invisible), 0 \ (visible)\}$.

For each route point r_x, the visibility of r_x is defined as the degree of exposure of r_x to other points in the Halo annulus. On the other hand, to determine a potential road location for attack, the red team may select the boundary of the viewshed, known as the escape adjacency (EA) [11], that simultaneously provides target observation and cover from return fires. Therefore, the actual exposure rate in considering the opponent's location choice is evaluated as $F_{exp}(r_x) = \sum_i \text{EA}(r_x, p_i)$, $r_x \in R$, $p_i \in H(r_x, d_{min}, d_{max})$, and EA: $R \times P \rightarrow \{1 \ (p_i \text{ is on EA}), 0 \ (p_i \text{ is not on EA})\}$, where d_{min} and d_{max} are respectively the inner and outer radii of the Halo annulus H [11].

Tactical reasoning can be largely grouped into two types: the environmental behavior (R→P) module and the route-based behavior (P→R) module. The former calculates the tactical value of a location r_x in R to get the potential Monitoring/Command locations in P. The latter one calculates the tactical values of R locations with respect to the given set of P points to derive the potential high risk locations on R. For R→P, risk points indicate the highly possible locations that have insurgents to monitor or control the timing of the attack on route. For P→R, high risk points indicate locations in R that are most exposed to P. These two modules can interactively execute to address to some specific tactical question. For instance, given the best long-term *Monitoring* location $p_y \in P$, where is the best location r_x to place an IED on R? To answer this question, R→P is introduced by taking the selected route as input R points to calculate the best P point to get the optimal monitoring point p_y. Subsequently, by taking p_y as the input of P→R module, we can calculate the best IED deployment locations along R. The differences between MECH-generated TRM and historical event locations allow analysts to interpret likely red team tactical behaviors.

3   Case Studies

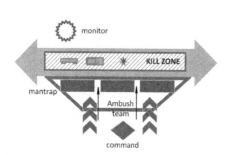

Fig. 2. An idealized ambush model of military activity

In this section, we present some case studies on the route-level assessment and the incident-level assessment to learn about visibility patterns of environs of past IED and DF attacks. Using the ASTER Global Digital Elevation Map [1] (DEM) of tan and some samples of the AC data set collected across Afghanistan in a time period of 19 months (Feb. 2011- Aug. 2012), a number of experiments were conducted to learn about the visibility properties under the MECH model. Using an idealized ambush model, an AC area used for ambush can be divided into the *kill zone* of the victim, the *mantrap* and blockage area for controlled access by the ambush team and its command. Based on the MECH model, we also added the *monitoring points* to watch the movement of the victims. The resulting model is illustrated in Figure 2. In the following discussions we manually interpreted and marked the probable areas for the kill zone, mantrap(s) and the monitoring point(s) of a study area, after the observability, route exposure and past incidents were generated by MECH.

First, we consider route-level assessment, where the red team is located at P, and the blue team R. Figures 3a and 3c show the locations along a long corridor that have the highest exposures. Here, the two thumb pins define the area of study, and points with the highest exposures are highlighted using heat maps. Past incidents, IED and DF events are highlighted respectively on the map as blue and red cross-marks after the exposure map was first computed. In these cases, past incidents were found to be near at or near some local maxima of route exposures. We note that when we change the route range, heat map locations may also change. A highway stretch near Kabul is used in the first study. In this case, we first selected a 743-meter long stretch of route to assess the observability of the adjacent P points, exposure rates of the route, as well as the past attack locations. Next, we manually marked the kill zone (the high incident stretch), and top candidate areas for mantraps (purple lined trapezoidal blocks), and short range control spots (red lined stars). It is worth noting that not all attacks occurred at the locations with the highest exposure rate. An anecdotal tactical interpretation is that these locations (with high exposure rate) might have been used as a signaling or overwatch point for the attack to prepare for imminent engagement.

To observe if different traffic and terrain would lead to different observability patterns, we considered two types of areas: busy traffic at flat regions, and rugged terrain on the mountains. For the busy traffic area, we used the area near Jalalabad Airport to analyze the studied routes as shown in Figure 3c. We observed that most attack locations are near

---

[1] https://lpdaac.usgs.gov/data_access

locations with high observabilities. For several studied corridors near rough terrains, both locations of frequent and infrequent attacks appear to be at or near high exposure road locations. It is also interesting to observe that several, if not most of the attack locations are also situated near the boundary of high visibility vs. no or low visibility areas.

We present two areas with rough terrain in Figures 3b and 3d to illustrate our observations. As these examples show, attack locations tend to have large viewsheds, and they tend to locate at nearby the boundary of the viewsheds. The rough shape of the terrain makes it easy to create asymmetry in intervisibility between the red and blue teams, making concealment power a critical factor in shaping of the attack. What particularly interesting is the area in Figure 3d, which is a mountain area near Esma'il Kalay. Unlike the area in Figure 3b, this area has no valleys of open spaces. The poor accessibility of the area may render mechanized blue teams ineffective in their operations.

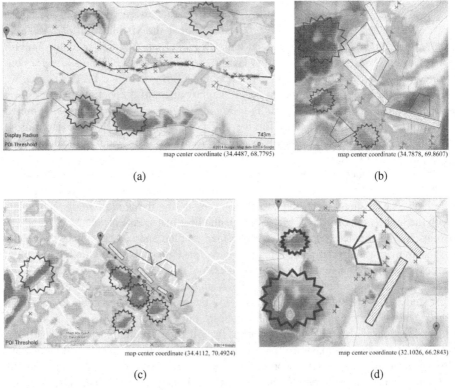

(a)

(b)

(c)

(d)

Fig. 3. (a) An example of route-level assessment in a valley facing heavy attacks along Kabul-Behsud Hwy: the red (blue) cross marks the IED (DF) events; (b) Mountain area in Nuristan Forest National Reserve; (c) A heavy attack area near the Jalalabad Airport; and (d) Mountain area near Esma'il Kalay

4   Conclusions

This paper presents a computing framework to create TRM based on the MECH model. Through some studies of some representative cases, we found that MECH model could be used for extraction of terrain factors that shape the attacks. For some cases, attack locations are strongly collocated with strong visibility, boundary of viewsheds, while in other cases the attack locations did not appear to have any relationship with nearby high visibility points. Given that the current model only considers different measures derived from terrain elevation, the result suggests that other features not considered or not visible as the map resolution used in this paper would have been used to provide cover. In conjunction with the statistical models presented in [11, 12], a wide range of experiments can be conducted for situational awareness analytics.

References

1. Lanchester, F.W.: Mathematics in warfare. In: Newman, J.R. (ed.) The world of mathematics, vol. 4, 2138–2157. Simon and Schuster, New York (1956)
2. Deitchman, S.J.: A lanchester model of guerrilla warfare. Operations Research **10**(6), 818–827 (1962)
3. Shakarian, P., Dickerson, J.P., Subrahmanian, V.S.: Adversarial geospatial abduction problems. ACM Trans. Intell. Syst. Techno. 3(2), 34:1–34:34 (2012)
4. Kahneman, D., Tversky, A.: Prospect theory: an analysis of decision under risk. Econometrica **47**(2), 263–292 (1979)
5. Roos, P., Carr, J.R., Nau, D.S.: Evolution of state-dependent risk preferences. ACM Trans. Intell. Syst. Techno. 1(1) 6:1–6:21 (2010)
6. Okada, I., Yamamoto, H.: Mathematical Description and Analysis of Adaptive Risk Choice Behavior. ACM Trans. Intell. Syst. Techno. 4(1) (2013)
7. Steinbach, M.C.: Markowitz revisited: Mean-variance models in financial portfolio analysis. SIAM Rev. **43**, 31–85 (2001)
8. Krokhmal, P., Zabarankin, M., Uryasev, S.: Modeling and optimization of risk. Surveys in Operations Research and Management Science **16**(2), 49–66 (2011)
9. Richbourg, R., Olson, W.K.: A Hybrid Expert System that Combines Technologies to Address the Problem of Military Terrain Analysis. Expert Systems with Applications **11**(2), 207 (1996)
10. Janlov, M., Salonen, T., Seppanen, H., Virrantaus, K.: Developing military situation picture by spatial analysis and visualization. In: Presented at the ScanGIS 2005: The 10th Scandinavian Research Conference on Geographical Information Science, Stockholm, Sweden (2005)
11. George, S., Wang, X., Liu, J.-C.: MECH: A Model for Predictive Analysis of Human Choices in Asymmetric Conflicts. In: Presented at the International Conference on Social Computing, Behavior-Cultural Modeling and Prediction 2015, Washington D.C (2015)
12. Wang, X., George, S., Lin, J., Liu, J.-C.: Quantifying Tactical Risk: A Framework for Statistical Classification Using MECH. In: Presented at the International Conference on Social Computing, Behavior-Cultural Modeling and Prediction 2015, Washington D.C (2015)

Styles in the Fashion Social Network: An Analysis on Lookbook.nu

Yusan Lin[1]($\boxtimes$), Heng Xu[2], Yilu Zhou[3], and Wang-Chien Lee[1]

[1] Department of Computer Science and Engineering,
Penn State University, State College, USA
yusan@psu.edu,wlee@cse.psu.edu
[2] Department of Information Science and Technology,
Penn State University, State College, USA
hxu@ist.psu.edu
[3] Schools of Business, Fordham University, Bronx, USA
yilu.zhou@gmail.com

Abstract. As birds of a feather flock together, so do people with similar interests and preferences befriend with each other. Numerous Social Network Analysis (SNA) researchers have investigated how individuals' identification affects their behaviors, such as ethnicity, education, political opinions and even musical tastes. What about one's fashion style? These days, online social networks provide rich resources for us to study these phenomena; however, no research has investigated people's styles/tastes within a social network. In this paper, we analyze the largest fashion social network, Lookbook.nu. By applying SNA techniques, we answer whether people with similar styles tend to connect with each other on the online social networks and whether people form communities based on their styles. We believe this is the first work studying people's fashion styles on an online social network empirically.

1 Introduction

Birds of a feather flock together is much more than an old adage. Our societal behavior is heavily linked to individuals' identification and association with groups, ideas, and trends, as well as visual aesthetics. We see this clearly in many forms, such as sports team loyalty [1], political ideologies, socioeconomic groups, and nationalistic tendencies [2,3] . People often join together in communities, groups, or at least via friendships to share in these common beliefs and interests.

The above-mentioned phenomenon has been widely studied in various fields of social science. However, none has done on the relationship between people's tendencies in choosing attire and their (online) social friends. Is it safe to assume that people with similar fashion tastes are also more likely to be friends? Research in the past has shown that people in the same environment affect each other's fashion taste on a small scale [4]. However, has the definition/phenomenon changed in this online social network era? And how can we observe people's fashion style in social networks? Lastly, how do people's interactions affect an individual's style? To the best knowledge of the authors, none

© Springer International Publishing Switzerland 2015
N. Agarwal et al. (Eds.): SBP 2015, LNCS 9021, pp. 356–361, 2015.
DOI: 10.1007/978-3-319-16268-3_43

of these questions have been addressed or explored. In this paper, we study the fashion styles identified in an online social network specifically geared toward fashion, from a data analysis perspective.

This project contributes to the field of Social Network Analysis (SNA) by studying styles on fashion social networks to determine how interaction may be connected to visual stimulation and, similarity, in personal preferences. Previous research has studied the interaction between the tastes and the interests of users in different types of social networks, e.g., how taste in music [5] affects friendship, but not in a fashion social network. In this research, we apply SNA techniques to the network of users on Lookbook.nu, the largest fashion social network. We aim to explore users' tastes and preferences of fashion styles within the social networks and study whether users' friendships/relationships are formed based on their fashion styles.

We aim to answer these questions by analyzing data we collected from an online fashion social network: (1) Are users with similar fashion styles more likely to follow each other and be more closely connected? (2) In the fashion network, how do social ties based on users' friendships/relationships and ties based on users' fashion styles relate to each other?

For the first question, we transform the network into a manageable structure, then apply community detection to visualize the result in order to answer it to some extent. As for the second question, based on the findings from answering the first question, a statistical test is conducted to show the significance level.

2   Related Works on Fashion Analysis

In the field of fashion, style is the foundation of every single design. The question is, how does one define style? Or, more specifically, how does one define style within the context of fashion? Also, how do different people behave toward different styles?

Articles within the literature discuss the potential mechanisms behind fashion trends. Pesendorfer believes that the fashion industry is driven by a fashion czar [6]. Alberti believes that fashion trends are led by the consumers selection process [7]. And Tassier proposes design models to capture the phenomenon by including information cascades [8]. However, these ideas are not verified empirically, not to mention via data analysis on a fashion social network to study the relationships and phenomenon within it.

In order to better understand how people with various styles interact with each other, we believe that performing a network data analysis based on the consumers' relationships will provide enormous value to the research community.

3   Data Collection: Lookbook.nu

Lookbook.nu, one of the largest online fashion social networks, provides a platform as a global community for users to share fashion tastes and to find inspiring fashions from others. Established in 2007, it has grown to 1.6 million users in 2014. The social function of Lookbook.nu comes from the following/follower relationship and giving hypes, loves, and comments on other users' looks.

Table 1. Labeled users network's profile

(a) Networks' summary

Metrics	Value
Number of nodes	797
Number of edges	7,588
Number of styles	32
Avg. clustering coef.	0.352

(b) Centrality metrics

Metric	Mean	Std	Min	Max
Degree	20.949	36.224	0	573
Indegree	11.848	23.137	0	229
Outdegree	9.100	22.913	0	513
Betweenness	0.002	0.010	0	0.247
Closeness	0.221	0.124	0	0.688

(a) Before Simmelian Backbones transformation   (b) After Simmelian Backbones transformation

Fig. 1. Labeled user network colored by styles (a) before and (b) after Simmelian Backbones transformation

Discover, one of the different ways to browse Lookbook.nu, allows users to browse other users based on 32 different fashion styles. According to the Lookbook.nu employee whom we contacted with, the approach they use to classify users into different styles is manually picked by the staff. The total number of users in each style is 1,268, while the actual total number of users classified is only 797. This is because between styles, certain users are overlap. We call these users labeled users. These labeled users form a social network with 7,588 edges, denoting the following/follower relationships among them. The basic summary of this data set is presented in Table 1 (a), and its basic network statistics are presented in Table 1 (b). Since the collected data is a small sample of the entire Lookbook.nus network, for clarity, we call it the labeled user network. As seen in Fig. 1(a), the labeled user network is a giant component of the network on its own; i.e., all of the labeled users are connected.

4 Experiment

In order to answer the research questions raised earlier, via our data analysis, we define the two questions more specifically here: (1) Given a fashion style social network G, do users who are strongly connected together form communities that have dominant styles? (2) Given a fashion style social network G, do the following social ties e_f and the style ties e_s correlate with each other?

Note that for question 2, if the relationship between two users is based on following/follower, we say they have a following tie e_f between them. If the relationship is based on two users belonging to the same fashion style, we say there is a style tie e_s between these two users.

In the rest of this section, we first examine question 1 by finding communities in the network and measure its quality. We then examine question 2 by using the Quadratic Assignment Procedure (QAP) correlation between the users following ties e_f and style ties e_s. Note that in order to analyze how different styles play different roles in the network due to the network structure, we ignore what fashion elements each style includes and simply focus on the network structure.

4.1 Network Transformation, Community Detection and Styles

Does the network form communities? Are those communities dominated by specific styles? To examine this, we apply a community detection algorithm on the labeled user network. Note that when encountering hairball-like networks, as shown in Fig.1(a), one should preprocess the networks with transformation algorithms to make the structure of network clearer. In our application, we choose Simmelian Backbones, which is based on the idea of local ranking and overlap calculations [9]. It can filter out redundant relations between nodes, leaving only the strong connections and amplifying the homophily structure within networks. A sample result of applying Simmelian Backbones is shown in Fig. 1(b). Since Simmelian Backbones requires two parameters for its two stages, we tune them by using grid search with objective function as the average Gini-index for assessing the impurity of the communities.

We begin by processing both networks with the transformation algorithm, Simmelian Backbones. After transforming with given α and β (the two parameters needed for Simmelian Backbones), we then apply Modularity Maximization [10] to partition the transformed network into communities. We then assess the purity of the communities by computing their Gini-index.

4.2 Correlation Between Following/Follower and Styles

For the second test, we construct another network, style-oriented network. It is purely based on styles users have in common, in which users link to all the other users who have the same styles with themselves. To distinguish from this network, here we refer to the original labeled user network as the following/follower-oriented network. The purpose of the style-oriented network is to calculate its correlation with the following/follower-oriented network. We use a Quadratic Assignment Procedure (QAP) correlation to perform this test. For our QAP test on the two networks, the null hypothesis is as follows:

H_0: The style relation and following/follower relation do not correlate among the same set of users; i.e., they have a correlation of 0.

For each network, we construct a full adjacency matrix to serve as the inputs of the QAP test, which includes two stages. In the first stage, correlation coefficients between the corresponding cells in two matrices are calculated. In the second stage, rows and columns in the two matrices are randomly permuted synchronously to recompute the correlation.

5 Result

We first present the result of our first experiment, examining whether users form communities based on their styles. Fig. 2 shows the visualization of the optimally transformed network and Table 2 shows the basic statistics of the network.

Fig. 2. Labeled user network after applying optimal Simmelian Backbones

Table 2. Statistics of the labeled network after optimal Simmelian Backbones transformation

Parameter/Metric	Value
α	16
β	9
Average Gini-index	0.6012
Number of connected nodes	83
Number of edges	188
Number of communities	10

Based on the results, we are able to answer the two questions raised earlier. As shown in Fig. 2, even though Simmelian Backbones has been applied to achieve the lowest impurity in each community, the community-like structure in such a network is still not obvious. For the generated 10 communities, 6 are dominated by the style effortless. Therefore the answer to our first question is "No."

To confirm our observation, we look into the result from the second experiment, which can be also be used to answer the second question. The result shows 0 correlation between the following/follower-oriented network and the style-oriented network. A p-value of 1 was returned by the QAP correlation test, which failed to reject the null hypothesis H_0. This contradicts our initial belief.

Why do we achieve these results? We believe it is because the labeled user network, with manually selected nodes, represents a highly biased sample of the whole Lookbook.nu's network. One of the reasons these users are labeled is because they are relatively more popular in the network; they have more followers and, therefore, more attention. This might create a phenomenon where, in the labeled user network, users tend to follow each other simply because of fame and regardless of style.

6 Conclusions and Future Work

In this paper, we explore the network structure of Lookbook.nu by looking into the role each style plays in the network. Two questions are raised: (1) Are people with similar styles more likely to form communities? (2) Do social ties and style ties correlated with each other? We identify and answer them using a range of observation and network data analysis techniques.

In terms of the likelihood of users with similar styles to connect and follow one another, we apply Simmelian Backbones and Modularity Maximization to reveal the interactions between Lookbook.nu's users. However, at this stage of our exploration, we discover that the way users connected to each other is not strongly correlated to the styles they personally own, according to the conducted QAP correlation test. We believe this result is caused by the fact that the labeled user network is a biased network, and the users tend to follow each other simply because they are all included in this network and regardless of their styles. While some of our expectations are upheld in the research, there are clearly more areas that should be explored to understand more about social network relationship interconnectivity based on fashion tastes and interests.

For future work, we aim to expand the scope of data and include other users who are not labeled with styles in Lookbook.nu to study the relationship between a network's community and users' fashion styles.

References

1. Lusher, D., Robins, G., Kremer, P.: The application of social network analysis to team sports. Measurement in Physical Education and Exercise Science **14**(4), 211–224 (2010)
2. Thelwall, M.: Homophily in myspace. Journal of the American Society for Information Science and Technology **60**(2), 219–231 (2009)
3. Oliver, M.L.: The urban black community as network: Toward a social network perspective. The Sociological Quarterly **29**(4), 623–645 (1988)
4. Morris, T.L., Gorham, J., Cohen, S.H., Huffman, D.: Fashion in the classroom: Effects of attire on student perceptions of instructors in college classes. Communication Education **45**(2), 135–148 (1996)
5. Steglich, C., Snijders, T.A., West, P.: Applying SIENA: An Illustrative Analysis of the Coevolution of Adolescents' Friendship Networks, Taste in Music, and Alcohol Consumption. Methodology: European Journal of Research Methods for the Behavioral and Social Sciences **2**(1), 48 (2006)
6. Pesendorfer, W.: Design innovation and fashion cycles. American Economic Review **85**(4), 771–792 (1995)
7. Alberti F (2013) A note on fashion cycles, novelty and conformity. Jena Economic Research Papers
8. Tassier, T.: A model of fads, fashions, and group formation. Complexity **9**(5), 51–61 (2004)
9. Nick B, Lee C, Cunningham P, Brandes U Simmelian Backbones: amplifying hidden homophily in Facebook networks. In: 2013 IEEE/ACM International Conference on Advances in Social Networks Analysis and Mining (ASONAM), pp. 525–532. IEEE (2013)
10. Blondel, V.D., Guillaume, J.-L., Lambiotte, R., Lefebvre E.: Fast unfolding of communities in large networks. Journal of Statistical Mechanics: Theory and Experiment **10**, P10008 (2008)

Beyond Mere Following: Mention Network, a Better Alternative for Researching User Interaction and Behavior

Minh-Duc Luu[1]([✉]) and Andrew C. Thomas[2]

[1] Living Analytics Research Center,
Singapore Management University, Singapore, Singapore
mdluu.2011@smu.edu.sg
[2] Department of Statistics, Carnegie Mellon University, Pittsburgh, USA

Abstract. In popular online social networks, there are various kinds of user behavior (e.g. retweeting, posting hashtags on Twitter; wall-posting, commenting on Facebook). At the same time there exist different kinds of relationships among users (e.g. follow relationship or mention relationship in Twitter). It is interesting to study how these relationships affect users' behaviors. Current literature already pointed out that the follow relationship is an insufficient metrics for users' interaction level. Thus, in this work we compare mention against follow relationship in several aspects, especially in terms of their correlations with hashtag usage of users. We propose a rigorous way to perform significance test for the correlations. Our results show that using mention can be a better alternative as it can provide stronger correlation to users' behavior with a smaller cost of obtaining data.

1 Introduction

The richness of modern online social networks allows us to examine many types of social behavior, rather than just simple ones. For example, a Facebook friend can post messages, affix notes to walls and play games; a Twitter user can retweet or modify an existing message along with direct messages between mutual followers. Some of those relationships are considerably more public than others; when Twitter releases its tweet database for academic use, they will release some but not all of this information, such as public mentions but not the active follower list for each user. As a result, the ability to access interesting relationships may be considerably more difficult, but can be mitigated using alternative, more readily available data such as mentions. This eases a considerable financial burden since we now can use another data set which is publicly available and cheaper to obtain.

To demonstrate these differences, we study the effective correlation of user expression on Twitter given their two relationship types: *mentions*, which are contained within the tweets themselves, and *follow* relationships, which are stored separately within Twitter database. We estimate correlations of expression with each of the two relationship types, where the expression is the behavior

© Springer International Publishing Switzerland 2015
N. Agarwal et al. (Eds.): SBP 2015, LNCS 9021, pp. 362–368, 2015.
DOI: 10.1007/978-3-319-16268-3_44

of adopting information, especially *hashtags*, which are keyed phrases (marked by the symbol "#") in tweets. Since it is well-known that using follow relationship alone provides very limited information about user interactions, we want to see if mention relationship can complement this. Using mention relationship is vastly more convenient than follow relationship as the former is often *publicly available* from tweets while the latter is not, especially for private accounts. Moreover, links obtained from mentioning provide us with *exact time stamp*, which is crucial for several tasks e.g. researching link dynamics over time. Finally, mention links can be a tell-tale signal of an event in progress while follow links cannot. It is because the former typically changes faster and in synchronization with events while the latter does not.

All these reasoning convince us that mention relationships may be a better alternative in user behavior studies and inspire us to go deeper in comparing the two relationships. More specifically, we aim to answer the following research questions.

1. Is there a statistically significant correlation between users expression and their *follow/mention* relationships?
2. Given that both relationships have significant correlation with hashtag usage, which of the two correlations are stronger?

2   Related Works

The works [3], [9], [8] pointed out that relying only on follow relationship not only provide limited information about user interactions but also can mislead our evaluations. It is because a huge portion of follow links is *inactive* but they are confused as actual interactions of users. On the contrary, mentions are allocated more parsimoniously and with more attention (e.g. see [12], [13]). Thus, they provide solid evidence of *active* and *quantifiable* interaction. Hence, we hypothesize that mention relationship should have a stronger correlation with adopting behavior of users.

To test the hypothesis, we adapt *permutation test*, which is a rigorous statistical method resampling theory ([2], [7]). Permutation test provides us with various important advantages. First, it requires fewer assumptions since it is a non-parametric approach ([5], [4]). For example, we do not need to assume a prior form for the distribution of hashtag usage frequencies. Second, permutation test is more versatile than traditional parametric and nonparametric tests in that it is available for various statistics such as mean, median, correlation ([11]). Finally, it is known that permutation testing is more robust than parametric tests when there are outliers and when the distribution does not conform to a theoretical frequency distribution e.g. t and F distributions ([10]).

3   Methodology

Given a relationship r, we build a graph $G_r = (V, E_r)$ whose nodes are Twitter users; a weighted edge connecting two nodes representing the strength of relationship r between them. Let $\mathbf{w}(G_r)$ be the vector of weights on all edges of

G_r, then $\mathbf{w}(G_r)$ represents the strength of the relationship among users in G_r. Now, for a hashtag h, denote $f_h(x)$ as the frequency with which node x used the hashtag. For each pair of nodes u and v, we measure the frequency $c_h(u,v)$ that each used h simply by $\min(f_h(u), f_h(v))$. Thus, the vector $\mathbf{c}_h(G_r)$ of co-use frequencies on all edges of G_r will represent the (co-)usage of hashtag h among users in G_r.

Under this setting, the correlation between the relationship r and the usage of hashtag h can be estimated by the correlation between two vectors $\mathbf{w}(G_r)$ and $\mathbf{c}_h(G_r)$. In other words, significance test for the correlation between relationship r and usage of h becomes the problem of testing significance of the statistics $cor\left[\mathbf{w}(G_r), \mathbf{c}_h(G_r)\right]$. Since we are considering follow and mention relationships we state two following null hypothesis for the two relationships.

1. (H_0^F): *Correlation between follow relationship and usage of hashtag h is zero.*
2. (H_0^M): *Correlation between mention relationship and usage of hashtag h is zero.*

We test these null hypotheses by first estimating the observed correlation coefficients $cor\left[\mathbf{w}(G_r), \mathbf{c}_h(G_r)\right]$ from data, then deriving the *null distribution* D_0 of the test statistics (correlation coefficient), which is the distribution of each test statistic under assumption that null hypotheses are true. Once these are known, we decide whether to accept or reject null hypothesis depending on the observed correlation coefficient (CC) is in the critical region of the null distribution. The main challenge here lies in estimating D_0 since we do not have any prior knowledge of D_0. A solution for this is the permutation test, which approximates the null distribution D_0 by calculating all possible values of CC under rearrangements of data. Specifically, we resample the frequencies of hashtag usage for each user while keeping the underlying social network unchanged. The interested CCs are then estimated by the standard Pearson formula.

However, performing permutation test on large data sets is very expensive as we need to do a huge number of computations for a large number of permutations. To mitigate this, we employ R package `data.table` [6], which provides very fast operations on large data sets.

4 Empirical Study

4.1 Collecting Data and Mention Networks

We first crawled the set of tweets from the United States Presidential Election (USPE) in October 2012 using Twitter's streaming API. A set of 56 seed users was then selected from popular politics-related accounts. These include major American politicians (e.g. Barack Obama and Mitt Romney); well known political bloggers (e.g. America Blog, Red State and Daily Kos); and political sections of mass media sites (e.g. CNN Politics and Huffington Post Politics). We then expanded from the seed users by identifying those who follow at least three of them and obtained a set of more than 408,000 users, their tweets and follow networks. We filtered out users who did not tweet frequently (less than 45 tweets or retweets) or active less than 15 days. This yields a smaller dataset of 104,000 users and about 80 million tweets.

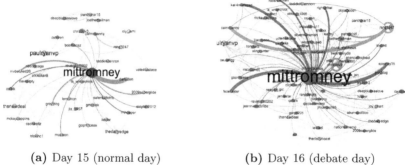

(a) Day 15 (normal day)   **(b)** Day 16 (debate day)

Fig. 1. Samples of ego mention network for candidate Mitt Romney in Oct. 15 and 16

Table 1. Top 10 hashtags in tweets at day 16

Hashtag	Explanation	Frequency
#tcot	Top Conservatives on Twitter	38326
#p2	Progressives	15394
#teaparty	Tea Party[1]	7100
#gop	Grand Old Party i.e. Republican Party	6676
#lnyhbt	Let not your heart be troubled[2]	3478
#getglue	A mobile app for TV fans	3290
#news	Common hashtag for retrieving news	3113
#fb	Facebook	1977
#sfgiants	San Francisco Giants baseball team	1974
#sgp	Smart girl politics	1967

Due to the volatility of mention links, we build mention networks on a daily basis. More specifically, we extracted ego mention networks whose centers are leading politicians. Each edge is weighted by the mention frequency between its two vertices. The networks were then visualized by *Gephi* ([1]), a popular open source for the task. From samples of these networks, Figures 1a and 1b, we can observe a distinct difference in mention networks on a normal day compared to a debate day. The latter is denser and more clustered than the former; the effective density of the subnetwork jumped from 0.058 to 0.099 between these days, and the relative number of "triangles" in the data more than tripled.

Finally, hashtags in our study are chosen from the most popular ones in debate days and periods of surging activity. Table 1 shows the top 10 hashtags in 16 October, which we divided into two groups: those closely related to the election, and those which are either loosely (e.g. #news) or not related to the election (e.g. #fb). The two groups are expected to exhibit a distinctive contrast in their significance test results. We expect the first group to pass the significance test while the second does not.

---

[1] A party focusing on government reforms
[2] A popular hashtag by conservative political commentator Sean Hannity

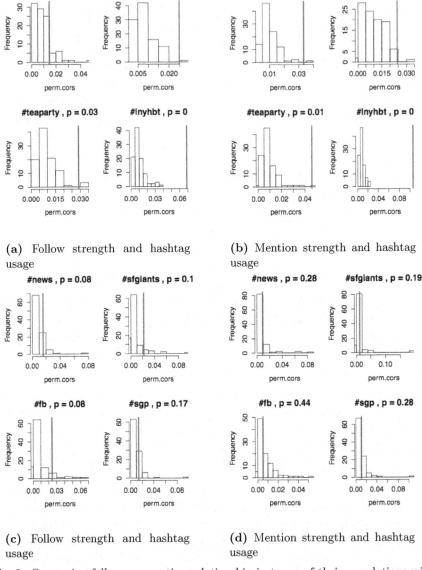

(a) Follow strength and hashtag usage

(b) Mention strength and hashtag usage

(c) Follow strength and hashtag usage

(d) Mention strength and hashtag usage

Fig. 2. Comparing follow vs. mention relationship in terms of their correlations with hashtag co-usage frequency. Red vertical line marks the original correlation coefficient (CC_0) estimated from data while bars show null distribution of CCs obtained from permutation test.

4.2   Results

We first look at significance of the two correlations. For the first group of political hashtags, (Figures 2a and 2b), all correlations between *mention* strength and hashtag usage are significant (p-values < 0.05) while most correlations between

Table 2. Correlations of *follow* (column 2) and *mention* (column 3) with hashtag usage on day 2012/10/16. G_f and G_m denote follow and mention graphs.

Hashtag	$cor\left[\mathbf{w}(G_f), \mathbf{c_h}(G_f)\right]$	$cor\left[\mathbf{w}(G_m), \mathbf{c_h}(G_m)\right]$
#p2	0.0146	0.0333
#tcot	0.0244	0.0254
#teaparty	0.0290	0.0461
#lnyhbt	0.0687	0.1059
#news	0.0090	0.0158
#sfgiants	0.0109	0.0217
#fb	0.0054	0.0242
#sgp	0.0111	0.0139

follow strength and hashtag co-usage are significant (except for hashtag #p2). For the second group (Figures 2c and 2d) of hashtags very loosely related to the election, we expect that correlation should be insignificant i.e. $p > 0.05$, which is actually proved in Figures 2c and 2d. Thus, the answer for the first two research questions is "yes".

To compare the two relationships, let us look at Table 2, which shows their correlations with hashtag usage on Debate Day October 16. From the table, it can be seen that, in terms of correlation with hashtag usage, mention is stronger than follow relationship. This answers our third research question.

5   Conclusion

Our results show that both follow and mention relationships have significant correlation with hashtag usage behavior. However, our work points out that using mentions can be a better alternative in many ways. First, mention data is free, publicly available in tweets. Secondly, they can capture events better than slow-changing follow relations. Most importantly, mention relationship have stronger correlation with hashtag usage. Thus, employing mention relationship can be easier in terms of collecting data and more effective as a signal for information adoption at the same time.

Acknowledgments. This research is supported by Living Analytics Research Center, a partnership between Singapore Management University and Carnegie Mellon University. The center is funded by the Singapore National Research Foundation and administered by the IDM Programme Office.

References

1. Bastian, M., Heymann, S., Jacomy, M., et al.: Gephi: an open source software for exploring and manipulating networks. ICWSM **8**, 361–362 (2009)
2. Berger, V.W.: Pros and cons of permutation tests in clinical trials. Statistics in medicine **19**(10), 1319–1328 (2000)
3. Cha, M., Haddadi, H., Benevenuto, F., Gummadi, P.K.: Measuring user influence in twitter: The million follower fallacy. ICWSM **10**, 10–17 (2010)

4. Chau, W., McIntosh, A.R., Robinson, S.E., Schulz, M., Pantev, C.: Improving permutation test power for group analysis of spatially filtered meg data. Neuroimage **23**(3), 983–996 (2004)
5. Collingridge, D.S.: A primer on quantitized data analysis and permutation testing. Journal of Mixed Methods Research **7**(1), 81–97 (2013)
6. Dowle, M., Short, T., Lianoglou, S.: data.table: Extension of data. frame for fast indexing, fast ordered joins, fast assignment, fast grouping and list columns. R package version 1(2)
7. Hesterberg, T., Moore, D.S., Monaghan, S., Clipson, A., Epstein, R.: Bootstrap methods and permutation tests. Introduction to the Practice of Statistics **5**, 1–70 (2005)
8. Huberman, B.A., Romero, D.M., Wu, F.: Social networks that matter: Twitter under the microscope. arXiv preprint :0812.1045 (2008)
9. Kwak, H., Lee, C., Park, H., Moon, S.: What is twitter, a social network or a news media? In: Proceedings of the 19th World Wide Web. pp. 591–600 (2010)
10. LaFleur, B.J., Greevy, R.A.: Introduction to permutation and resampling-based hypothesis tests. Journal of Clinical Child & Adolescent Psychology **38**(2), 286–294 (2009)
11. Ludbrook, J., Dudley, H.: Why permutation tests are superior to t and f tests in biomedical research. The American Statistician **52**(2), 127–132 (1998)
12. Romero, D.M., Meeder, B., Kleinberg, J.: Differences in the mechanics of information diffusion across topics: idioms, political hashtags, and complex contagion on twitter. In: Proceedings of the 20th World Wide Web, pp. 695–704 (2011)
13. Yang, J., Counts, S.: Predicting the speed, scale, and range of information diffusion in twitter. International Conference on Weblog and Social Media (ICWSM) **10**, 355–358 (2010)

Agent-Based Models of Copycat Suicide

Patrick N. Morabito$^{(\boxtimes)}$, Amanda V. Cook, Christopher M. Homan,
and Michael E. Long

Rochester Institute of Technology, Rochester, NY, USA
{pnm5379,melsch,avc9971}@rit.edu, cmh@cs.rit.edu

Abstract. In 1774, soon after Goethe published his novel *The Sorrows of Young Werther*, there were reports of suicides commited in the protagonist's distinct manner. These may have been the first noted instances of copycat suicide. This phenomenon results in geographic, temporal, and/or social clustering, and has become known as the Werther effect. However, suicide is very difficult to study experimentally, not only because it is under reported, but also because it is extremely difficult to predict the influence of social contact in real-life situations. We use agent-based modeling to study the effect of social influence on suicide rates. Our results demonstrate that both the scale of an individual's social group and the presence of celebrity suicides influence aggregate suicide rates by small, but measurable, amounts.

1 Introduction

Suicide is the tenth overall leading cause of death in the United States and the second leading cause of death in people 10 to 24 years of age [5]. Additionally, the after effects can be devastating, and reach far beyond just the victim. Suicide is difficult to study experimentally because of the infrequency and unpredictability of cases. Despite these difficulties, suicide is an important topic to study and each piece of knowledge gained can help to aid in its prevention.

This paper studies how suicide rates for a population are affected by four key social factors. *Social learning* is the concept that individuals exposed to the suicide of another have an increased chance of committing suicide themselves. *Prestige bias* is the concept that individuals tend to copy the behavior of prestigious figures, such as celebrities. This is enabled by *one-to-many cultural transmission*, which means that celebrity suicides are highly publicized and reach a much larger number of people than the suicide of a regular person. *Similarity bias* means that individuals are more likely to adopt the behavior of others who are similar to themselves. *Homophily* is the phenomenon that individuals who are similar to each other tend to be friends [6].

Our work is an extension of an agent-based model by Mesoudi [6], who showed that social learning, prestige bias, one-to-many transmission, similarity bias, and homophily all affect suicide rates to some degree. However, Mesoudi did not use

This material is based upon work supported by a GCCIS Kodak Endowed Chair Fund Health IT Strategic Initiative Grant

© Springer International Publishing Switzerland 2015
N. Agarwal et al. (Eds.): SBP 2015, LNCS 9021, pp. 369–374, 2015.
DOI: 10.1007/978-3-319-16268-3_45

realistic values for these factors. Additionally: the iterations of his model were not bound to physical time intervals; social learning—a complex phenomenon—was linked in a categorical fashion to similarity bias; and the number of agents used (1000) was too small to produce noticeable effects under real-world conditions. This resulted in unrealistic suicide rates, including some scenarios where nearly all agents committed suicide.

Our focus in this paper is to use a similar agent-based framework to compute realistic suicide rates based on observed values [1,8] for the aforementioned suicidal factors. Additionally, we bind each of our model's iterations to specific physical time intervals. We also change similarity bias from a categorical to a continuous variable. Finally, we use multiple combinations of social group size, world scale, and total population in our model, and show that suicide rates increase as the size of individuals' social groups or the prevalence of celebrity or prestige suicides increases.

2   The Model

The model was implemented in Netlogo [11] and consists of a two-dimensional, fixed-size square grid, representing a landscape, and an *agent density* parameter representing the number of agents per *patch*, or grid location. For example, if the landscape is 10 patches wide there would be 100 patches total and with an agent density of 100 there would be 100,000 total agents. For each model run, we fixed a single iteration, or *tick*, as one month. Each run lasts for 130 ticks. We ignore the first 10 ticks, in order to initialize the model [6], and collect data from the 11th through the 130th tick, which corresponds to 10 years of activity.

Each agent has a set of six binary traits, assigned randomly by fixed probabilities, which simulate individual-level suicide risk factors [6]. For the six traits, five were risk factors and one was a protective factor. These traits are: age, gender, race, mental disorder, and medical disorder. We use observed values for these traits and for suicide rates, as shown in Table 1. Since the simulated time was only 10 years, we assume that these values would remain constant for an individual during this time frame.

Table 1. Suicide Risk Factors

Risk Factors	Value	Probability	Reference
Age	Adult	0.774	
Gender	Male	0.483	[1]
Race	White	0.767	
Mental Disorder	Have	0.186	[8]
Medical Disorder	Have	0.1	**1
Support Factor	Have	0.2	

These traits are used to modify agents' base probabilities of committing suicide. Each agent is initially assigned a base probability of $p_0 = 1.1 \times 10^{-5}$ for committing suicide per month. So, for 100,000 agents the expectation is 1.1 suicides per month, or 13.2 suicides per year. We then modify this base

---

[1] These values were assumed based on internal discussion.

probability using the same formula as Mesoudi [6], where each of the six factors is expressed by $i = \{1, 2, \ldots, 6\}$, $k_i \in \{1 - q, 1 + q\}$ and $q = 0.2$. An agent's probability of committing suicide is then modified from p_0 to p_1 as defined in Equation 1.

$$p_1 = p_0 \Pi_{i=1}^{i=6}(1 \pm q_i) \tag{1}$$

Thus, if an agent had the higher value for all six traits their new probability would be $p_1 = (1+q)^6 p_0 = 2.99 p_0$, and if it had the lower risk value for all traits its new probability would be $p_1 = (1 - q)^6 p_0 = .26 p_0$.

We define two distinct types of agents: normal and celebrity. An agent has a probability of 1.0×10^{-4} of being a celebrity [3]. Celebrity agents function the same in every way as normal agents, but are used to model prestige bias with one-to-many transmission, while normal agents are used to model social learning. The difference is that social learning only affects agents in the patches where the suicides occur, but when a celebrity commits suicide all agents in all patches are affected.

We model homophily as follows. In each patch we order the agents and, according to this ordering, assign them their six risk factor traits as described above, except that, for all but the first agent, with probability $h = 0.3$ we instead copy the traits of the first agent into the current agent.

We simulate social learning as follows. Whenever an agent commits suicide, we replace the suicide probability p_1 of each surviving agent in the victim's patch with p_2, defined as:

$$p_2 = p_1\Big(1 + \sum_{i=1}^{x_n}(s * t_i) + \sum_{j=1}^{x_s}(c_s * t_j)\Big), \tag{2}$$

where x_n is the number of normal agents that committed suicide in the survivors' patch within the past 3 time ticks, $s = 1.5$ [4,7], and $t_i = 1.01^{m^2}$, where m is the number of traits shared by the victim and the survivor. The second term reflects the impact of celebrity suicides, where x_s is the number of celebrity agents that committed suicide, $c_s = 8.0$ [9,10], and t_j is defined in the same way as t_i.

3   Simulation Scenarios

We modeled a total of three different scenarios and for each case the change in suicide rate per 100,000 agents per year was measured. The first scenario used a constant total population of 100,000 agents and the size of the world was varied from a 10×10 grid up to a 100×100 grid through a total of 11 different sizes. By keeping the population constant and varying the scale of the word, the density of agents was forced to change.

The second scenario used a constant agent density of 100, but varied the scale of the world from a 10×10 grid to a 100×100 grid through a total of 10 different sizes. By keeping the density constant in this scenario and varying the scale of the world, the total population was forced to change.

For the third scenario, the size of the world was kept constant at 100×100, but the density was varied from 5 to 150 through 8 different values. By keeping the scale of the world constant and varying the density of agents, the total population was forced to change. This would therefore result in a combination of the effects from the first two scenarios because both the density of the agents and the total number agents were changed.

4   Results

The first scenario, which has a fixed population size but varying world size, showed that the suicide rate varied linearly with respect to the density. We ran the model 60 times for each different world size. The average values for suicide rate per 100,000 agents per year versus the density of agents for the 60 runs are presented in Figure 1. The error bars for each plot correspond to standard error.

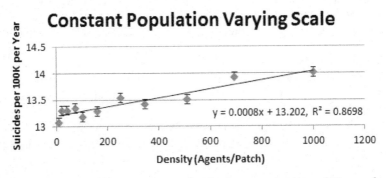

Fig. 1. Suicides per 100,000 agents versus density for a constant population and varying world scale

The second scenario, which had the constant density and varying world size, showed that the suicide rate varied linearly with respect to the population. We ran the model 80 times for each different world size. The average values for suicide rate per 100,000 agents per year versus the density of agents for the 80 runs are presented in Figure 2. However, the four smallest world sizes that were modeled did not result in a single celebrity suicide, so only the results from the other six runs are presented.

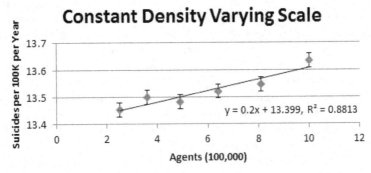

Fig. 2. Suicides per 100,000 agents versus agents for a constant density and varying world scale

The third scenario, which had the constant world size and varying density, which was a combination of the effects from the first two scenarios, showed that the suicide rate varied linearly with respect to the density and also linearly with respect to the total population. We ran the model 40 times for each different density. The average values for suicide rate per 100,000 agents per year versus the density of agents for the 40 runs are presented in Figure 3 (a), and versus population in Figure 3 (b).

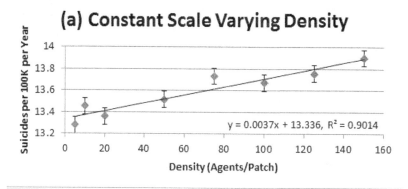

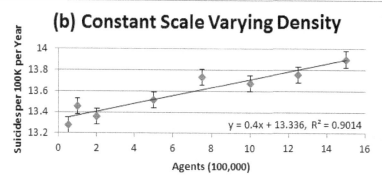

Fig. 3. (a) Suicides per 100,000 agents versus density for a constant world scale and varying population. (b) Suicides per 100,000 agents versus agents for a constant world scale and varying density.

5   Discussion and Conclusion

We showed that the density of agents and the total population both have small, but measurable, linear effects on overall suicide rates for a population. We also generated realistic suicide rates[2] , ranging from approximately 13 to 14 suicides per 100,000 people per year. Additionally, it should be noted that when these two factors were combined, the suicide rate still varied linearly with respect to each variable, but not as a simple addition of each. From Figures 1 and 3 (a), the suicide rate varies with density with a slope of 4.625 times greater when both density and population are varied than when only density is varied.

---

[2] US suicide rate for 2011 was 12.64 per 100,000 [2].

Also, from Figures 2 and 3 (b), the suicide rate varies with population with a slope 2 times greater when both density and population are varied than when only population is varied. Therefore, the effects of the density and population are not merely additive when both are present.

There are several additional possible scenarios that could also be modeled. The first suggestion is the introduction of local celebrities who would be an important member of a community, such as a mayor, who have a greater range of influence than a normal individual and less than a traditional celebrity. For this case, the suicide of a local celebrity would be viewed by agents within a certain radius of themselves, such as the eight patches immediately adjacent to their own. Another suggestion for future work is to increase the total number of risk and protective factors to better capture the factors that affect suicide rates. These values could also be varied to simulate different regions, based on their own specific demographics.

References

1. United States Census, Monroe County, NY. http://www.census.gov/easystats/
2. National violent death reporting system. Center for Disease Control. http://wisqars.cdc.gov:8080/nvdrs/nvdrsDisplay.jsp
3. Arbesman, S.: The fraction of famous people in the world. Wired (2013). http://www.wired.com/2013/01/the-fraction-of-famous-people-in-the-world/
4. Cutler, D.M., Glaeser, E.L., Norberg, K.: Explaining the rise in youth suicide, pp. 219–269. Chicago University Press Risky Behavior Among Youths: An Economic Analysis (2001)
5. Heron, M.: Deaths: Leading causes for 2010. National Vital Statistics Report **62**(6), 9–10 (2013)
6. Mesoudi, A.: The cultural dynamics of copycat suicide. PLoS One 4(9) (2009)
7. Gould, M.S., Wallenstein, S., Kleinman, M.: Time-space clustering of teenage suicide. Am. J. Epidemioly **131**, 71–78 (1990)
8. National Institute of Mental Health. http://www.nimh.nih.gov/health/statistics/prevalence/any-mental-illness-ami-among-adults.shtml
9. Stack, S.: Media coverage as a risk factor in suicide. Journal of Epidemiology and Community Health **57**, 238–240 (2003)
10. Niederkrotenthaler, T., Fu, K.W., Yip, P.S., Fong, D.Y., Stack, S., Cheng, Q., Pirkis, J.: Changes insuicide rates following media reports on celebrity suicide: A meta-analysis. Journal of Epidemiology and Community Health **66**, 1037–1042 (2012)
11. Wilensky, U.: Center for Connected Learning and Computer-Based Modeling, Northwestern University (1991). http://ccl.northwestern.edu/netlogo/

Analyzing Second Screen Based Social Soundtrack of TV Viewers from Diverse Cultural Settings

Partha Mukherjee[✉] and Bernard J. Jansen

College of Information Science and Technology, Pennsylvania State University,
University Park, PA, USA
pom5109@ist.psu.edu, jjansen@acm.org

Abstract. The presence of social networks and computing device in form of secondary screens used in conjunction with television plays a significant role in the shift from traditional television to social television (sTV). In this research, we explore the overall secondary screen usage in terms of four conversation patterns each over different cultural social soundtrack related to different TV shows. We process more than 469,000 tweets from second screens for four popular TV shows from four different countries. The ANOVA test results identify significant differences in overall usage of conversational patterns among viewers for TV broadcast across diverse cultures using second screens.

Keywords: Second screen · Social television · Social soundtrack · Cultural bias

1 Introduction

With the advent of Internet technology and the emergence of online social networking, the social possibility of TV has greatly expanded, as the merging of these technologies now allow a number of social activities and conversation concerning TV content via social networks (e.g., Facebook, Twitter, Weibo, etc.). This combination of TV and online social networks has forged a *social TV* that imparts feelings of togetherness and facilitates communication among people in dispersed locations. The social network has embedded itself within the modern TV culture and acts as a *social soundtrack* for TV content. The social soundtrack is the social commentary that results from integrating social networks as interactive tools with TV broadcasts.

The integration of Twitter (or other online social network) as the social soundtrack with televised broadcasts both in real time and non-real time marks the emergence of a new phenomenon augmenting the prior limited social aspect of TV. This new usage phenomenon is referred to as *second screen* (TV and computing device). The secondary screen allows the social soundtrack, the conversation with others, regarding the particular TV program.

In this research, we investigate the relative usage of four different patterns of social soundtrack conversations used by the audiences of TV shows from four different cultural biases, specifically examining if viewer's overall usage of each second screen interaction patterns differ for different cultural settings. This research is important as the variation in conversation patterns in distinct cultural social soundtracks can

© Springer International Publishing Switzerland 2015
N. Agarwal et al. (Eds.): SBP 2015, LNCS 9021, pp. 375–380, 2015.
DOI: 10.1007/978-3-319-16268-3_46

facilitate the understanding of the intent of viewer's communication via second screen on entertainment and TV commercials belong to divergent societies.

2   Related Work

2.1   Social TV (sTV) and Second Screen Usage

There is existing literature concerning socializing aspect of TV. Abreu, Almeida and Branco [1] developed the 2BeOn sTV system with the goal of making users go online to ensure interpersonal communication. Mukherjee and Jansen [2] investigate the engagement of second screen in sTV for sharing information during real time and non-real time transmission of TV shows. The degree of sociability increased significantly via second screen based online social networks services for TV viewing. The conversation interactions in the social soundtrack can be considered as a form of end-user enrichment of the TV content [3]. Such end user enrichment enhances the sTV socialization via the user generated content and converts sTV into a community [4].

2.2   Conversation in Social Soundtrack

There are prior studies on social network interaction framework that focus on aspects of second screen based social conversation. Mukherjee, Wong, and Jansen [5] examined predominant conversation pattern in TV show related social soundtrack during live and after live broadcast of TV shows. Honeycutt and Herring [6] examined the tweets to find specific purposes of interlocution (i.e., '@' symbol) in directed communication and referencing. boyd, Golder, and Lotan [7] studied the conversational aspects of retweet and investigated the reasons of retweeting in Twitter, while Naaman, Boase and Lai [8] introduced an item list of broadcast statements including information sharing, personal opinion along with random thoughts and observations in an undirected manner as observed by Jansen, Zhang, Sobel and Choudhury [9].

Though the aforementioned research speaks for social soundtrack conversations none of the above studies examined the relative usage of conversation patterns of varied cultural social soundtrack.

3   Research Question

Social networks allow for TV programs to be accessed and shared by viewers asynchronously using second screen. We evaluate the relative usage of second screen based interaction patterns in overall social soundtrack conversations over different cultural biases carried out by the viewers via second screen. The overall social soundtrack conversation is accounted for including the patterns of second screen interactions during real time (i.e. live TV broadcast) and non-real time transmission (after live broadcast) of TV shows. The notion that divergent societies may have different prevalent conversation patterns leads to frame our research question:

RQ. *Is there any change in relative usage of conversation pattern in viewers' overall social soundtrack conversations in diverse cultural settings related to TV shows?*

To investigate our research question, we have segregated viewers' tweets from three TV shows into four categories such as: 1) Response (RS), 2) Referral (RF), 3) Retweet (RT) and 4) Broadcast (BC). We categorized the queries based on results from prior literature [5; 6; 8]. Table 1 describes the communication patterns for the categories. We inquire the existence of such patterns as described in Table 1 in the tweets posted by viewers to classify the collected tweets into four categories. The categories are mutually exclusive by introducing a priority order: RT > RF > RS > BC. This means that any tweet that includes characteristics of retweets will inevitably categorize to RT even if it contains the characteristics of other three categories and so on.

Table 1. Categories of social soundtrack conversation patterns

Categories	Description
Retweet (RT)	Any retweet as recognized by "'RT: @', 'retweeting @', 'retweet @', '(via @)', 'RT (via @)', 'thx @', 'HT @' or 'r @' ". It is immaterial whether the tweet contains the characteristics of other categories.
Referral (RF)	Any full length or shortened URL directed at another user. It does not contain any pattern contained in RT category.
Response (RS)	Tweets intentionally engaging another user by means of '@' symbol which does not meet the other requirements of containing retweets or referrals
Broadcast (BC)	Undirected statements (i.e., does not contain any addressing) which allow for opinion, statements and random thoughts to be sent to the author's followers.

Based on the notion of the research question, the research hypothesis is formulated as:

Hypothesis: There is a significant difference in usage of each conversation pattern in viewers' overall social soundtrack conversations in diverse cultural settings.

4   Data Collection and Analysis

We selected four popular TV shows, each from a different country and collected users' interactions in form of tweets from Twitter. The TV shows selected for this research are: 1) *True Blood* from US, 2) *Bienvenidos al Lolita* from Spain, and 3) *Mahabharat* from India, and 4) *Doce de Mae* from Brazil. The tweets for all non-US TV shows were collected starting from 6th Feb 2014 to 5th March 2014, while for *True Blood* we collected tweets from 9th June to 29th June 2013. The number of tweets for *True Blood, Bienvenidos al Lolita, Mahabharat* and *Doce de Mae* are 220390, 98230, 76443 and 74328 respectively. The queries used are the TV show names.

Once the tweets are collected, we extract number of unique users from geo-enabled tweets (i.e., < 10% of the collected tweets) participated in TV show related social soundtrack conversation. The number of unique users for *True Blood, Bienvenidos al Lolita, Mahabharat* and *Doce de Mae* are 4256, 1876, 1356 and 1296 respectively. We analyze our research hypothesis based on the overall second screen interaction that includes both real time (rtSS) and non-real time (nrtSS) second screen interaction of the users related to four TV shows. The coordinates extracted from the location features via Twitter geo REST api are mapped into time zones using timezoneDB api.

Table 2. Show times with time zones for *True Blood* on Sunday

Eastern	Central	Mountain	Pacific	Alaska	Hawaii
9 PM	8PM	10 PM	9 PM	8 PM	3 PM

The show timings of *True Blood* are displayed in Table 2, while Table 3 displays show timings of the non-US TV programs. The running time for *True Blood* is 60 minutes and was broadcasted by HBO in US, while *Bienvenidos al Lolita, Mahabharat* and *Doce de Mae* are televised by Antena3, Star Plus and Rede Globo TV networks in Spain, India and Brazil, respectively.

Table 3. Show times with time zones for non-US TV shows

TV show name	Day	Show Time (PM)	Running time	Time zone
Bienvenidos al Lolita	Tuesdays	10:40	75 minutes	Central European
Mahabharat	Monday to Saturday	8:30	30 minutes	Indian Standard
Doce de mae	Thursday	11:15	30 minutes	Brasilia

Regarding investigation on relative overall usage of social soundtrack conversation patterns in diverse cultural settings, we classify taking both rtss and nrtSS tweets as a whole posted by every interactant for each show into four mutually exclusive categories (i.e.,RT, RF, RS and BC), as described in Table1. The overall tweet counts formed by identification of four conversation patterns against the interactants from different countries are used as units of analysis for evaluation of our first and second research questions respectively.

5   Results

The data is imported into SPSS platform for investigation. We approximately normalized the data using log transformation function *log (variable + 1.0)*, as the data is not normal. In SPSS, we ran one way ANOVA tests for examining both the research questions where $F_{critical(3, > 120)}$ = 2.60 at 95% confidence level. We use the Games

Howell test as the post-hoc test since it is not sensitive to unequal variances and unequal group sizes. To perform the post hoc test, we lower the level of significance (α) from 0.05 to 0.008 with the help of Bonferroni correction to reduce the chance of false positive error.

While investigating our research question, we perform ANOVA on TV show related social soundtrack by users of four different countries for each of the four different conversation patterns presented in Table 1. The F-statistics over four different cultural biases for each conversation pattern is given by: $F_{(3,8781)} = 58.49$ for RT, $F_{(3,8781)} = 16.48$ for RF, $F_{(3,8781)} = 42.63$ for RS and $F_{(3,8781)} = 42.22$ for BC. The F values are significant and thus support the hypothesis.

Table 4. T-values (* denotes significance) for four different TV shows for each conversation pattern

	True Blood	Bienvenidos al Lolita	Mahabharat	Doce de Mae
RT	12.66*		9.53*	19.47*
RF		6.32*	5.53*	5.44*
RS	9.76*	8.88*		11.34*
BC	2.81	16.21*	15.22*	

To identify the specific cultural biases that have the significant differences among the remaining cultural settings, the Games-Howell test is performed. From Table 4, Spanish TV show related social soundtrack becomes relatively predominant in using retweets (RT); US TV viewers use relatively more referral or URL based recommendation (RF) than the viewers from other societies; Indian TV show related social soundtrack involves more mention or reply based conversation, while undirected broadcast (BC) is relatively prevalent in overall discussions related to Brazilian TV show. It is also understood from Table 4, that though undirected broadcast in Brazilian TV show is significantly different over that in other non-US TV show related social soundtracks; it does not infer significance over US TV show related conversations.

6   Discussion

In this research, the analysis of each of interaction patterns in distinct cultural social soundtrack conversations shows (from Table 4) that US TV viewers rely comparatively more on URL based recommendation (RF) while Spanish, Indian and Brazilian audiences use relatively more pass along tweets (i.e. retweets (RT)), replies (RS) and undirected messages (BC)) respectively in social soundtrack conversations via second screen. The findings are important as they indicate the variations over each communication pattern and secondary screen usage in multiple cultures by its viewers in terms of information sharing behavior and social interactivity.

Regarding practical implication analyzing the reactions from undirected and directed recommendation will help cable providers and advertisers to identify the positive and negative effects of the televised shows and ads respectively for better personalization of TV shows and ads.

7   Conclusion

Though our data set spans for three/four weeks for each TV show, the results indicate a variation in social soundtrack conversation patterns for viewers from diverse cultural settings. We believe that our research provides valuable contribution concerning user's behavior and interaction while viewing of mass media from diverse cultural biases in a relatively new but emerging avenue of user behavior research in social soundtrack using second screen.

For future work, we will evaluated relative usage of second screen technology across diverse biases and also conduct content analysis of the social soundtrack data collected over a lengthier period from diverse cultures to determine the sentiment of the conversation occurring via user interactions with the second screen.

References

1. Abreu, J., Almeida, P., Branco, V.: 2BeOn—interactive television supporting interpersonal communication. In: Multimedia 2001, pp. 199–208. Springer (2002)
2. Mukherjee, P., Jansen, B.J.: Social TV and the social soundtrack: significance of second screen interaction during television viewing. In: Kennedy, W.G., Agarwal, N., Yang, S.J. (eds.) SBP 2014. LNCS, vol. 8393, pp. 317–324. Springer, Heidelberg (2014)
3. Cesar P., et al. Enhancing social sharing of videos: fragment, annotate, enrich, and share. In: Proceedings of the 16th ACM International Conference on Multimedia. ACM Press (2008)
4. Alliez, D., France, N.: Adapt TV paradigms to UGC by importing social networks. In: Adjunct Proceedings of the EuroITV2008 Conference (2008)
5. Mukherjee, P., Wong, J.-S., Jansen, B.J.: Patterns of social media conversations using second screens. In: Sixth ASE Conference on Social Computing (SocialCom), Stanford, CA, USA (2014)
6. Honeycut, C., Herring, S.C.: Beyond microblogging: Conversation and collaboration via Twitter. In: Proceedings of the 42nd Hawaii International Conference on System Sciences, pp. 1–10 (2009)
7. Boyd, D., Golder, S., Lotan, G.: Tweet, tweet, retweet: Conversational aspects of retweeting on twitter. In: Proceedings of the 43rd Hawaii International Conference on System Sciences, pp. 1–10 (2010)
8. Naaman, M., Boase, J., Lai, C.-H.: Is it really about me?: message content in social awareness streams. In: Proceedings of the 2010 ACM Conference on Computer Supported Cooperative Work (CSCW), pp. 189–192 (2010)
9. Jansen, B.J., et al.: Twitter power: Tweets as electronic word of mouth. Journal of the American Society for Information Science and Technology **60**(11), 2169–2188 (2011)

Error-Correction and Aggregation in Crowd-Sourcing of Geopolitical Incident Information

Alexander G. Ororbia II[1]([✉]), Yang Xu[1], Vito D'Orazio[2], and David Reitter[1]

[1] Pennsylvania State University, University Park, USA
ago109@ist.psu.edu
[2] Harvard University, Cambridge, USA

Abstract. A discriminative model is presented for crowd-sourcing the annotation of news stories to produce a structured dataset about incidents involving militarized disputes between nation-states. We used a question tree to gather partially redundant data from each crowd worker. A lattice of Bayesian Networks was then applied to error correct the individual worker annotations, the results of which were then aggregated via majority voting. The resulting hybrid model outperformed comparable, state-of-the-art aggregation models in both accuracy and computational scalability.

1   Introduction

Crowd-sourcing has challenged the notion that complicated problems call for great expertise. Instead, it parallelizes the solution-finding process: many untrained individuals contribute to a joint solution. Tasks ranging from natural language processing (NLP) [14] to image recognition [15] have been shown to be amenable to the use of crowds in lieu of experts. Given these successes, the use of crowd-sourcing has proliferated to other fields of study, notably the quantitative social sciences [1,7].

In social science research, where metrics for studying social phenomena are often derived by expert judgment and analysis, crowd-sourcing has the potential for ubiquitous application. For example, *militarized conflict* has traditionally been measured by the expert analysis of text documents [10]. In sufficient numbers, however, non-experts should be able to analyze these documents as effectively as experts. This leaves a problem of aggregation: how can redundant work be most effectively combined?

In this paper, we evaluate methods for aggregating partially redundant information from crowd workers to code geopolitical incidents using the criteria defined by the Militarized Interstate Dispute (MID) project [10]. To begin, we deconstructed the task into several simple and objective questions, the answers to which provided us with sufficient information to annotate the document. Unlike previous approaches, we asked partially redundant questions that do not follow a one-to-one mapping to target variables. The data gathered from workers inform

© Springer International Publishing Switzerland 2015
N. Agarwal et al. (Eds.): SBP 2015, LNCS 9021, pp. 381–387, 2015.
DOI: 10.1007/978-3-319-16268-3_47

a set of Bayesian Networks, which are trained to error-correct individual worker results. The models are combined with a voting scheme that aggregates multiple worker inputs to perform different classification tasks. Finally, we compare the accuracy of our approach with competing models and find that our hybrid approach outperforms all others.

2   Related Work

The coupling of human and machine intelligence is emerging as a critical tool in utilizing large-scale datasets where manual labeling is expensive [3]. Often, machine learning algorithms are trained to emulate collective human intelligence through the interaction with human users [8]. Both applications [2] and evaluations of efficiency and cost effectiveness [11] are available. Hybrid methods have proven effective in handling issues of global interest, such as early stage tracking of disease outbreaks [9]. The successes reported in these studies motivate the current study.

Benchmark platforms such as SQUARE [13] have made available representative worker aggregation methods that improve upon simple majority voting. We compete with two aggregation models, the state-of-the-art ZenCrowd [5] and the established Dawid and Skene & Naive Bayes method [4] (DS/NB). [1] Both ZenCrowd and DS/NB model worker behavior and problem difficulty to vertically aggregate responses. In contrast, our method circumvents the computational overhead imposed by models of behavior and task complexity. While task-agnostic, it can still incorporate domain heuristics.

3   Methodology

3.1   Crowd-Sourcing

Workers on Amazon Mechanical Turk (AMT) read a news story and answered a set of simple, objective questions about it implemented using Qualtrics. Questions were designed to address the coding criteria defined by the MID project, a well-known, ongoing effort that collects data on international conflict [10]. News stories were randomly sampled from a set of "potentially relevant" MID documents in equal portions from years 2007, 2009, and 2010.[2] Each news story read by the workers was either irrelevant or about a *threat*, *display*, or *use* of military force. There are three primary coding tasks, each specifying a target variable: 1) the hostility level (threat/display/use of force), 2) the initiator and target nation-states, and 3) the type of actions taken by these countries.[3]

---

[1] A third, and similar to our own work, is [12], where the DALE model was proposed to solve the object localization task. No public implementation was available.

[2] The algorithm in [6] was used to create the set of "potentially relevant" source documents, from which 150 were randomly sampled per year. After discarding some for formatting reasons, 446 total were left.

[3] See [10] for additional details about the MID coding ontology.

Table 1. Sample features provided by workers

Meaning	Value Examples
Hostility level of MII	Non-incident, Threat to use force, Display of force, Use of force
Type of action taken by country that started incident	Alert, Seizure, Attack, Join interstate war etc.
Is action just verbal, or material?	Verbal, Material
Whether or not a story is conflictual or cooperative	Cooperative, Conflictual
State entity first taking action (the "initiator")	Afghanistan, Armenia, etc.
State entity opposing the initiator	Afghanistan, Armenia, etc.

3.2   The Question Tree

A question tree, comprised of blocks of multiple-choice and yes-or-no questions, was used to guide the workers to finish the coding tasks.[4] Workers are branched to different sub-blocks depending on their answers to previous questions. The answers to many of these questions provide information necessary for completing the underlying coding tasks and building feature representations for horizontal integration. Additional questions extract partially redundant information that is later used for error correction. Examples of the features provided by the questionnaire are shown in Table 1.

1644 workers on Mechanical Turk coded the 446 documents. While 1251 workers did this task only once, some completed many dozens of annotation tasks. On average, each document was coded by 8.47 different workers (range: 6–10).

3.3   Prediction Targets

There are 5 labels that are most informative of the nature of an MII and hence are the target variables that we predict. *Initiator* refers to the country that took the first action in the dispute. *Type* is the primary action type taken by the initiator. *Target* refers to the country that is the target of the initiator's action. *Level* refers to the hostility level of the action taken by the initiator. *Incident* is a binary variable that distinguishes a story about a militarized conflict event between nation-states from other articles. All target variables but *Incident* are directly answered by corresponding regular questions given to the workers. *Incident* is built from answers to these questions. The error-correction approach will, however, use all available information to revise those choices.

For our gold standard to compare each model's predictions against, each document was independently labeled by three subject matter experts, each of whom were graduate students of political science experienced with the MID coding scheme. Disagreements among the experts were resolved by majority

---

[4] A copy of the Qualtrics questionnaire is available at http://goo.gl/TZnkVd.

vote. In 19 cases, all three disagreed and these were resolved by subsequent discussions of MID coding rules.[5]

3.4   Approaches to Predicting Targets

Several approaches for predicting the target classes were evaluated. The most frequent class label served as a baseline (*Baseline*) while the first model, *Voting*, performs a vertical aggregation across worker annotations via their modal response. This commonly-used approach only aggregates direct responses for single variables. The same limitation holds for the two previously proposed approaches we also compare our methods against, ZenCrowd (*Zen*) and DS/NB (*Bayes*).

The *Horizontal* approach consists of a classifier trained on annotated story features as determined by the workers. Information from all features is integrated to predict all variables, separately for each worker.

Finally, our hybrid method *Hori+Vert* first trains a set of classifiers on a training set of worker-annotated story features. The resulting model is applied to each worker's annotations of a single story. This produces a prediction matrix for each unseen story, where columns represent the five predicted variables and rows correspond to error-corrected annotations of the workers. The most-frequent choice is then computed for each column, yielding the predictions for each story. At this stage, simple domain heuristics may be applied to guarantee plausibility. (For example a "non-incident" never has initiators or targets.) Ties are resolved at random.

The Bayesian (Belief) Network, a probabilistic directed graphical model, was chosen as the base classifier for both the *Horizontal* and *Hori+Vert* models [6]. Such a model does not require explicit hyper-parameter tuning and can be constructed efficiently, advantages we exploit in composing our lattice of discriminative experts.

4   Results

Evaluation results were produced on non-overlapping test sets, separate from the training data. We used 40-fold cross-validation by story (i.e., classifiers were never trained on the same story used to evaluate them). As shown in Table 2 and Figure 1, the results of our experiments indicate that a combined approach, which leverages the power of crowd-sourcing aggregation and supervised machine learning integration, yields the best predictive model for each of the

---

[5] Inter-annotator agreement was acceptable (Cohen's Kappa across years: 0.7-0.78). While this indicates a well-defined coding scheme, it also shows that MIDs remain difficult to code.

[6] Waikato Environment for Knowledge Analysis (3.7.10) was used to build the models. This model performed best in comparison to other algorithms we tuned, such as the Support Vector Machine (linear & Gaussian kernels).

Table 2. Predicting target variables via different approaches

	Baseline	Hori+Vert	Horizontal	Voting	Worker	Bayes	Zen
Initiator	11.38	**73.75**	64.99	69.06	57.18	67.43	64.10
Target	14.49	**71.25**	60.94	68.51	56.28	64.87	63.58
Type	46.47	**73.33**	60.45	68.97	51.82	64.10	67.50
Level	33.05	**83.33**	69.99	71.26	59.81	69.23	68.21
Incident	53.53	**87.50**	77.27	$--$	$--$	$--$	$--$

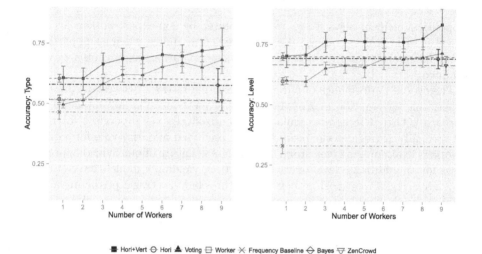

■ Hori+Vert ⊖ Hori ▲ Voting ⊟ Worker ✕ Frequency Baseline ◇ Bayes ▽ ZenCrowd

Fig. 1. Accuracy vs. Number of Workers. Some data from workers annotating very few or many stories were removed for this analysis.

target variables.[7] Furthermore, the performance of the *Zen* and *Bayes* aggregation models worsened with task complexity. Specifically, the results indicate that these models do not scale well with the size of the category set (confirming a hypothesis stated in [13]). Run-time performance similarly worsens as problem difficulty increases. For example, when predicting *Initiator*, the average model training time is 18.97 seconds for *Zen* and 6.62 for *Bayes*, as compared to 0.008 for *Hori+Vert*. Similar trends are exhibited for each target variable.[8]

Our hybrid model continues to outperform the non-hybrid ones as additional workers are employed. This is seen in Figure 1 (*Hori+Vert*). Intuitively, this makes sense since our model leverages both the horizontal *and* vertical features of the data, while the other models are restricted to one or the other.

---

[7] Since workers were not asked to directly classify a story as MID or non-MID, for *Incident* our hybrid model was only compared to the *Horizontal* and *Baseline* models.

[8] For *Target*, average model training time is 19.17 seconds for *Zen*, 4.73 for *Bayes*, and 0.009 for *Hori+Vert*. For *Type*, it is 1.76 seconds for *Zen*, 0.56 for *Bayes*, and 0.005 for *Hori+Vert*. For *Level*, it is 0.14 for *Zen*, 0.12 for *Bayes*, and 0.004 for *Hori+Vert*.

One explanation for why the *Hori+Vert* model outperforms the others pertains to the influence of erroneous annotations. Specifically, some workers will very likely make erroneous annotations, and the modal computation step helps to mitigate the impacts of such mistakes. In basic voting or other aggregation models (Bayes, Zen), however, erroneous annotations are still leveraged for cross-feature prediction.

5   Conclusion

Geopolitical incident news stories were annotated by non-expert workers according to the MID project coding rules. The prediction ability, when using partially redundant information provided by the workers, of various algorithms was then evaluated.

The overall performance of our error-correcting lattice of Bayesian Networks outperforms aggregation or classification-only methods. The advantage of our approach is that it integrates annotations horizontally via supervised learning, and vertically aggregates the results via the predicted majority vote for a group of workers examining a given story. The ensemble nature of our hybrid approach allows for an additional level of error-correction, yielding a model that not only takes into account relationships between features but also target predictor values. In future work, simple rule-mining could be used to "tune" these higher-level correction heuristics to the target task. Our method exploits even workers that make mistakes by integrating across their answers to both direct and indirect questions without the need for modeling worker behavior.

Acknowledgments. We acknowledge seed funding from the Social Sciences Research Institute. We thank Jessie Li, C. Lee Giles, project collaborators Glenn Palmer, Michael Schmierbach, & data annotators Matthew Lane, & Michael Kenwick.

References

1. Benoit, K., Conway, D., Laver, M., Mikhaylov, S.: Crowd-sourced data coding for the social sciences: Massive non-expert human coding of political texts. Presentation at the 3rd Annual New Directions in Analyzing Text as Data Conference. Harvard University (2012)
2. Boia, M., Musat, C.C., Faltings, B.: Acquiring commonsense knowledge for sentiment analysis through human computation. In: 28th American Association for Artificial Intelligence (2014)
3. Culotta, A., McCallum, A.: Reducing labeling effort for structured prediction tasks. In: Proceedings of the 20th National Conference on Artificial Intelligence, AAAI 2005, vol. 2, pp. 746–751. AAAI Press (2005)
4. Dawid, A.P., Skene, A.M.: Maximum likelihood estimation of observer error-rates using the EM algorithm. Journal of the Royal Statistical Society. Series C (Applied Statistics) **28**(1), 20–28 (1979)
5. Demartini, G., Difallah, D.E., Cudr-Mauroux, P.: ZenCrowd: Leveraging probabilistic reasoning and crowdsourcing techniques for large-scale entity linking. In: Proc. 21st International Conference on World Wide Web, pp. 469–478. ACM (2012)

6. D'Orazio, V., Landis, S.T., Palmer, G., Schrodt, P.: Separating the wheat from the chaff: Applications of automated document classification using support vector machines. Political Analysis **22**(2), 224–242 (2014)
7. Gao, H., Wang, X., Barbier, G., Liu, H.: Promoting coordination for disaster relief – from crowdsourcing to coordination. In: Salerno, J., Yang, S.J., Nau, D., Chai, S.-K. (eds.) SBP 2011. LNCS, vol. 6589, pp. 197–204. Springer, Heidelberg (2011)
8. Lughofer, E.: Hybrid Active Learning for Reducing the Annotation Effort of Operators in Classification Systems. Pattern Recognition **45**, 884–896 (2012)
9. Munro, R., Gunasekara, L., Nevins, S., Polepeddi, L., Rosen, E.: Tracking epidemics with natural language processing and crowdsourcing. In: 2012 American Association for Artificial Intelligence Spring Symposium, Toronto, Ontario, Canada (2012)
10. Palmer, G., D'Orazio, V., Kenwick, M., Lane, M.: The MID4 Data Set, 2002–2010: Procedures, Coding rules, and Description. Conflict Management and Peace Science (Forthcoming, 2015)
11. Ramirez-Loaiza, M.E., Culotta, A., Bilgic, M.: Anytime active learning. In: 28th American Association for Artificial Intelligence (2014)
12. Salek, M., Bachrach, Y., Key, P.: Hotspotting - a probabilistic graphical model for image object localization through crowdsourcing. In: DesJardins, M., Littman, M.L. (eds.) Proc. 27th American Association for Artificial Intelligence, July 14-18, Bellevue, Washington, USA. AAAI Press (2013)
13. Sheshadri, A., Lease, M.: Square: A benchmark for research on computing crowd consensus. In: First AAAI Conference on Human Computation and Crowdsourcing (2013)
14. Snow, R., O'Connor, B., Jurafsky, D., Ng, A.Y.: Cheap and fast-but is it good? evaluating non-expert annotations for natural language tasks. In: Proceedings of the Conference on Empirical Methods in Natural Language Processing, pp. 254–263. Association for Computational Linguistics (2008)
15. Von Ahn, L., Blum, M., Hopper, N.J., Langford, J.: CAPTCHA: Using hard AI problems for security. In: Biham, E. (ed.) EUROCRYPT 2003. LNCS, vol. 2656, pp. 294–311. Springer, Heidelberg (2003)

Socio-Spatial Pareto Frontiers
of Twitter Networks

Brandon Oselio[✉], Alex Kulesza, and Alfred Hero

Department of Electrical Engineering and Computer Science,
University of Michigan, Ann Arbor, MI, USA
{boselio,kulesza,hero}@umich.edu

Abstract. Social media provides a rich source of networked data. This data is represented by a set of nodes and a set of relations (edges). It is often possible to obtain or infer multiple types of relations from the same set of nodes, such as observed friend connections, inferred links via semantic comparison, or relations based off of geographic proximity. These edge sets can be represented by one multi-layer network. In this paper we review a method to perform community detection of multi-layer networks, and illustrate its use as a visualization tool for analyzing different community partitions. The algorithm is illustrated on a dataset from Twitter, specifically regarding the National Football League (NFL).

1 Introduction

Social networks increasingly comprise multiple types of connectivity information. The set of users form nodes, while the connectivity information form sets of edges. This information can either be observed directly from the data — *relational* edge sets — or inferred using ancillary data that describes users — *behavioral* edge sets [1]. A multi-layer network [1–3] is a framework that allows for nodes with multiple edge sets. Multi-layer networks place each type of connectivity in its own layer; the goal is to then analyze this structure.

In performing tasks like community detection on multi-layer networks, we seek a flexible method that allows for each layer to contribute to the overall community structure at varying levels of strength. One method to do this is to incrementally combine single layer solutions while visualizing each community partition.

Prior work has analyzed two-community networks using Pareto frontiers [1]. In this paper we generalize those methods to handle complex networks with many communities, applying the resulting algorithm to socio-spatial Twitter networks focusing on the 2013 NFL playoffs.

This work was partially supported by ARO under grant #W911NF-12-1-0443. We are grateful to Qiaozhu Mei who provided the Twitter data stream through his API gardenhose level access.

© Springer International Publishing Switzerland 2015
N. Agarwal et al. (Eds.): SBP 2015, LNCS 9021, pp. 388–393, 2015.
DOI: 10.1007/978-3-319-16268-3_48

2   Methods

2.1   Pareto Optimality

The development of the algorithm will follow [1]; it is briefly described here. First, the concept of Pareto optimality must be introduced. Pareto optimality stems from multi-objective optimization, also known as vector optimization. This field of research deals with problems where the goal is to optimize multiple objective functions simultaneously.

Let $f_i : X \to \mathbb{R}$, with X being our solution space, and consider the following minimization problem:

$$\min_{x}[f_1(x), \ldots, f_k(x)]. \tag{1}$$

The key definition is dominating and non-dominating points:

Definition 1. *Let x_1 and x_2 be in the solution space X. We say that x_1 dominates x_2 if for all $i = 1, \ldots, k$, $f_i(x_1) \leq f_i(x_2)$, and at least for one index j, $f_j(x_1) < f_j(x_2)$. A solution y in the solution space that is not dominated by any other point in the solution space is said to be non-dominated.*

The **Pareto front** of solutions is the set of solutions which are non-dominated. So, if a solution is on the Pareto front, it's not possible to achieve a lower value for a particular objective function without increasing the value of at least one other. The Pareto front defines a type of optimal set for vector optimization problems.

2.2   Pareto Front Algorithm

Many single-layer community detection algorithms are objective based. We can adapt these techniques to multi-layer networks by defining an objective function for each layer, and then exploring the Pareto front of community partitions.

A multi-layer network $G = (\mathcal{V}, \mathcal{E})$ consists of vertices $\mathcal{V} = \{v_1, \ldots, v_p\}$, common to all layers, and edges $\mathcal{E} = (\mathcal{E}_1, \ldots, \mathcal{E}_L)$ in L layers, where $\mathcal{E}_l$ is the edge set for layer l, and $\mathcal{E}_l = \{e^l_{v_i v_j}; \quad v_i, v_j \in V\}$. Each edge is undirected. Further, a series of adjacency matrices are defined, one for each layer, where we have:

$$[[A^l]]_{ij} = e^l_{v_i v_j}. \tag{2}$$

These adjacency matrices are important in order to evaluate the objective function for each layer. The goal is to find a community partition C which divides the nodes into k communities $B_1, \ldots B_k$, where $B_i \subseteq V$. In this paper, RatioCut [4] is used as the objective function for each layer, defined as:

$$f_l(C) = \frac{1}{2} \sum_{i=1}^{k} \frac{\mathrm{cut}_l(B_i, \bar{B}_i)}{|B_i|} \qquad \mathrm{cut}_l(B_i, \bar{B}_i) = \sum_{i \in B_i, j \notin B_i} [A^l]_{ij} \tag{3}$$

In describing the algorithm, we specialize to $L = 2$ layers. First we perform single-layer network community detection on both layers, resulting in two community partitions C_1 and C_2. Note that normalized spectral clustering is a relaxation of the problem of minimizing RatioCut, which we perform on each layer, and obtain C_1 and C_2. We assume that these solutions are optimal in their respective objectives. Thus, C_1 and C_2 both lie on the Pareto front; in order to find other points on an approximate Pareto front between the given community partitions, a node swapping technique is used, similar to the KL algorithm [5]. Starting at one solution C_1 and ending at solution C_2, one node i changes its community membership from $C_1[i]$ to the membership $C_2[i]$ at each step of the algorithm. In order to determine which nodes to swap, the RatioCut is calculated for each possibility. This is continued until the ending solution C_2 is reached. Finally, all the traversed partitions are filtered to find non-dominated solutions; these solutions form an approximate Pareto front. Algorithm 1 describes the traversal in more detail. This algorithm improves on [1] in that it allows for unequal communities in the resulting partition and it is effective for more than two communities, which is useful when applying to real social network datasets.

Algorithm 1. Pareto Frontier Algorithm

1: **procedure** POSSIBLEFRONTIERPOINTS(A_1, A_2, C_1, C_2)
2:     $C_{cur} \leftarrow C_1$
3:     $t \leftarrow 0$
4:     Cost $\leftarrow \infty$
5:     **while** $C_{cur} \neq C_2$ **do**
6:         **for** i where $C_{cur}[i] \neq C_2[i]$ **do**
7:             $C_{cur}[i] \leftarrow C_2[i]$
8:             Cost$[i] \leftarrow$ RatioCut(A_2, C_{cur})
9:             $C_{cur}[i] \leftarrow C_1[i]$
10:        $i^* \leftarrow \min_i$ Cost$[i]$
11:        $C_{cur}[i^*] \leftarrow C_2[i^*]$
12:        Memberships$[t] \leftarrow C_{cur}$
13:        Cuts$[t] \leftarrow$ (Ratio-Cut(A_1, C_{cur}), Ratio-Cut(A_2, C_{cur}))
14:        $t \leftarrow t + 1$
15:    **return** Memberships, Cuts

---

3   Twitter Dataset

Data to create a multi-layer network was obtained from the Twitter stream API at gardenhose level access during January of 2013. Tweets were filtered based on the availability of geolocation data. This geolocation information allowed for the creation of the first layer of the multi-layer network. For every pair of users i and j, they were connected ($A_{ij} = A_{ji} = 1$) if the users were closer than a certain distance threshold δ. The δ parameter changed based on the density of users and size of area that was being observed. This layer is called the coordinate network layer.

The second layer that is created utilizes hashtags to connect users. Hashtags are any words beginning with a # sign. In this layer, a user i and j are said to be connected if they use the same hashtag from a specified set of hashtags over the one month period. In order to focus on a smaller set of users, specific hashtags were chosen that applied to an event or set of events that were occurring in this period; in this case the events were the National Football League (NFL) playoffs. The dataset was created by first filtering on four of the most popular pertinent hashtags in the three month time period: #Ravens, #49ers, #Falcons, and #Patriots. These correspond to the four NFL football teams that reached the end of the NFL playoffs for that year. A two-layer network consisting of the hashtag network layer and coordinate network layer ($\delta = 50$) is analyzed. The resulting dataset contains 3456 nodes (Twitter users).

4   Results

We first perform some single-layer community detection. The partition resulting from spectral clustering on the hashtag network does a good job at stratifying the popular hashtags into communities, as seen in Table 1. Community 1 is mostly the #Ravens hashtag, while community 4 is the #49ers. Community 2 sees the #Patriots and #Falcons hashtags grouped together, while community 3 is a mixture of all four. Figure 1 shows a false color map of the densities of users in each community. It is surprising that while there is strong community structure in this network, it is less correlated with geography than one might expect.

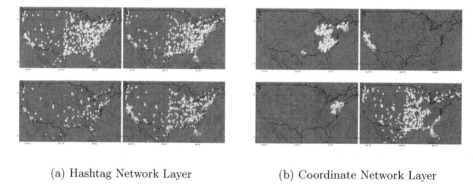

(a) Hashtag Network Layer   (b) Coordinate Network Layer

Fig. 1. Density Plot by Community. For the hashtag network layer, the communities correspond to the numbers in Table 1 going from left to right and subsequently from top to bottom. Communities based on the discussion of NFL teams are less localized than the communities based on geographic proximity.

As expected, the coordinate network layer partitions according to high population density. Specifically, it clusters the San Francisco and LA area together,

Table 1. Hashtags per Community for Hashtag Network Layer Solution

	Community 1	Community 2	Community 3	Community 4
#Ravens	1232	0	170	0
#49ers	57	0	155	762
#Patriots	45	291	29	10
#Falcons	49	273	29	7

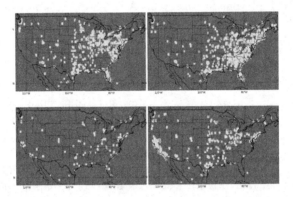

Fig. 2. Density Plot for Pareto Combined Community. This community partition retains attributes from both layers, while still giving a visual sense of the overall community structure. The communities in the upper left and lower right have become more concentrated about east coast and west coast, respectively. Further, the community in the upper right shows high concentration in Atlanta up to both the Maryland and Massachusetts area.

the Maryland area by itself, and the Atlanta and Boston area together. The last community seems to be a catch-all for everywhere else, i.e., those places with less density.

Using the described algorithm, a Pareto solution is found for the multi-layer network; the community partition is shown in Figure 2. The communities are more geographically localized when compared with the mention network layer solution, while still visually resembling its structure. For instance, the last community picks out the San Francisco/LA area in a single community, which the original mention network did not. Further, the second community groups the Atlanta area with the Massachusetts area, though not as well as the coordinate network layer. The Pareto community partition, however, still contains some of the interesting patterns of the hashtag network layer and is not completely given to geographic localization.

5   Related Work and Conclusion

Work on multi-layer networks continues to increase; a more comprehensive overview of the techniques and theoretical background for the multi-layer structure can be found in [2]. Other methods have been proposed to utilize multiple types of data on nodes, including a mean approximation [6].

[7] provides an overview of the types of algorithms used for community detection. For multi-layer networks, the extension of modularity-type algorithms have been proposed [8]. There have also been papers that analyzes single layer network community partitions when attempting to understand the multi-layer structure, as we are doing in this paper. [9] compares single-layer communities via normalized mutual information (NMI), although does not try to combine the solutions in any way. The particular single-layer algorithm used in this paper, normalized spectral clustering, has been well studied [4]. The Pareto front algorithm is an extension of the one found in [1].

This paper revisited and expanded upon the Pareto frontier algorithm which was detailed in [1]. Extending the algorithm to multiple communities, the main purpose of this approach was to evaluate community partitions for varying levels of involvement by the layers of the network. The algorithm was applied to data from Twitter. The Pareto front method is useful to visualize community partitions for multi-layer datasets; this type of visualization is key to understanding how the layers are similar to each other, and how their community structures can interact. For future work, an explanation of how well we can approximate the Pareto front, as well as quantitative measurements on the similarity of community structure between layers would be helpful in the analysis of multi-layer networks.

References

1. Oselio, B., Kulesza, A., Hero, A.: Multi-objective optimization for multi-level networks. In: Kennedy, W.G., Agarwal, N., Yang, S.J. (eds.) SBP 2014. LNCS, vol. 8393, pp. 129–136. Springer, Heidelberg (2014)
2. Kivelä, M., Arenas, A., Barthelemy, M., Gleeson, J.P., Moreno, Y., Porter, M.A.: Multilayer networks. Journal of Complex Networks **2**(3), 203–271 (2014)
3. Magnani, M., Rossi, L.: Pareto distance for multi-layer network analysis. In: Greenberg, A.M., Kennedy, W.G., Bos, N.D. (eds.) SBP 2013. LNCS, vol. 7812, pp. 249–256. Springer, Heidelberg (2013)
4. Luxburg, U.: A tutorial on spectral clustering. Statistics and Computing **17**(4), 395–416 (2007)
5. Kernighan, B.W., Lin, S.: An efficient heuristic procedure for partitioning graphs. Bell System Technical Journal **49**(2), 291–307 (1970)
6. Tang, L., Wang, X., Liu, H.: Community detection via heterogeneous interaction analysis. Data Min. Knowl. Discov. **25**(1), 1–33 (2012)
7. Fortunato, S.: Community detection in graphs. Physics Reports **486**, 75–174 (2010)
8. Mucha, P.J., Richardson, T., Macon, K., et al.: Community structure in time-dependent, multiscale, and multiplex networks. Science **328**(5980), 876–878 (2010)
9. Barigozzi, M., Fagiolo, G., Mangioni, G.: Identifying the community structure of the international-trade multi-network. Physica A **390**(11), 2051–2066 (2011)

SOR: A Protocol for Requests Dissemination in Online Social Networks

Salem Othman[(✉)] and Javed I. Khan

Networking and Media Communications Research Laboratories,
Department of Computer Science, Kent State University
Kent OH 44242, USA
{sothman,javed}@kent.edu

Abstract. Routing over online social networks (OSN) requires a new routing mechanism that can adhere to individual privacy needs. In this paper, we explore a social online routing protocol (SOR) that can find best paths for request dissemination, using friends' network with privacy aware route messaging. We use an OMNeT++ simulator to study the proposed SOR at three cognoscenti' levels over real social circles from Google+ to study their routing efficacy.

Keywords: Online social networks · Social routing protocol · Social requests

1   Introduction

Despite the emergence of online social networks, there is no satisfactory social routing protocol. In general, routing requires sharing of some global information by all parties involved. On the other hand, social routing is highly sensitive to privacy concerns. In this paper, we explore a protocol for social request routing. Requests routing is a rather universal application in real-life social network [6]. It has also been studied by social scientists since the 1980s [2, 3]. In the online world, a few elementary current examples include LinkedIn's request for endorsements, and Facebook's request to join in a Cause, etc. However, there is no mechanism to route them in regards to privacy concerns. It can be a major help for future online philanthropy communities [1]. We identify two important aspects of real-life request circulation. First is the individualized social priority. People respond or route requests based on highly varying and individualized priorities. Varying social characteristics (such as gender, fame, centrality, closeness, etc.) could be weighed by individuals with varied importance to estimate routing priorities between two adjacent neighbors. Secondly, complex privacy considerations are used to share various pieces of information needed to make the routing happen in a social network. In this research, we outline a *social online routing* (SOR) protocol that can help social request dissemination in OSNs by considering the individual's social characteristics and conforming to a multi-circle privacy domain system. In this paper, the next section presents the social priority framework. Section 3 then introduces the multi-circle privacy domain, and presents the proposed protocol SOR. An evaluation of several possible versions of SOR is then

© Springer International Publishing Switzerland 2015
N. Agarwal et al. (Eds.): SBP 2015, LNCS 9021, pp. 394–399, 2015.
DOI: 10.1007/978-3-319-16268-3_49

presented using several online social networks from Google+ using the OMNeT++ simulator.

2 General Social Priority Framework

SOR assumes a quantity called *In-Social Priority (iSP)* for routing. If there is a friendship edge e_{uv} (from node u to v) with *iSP* value p_{uv}, it represents a form of proportionate priority with which v will treat a request arriving from u. A proportionate priority is defined as following: If v has a current task queue of length L (1 means top of the queue and L means end of the queue), u's request will be placed at $L*(1-p_{vu})$ location of the queue. Of course, the value of p_{uv}, is known to v, but we cannot expect v to reveal it candidly to u. Thus, u continually learns *Out-Social Priority (oSP)* p^*_{uv} which is an estimate of p_{uv}. If node u makes a correct estimation then $p^*_{uv}=p_{uv}$. SOR nodes use p^*_{uv} to determine the best path to forward.

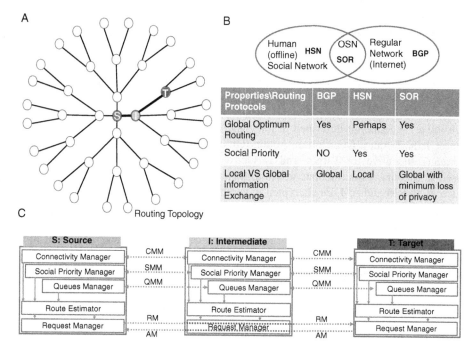

Fig. 1. (A) Routing Topology. (B) Protocols comparison. (C) SOR layout when node S sends request to node T through node I. The messages CMM, SMM, QMM, RM, and AM (Section 3.1). The managers Connectivity, Social Priority, Route and Request (section 3.2). The dashed line represents encryption (Section 3.3).

An individual (node) can use any set of social factors to generate and estimate p_{uv} of the senders and p^*_{uv} of the targets in the personal circle. (In our simulation, we factor namely gender, degree, closeness, betweenness, eigenvector centralities, etc.). In order to generate social priorities for all potential senders (receivers), each SOR node uses its own set of factors and uses singular value decomposition (SVD) [4], to generate a SP vector for the immediate in(out) circle made of adjacent neighbors. Besides, SPs additionally each node requires knowledge about the network connectivity as well as availability and load information of other nodes in the network to make optimum routing decision. Unlike computer networks - due to privacy concerns, no node is expected to know all these pieces of information. We propose three cognoscenti types for the SOR nodes.

3   SOR Overview

3.1   SOR Messages

SOR uses three different update messages. These are 1) Connection Metric Message (CMM) for carrying connectivity information, 2) Social Priority Metric Message (SMM) for carrying social priorities, 3) Queue Metric Message (QMM) for carrying queues statistics. Normally, in regular networks, update messages are sent in a single message, but in SOR (for privacy and security), each message has a separate encryption key. The CMM's key is public, but the SMM's key and QMM's key are private keys for specific nodes in OSN. Source field (Scr CIM) refers to community based identity mechanism. Request Message (RM) contains the actual request. Acknowledgment Message (AM) carries the replies. RM header contains fields Next Hop, Request ID, Request Type, Request Scope (Private, Public), Request Deadline, and Incentive (schema not described in this paper) for forwarder. RM body contains Request Title, Request Description, Action (Online, Offline) Request Delivery (Normal, Urgent), Incentive for Target, and an Optional fields. Separate incentive fields are used to encourage intermediate nodes to forward requests and also target to give service. Both RM and AM are encrypted in another way similar to Onion Routing where source and destination are hidden. Nearly all messages share a common header which contains Hop Counter, Creation Time, Propagation Rules, and TTL. The Propagation Rules field carries specific rules chosen by source to control message propagation directions. A node may share its values with other trusted friends. Conversely a node may guess these values of others whether friend or not.

3.2   SOR Architecture

SOR has five components. 1) *Connectivity Manager,* which is responsible for propagating, receiving, and managing connectivity information using CMM. 2) *Social Priority Manager,* which caculates social priorities for incoming adjacent neighbors and estimates social priorities given by outgoing adjacent neighbors as well as propagates the latter using SMM. 3) *Queues Manager,* which propagates queues information using QMM. 4) *Routing Estimator,* which combines the information collected via three above components to understand the real situation in the OSN and calculates the best path for a given request. 5) *Request Manager*: users intract with this component to generate requests and recive replies.

3.3   Protocol

A node may share its values with other trusted friends. Conversely a node may guess these values of others, whether friend or not. SOR is different from internet routing protocols. In SOR, each individual is a router (forwarder) but has multiple circles of trusts. This is because any advertisement affects the privacy and other social factors. Individuals in OSN have to feel that choosing the best path is under their hands, not totally in the routing protocol's hands. Thus its circle consists of the circle of adjacent neighbors, the circle of social priority sharing, and the circle of load information sharing as explained In Figure 1 (B).   SOR uses *path vector* type mechanism to propagate SMMs, link state type mechanism to distribute QMMs, and flooding type mechanism to propagate CMMs. These propagation mechanisms are encrypted in a way similar to the one implemented in Tor and Ripple [5]. Moreover, the Request Message (RM) is encrypted in a way that each intermediate node can only see relevant portions (The RM's Header, see figure 2) of its fields. Each intermediate node uses its private key to decrypt the RM in order to read this portion and then forward the RM to the next neighbor. Figure 1 (C) illustrates, with an example, how SOR works in the following six steps: 1) Initially, nodes S, I, and T exchange messages (CMM, SMM, QMM) by using three different managers with different propagation and encryption techniques; 2) S determines T as target (out of the scope of this paper); 3) S's Request Manager asks Route Estimator for best path to target T; 4) Route Estimator requires information from other three managers; 5) Based on the response, which determines how much information is available, Route Estimator decide which routing algorithm must be used (section 3.4) and what is the best path to T; 6) S's Request Manager receives the path to T, and forward the request to next node.

Community Structure and ID Mechanism: Individuals inside OSN are organized in communities in SOR. We propose a hierarchical *community based identity mechanism* (CIM) which is used to determine individual's membership to specific communities. Each individual can be a member of multiple communities, but one identity is allowed to send in a message. The idea is simple, but it is efficient and effective in social routing. We omit the details of this mechanism, due to lack of space.

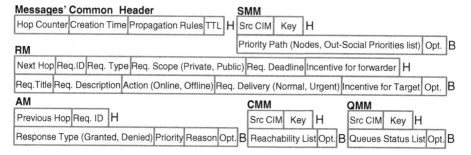

Fig. 2. SOR Messages where H (Header) and B (Body) and keys (discussed in Section 3.1)

3.4   SOR Routing Algorithms

In SOR, forwarding process is based on Social Priority (SP) and request queue. The routing is computed by using information received via CSP, SPB_s, or SPB_D. However, due to limited trust domains - a node may estimate paths for their requests with best available partial information. As a result, there can be algorithms at three cognoscenti' levels:

Conventional Shortest-Path-Based routing algorithm (CSP): is an algorithm that uses only connectivity information collected by Connectivity Manager. It is based on Dijkstra's algorithm. In our simulation it uses only hop count. It emulates classical routing.

Static Social-Priority-Based routing algorithm (SPB_S): is an algorithm that adds SP which is collected by Social Priority Manager. It uses Prioritized metric (the minimum sum of the SP between source and destination) to evaluate the best path for a request to travel. It is also based on Dijkstra's algorithm. In this static version, the queuing load is assumed to be zero (or constant at snapshot). Same network knowledge is used while estimating requests r_1, r_2, and r_3.

Dynamic Social-Priority-Based routing algorithm (SPB_D): is an algorithm that also adds dynamic queue status information with special priority collected by Queue Manager and Social Priority Manager respectively to evaluate the best path for a request to be assigned. It is simply a modified dynamic version of Dijkstra's algorithm. After assignment of each request (r_1), the queue load is updated between computing best path for the subsequent request (r_2).

4   Experimental Results

For our experiment[1] we use real-world Google+ circles from Jure Leskovec's Website [7], in order to simulate real social environment. See Table 1 for some statistical information of the five used datasets. An experiment has been conducted for the three routing algorithms (CSP, SPB_S, and SPB_D). Figure 3 (a) shows the percentage distribution of social priority for five datasets. We grouped the social priorities values in ten intervals 0-0.1, 0.1-0.2, 0.2-0.3 and so on. The horizontal axis is labeled with the class intervals. The shape of DS-1 distribution has a bell-shaped curve, asymmetrical about its highest point 0.6-0.7. In addition, DS-2 and DS-3 have skewed left "negatively" distributions and DS-4 and DS-5 have other distributions. Figure 3 (b) compares the average end-to-end routing delay of the three algorithms over five different social graphs of google plus. In general, SPB_D algorithm performs the best and CSP algorithm performs the worst among the three algorithms. While in case of small scale social graphs SPB_S algorithm is very close to the performance of CSP, it performs better in large scale social graphs. That is probably because of the lack of paths between sources and destinations in small social graphs, and large number of paths in case of large scale social graphs.

---

[1] We have designed a new OSN simulator using OMNeT++.

Table 1. Statistical information of Google+ datasets

Statistical Information	DS-1	DS-2	DS-3	DS-4	DS-5
#nodes	54	116	342	1648	2211
#directed edges	252	1255	4176	166291	93509

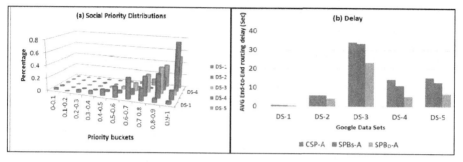

Fig. 3. (a) Percentage of Social Priority and (b) delay metric

5   Conclusion

This paper discusses SOR - a novel social routing protocol for OSNs. In order to estimate social priority, SOR uses social characteristics based metrics such as Gender-based social priority, Degree-based social priority, and so on. It also offers individual nodes the opportunity to disseminate basic connectivity, social priority and operation states separately, and to varying trust circles. Like real social network - the optimality (effort spent) of communication will depend on the overall trust level in the society.

References

1. Tim, A., Danescu-Niculescu-Mizil, C., Jurafsky, D.: How to Ask for a Favor: A Case Study on the Success of Altruistic Requests. arXiv preprint arXiv:1405.3282 (2014)
2. Goldschmidt, M.M.: Do me a favor: A descriptive analysis of favor asking sequences in American English. Journal of Pragmatics **29**(2), 129–153 (1998)
3. Goldschmidt, M.M.: For the favor of asking: an analysis of the favor as a speech act, pp. 35–49 (1980)
4. Virginia, K., Laub, A.J.: The singular value decomposition: Its computation and some applications. IEEE Transactions on Automatic Control 25(2), 164–176 (1980)
5. Jan, M.: Security and routing in the ripple payment network. Master's thesis, Masaryk University, Brno (2011)
6. Modern etiquette: asking for a favor. http://www.designsponge.com/2013/01/modern-etiquette-asking-for-a-favor.html
7. Stanford Large Network Dataset Collection. http://snap.stanford.edu/data/index.html

Resilience of Criminal Networks

Fatih Ozgul[1]([⊠]) and Zeki Erdem[2]

[1] Faculty of Security Sciences, National Police Academy,
06100 Golbasi, Ankara, Turkey
fatih.ozgul@istanbul.com
[2] TUBITAK BILGEM, Information Technologies Institute,
41470 Gebze, Kocaeli, Turkey
zeki.erdem@tubitak.gov.tr

Abstract. In this paper we presented resilience measure for criminal networks which is tested on two real criminal networks. We investigated resilience results in parallel with their activities, their recruitment policy, and growth of network, their survival strategy and secrecy after they are prosecuted.

Keywords: Criminal networks · Network resilience · Secrecy · Survival of networks

1 Introduction

Average Path Length (APL)[5] is a concept in network topology that is defined as the average number of steps along the shortest paths for all possible pairs of network nodes. It is a measure of the efficiency of information or mass transport on a network. APL of a criminal network shows average walk distance between every distant member. This measure shows the diffusion of communication and information sharing. APL is good for measuring information flow in the network. Clustering coefficient (CC)[5] is a measure of the degree to which nodes in a graph tend to cluster together. Clustering coefficient shows how much a network is tend to be in cliques of members. This is desirable for a network in terms of secrecy because in a clustered topology less individual members are exposed to information and communication. Centrality [3] refers to indicators which identify the most important vertices within a graph. Average Centrality of Leaders (ACL) in a criminal network shows how much authority is in the hands of leaders in a network setting. This measure shows the degree of control is in the hands of leaders. Since Number of Members (NM) and links expand in a robust network, growth of a social network shows the robustness in time. Average Centrality of Leaders (ACL), Average Path Length (APL), Number of Members in criminal network and Clustering Coefficient (CC) [5] are the measures required for calculating network resilience. Maximum number of members, low path length, high authority and high clustering co-efficient are desirable scores for better secrecy versus information and authority trade-off. Resilience of a network is dependent on qualities of high authority, robustness, reachability, and secrecy. Using these qualities, Criminal Network Resilience (CNR) can therefore be defined as;

$$CNR = log\ [(ACL*CC*NM)\ /\ (APL)]$$

© Springer International Publishing Switzerland 2015
N. Agarwal et al. (Eds.): SBP 2015, LNCS 9021, pp. 400–404, 2015.
DOI: 10.1007/978-3-319-16268-3_50

2   Operation Cash Network

We conducted experiments on two theft networks, Operation Cash and Ex-inmates Networks. We used an R package called Tnet [4] for measuring their network resilience in seven years' time. The first network, Operation Cash was finally prosecuted by Bursa Police Department, Turkey in 2007 [2]. This network was first detected in 1999 (figure 1 left) with two members (i.e. #12113, #41211) who are one ex-inmate and one handler of stolen goods. In 2000 and 2001 a new recruit is added and they conducted several thefts together (figure 1). From 2005 to 2007 they also recruited A new young member in 2005 but in 2006 there were also new five members and in 2007 one of the creators of this network was not as influential and a new recruit (i.e.#220868) was more central then any other network member. After many operations in 1999, 2001, 2003, 2005, 2006 a final operation was made by the police in 2007 and the network is collapsed.

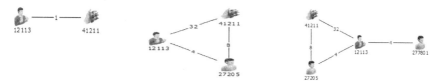

Fig. 1. Operation Cash Network in 1999 (left) in 2001(centre) and in 2005(right)

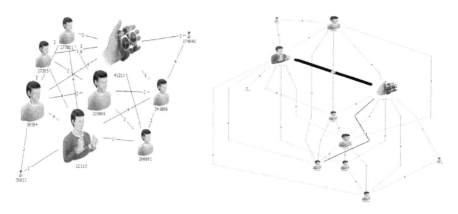

Fig. 2. Operation Cash Network in 2006 (left) and in 2007 (right)

Possible reason of the police operations were high visibility and intelligence about the amount of cash they had from selling stolen goods. The leaders (i.e. #12113, #41211) had loose connections of new recruits; they have lost the control in time, after the police raids. Based on the experiment results, this theft network was weak until 2005 and it is evolved from 2005 to 2007 with a better resilience. The best action for network resilience was creating more clusters and using some recruits to manage these clusters while taking control of network by initial leaders. This ensured the layered secrecy and gave higher resilience.

3   Ex-Inmates Network

The second criminal network is a theft network from Bursa, Turkey, operated between 1997 to 2005 [1]. It started with two ex-inmates, who were released and they were back to business. They developed weak connections with other local theft networks and finally they succeeded to connect all these small networks into a big hub. Figures 3, 4, 5 and 6 (from left to right) show how they emerged and got bigger in time.

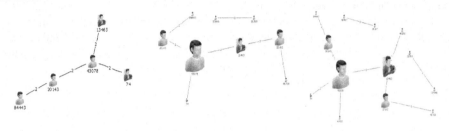

Fig. 3. Ex-Inmates Network in 1997(left) in 1998 (centre) and in 1999(right)

Two ex-inmates (i.e. #74 and #13463) started working with different new recruits. Both preffered working with the same member (i.e. #43078) in 1997 but these new recruits also worked with other new members. Ex-inmates never put off semi-independent working of new recruits. One of the ex-inmates (i.e. #13463) preserved his central role in the network while the other (i.e. #74) was more peripheral, the first recruit (i.e. 43078) was one of the two leaders in the network. In 1999 and 2000 these two leaders tried to get connected to small pairs of thieves into the network. In 2001 and 2002, two ex-inmates regained their authority together, while another recruit (i.e. #55788) played a leader role for the half of the network and continued recruitment fast. By 2003 and 2004 the network growed so fast even the ex-inmates were getting more peripheral and lost control of the new recruits. Instead new leaders emerged from previuos members.

Fig. 4. Ex-Inmates Network in 2000 (left) in 2001(centre) and in 2002 (right)

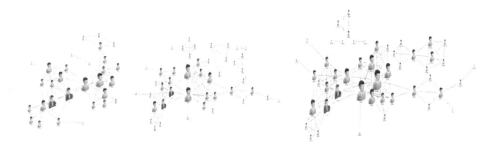

Fig. 5. Ex-Inmates Network in 2003(left) in 2004 (centre) and in 2005 (left)

Fig. 6. Ex-Inmates Network in 2006(left) and in 2007 (right)

By 2005 and 2006, there were many subgroups and smaller cliques but most of them are connected to the big network anymore. As shown in figure 6, there were five subgroups and many cliques in the network but ex-inmates were in the central position of overall network.

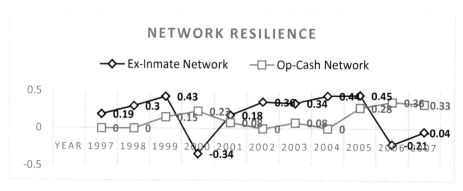

Fig. 7. Resilience of two criminal networks in time

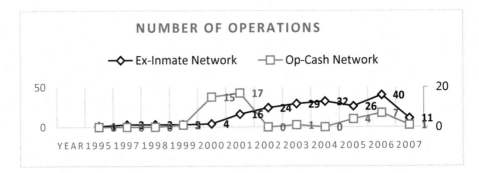

Fig. 8. Number of Police Operations made againist two criminal networks

4   Conclusion

Based on the experimental results, criminal network resilience for two selected networks is presented in Figure 7 and number of operations made by the police againist the network members in Figure 8. It is clear to see how the operations push down the network resilience. Operation Cash network revived itself in time and has kept its trade-off balance between robustness, recruitment and secrecy. Ex-inmate Network was not stable for its recruitment policy. Due to risk of uncontrolled enlargement and lack of control. Operation Cash Network is smaller but it survived after police raids and several prosecutions against itself. It never dropped to minus zero level. After operations it developed its resilience after 2005. It was almost collapsed in 2000 and 2006 after harsh operations made against them. By the end of 2007, it was still in crisis, resilience level was still minus. Its vulnerability for unplanned recruitment and lack of secrecy caused more and more police raids and operations and finally the collapse of the network in 2007.

References

1. Ozgul, F.: Mining Crime Data for Detection, Prediction and Similarity of Criminal Networks, Ph.d. Thesis. University of Sunderland, UK (2010)
2. Ozgul, F., Bondy, J., Aksoy, H.: Mining for offender group detection and story of a police operation. In: Proceedings of the Sixth Australasian Conference on Data Mining and Analytics, vol. 70, pp.189-193. Australian Computer Society, Inc. (2007)
3. Everett, M.G., Borgatti, S.P.: The Centrality of Groups and Classes. Journal of Mathematical Sociology (1999)
4. Opsahl, T.: R Tnet Package. [http://toreopsahl.com] (accessed on November 28, 2014)
5. Wasserman, S., Faust, K.: Social Network Analysis Methods and Applications. Cambridge University Press, Cambridge (1994)

Modeling Group Dynamics for Personalized Group-Event Recommendation

Sanjay Purushotham$^{(\boxtimes)}$ and C.-C. Jay Kuo

University of Southern California, Los Angeles, California, USA
spurusho@usc.edu, cckuo@sipi.usc.edu

Abstract. Event-Based Social Networks (EBSNs) such as Meetup, Plancast, etc., have become popular platforms for users to plan and organize social events with friends and acquaintances. These EBSNs provide rich *online* and *offline* user interactions, and rich event content information which can be leveraged for personalized group-event recommendations. In this paper, we propose collaborative-filtering based Bayesian models which captures group dynamics such as user interactions, user-group membership etc., for personalized group-event recommendations. We show that modeling group dynamics learns the group preferences better than aggregating individual user preferences, and that our approach out-performs popular state-of-the-art group recommender systems. Moreover, our model provides interpretable results which can be used to study the group participations and event popularity.

Keywords: Personalized group recommendation · Group dynamics · Event-based Social Networks · Bayesian models · Collaborative filtering

1 Introduction

Growth and popularity of online social networks has changed how users connect and interact with friends and family in today's internet age. Event-based Social Networks (EBSNs) such as Meetup, Plancast, Eventbrite etc., have become popular convenient platforms for users to co-ordinate, organize and participate in social meet-ups/events and share these activities with their friends and family. Event recommendation in EBSNs has been recently studied in the past couple of years [3], [8], [11], for recommending events or to recommend event-sponsoring groups to the EBSN users. Most of the previous works were designed for single user recommendations by making use of user event participation information. However, many users use EBSNs to organize personal group activities (such as dining with friends, etc.) since EBSN services provide easy to use online interfaces, and they provide rich user interaction and networking options. We believe that EBSN provides a natural platform for studying personalized recommendation of events to group of users i.e. group-event recommendations. Recommendation to groups (for example, recommending a movie for friends to watch together), in general, is a very challenging problem, since the users of the group

© Springer International Publishing Switzerland 2015
N. Agarwal et al. (Eds.): SBP 2015, LNCS 9021, pp. 405–411, 2015.
DOI: 10.1007/978-3-319-16268-3_51

may or may not share similar tastes, and user preferences may change due to other users in the group. Therefore, it is important for recommender systems to capture group dynamics such as user interactions, user group membership, user influence etc. for personalizing group recommendations. Personalized event recommendation for groups is possible for EBSNs since they provide rich social network information in terms of *online* user interactions and *offline* user participations, and rich event information (location tags, time-stamps, group sponsoring the event, etc.) which help in accurate modeling of group dynamics. Thus, modeling and mining of EBSNs help us to study research issues of group dynamics and group recommendation by leveraging the social and event characteristics of users and locations.

In this paper, we present a novel collaborative filtering based Hierarchical Bayesian model for personalized group-event recommendation in EBSNs, and study how modeling group dynamics affects group-event recommendation. Our contributions include: (1) proposing novel probabilistic modeling framework for capturing group dynamics for group-event recommendation, (2) Studying the group and event characteristics of a large EBSN, and (3) Handling data sparsity and cold-start recommendation challenges associated with group recommenders.

2   Related Work

There is abundant body of research on Group recommendation, however, there is limited previous work on personalized group-event recommendations in EBSNs. Recently, [10] proposed a probabilistic approach to model group activities of online social users using generative processes; though the final group recommendation is done by aggregation of user preferences without directly learning group preferences. In our previous works [6], [7], we proposed a probabilistic approach, where we model groups and activities as generative processes to capture user-group interactions and location semantics respectively, to perform group-activity recommendations in Location-Based Social Networks (LBSNs). However, in our previous works, we investigated group recommendation in LBSNs, and we did not study group-event recommendation in EBSNs which we feel requires dedicated study since EBSNs and LBSNs are quite different and have different characteristics [4].

In this paper, we present a hierarchical Bayesian model to incorporate group dynamics such as user's offline social interactions and user-group membership into our probabilistic framework, and study their effect on personalized group-event recommendation in EBSN. In the following sections, we present our model, and report our experimental results on a real-world large EBSN (Meetup) dataset.

3   Our Approach

We first define the group-event recommendation problem in section 3.1 and then we present our proposed Personalized Group-Event Recommender model in section 3.2.

3.1   Problem Statement

Let U, E, G_{on}, G_{off} represent the set of Users, Events, Online Social Groups and Offline Social Groups of the EBSNs. As the name indicates 'Online Social Groups' correspond to groups whose users interact online, i.e. these groups arise due to the online social interactions. On the other hand, 'Offline Social Groups' corresponds to groups of users who physically meet and participate in events organized by members of online social groups. Offline social group users interact at a location during a particular interval of time while participating in a social event. Mathematically, Online and Offline social groups correspond to the connected components of the Online and Offline Social Network Graphs [4].

The problem of group-event recommendation is defined as recommending a list of events that the users of offline social groups may participate in. It is related to the group-event rating prediction task where the group's event participation is predicted as (implicit) group-event rating. Throughout this paper, we consider the 'Offline social groups' as 'groups' for our group-event recommendation task since they provide rich event content and user interactions to model and study group dynamics for personalization of group recommenders.

3.2   Personalized Group-Event Recommender (PGER)

In this section, we briefly discuss our proposed model- Personalized Group-Event Recommender (PGER), shown in Figure 1. Our model is a generalized hierarchical Bayesian model which jointly learns the group and event latent spaces. We use topic models based on Latent Dirichlet Allocation (LDA) [1] to model the descriptions of events and the groups, and we use matrix factorization to match the latent features of group to the latent features of events. Our model fuses topic models with matrix factorization to obtain a consistent and compact feature representation. We introduced this model in our previous work [7]. In this paper, we adopt the same model for studying Group-event recommendation in EBSN. We urge the readers to refer to our previous work [7] for detailed description of our algorithm, the parameter learning and the definitions for prediction tasks (in-matrix and out-matrix prediction).

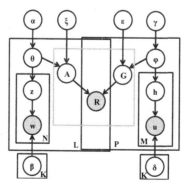

Fig. 1. Our proposed model -Personalized Group Event Recommender (PGER)

4   Experiments

We conduct several experiments to evaluate our model on a EBSN dataset for group-event recommendation. Our experiments help us to answer the following key questions: (a) What are the group and event characteristics of EBSNs? (b) How does our model perform when compared to the state-of-the-art group recommenders? (c) How to interpret the group preferences learned by our model?

4.1   Dataset Description

Our experiments were conducted on Meetup dataset [4] which is a real-world EBSN. We study the event & offline group characteristics and group-event locality properties of EBSNs on the Full Meetup dataset, while we do performance comparsions using a small Meetup dataset (Small dataset was chosen due to limited computing power, however, our model works for larger dataset). Table 1 shows the description of these datasets. Due to space constraints, we don't discuss the group and event characteristics of Meetup dataset[1]. To make our

Table 1. Meetup Dataset Description

Dataset	Meetup (Full)	Meetup (Small)
#Users	4,111,476	3,650
#Events	2,593,263	27,244
#Online Groups	70,604	454
#Offline Groups of size 2	345,998	2,000
#Offline Groups of size 2 at atleast 10 events	126,141	511
#Events attended by Offline groups of size 2	790,547	9510

analysis and experiments on group-event recommendation more sound, we considered only the offline social groups who have participated in atleast 10 events.

4.2   Experimental Settings

We split Meetup dataset into three parts - training ($\sim$80%), held-out ($\sim$5%) and test datasets ($\sim$15%). The model is trained on training data, the optimal parameters obtained on the held-out data and ratings are predicted for the test data. We ran 5 simulations for all our comparison experiments. For performance evaluation, we consider three metrics, namely: (1) Average Group-Event Rating Prediction Accuracy (Avg. Accuracy), (2) Average Root Mean Squared Error (Avg. RMSE) and (3) Average Recall. We define the avg. accuracy for test data as the ratio of correctly predicted ratings compared to the total ratings in test data. RMSE is given by: $RMSE = \sqrt{\frac{1}{|T|}\sum_{(i,j)\in T}(\hat{R}_{ij} - R_{ij})^2}$ Where $\hat{R}_{ij}$ is predicted ratings of group-event pairs (i,j) for a test set T, and R_{ij} are

---

[1] The group and event characteristics is discussed in the longer version of this paper which is available at $http : //www-scf.usc.edu/\sim spurusho/$

true ratings. 'Recall' only considers the events participated within the top M suggestions. For each group, we define the *recall@M* as the ratio of number of events the group participates in Top M suggestions to the total events the group participates in.

4.3   Evaluated Recommendation Approaches

We compare our PGER with the following popular and state-of-the-art recommendation systems: (1) Matrix Factorization (MF) [5](2) Collaborative Topic Regression (CTR) [9] (3) Aggregation methods [2]: We considered the following popular aggregation methods: (a) Least Misery method (b) Most Pleasurable method (c) Averaging method, (d) Plurality Voting.

4.4   Results

In this section, we present our experimental results and answer the questions raised in section 4.

Table 2. Performance Comparison on Test Data

	Best Aggregation method (Averaging)	MF [5]	CTR [9]	PGER (Ours)
In-matrix Avg. Accuracy	0.870(0.10)	0.881(0.09)	0.915(0.05)	**0.955(0.03)**
In-matrix Avg. RMSE	0.359(0.12)	0.329(0.11)	0.282(0.08)	**0.205(0.06)**
Out-matrix Avg. Accuracy	-	-	0.794(0.16)	**0.908(0.08)**
Out-matrix Avg. RMSE	-	-	0.441(0.14)	**0.298(0.1)**
Avg. Recall@10 (K=50)	0.49	0.532	0.583	**0.792**

Performance Comparisons

Table 2 shows that our approach (PGER) outperforms all the state-of-the-art models in both in-matrix and out-matrix prediction tasks. Our model performs better than the state-of-the-art models by an impressive 20% in terms of recall@10 metrics. Variance of avg. accuracy and RMSE is shown in brackets. All our experiments were run on intel quad-core 2.2 GHz CPUs with 8 GB RAM.

4.5   Learned Group Preferences vs. Aggregating User Preferences

We compare our learned group preferences w.r.t aggregating user preferences using two scenarios. In first scenario, Group I has users who have similar user preferences and in second scenario, Group II has users who have different user preferences. From table 4.5 we observe that aggregating user preferences may recommend events that may not be relevant for their groups, while directly learning group preferences (by capturing group dynamics) will recommend events that are similar to the events that the group has participated in. Note: Event tags (provided in the dataset) are used to interpret the event-topics and the learned group preferences.

4.6   Examining Latent Spaces of Groups and Events

We can interpret our model's group recommendations by studying the latent spaces of the groups and events. Table 4 shows top 3 group preference topics for an example offline social group (say Group I). Group I has users who are interested in fantasy literature, fitness and adventure related events, and most of the recommended events belong to these topics. One advantage of our model is that it can predict if an event will become popular for groups (i.e. if an event will have more group participants) by studying the offsets of the event-topic proportions. An event whose topics have large offsets indicates that many groups (of different group sizes) will take part in that event. An example for such an event is a hiking trip organized by a school's travel club, or an author book reading session hosted by a book club.

Table 3. Learned Group Preferences vs. Aggregating User Preferences

	Group I	Group II
Learned group preferences	Book-club, Spirituality, Adventure	Sports, Fitness, Business Networking
Aggregating user preferences	Board games, Religion, Spirituality	Politics, Movies, Book club
Events participated by Groups	Harrypotter reading session, Meditation, Hiking	Baseball, Yoga, Conference

Table 4. Latent Topics for an Offline Social Group. We list the top 5 events recommended by PGER. Last column shows whether the group participates in the event.

	Group I	Event participation
Top 3 topics (top 5 words)	1. bookclub, sci-fi, harrypotter, kidlit, fantasy-literature 2. wellness, spirituality, self-empowerment, Yoga 3. nightlife, travel, adventures, dance, singles	
top 5 event recommendations	7466, 8994, 10298, 13200, 16194 (harrypotter, sci-fi, Yoga, book-club, hiking)	Yes,Yes,Yes,Yes,Yes

5   Summary and Future Work

In this paper, we presented Collaborative filtering based Hierarchical Bayesian Model that exploits event tag information, user interactions and user group memberships to learn group preferences and to recommend personalized group events. Our experiments on Meetup datasets showed that our model consistently outperforms the state-of-the-art group recommmender systems. Our framework models the group dynamics and allows us to address cold-start recommendation for new events. For our future work, we will study how to leverage online social network structure for improving group-event recommendations. We will study the impact of model parameters and investigate how to make our algorithms scalable to the ever-growing EBSNs.

References

1. Blei, D.M., Ng, A.Y., Jordan, M.I.: Latent dirichlet allocation. JMLR (2003)
2. Cantador, I., Castells, P.: Group recommender systems: new perspectives in the social web. In: Pazos Arias, J.J., Fernández Vilas, A., Díaz Redondo, R.P. (eds.) Recommender Systems for the Social Web. ISRL, vol. 32, pp. 139–157. Springer, Heidelberg (2012)
3. de Macedo, A.Q., Marinho, L.B.: Event recommendation in event-based social networks. In: Hypertext, Social Personalization Workshop (2014)
4. Liu, X., He, Q., Tian, Y., Lee, W.-C., McPherson, J., Han, J.: Event-based social networks: linking the online and offline social worlds. In: ACM SIGKDD (2012)
5. Mnih, A., Salakhutdinov, R.: Probabilistic matrix factorization. In: NIPS (2007)
6. Purushotham, S., Kuo, C.-C.J.: Studying user influence in personalized group recommenders in location based social networks. In: NIPS Personalization (2014)
7. Purushotham, S., Shahabdeen, J., Nachman, L., Kuo, C.-C.J.: Collaborative group-activity recommendation in location-based social networks. In: ACM SIGSPA-TIAL, GeoCrowd (2014)
8. Qiao, Z., Peng, Z., Zhou, C., Cao, Y., Guo, L., Zhang, Y.: Event recommendation in event-based social networks. In: AAAI (2014)
9. Wang, C., Blei, D.M.: Collaborative topic modeling for recommending scientific articles. In: ACM SIGKDD (2011)
10. Yuan, Q., Cong, G., Lin, C.-Y.: Com: A generative model for group recommendation. In: ACM SIGKDD (2014)
11. Zhang, W., Wang, J., Feng, W.: Combining latent factor model with location features for event-based group recommendation. In: ACM SIGKDD (2013)

Determinants of User Ratings in Online Business Rating Services

Syed A. Rahman, Tazin Afrin, and Don Adjeroh[✉]

West Virginia University, Morgantown, WV 26506, USA
donald.adjeroh@mail.wvu.edu

Abstract. We investigate the key determinants of user ratings in social media-based business rating services. Our hypothesis is that beyond factors internal to a user, external factors, such as the direct and indirect influence of other users, and environmental factors beyond the control of the user have a significant role in determining the actual rating assigned by a user for a given service. To test this hypothesis, we used data from Yelp, and attempted to predict user ratings on location-based services, in particular food and restaurant business. Our results show improved prediction performance over the baseline, with improved robustness to rating variability and rating sparsity.

Keywords: Review rating · Weather · Random forest · Yelp · Business rating

1 Introduction

Social learning and online review databases are increasingly supplementing expert opinions as the basic source of data in assessing the quality of a product, service, or business entity see [3,7,16]. This in turn has led to an exponential growth in the quantity and diversity of available rating data, and hence the ability for machine learning and data mining algorithms to improve on their rating predictions e.g., using collaborative filtering [1,7,16]. At the same time, with increased user participation comes the question of quality of the data, credibility of the ratings, and difficulty in dealing with the diversity in user-generated content. There is a need for a careful consideration of the factors that could impact the rating that a given user will assign for a given service or business.

Various factors can have an impact on the rating a given user assigns to a service provider (business). These can range from internal factors directly connected to the specific encounter or experience and satisfaction (otherwise) with the provided service, to external factors not directly connected to the experience the user had. We are motivated by recent efforts how a careful analysis of available data can expose unexpected connections between people, human behaviors, or other attributes [4,5,8,15].

Table 1 shows the correlation coefficients for the ratings for the three business categories (Food and Restaurants, Health, and Shopping) in 8 cities (with large colleges) in the US (within and between cities) using data from Yelp.com.

© Springer International Publishing Switzerland 2015
N. Agarwal et al. (Eds.): SBP 2015, LNCS 9021, pp. 412–420, 2015.
DOI: 10.1007/978-3-319-16268-3_52

Ratings from users in the same city seem to be correlated, independent of the particular business category. The observed correlation, however, is not universal between every pair of business categories. We also observe the correlation between ratings of similar businesse categories between-cities. Thus, there is an underlying process that drives the ratings, independent of the specific experience of the user with the particular service being rated.

Table 1. Corr. in avg user rtngs for 3 biz categories(above diag: Pearson ρ, below diag:Kendall τ)

	CMU_F	CMU_H	CMU_S	UCLA_F	UCLA_H	UCLA_S	Col_F	Col_H	Col_S	PU_F	PU_H	PU_S	WAS_F	WAS_H	WAS_S	UTA_F	UTA_H	UTA_S	GT_F	GT_H	GT_S	MIT_F	MIT_H	MIT_S
CMU_F	1	0.38	0.31	0.28	0.08	0.08	0.42	0.02	0.55	0.28	0.47	0.66	-0.38	0.02	-0.05	0.17	-0.36	-0.26	-0.19	0	-0.44	0.06	-0.05	-0.09
CMU_H	0.16	1	0.3	-0.36	-0.08	-0.49	0.19	0.18	0.61	-0.31	0.2	0.21	-0.03	-0.1	0.13	-0.06	-0.18	-0.19	-0.15	0.5	-0.37	0.3	-0.33	-0.59
CMU_S	0.09	0.25	1	-0.39	-0.2	-0.28	0.18	0.26	0.27	0.13	-0.12	0.05	-0.23	0.26	0.44	0.42	-0.52	0.55	-0.19	-0.19	-0.19	-0.06	-0.47	-0.2
UCLA_F	0.12	-0.25	-0.24	1	0.36	0.66	0.24	0.02	-0.17	0.48	0.44	0.42	-0.23	0.29	-0.65	-0.26	0	-0.14	0.14	-0.04	-0.35	-0.16	0.68	0.57
UCLA_H	0.29	0.02	-0.14	0.11	1	0.66	-0.17	0.21	0.08	0.44	0.33	0.45	-0.47	-0.22	-0.18	0.45	-0.27	-0.1	0.26	0.24	0.02	-0.01	0.4	0.28
UCLA_S	0.12	-0.22	-0.3	0.58	0.41	1	-0.16	0.27	-0.54	0.41	0.17	0.47	-0.35	0.15	-0.43	0.17	-0.17	0.2	0.39	0.03	0.08	-0.17	0.59	0.5
Col_F	0.42	0.32	0.24	0.09	-0.05	-0.15	1	-0.24	0.42	-0.28	0.36	0.28	-0.12	-0.18	0.33	0.04	0.13	0.02	-0.2	-0.07	-0.54	0.55	0.39	0.08
Col_H	-0.09	0.39	0.16	-0.06	0.17	0.16	-0.13	1	-0.16	0.33	-0.25	0.46	-0.51	0.24	-0.39	0.26	0.13	0.22	0.27	0.19	-0.13	-0.07	0.03	0.11
Col_S	0.42	0.44	0.24	-0.15	0.08	-0.39	0.52	-0.09	1	-0.05	0.12	0.04	-0.12	-0.28	0.35	0.09	-0.12	-0.49	-0.41	0.29	-0.46	0.14	-0.14	-0.07
PU_F	0.12	0	0	0.45	0.29	0.39	-0.03	0.25	-0.03	1	0.06	0.32	-0.08	0.58	-0.44	0.05	-0.25	0.11	0.39	0.06	-0.03	-0.33	0.11	0.28
PU_H	0.34	0.04	-0.18	0.34	0.41	0.3	0.34	-0.23	0.14	0.02	1	0.24	-0.09	-0.23	-0.15	-0.24	-0.11	0.06	0.17	-0.01	-0.02	0.12	0.37	-0.13
PU_S	0.53	0.1	0.11	0.29	0.22	0.41	0.32	0.17	0.02	0.17	0.27	1	-0.3	0.27	-0.34	0.19	-0.05	-0.06	0.34	0.27	-0.35	0.37	0.24	-0.09
WAS_F	-0.18	-0.13	-0.12	-0.03	-0.38	-0.15	-0.03	-0.41	-0.21	-0.09	-0.18	-0.17	1	0.25	0.2	-0.59	0.16	-0.04	0.41	0.34	0.4	0.21	-0.17	-0.49
WAS_H	-0.09	0.06	0.33	0.24	-0.23	0.12	-0.12	0.31	-0.18	0.3	-0.18	0.26	0	1	-0.45	-0.16	-0.34	0.24	0.19	0.02	-0.28	-0.13	-0.25	-0.11
WAS_S	-0.06	0.03	0.3	-0.39	-0.05	-0.33	0.27	-0.22	0.21	-0.27	0.02	-0.17	0.09	-0.24	1	0.34	-0.12	0.25	-0.12	0.03	0.22	0.31	-0.16	-0.2
UTA_F	0.06	-0.06	0.24	-0.21	0.23	-0.21	0.03	0.25	0.15	-0.03	-0.02	0.02	-0.05	0.09	0.3	1	0.27	-0.12	-0.35	0.29	-0.22	-0.26	-0.2	0.24
UTA_H	-0.27	0	-0.39	0	-0.23	-0.06	0.12	0.03	-0.06	-0.12	-0.18	-0.2	0.18	-0.21	0	0.27	1	-0.12	-0.06	0.12	0.43	1	0.19	-0.1
UTA_S	-0.3	-0.22	0.3	0.09	-0.2	-0.03	0.09	0.09	-0.33	0.03	0.02	-0.05	0.3	0.27	-0.06	-0.12	0	1	-0.28	0.25	0.05	0.23	0.25	0.12
GT_F	-0.27	-0.06	-0.21	0.06	0.02	0.24	-0.24	0.25	-0.42	0.24	-0.02	0.08	0.24	0.09	0	-0.4	-0.06	-0.12	1	0.35	-0.23	0.23	0.09	-0.1
GT_H	0.09	0.36	0	-0.03	0.11	0.15	0.12	0.16	0.03	0.25	0.04	0.26	0.12	0	0	-0.4	0.12	0.21	0.18	1	0.5	0.52	0.26	-0.19
GT_S	-0.42	-0.22	-0.24	-0.15	0.05	0.15	-0.39	0.06	-0.33	-0.03	-0.06	-0.32	0.15	0.06	0.09	-0.09	0.24	0.15	0.3	-0.06	1	-0.24	-0.05	-0.15
MIT_F	0.12	0.06	0.06	-0.09	0.02	-0.09	0.21	-0.03	-0.15	-0.09	0.06	0.26	0.15	-0.3	0.21	0.15	0.06	0.15	0.3	0.31	-0.27	1	0.19	-0.45
MIT_H	-0.03	-0.16	-0.25	0.55	0.14	0.43	0.25	0.1	-0.12	0.09	0.31	0.29	-0.06	-0.09	-0.06	-0.03	0.34	0.22	0.25	0.09	0.06	0.15	1	0.6
MIT_S	-0.03	-0.19	0.03	0.36	0.08	0.24	0.06	0.03	0.12	0.06	-0.06	0.11	-0.3	0.03	0	0.12	0.15	0	-0.27	-0.25	-0.06	-0.24	0.46	1

We focus on the role of environmental factors on user ratings. In particular, in modeling user ratings for a business or service, the specific location of the business, the location of the user, and environmental factors at these locations become important considerations. Ideally, such a model should consider the environmental factors at the time the user experienced the service, the location of the service provider and user, the location of the user at the time of rating, and the environmental factors at each of the locations mentioned. In practice, we do not always have data on all these factors. Thus, we assume that both users and the service/business being rated are in the same geographic location, and that the environmental factors on the date of rating are similar to those on the date service was received.

2 Background and Related Work

User rating is an essential procedure in recommender systems (see [1] for a survey). Let $U = \{U_1, U_2, \ldots, U_n\}$ be the set of users, and $B = \{B_1, B_2, \ldots, B_m\}$ be the set of businesses, products, or services being rated by the users. Each user U_i can be described using a corresponding user profile,P_i which indicates certain characteristics of the user, such as a user ID, gender, location, income, race, age, etc. Similarly, each product B_j in B is described using a business profile Q_j, based on certain attributes, depending on the specific product or business. For instance, for a restaurant, this could include its location, types of food served, times of operation, etc. The user feedback is then often captured using a rating matrix R, whereby the element R_{ij}, $(i = 1, 2, \ldots, n, j = 1, 2, \ldots, m)$, indicates the rating user U_i assigned to product or business B_j. In reality, not all users provide a rating for every product or service. Thus, the matrix R is typically sparse, and the key challenge in recommender systems is to use available information in R to estimate the unknown ratings, and thus populate the empty cells. That is, compute the estimated rating: $\hat{R}_{ij}, \forall i, j, R_{ij} = $"unknown". Once the rating is estimated, item recommendation becomes a simple matter of suggesting the item with the highest rating to the user, or suggesting the top-k items, for the user to make the final selection.

Recommender systems have used various methods to estimate the unknown ratings, given the rating matrix R. These approaches are often classified as content-based filtering [10], collaborative filtering [20], or their combination. More recently, collaborative filtering methods have been shown to perform better than content-based filtering approaches [7], and provide better support for estimating ratings for new items, or ratings from new users [1,13]. Our work aims at generating more features which can be combined with the often limited information in the rating matrix for improved prediction. In a sense, the major purpose of this so called "feature engineering" approach [2,9] is to alleviate the problem of sparsity of the rating matrix [19]. This is akin to the idea of feature expansion used in traditional information retrieval and pattern recognition. Recent work include graph-based methods, e.g., using random walks [19], tripartite and bipartite graphs using meta-data [18], multimodal social networks [17], sentiment analysis-based methods, e.g., bag of opinions models [11], factorization machines using both ratings and review data [12], and multiple aspect ranking [6]. None of these methods seem to have paid much attention to issues such as environmental factors that can affect user ratings, the potential impact of emotional contagion, and the impact of socio-economic factors in user ratings.

3   Materials and Methods

3.1   Data Sets

Yelp Academic Dataset is a public dataset released by the business review site Yelp.com. It has about 67 million reviews for local businesses in different cities. Yelp academic dataset contains information about 30 universities - 29 in USA and 1 in Canada. It has total 330072 reviews, 13490 businesses and 130873 user information. Each business can be classified to one or more categories, e.g., food, nightlife, health, shopping, beauty etc. The dataset has information from year 2004 to 2012. In this work, we have considered 8 universities in 8 cities. CMU (Pittsburgh), Columbia (New York), MIT (Cambridge), GeorgiaTech (Atlanta), Purdue (Lafayette), UT Austin (Austin), UCLA (Los Angeles) and University of Washington (Seattle). We have considered 3 business categories Food and Restaurants, Health and Medical and Shopping, between the year 2008 to 2012.

Figure 1 shows the statistics of Yelp data for the 8 cities considered. The distribution of ratings show that the modal rating is 4, followed by a rating of 5. Also more users tend to give higher ratings, while fewer users give lower ratings.

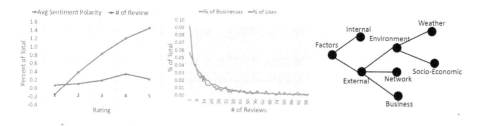

(a) #reviews & sentiments.   (b) %Biz-Usr vs #review   (c) Feature Categories.

Fig. 1. Yelp statistics and proposed feature categories

Weather Data was collected from Quality Controlled Local Climatological Data (QCLCD) from National Climatic Data Center (NCDC). NCDC has data archive for every sector of United States and is publicly available. We have used daily summary data for the 8 cities.

Socioeconomic indicators was collected from Model-based Small Area Income & Poverty Estimates (SAIPE) for School Districts, Counties, and States data from US Census Bureau. This dataset has updated estimates for poverty and income levels for local jurisdiction, at county level.

3.2 Factors Influencing User Ratings

Our view is that the rating given by a user depends primarily on the users opinion of a product, or of the service received from a business. This rating also depends on other factors, such as the mood of the user on the day of service, and on the day the service was rated, influence from friends and relatives, seasonal variations which could affect the mood of raters in the city, and socio economic status of the user, etc. We consider two broad groups of factors – internal and external factors.(See Figure 1c).

Internal Factors are directly related to the user's experience of the service provided. These capture information that are internal to the user, which we assume are encoded in the overall rating a user gives to a product or service, and by the text of the user's review, when available. We use the following features from Yelp to describe the internal attributes about a user: user's average rating ([1.0, 5.0], mean 3.6), total number of reviews from the user (mean 117) and review polarity ([-1, 1], mean 0.22). The review texts were analyzed to calculate sentiment polarity using TextBlob Python library (http://textblob.readthedocs. org/en/dev/).

External Factors are those that could affect a user rating but are not under the control of the user. We consider three broad categories of external factors - business-related, network, and environmental. Environmental factors are further divided into weather and socio-economic factors.

Business attributes are related to the business or product that is being rated or considered. These are external to the user, and are often beyond the control of the individual. From Yelp data, we use the following features to describe the business attributes: the business' average rating, total review count, and binary flag on whether the business is still operating or not, review votes for 3 categories (funny, useful, and cool), each users total votes for those categories, daily changes in average ratings, and the geometric mean of the average user rating for a given business.

Network attributes capture the influence of the users social networks and other social ties on the user's ratings. Features in this category also attempt to capture the impact of affect and mood changes attributed to the user's social contacts, for instance the issue of emotional contagion in large scale social networks [8]. A user may give a business or service a high rating just because most of his friends or friends-of-a-friend in the user's social network were doing the same. We include the cumulative rating for each business, and review count for all previous reviews of the business, for the past month, week and day from the date of individual review. Further, we construct a network of users by connecting any two users that have rated or reviewed the same business or service. We assign edge weights in the network based on the number of times two users rated the same business. We record the degree of each node (user), the average ratings from nearest neighbors, and weighted average ratings for the k-hop neighbors. Ratings are weighted using the inverse square law on the hop count following the shortest path between two nodes.

Weather attributes capture the weather and seasonal changes, including changes in length of daylight that have an impact on the peoples mood, even across cultures [5]. Given that the mood of the user can affect his or her ratings, it is therefore possible that the variations in weather can also affect user ratings. To capture potential impact from variations in weather, we used key weather-related attributes of the city, namely min, max, and average daily temperatures, temperature differences, wet-bulb temperature (indirect measure of humidity), and wind speed.

Socioeconomic attributes in this group are median household income, average poverty rates for all ages, and for two age categories. Given that we do not have direct information of this type of data from Yelp, we used data about the socioeconomic factors in the city. This is based on the assumption that for business such as restaurants and fast food, most of the customers (raters) will be local to the business. Thus, we can use the average of these attributes for the given city.

Figure 2 and Figure 1a shows the empirical evidence of the impact of selected attributes on user ratings or on user review counts. An observation is the interplay between the number of ratings, average ratings, and the sentiment polarity from user reviews (Figures 1a and 1). We notice that with negative sentiments or low positive sentiments, the ratings are also low, and only few people are providing ratings with very negative sentiments. Higher sentiments tend to correspond to situations when more people are providing reviews, and often with higher ratings. This is not surprising, given that the sentiment can be used as a surrogate for the user mood, and users in a good mood (more positive sentiments) are more likely to provide a review, and are likely to give a high rating

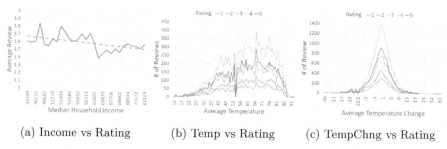

Fig. 2. Feature classes and influence on user ratings

when they do. We can also observe the subtle impact of socioeconomic factors (see Figure 2 (a)) and of weather (see Figure 2 (b, c)) on both the number of ratings, and the actual ratings. In general, the number of ratings is highest when the weather change is minimal, but decreases quickly with increasing changes in the weather (temperature, or wetbulb temperature). Similarly, the highest number of reviews occur at temperatures close the 60-70 degrees (Fahrenheit), and tend to decrease at very low temperatures or very high temperatures (see Figure 2). A similar trend is observed for the ratings, since the average ratings tend to mirror the number of ratings (Figure 1).

3.3   Modeling User Ratings

Initially we applied **Linear Regression (LR)** to predict the review ratings. It has a reasonable performance with 0.681 correlation with the predicted ratings and the mean absolute error is 0.692. But there might exist non-linear relationship between different attributes and review. Thus we applied **Generalized Additive Model (GAM)**. The result is slightly better ($\rho = 0.682$) for some cases but does not improve significantly with GAM. We then applied **Random Forest (RF)** regression to predict the user ratings. Random Forest [14] is an ensemble learning algorithm that de-correlates the decision trees at each split. The key is that, trees are repeatedly fit to subsets of the observations. A portion of the dataset is used for training, while the remaining data (called out-of-bag observations) are used for testing. In general, when more trees are considered, the correlation of the predicted model with the true value increases and the error decreases. We observed that beyond 13 attributes, the error does not really improve with adding more attributes for building the trees. Thus we set the number of random attributes to be selected at each decision node to 13. After careful considerations we used 100 as the number of trees. Random Forest regression has better results for all the datasets ($\rho = 0.70$, $MAE = 0.65$, which is reasonably better than the other two regression models).

4   Results

To evaluate the impact of the features identified, we performed rating prediction using Yelp data, based on the prediction models described in the previous section.

We used the results based on only user ratings and business ratings as the baseline, and sequentially evaluated the impact of the feature classes identified. We evaluate the prediction performance using correlation, MAE, RMSE. Our experiments were performed using the R package (http://www.r-project.org/).

4.1 Effectiveness of Identified Feature Classes

Table 2 shows the results obtained for different combinations of the identified feature categories. Individually, the internal feature category and the business feature category produced the best results. The other feature categories were much less effective, when taken singularly. However, an overall best result was obtained when combining all the feature categories. The table also shows the effect of different prediction methods, indicating that the Random Forest approach produces the best overall results over the baseline model. We can see that the new features introduced have a definite impact in reducing the rating prediction errors, while improving the correlation.

Table 2. Selected Features

Table 3. Impact of weather attributes on RMSE (All-using all features, woW-Prediction without weather attributes)

Features	Corr.	MAE	RMSE
Internal	0.604	0.761	0.951
Business	0.605	0.747	0.950
Network	0.187	0.965	1.172
Weather	0.040	0.996	1.192
SocioEcon	0.057	0.994	1.191
All(RF)	0.707	0.657	0.842
All(LR)	0.681	0.692	0.873
Baseline	0.576	0.765	0.975

	UTA			CMU			MIT			PUR			UCLA			WAS		
	All	woW	%diff	All	woW	%diff	All	woW	%diff	All	woW	%diff	All	woW	%diff	All	woW	%diff
Jan	0.74	0.83	11.24	0.8	0.84	5.16	0.89	0.89	0.86	0	0.33	-	0.75	0.77	1.87	0.77	0.8	4.23
Feb	0.76	0.78	1.94	0.81	0.83	2.62	0.82	0.83	1.39	0	0.34	-	0.84	0.84	1.01	0.79	0.81	2.52
Mar	0.82	0.83	1.78	0.72	0.77	6.44	0.84	0.85	0.6	0	0	-	0.81	0.81	1.07	0.8	0.84	5.01
Apr	0.71	0.73	3.9	0.75	0.79	4.29	0.86	0.87	1.74	0.44	0.57	29.3	0.84	0.85	0.85	0.74	0.76	2.56
May	0.79	0.82	3.36	0.65	0.7	7.33	0.83	0.84	0.78	0.54	0.68	26.17	0.85	0.87	2.23	0.75	0.78	3.96
Jun	0.86	0.87	1.75	0.75	0.78	4.02	0.88	0.88	0.39	0.51	0.64	24.5	0.9	0.92	1.75	0.88	0.91	3.38
Jul	0.8	0.81	2.14	0.73	0.75	3.82	0.88	0.88	0.38	0.44	0.52	18.37	0.84	0.85	1.85	0.81	0.84	3.19
Aug	0.8	0.82	2.25	0.72	0.76	5.58	0.86	0.87	1.22	0.5	0.67	32.76	0.81	0.82	1.49	0.77	0.78	2.19
Sep	0.76	0.79	4.1	0.77	0.81	5.2	0.84	0.85	1.07	0.53	0.57	7.21	0.8	0.83	3.59	0.87	0.89	1.71
Oct	0.86	0.87	2.12	0.79	0.81	2.36	0.89	0.9	1.23	0.38	0.51	36.65	0.79	0.8	1.3	0.83	0.84	1.35
Nov	0.84	0.9	6.7	0.74	0.77	4.38	0.89	0.9	0.86	0.44	0.65	47.11	0.83	0.85	2.68	0.76	0.79	3.72
Dec	0.73	0.82	11.82	0.54	0.59	9.04	0.82	0.82	0.79	0.5	0.61	23.09	0.81	0.83	2.34	0.77	0.81	4.36
Avg			4.42			5.02			0.94			27.24			1.84			3.18

4.2 Feature Class Significance

While the internal and business feature categories were significant in each city, the environmental factors did not seem to have a universal impact in each city (data not shown). For instance, weather factors seemed to be more significant in Lafayette, IN (Purdue University) and Pittsburgh, PA (CMU and University of Pittsburgh), while social economic factors were significant only in New York City (Columbia University). A clearer picture on the impact of weather is shown in Table 3. When we consider the data on monthly basis for some months, the impact of weather could be very significant. The challenge is in how we exploit the observed impact.

4.3 Robustness

Robustness to Variability. We evaluated the robustness to variability using the method in [18], by considering variability bands in the user rating, and assessing the prediction error within each band. Thus, for each business we computed the standard deviation of its ratings, and then sorted the business based on this value. We then divided the businesses into 10 groups, and computed the average RMSE values for the business in each group, for the three prediction approaches. The results are shown in Figure 3. The figure shows that, while the prediction errors generally increase with increasing standard deviation, the proposed methods still maintained relatively lower prediction errors across the range, when compared with the baseline model.

Robustness to Sparsity – User. Using a similar approach, we investigate the impact of sparsity in the rating matrix. Here, the users are considered in groups, based on their number of ratings. Thus, those with low number of ratings should be more difficult to predict, when compared with those with a larger number of ratings. The results of this experiment are presented in Figure 3. Again, the proposed method performed better than the baseline on every sparsity band.

Fig. 3. Robustness to variability and sparsity. (Results based on RMSE)

Robustness to Sparsity – Business/Product. Figure 3 shows the results of a similar experiment as the above, but grouping based on the total number of business ratings rather than user ratings. The proposed method using the combined internal and external features under the random forest regression model produced the best overall result.

5 Discussion and Conclusion

We have argued for a consideration of different categories of features in analyzing the determinants of user ratings, especially in location-based businesses and services. In particular, we consider environmental and social network related factors as being important in accurate modeling of user ratings for such services. Our initial results show that this could be a promising direction for further studies. However, various issues still need to be considered. We assumed that the impact of weather changes could be modeled using relatively simple regression schemes, such as the linear model or the random forest. Given the observed impact of weather factors on the number of ratings, and the relationship between number of reviews and the rating, more sophisticated models that can exploit this relationship could lead to further improvements. In the current work, we only considered the environmental factors at the site of the business or service provider, on the assumption that both users and the providers are co-located. More refined information on the environmental attributes at the location of the user when the rating was made could further reduce the prediction errors.

420   S.A. Rahman et al.

References

1. Adomavicius, G., Tuzhilin, A.: Toward the next generation of recommender systems: A survey of the state-of-the-art and possible extensions. IEEE TKDE **17**(6), 734–749 (2005)
2. Anderson, M., Antenucci, D., Bittorf, V., Burgess, M., Cafarella, M.J., Kumar, A., Niu, F., Park, Y., R, C., Zhang, C.: Brainwash: A data system for feature engineering. CIDR (2013). www.cidrdb.org
3. Anderson, M., Magruder, J.: Learning from the crowd: Regression discontinuity estimates of the effects of an online review database. The Economic Journal **122**(563), 957–989 (2012)
4. Crandall, D.J., Backstrom, L., Cosley, D., Suri, S., Huttenlocher, D., Kleinberg, J.: Inferring social ties from geographic coincidences. PNAS **107**(52), 22436–22441 (2010)
5. Golder, S.A., Macy, M.W.: Diurnal and seasonal mood vary with work, sleep, and daylength across diverse cultures. Science **333**(6051), 1878–1881 (2011)
6. Gupta, N., Di Fabbrizio, G., Haffner, P.: Capturing the stars: Predicting ratings for service and product reviews, pp. 36–43. SS 2010. ACL, Stroudsburg (2010)
7. Koren, Y., Bell, R., Volinsky, C.: Matrix factorization techniques for recommender systems (2009)
8. Kramer, A.D., Guillory, J.E., Hancock, J.T.: Experimental evidence of massive-scale emotional contagion through social networks. PNAS, p. 201320040 (2014)
9. Markovitch, S., Rosenstein, D.: Feature generation using general constructor functions. Machine Learning **49**, 59–98 (2002)
10. Pazzani, M.J., Billsus, D.: Content-Based Recommendation Systems. In: Brusilovsky, P., Kobsa, A., Nejdl, W. (eds.) Adaptive Web 2007. LNCS, vol. 4321, pp. 325–341. Springer, Heidelberg (2007)
11. Qu, L., Ifrim, G., Weikum, G.: The bag-of-opinions method for review rating prediction from sparse text patterns. COLING 2010, pp. 913–921. ACL (2010)
12. Rendle, S.: Factorization machines with libfm. ACM Trans. Intell. Syst. Tech. **3**(3), 1–22 (2012)
13. Seroussi, Y., Bohnert, F., Zukerman, I.: Personalised rating prediction for new users using latent factor models. HT 2011, pp. 47–56. ACM, New York (2011)
14. Statistics, L.B., Breiman, L.: Random forests. In: Machine Learning, pp. 5–32 (2001)
15. Sun, L., Axhausen, K.W., Lee, D.H., Huang, X.: Understanding metropolitan patterns of daily encounters. PNAS **110**(34), 13774–13779 (2013)
16. Sun, M.: How does the variance of product ratings matter? Management Science **58**(4), 696–707 (2012)
17. Symeonidis, P., Tiakas, E., Manolopoulos, Y.: Product recommendation and rating prediction based on multi-modal social networks. RecSys 2011, pp. 61–68. ACM (2011)
18. Tiroshi, A., Berkovsky, S., Kaafar, M.A., Vallet, D., Chen, T., Kuflik, T.: Improving business rating predictions using graph based features. IUI 2014, pp. 17–26. ACM, New York (2014)
19. Yildirim, H., Krishnamoorthy, M.S.: A random walk method for alleviating the sparsity problem in collaborative filtering. RecSys 2008, pp. 131–138. ACM, New York (2008)
20. Yu, K., Schwaighofer, A., Tresp, V., Xu, X., Kriegel, H.P.: Probabilistic memory-based collaborative filtering. IEEE TKDE **16**(1), 56–69 (2004)

Social Support and Stress in Autism Blogging Community on Twitter

Amit Saha[1](✉) and Nitin Agarwal[2]

[1] University of Arkansas Medical Sciences, Little Rock, AR, USA
asaha@uams.edu
[2] University of Arkansas at Little Rock, Little Rock, AR, USA
nxagarwal@ualr.edu

Abstract. At present about one percent of the world population is diagnosed with autism spectrum disorder (ASD). With the rising prevalence of children diagnosed with ASD, the United States faces shortage of autism expertise and resources for delivery of effective therapy to the people who need services. For families dealing with autism, social media has become a crucial part of life. Parents of children with autism use social media platforms to build social bonding with other parents of children with special needs, access information, and share experiences. Content from social media such as blogs and Twitter can be analyzed to extract knowledge that could help develop practical and useful learning tools for other families with autistic kids. The current study is aimed to provide a research-based understanding of conversations in social media among families with autism. We systematically analyzed relevant interactions among the autism blogger community on Twitter. The study revealed that the autism blogger community provides quite significant social support to its members in the form of information sharing and reaching out to other members. We further found that perceived social support facilitated by the community members could be very effective in reducing psychological distress for parents of children with autism and help them adapt to the challenges.

Keywords: Autism · Social support · Stress · Twitter · Social media · Autism spectrum disorder · Community · Bloggers · Sentiment analysis

1 Introduction

Recent estimate by the Centers for Disease Control and Prevention (CDC, 2014) shows about 1 in 68 children in the USA were diagnosed with Autism Spectrum Disorder (ASD). Research activities in the field of ASD have increased significantly in recent years and has answered many questions but the larger group of public and private organizations are still working on getting a better understanding of ASD through research. Children or adults with ASD are generally characterized by communication difficulties, social deficit, repetitive behaviors, sensory issues and cognitive delays. They mostly indulge in social interactions only to fulfill their immediate needs such as to get food when hungry.

© Springer International Publishing Switzerland 2015
N. Agarwal et al. (Eds.): SBP 2015, LNCS 9021, pp. 421–427, 2015.
DOI: 10.1007/978-3-319-16268-3_53

With significant increases in coverage both online and traditional media on autism awareness, Internet, especially social media has become a critical means to gather and share information about autism for families handling ASD. Parents with kids diagnosed with autism share detailed accounts about their daily life experiences and information about autism in many social media platforms like blogs and Twitter that can be tapped as excellent sources of knowledge for other families dealing with autism. Studies show that caregivers and patients use social media to exchange and gather information related to health domains (Hamm, et al., 2013). Blogs, Twitter, Facebook and discussion forums have become new sources of information for health related topics. There are at present thousands of blogs about autism where person dealing with autism shares tips and daily experiences. Information shared over the Internet has a significant role in creating awareness of autism and eradicate misconceptions about various causes of autism. The premier US non-profit organization for autism "Autism Speaks" together with babble, which is an influential online blog network targeting young, educated, urban parents together publishes a list of top 30 autism blogs every year to raise autism awareness. Among the autism bloggers there is quiet a race and prestige to be included in the list. The top 30 autism blogs share their viewpoints on what needs to be heard in the autism community and raise awareness on various issues. Selection is based on the recommendation by parents of autistic kids who get invaluable practical advice and support from these blogs.

The purpose of this study is to provide a research-based understanding of conversations in the social media platform among families dealing with autism and to shed light on characteristics of autism community for providing support towards its members. The study is aimed to deduce the efficacy of social support provided by the autism community towards its members to cope with daily life challenges. For parents with children with autism, social support plays a key role to reduce stress among the parents. In this study, sentiments in the conversations among the members of the autism blogging community are quantitatively analyzed to shed light on the fact whether autism blogger community produces support and stress-free information to other members of the community on Twitter. For this study, microblogging activities of popular autism bloggers are analyzed.

2   Related Work

There is a huge body of research on autism studying the etiology of ASD, effectiveness of various intervention strategies, and several other aspects (Schaefer, et al., 2008). These studies are primarily conducted through clinical trials. Very few studies examine the vastly available social media interactions on autism. Study by Burton et al. (2014) used widely and freely available social media content in the form of mommy blogs and found health topic mentioned in 18% of the blogs and autism being one of the top topics discussed on those blogs. A recent study found 84.7% of 1039 participating parents use the Internet for searching information regarding their own medical conditions or those of family members or relatives (Bianco, et al., 2013). Many studies and survey results indicate social media, such as blogs, Twitter, and the like as new and convenient platform of communication within healthcare communities that affords to build social bonding with other patient families. The study found social support has a significant positive effect in depression, anxiety in mothers of children with autism (Weiss, 2002). Zhang & Yang (2013) in their study of the social media data found that online smoking support communities help establishing positive confidence among individuals to achieve better intervention outcome. Using freely available social media data, Carlisle (2014) found that using pets, especially dogs, could help improve the life of autism families. As indicated in the aforementioned studies, there is a huge potential of harvesting knowledge from the interactions among healthcare communities facilitated by social media.

3   Methods and Data

To evaluate social support and stress between members of autism blogging community, social network analysis along with sentiment analysis technique is conducted on the community interactions. Selection of autism blogger community members is based on the popularity of the blogger within the autism community. The popularity of an autism blogger is assessed by the fact that the member blog was selected among the top 30 autism blogs by "Autism Speaks" – a premier nonprofit autism awareness organization in the United States. As part of raising autism awareness, Babble – the website for promoting parenting in association with Autism Speaks – published the Top 30 Autism blogs of 2013 (*www.babble.com/baby/babbles-top-30-autsim-spectrum-blogs-of-2013*). The list for 2014 was not released by the time of writing this study. Presently 28 out of those 30 blogs are active. These 28 bloggers constitute the subjects of our study. Twitter IDs of the 28 popular autism bloggers are extracted from their blog profiles. We collect the interaction data posted on Twitter for the 2009-2014 period, which is publicly available. The dataset includes tweets, friends, followers, hashtags, timestamp and mentions among other publicly available information. NodeXL(*nodexl.codeplex.com*) is used to analyze the network structural characteristic of the autism blogger community. To determine the sentiments reflected in the interactions, Semantria (*www.semantria.com*), a sentiment analysis API is used.

For each of the 28 autism blogs, we are able to identify Twitter handles mentioned on the blog profiles of the corresponding bloggers. Three out of the 28 Twitter handles were found to be dormant on Twitter. We retrieved the most recent permissible tweets (up to 3200) for the 25 autism bloggers, resulting in 65,934 tweets. All the tweets were in English language. We analyzed their networks (tweet-retweet and friend-follower) and sentiments. All the friends' and followers' Twitter IDs of the 25 autism bloggers is downloaded to determine their social network. Twitter IDs of dormant users were excluded in the social network analysis. The 25 autism bloggers were grouped in three categories, viz., parent bloggers with autistic kids, autistic individuals who blogged themselves, and autistic bloggers who blogged as a group. The classification is shown in Table 1.

Table 1. Different categories of autism bloggers

Categories	Numbers
Parent Autism Bloggers	19
Autistic Bloggers	4
Group Autism Bloggers	2

4   Results

For the three distinct categories of bloggers, different analyses were conducted to get a richer insight into the autism blogger community as a whole. The Twitter activities of the popular autism bloggers were analyzed to extract the network characteristics of the autism blogger community.

The autism blogger community network shows distinct characteristics where any member of the community can reach a fellow member within an average of 4 hops (average geodesic distance), as compared to the average 6 degrees of separation for most social networks. This indicates the community members are well connected and more accessible than conventional social networks. The maximum value of outdegree is 105 that shows the reaching out tendency of the community. The friend and follower network of the autism blogger community is shown in Figure 1.

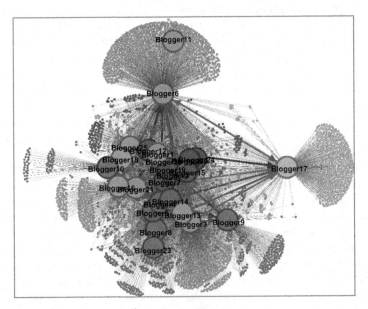

Fig. 1. Friend and follower network of autism blogger community. The 25 top autism bloggers are annotated with the prefix "Blogger". Their real identity is anonymized. Colors indicate different communities based on the network structure.

The mention and reply network of autism blogger community network is shown in Figure 2, where a thick edge indicates more interactions. With "autism" and "autistic" being the dominant words used in the tweets, it is evident that the autism blogger community's tweet network is focused on autism-related discussion. The entire autism blogger community network is one strong component, which indicates that the autism blogger community is strongly connected as one unit and any member of the community can reach any other member. Different colors indicate various sub-communities based on the network structure. Although, the community network looks fragmented, it is highly concentrated in its efforts of information dissemination and sharing. The reply and mention network of the popular autism bloggers reflects a spoke and hub nature of communication, where each popular blogger replies to and mentions several other members of the community – a behavior which is unusual to microblogging activity. The highly concentrated support provided by the autism blogger community to its members is also evident by the network's high connectedness.

To study the community characteristics for support and stress, autism bloggers' tweets are analyzed. Figure 3 shows a baseline comparison of emotions expressed in the tweets of the autism blogger community with other classes of text. We used LIWC (www.liwc.com) for assessment of stress, social support, positive emotion, and negative emotion. Stress is estimated using the LIWC scores for anger, anxiety and sadness. Social support is estimated using the scores for assent, positive sentiment, and social process for family, friends and humans. LIWC provides several baselines for comparison, e.g., emotional writings, control writings, science articles, and blogs.

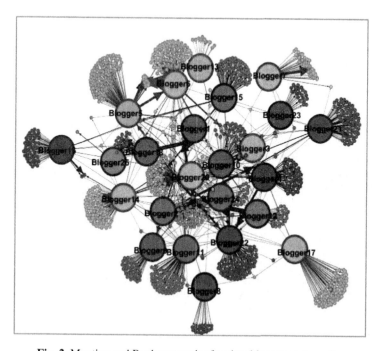

Fig. 2. Mention and Reply network of autism blogger community

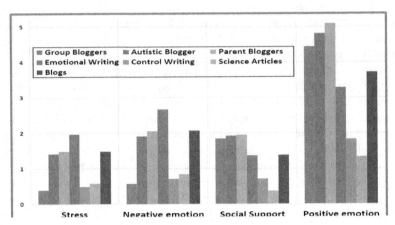

Fig. 3. Baseline comparison of emotion expressed in the tweets of autism blogger community

Baseline comparison shows that the positive sentiment and support expressed in the interactions among autism blogger community is way more than that expressed in the standard text. On the contrary, stress and negative sentiments are observed far less in the interactions among autism blogger community as compared to emotion writing and blogs, provided by LIWC. This indicates the less stressful and rather more upbeat nature of autism blogger community's Twitter interactions.

5   Conclusion and Future Works

In this paper, we study social support and stress in autism blogger community by identifying the popular bloggers and the members. The study considers top bloggers' Twitter activity and analyzes their friends, followers, tweets, retweets, mentions, and hashtags. We observed that the autism blogging community is tightly knit. The autism community provides extensive support to its members, both in the form of information sharing and emotional support. Families with autism connect to other members of the community to gain information for tackling disparate challenges with autism. The autism community empathetically provides support to other members by sharing a feeling of solidarity among its members. While stress and negative sentiments (anger, anxiety, and sadness to be precise) are reflected in some tweets by autism bloggers, an overwhelming majority of interactions exhibit positive sentiments. Another significant finding of the research is, the interactions of the autism blogger community on Twitter reflected more social support than any other forms of texts, including emotional writings, blogs, control writing, etc.

We envision that the study would create avenues to further explore the assessment of social support mechanisms in online healthcare communities. The research will facilitate data-driven studies to assess the efficacy of therapies for various disorders and evaluate success of different interventions as perceived by the caregivers based upon their shared experiences. This will help build a knowledge base for interventions and experiences, which in turn could help the clinical research on better understanding of behavioral interventions and developing new ones.

References

1. Bianco, A., Zucco, R., Nobile, C.G.A., Pileggi, C., Pavia, M.: Parents seeking health-related information on the Internet: cross-sectional study. Journal of medical Internet research **15**(9) (2013)
2. Burton, S., Tew, C., Thackeray, R.: Social moms and health: a multi-platform analysis of mommy communities. In: Proceedings of the 2013 IEEE/ASONAM. ACM. (2013)
3. Centers for Disease Control and Prevention (CDC): Community Report on Autism. http://www.cdc.gov/ncbddd/autism/states/comm_report_autism_2014.pdf/(Online accessed 20 November, 2014)
4. Carlisle, G.: Pet dog ownership decisions for parents of children with autism spectrum disorder. Journal of pediatric nursing **29**(2), 114–123 (2014)
5. Hamm, M., Chisholm, A., Shulhan, J., Milne, A., Scott, D., Given, M., Hartling, L.: Social media use among patients and caregivers: a scoping review. BMJ open **3**(5) (2013)
6. Zhang, M., Yang, C., Gong, X.: Social support and exchange patterns in an online smoking cessation intervention program. In: Proceedings of the 2013 IEEE ICHI 2013, pp. 219-228. IEEE Computer Society, Washington, DC (2013)
7. Schaefer, G., Mendelsohn, N., and Professional Practice and Guidelines Committee: Clinical genetics evaluation in identifying the etiology of autism spectrum disorders. Genetics in Medicine, **10**(4), 301-305 (2008)
8. Weiss, M.: Hardiness and social support as predictors of stress in mothers of typical children, children with autism, and children with mental retardation. Autism **6**(1), 115–130 (2002)

Dynamics of a Repulsive Voter Model

Farshad Salimi Naneh Karan and Subhadeep Chakraborty$^{(\boxtimes)}$

Department of Mechanical, Aerospace, and Biomedical Engineering,
University of Tennessee, Knoxville, TN, USA
schakrab@utk.edu

Abstract. In contrast to most conventional social interaction models, in this paper, the dynamics of repulsive interaction between nodes in a network is investigated. Such a model may be considered to be a first-order approximation for drivers making lane-changing decisions when confronted with other cars in the same lane. The model has been formulated and solved over a complete graph both with and without the influence of biased nodes. Analytical solutions have been derived for the model to determine its equilibrium distribution. Monte Carlo simulations have been performed to validate the accuracy of the model and solution. Interesting behavior of the model regarding dependence on population size, convergence time, and the effect of biased nodes are presented.

Keywords: Behavioral dynamics · Repulsive voter model · Influence

1 Introduction

Increase in online social interactions due to the proliferation of social media has recently led to accentuated efforts to understand the topology, mechanisms and characteristics of interactions among individuals in networked societies.

In this paper we consider the simplest case, where at any time, each node has only two discrete states $+1$ and -1; thus, a single variable $s_i = \pm 1$ fully specifies the state of node i. This Ising-based approach is not intended to model any real phenomena, rather allow a mathematical treatment of emergent phenomena where simple interactions among agents lead to complex global behavior. The zero temperature Glauber dynamics on complex networks has been widely studied with mean-field (MF) theory [1], and a comparison of the voter and Glauber ordering dynamics has been performed by Castellano and Loreto [2]. It has been shown that the voter dynamics always reaches the fully ordered state when the system size is finite.

In this paper, we discuss a modified scenario where in contrast to most conventional social interaction models, each node is repealed, rather than attracted by its neighbors. A direct application of such model is in the study of traffic flow patterns, where a car blocking a lane instigates lane change behavior in its immediate neighborhood. We also investigate the effect of including an additional I nodes, each with fixed pre-conceived allegiance to either of the states ± 1.

© Springer International Publishing Switzerland 2015
N. Agarwal et al. (Eds.): SBP 2015, LNCS 9021, pp. 428–433, 2015.
DOI: 10.1007/978-3-319-16268-3_54

We derive the equilibrium solution of the Master equation which reveals some interesting facts about the role of the influencing nodes on the convergence state, time to convergence, etc. This is then validated by numerical simulations. Section 2 presents the derivation of the master equation for the voter model with biased nodes. In section 3, we compare the theoretical results with numerical simulations and point out several interesting characteristics of the system. We conclude with a summary of our findings and some indications of possible future research directions.

2   Repulsive Voter Model Dynamics

In this paper, a complete graph is considered. The complete graph, being a fully connected random graph in the limit $p \to 1$, can be conveniently characterized by a single parameter, the magnetization factor, defining as $m = (N_+ - N_-)/N$, where, at any instant, N_+ represents the number of nodes (agents) in state $+1$, and N_- represents the number of nodes of the network in state -1. $N = N_+ + N_-$ nodes make up the vertices of the graph.

The repulsive Voter model dynamics is a straightforward modification of the standard Voter model. At each time step, a node i and one of its neighbors j is randomly selected, and s_i is set equal to $-s_j$, i.e. node i chooses the opposite state as its neighbor. Several social behaviors finds parallel in this anti-conforming tendency in a group, a typical example being lane selection in heavy traffic. Although not accurate at a fine level of granularity, an analysis of the dynamics can provide valuable information regarding the steady state equilibrium distribution, its dependence on group size and stabilization time.

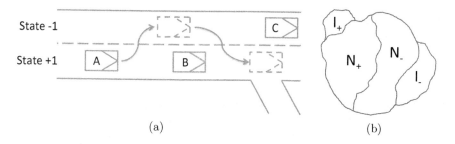

(a)   (b)

Fig. 1. (a) An example demonstrating traffic flow modeled with repulsive Voter dynamics, (b) Population composition

Figure 1 demonstrates the instantaneous composition of the population, which, in addition to N_+ and N_- nodes comprises of $I = I_+ + I_-$ external agents with respective pre-conceived allegiance to states $+1$ and -1. In the repulsive interaction scenario, the influence groups represent decentralized control inputs, which aim at repealing nodes away in order to attain a target state distribution. A natural example of such repulsive influencers in a traffic network are police vehicles stationed near accident or construction sites in order to attain smooth uninhibited traffic flow through reduced number of lanes.

In such a system, defining the control variable $u = \frac{I_+ - I_-}{I}$, we can proceed to compose the master equation as

$$\dot{P}_m = r_{m+\frac{2}{N}} P_{m+\frac{2}{N}} + g_{m-\frac{2}{N}} P_{m-\frac{2}{N}} - (r_m + g_m) P_m, \quad \text{where,} \tag{1}$$

$$r_m = P(m \rightarrow m - \frac{2}{N}) = \left(\frac{N_+}{N}\right)\left(\frac{N_+ + I_+ - 1}{N + I - 1}\right)$$

$$g_m = P(m \rightarrow m - \frac{2}{N}) = \left(\frac{N_-}{N}\right)\left(\frac{N_- + I_- - 1}{N + I - 1}\right)$$

$$r_{m+\frac{2}{N}} = \left(\frac{N_+ + 1}{N}\right)\left(\frac{N_+ + I_+}{N + I - 1}\right) \tag{2}$$

$$g_{m-\frac{2}{N}} = \left(\frac{N_- + 1}{N}\right)\left(\frac{N_- + I_-}{N + I - 1}\right)$$

Using Eqn. 2 in Eqn. 1, we get,

$$\dot{P}_m = \left(\frac{N_+ + 1}{N}\right)\left(\frac{N_+ + I_+}{N + I - 1}\right) P_{m+\frac{2}{N}} + \left(\frac{N_- + 1}{N}\right)\left(\frac{N_- + I_-}{N + I - 1}\right) P_{m-\frac{2}{N}}$$
$$- \left[\left(\frac{N_+}{N}\right)\left(\frac{N_+ + I_+ - 1}{N + I - 1}\right) + \left(\frac{N_-}{N}\right)\left(\frac{N_- + I_- - 1}{N + I - 1}\right)\right] P_m \tag{3}$$

After some manipulations, in the limit $N \rightarrow \infty$, the master equation can be expressed as:

$$\dot{P}_m = \frac{1}{N(N + I - 1)} \left[(1 + m^2)\frac{\partial^2 P_m}{\partial m^2} \right.$$
$$\left. + (2mN + 2m + Iu + Im)\frac{\partial P_m}{\partial m} + 2NP_m + IP_m\right] \tag{4}$$

With the proper time scaling as $\tau = t/N$, the final form of the influenced repulsive voter model can be expressed as:

$$\frac{\partial P_m}{\partial \tau} = \frac{\partial}{\partial m}\left[(1 + m^2)\frac{1}{N}\frac{\partial P_m}{\partial m} + 2mP_m + \frac{I}{N}P_m(m + u)\right] \tag{5}$$

If we assume that an equilibrium density function exists, then, $\lim_{\tau \to \infty} \frac{\partial P_m}{\partial \tau} = 0$, and by assuming that constants of integration are equal to 0, the equilibrium density $P_e(m)$ has to satisfy

$$(1 + m^2)\frac{1}{N}\frac{\partial P_e(m)}{\partial m} + 2mP_e(m) + \frac{I}{N}P_e(m)(m + u) = 0 \tag{6}$$

whose solution is,

$$P_e(m) = \frac{1}{C}\frac{e^{-Iu}e^{tan^{-1}(m)}}{(1 + m^2)^{(N + I/2)}} \tag{7}$$

where C is the constant that ensures proper normalization of the probability density function.

The effect of influence is apparent both in the exponential term in the numerator and in the power of the denominator. Putting $I = 0$ results in the solution of the repulsive Voter model without influence.

3   Discussion and Numerical Results

Performing Monte Carlo simulations and treating the magnetization parameter m as a random variable, we can estimate the probability density functions at each time (decision) point, simply with a rescaled histogram. We expect Eqn. 7 to accurately represent the density estimates at steady state and provide us with insight into some characteristics of this interesting system, such as the expected equilibrium magnetization, effect of N, I and u, etc.

3.1   The Equilibrium Density

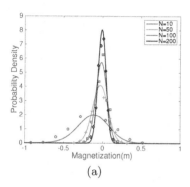
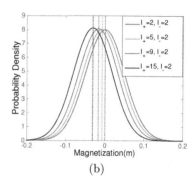

(a)   (b)

Fig. 2. (a) Effect of the population size (N) on the equilibrium density $P_e(m)$, and (b) Effect of size of influence groups on $P_e(m)$

It may be noted that $P_e(m)$ is not independent of N for large but finite population sizes. Thus, for bigger populations to reach similar expected magnetization levels, more control (bigger u) has to be exerted. This is in stark contrast to the regular Voter model with influence, where the mean magnetization was shown to be independent of the population size [3].

To investigate the dependence of $P_e(m)$ on N, we plotted probability densities for a variety of population sizes. Numerical results from MC simulations are overlayed on the analytical solutions for each N in Fig. 2a. Larger populations are characterized by lower variance and mean magnetization close to 0. The assumption of $N \to \infty$ lower bounds the population size. High degree of fidelity between MC simulation data and analytical result is observed for $N > 100$.

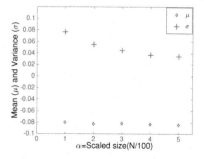

Fig. 3. Effect of scale factor on the mean and variance of $P_e(m)$

Expected Magnetization: The effect of different influence groups on the equilibrium density is shown in Fig. 2b along with the expected magnetization $\langle m \rangle$ plotted as vertical lines. It may be noted that the repulsive voter model is extremely stable in the sense that even high values of u are not be able to cause any noticeable shift in $\langle m \rangle$. In smaller populations, however, the effect is much more pronounced.

An interesting behavior for the system is observed if the influences I_+ and I_- are scaled proportionally with the population size. Figure 3 shows the effect of the scale factor α. Interestingly, $\langle m \rangle$ is independent of α, i.e.

$$\langle m(N, I_+, I_-) \rangle = \langle m(\alpha N, \alpha I_+, \alpha I_-) \rangle$$

This implies that a population of $N = 100$ nodes, with $I_+ = 5$ and $I_- = 1$ will have the same expected distribution mean as a population of $N = 200$, $I_+ = 10$ and $I_- = 2$. However, the variance monotonically decreases with scale.

3.2 Progression of the Density Function

So far, only aspects of the system at its equilibrium point have been presented. we now investigate the dynamic characteristics of the probability density function; specifically the effect of the influence groups I_+ and I_-.

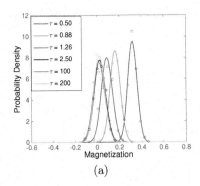
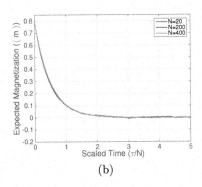

(a)   (b)

Fig. 4. a) Progression of the pdf with time for $N = 200$, $I_+ = 5$ and $I_- = 2$, b) Effect of N on the time to convergence

Figure 4a plots the progression of the pdf for such a numerical experiment with simulation parameters $N = 200$, $m_0 = 0.8$, $I_+ = 1$ and $I_- = 9$. Starting from a delta distribution $p(m)|_{\tau=0} = \delta(m - 0.8)$, the figure gives a snapshot of the distribution at time instants spaced in a geometric sequence. The choice of the time instants displayed are purely based on aesthetic considerations. The plots with the empty circles are scaled histograms from 1000 simulation runs at the corresponding scaled time points, while the solid lines are the normal distributions fit to the data.

Figure 4b plots the expected magnetization trajectory $\langle m \rangle$ with τ. It can be readily seen that the convergence time is independent of the population size N if time is scaled as $\tau = t/N$. This implies that the number of interactions it takes for a repulsive dynamical system to reach equilibrium is proportional to the size of the population. Even at this very low level of fidelity, this information can be potentially useful in a variety of situations, including the design of merge sections in highways.

4   Conclusions

In this paper, a new interaction model has been introduced, where each node repeals rather than attracts its neighbors. The model, is formulated on a complete graph and solved under the effect of influences at equilibrium. The equilibrium density $p_e(m)$ is derived as a function of the population size N, total influence size I and the control input $u = (I_+ - I_-)/I$.

Even though the repulsive dynamics proved to be very stable around the zero magnetization state, some interesting characteristics were observed. The expected equilibrium magnetization is dependent on the population size as well as the size of the positive and negative influences, but is invariant when all three are scaled identically. The distribution variance is a monotonically decreasing function of the population size. The time to convergence scales proportionately with the population size and is independent of influence.

The validity of analytical results has been checked by means of extensive numerical simulations. We found good agreement between the results from numerical experiments and analytical results for network size exceeding $N = 100$. It is important to note that the nontrivial phenomenology we have witnessed in this paper may lead to decision aids in traffic system designs. However this is only valid under the strict assumptions of the voter dynamics and simplified graph topology considered in this paper.

To address this issue and expand the applicability of the results, future studies along the same lines will adopt the solution for more complex decision space and more realistic diffusion characteristics of opinions. These problems are currently under investigation.

References

1. Castellano, C., Pastor-Satorras, R.: Zero temperature glauber dynamics on complex networks. Journal of Statistical Mechanics: Theory and Experiment **2006**(05), 5001 (2006)
2. Castellano, C., Loreto, V., Barrat, A., Cecconi, F., Parisi, D.: Comparison of voter and glauber ordering dynamics on networks. Physical review E **71**(6), 066107 (2005)
3. Chakraborty, S.: Analytical methods to investigate the effects of external influence on socio-cultural opinion evolution. In: Greenberg, A.M., Kennedy, W.G., Bos, N.D. (eds.) SBP 2013. LNCS, vol. 7812, pp. 386–393. Springer, Heidelberg (2013)

Ride Sharing: A Network Perspective

Erez Shmueli[1,2]([✉]), Itzik Mazeh[1], Laura Radaelli[1], Alex (Sandy) Pentland[2], and Yaniv Altshuler[3]

[1] Department of Industrial Engineering, Tel-Aviv University, Tel-aviv, Israel
[2] MIT Media Lab, Cambridge, USA
shmueli@tau.ac.il
[3] Athena Wisdom, Athens, Greece

Abstract. Ride sharing's potential to improve traffic congestion as well as assist in reducing CO_2 emission and fuel consumption was recently demonstrated by works such as [1]. Furthermore, it was shown that ride sharing can be implemented within a sound economic regime, providing values for all participants (e.g., Uber). Better understanding the utilization of ride sharing can help policy makers and urban planners in modifying existing urban transportation systems to increase their "ride sharing friendliness" as well as in designing new ride sharing oriented ones. In this paper, we study systematically the relationship between properties of the dynamic transportation network (implied by the aggregated rides) and the potential benefit of ride sharing. By analyzing a dataset of over 14 Million taxi trips taken in New York City during January 2013, we predict the potential benefit of ride sharing using topological properties of the rides network only. Such prediction can ease the analysis of urban areas, with respect to the potential efficiency of ride sharing for their inhabitants, without the need to carry out expensive and time consuming surveys, data collection and analysis operations.

1 Introduction

The increasing availability of portable technologies and ubiquitous connectivity makes the dream of having smart cities closer [2]. Easier collection of data on the way people live in a city and data analysis methods empower city administrators and policy makers that can better manage a city and improve the life in it.

Availability of large-scale datasets gives rise to new possibilities to study urban mobility. Calabrese et al. [3] show that mobile phone data can be used as a proxy to examine urban mobility and Noulas et al. [4] analyze social network data of different cities to find that mobility highly correlates with the distribution of points of interest. Mobile technologies support also successful applications, such as Waze [5], that provide traffic-aware city navigation by using data provided by the community. Many of the most urgent problems of big cities relate to cars. Alternative ways of moving in the city, such as autonomous mobility-on-demand and short-term car rental have been identified among the possible solutions to the transport headaches [6].

© Springer International Publishing Switzerland 2015
N. Agarwal et al. (Eds.): SBP 2015, LNCS 9021, pp. 434–439, 2015.
DOI: 10.1007/978-3-319-16268-3_55

Altshuler et al. [7] show that on-demand route-free public transportation based on mobile phones provides better traveling times than standard fix-route methods. Cici et al. [8] use mobile phone and social network data to show that traffic in the city of Madrid can be reduced by 59% if people are willing to share their home-work commute ride with neighbors. The authors study the effect of friendship on the potential of ride sharing, showing that if people are willing to ride with friends of friends the savings are close to the case of riding with strangers. Recent work by Santi et al. [1] introduces a new way to quantify the benefits of sharing, and studies a GPS dataset of taxi rides in New York City. The findings show that when passengers have a 5 minutes flexibility on the arrival time, and they are willing to wait up to 1 minute after calling the cab, over 90% of the sharing opportunities can be exploited and 32% of travel time can be saved. These results encourage the deployment and policies supporting ride sharing in urban settings.

Data analysis applied to large-scale datasets can reveal patterns of individual and group behaviors [9], and methods from social network analysis and graph theory show the structure and dynamics of social and communication networks [10]. A network can often be built on easily available data and becomes useful to predict apparently unrelated events or facts. A phone call network can signal an emergency situation [11], and social network of a Twitter account can identify a spammer [12]. Network analysis has been used to predict installation rates of mobile applications [13], spending behaviors of couples [14], and personality of individuals [15].

In this work, we propose a network-based approach to the prediction of benefit of ride sharing based on features of the transportation network of a city. We study its efficacy on a dataset of over 14 Million taxi trips taken in New York City during January 2013. First, we compute the benefits as a function of the maximum delay experienced by a rider, and we find encouraging results, e.g., more than 70% of the rides can be shared when riders are willing to accept a delay up to 5 minutes. Second, we apply the network-based approach to predict the benefit for a given maximum delay. We use topological features of the dynamic rides network to predict the benefit, and find that the combination of selected features has a high predictive power. Such prediction can enable analysis of urban areas, with respect to the potential efficiency of ride sharing for their inhabitants, without the need to carry out expensive and time consuming surveys or data collection.

The remainder of this paper is structured as follows: Section 2 describes the dataset; Section 3 presents the results of our study; and Section 4 summarizes our contributions and indicates future directions.

2   Datasets

We analyze a dataset of 14,776,615 taxi rides collected in New York City during January 2013 [16]. Each ride record consists of five fields: origin time, origin longitude, origin latitude, destination longitude, destination latitude. Removing rides with missing or erroneous GPS coordinates, we ended up with 12,784,243 rides.

Figure 1a reports the distribution of rides per day of the week and per hour of the day. The time distribution is far from uniform: the number of rides is higher in the middle of the week and it is lower during the weekend, and the daily distribution has peaks in the morning and around 6-7pm corresponding to start/end of working hours.

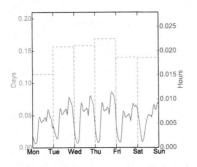

(a) Probability Density Function (PDF) of the number of rides per day of week/hour of day

(b) Illustration of the network nodes on the map of NYC

Fig. 1. Temporal and geographical distribution of data

3   Results

We construct the rides network from the dataset, where the set of nodes represents equally divided regions of New York City, and an edge (u, v) connects the two regions u and v if there exists at least one ride from region u to region v. The resulting network comprises 813 nodes and 58,014 edges. Figure 1b shows the geographical distribution of the nodes on a map, where New York is densely covered by the nodes with only a few empty spots in Staten Island.

We applied a simplified version of the methodology used by Santi et al. [1] to calculate the potential benefits of ride sharing. Benefits are expressed in terms of how many rides can be shared (therefore, how many rides can be saved), and are computed as a function of the guaranteed quality of service. The quality of the service is measured by the maximum time delay in catching a ride and arriving at destination, and it represents the maximum discomfort that a passenger can experience using the service.

Our analysis aims at finding pairs of rides, which are represented in the network by the same edge (i.e., have the same origin and destination), that can be shared. For each edge, we examine its corresponding set of originating rides, and count the number of ride pairs that can be merged, taking into consideration the maximum time delay parameter. Figure 2a shows the probability density function (pdf) of the number of rides per edge. As can be seen from the Figure, the distribution is heavy tailed and seems to follow a power-law. In other words, most of the edges (i.e., pairs of origin-destination) induce a small number of rides while a small number of edges induce an extremely high number of rides.

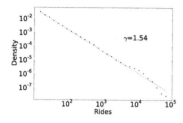 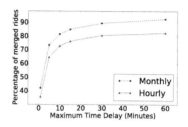

(a) Probability Density Function of the number of rides per edge

(b) Percentage of merged rides for the entire network (blue dashed line) and averaged over all sub-networks (green solid line)

Fig. 2. Analysis of rides

Figure 2b (blue dashed line) reports the percentage of shareable rides as a function of the maximum time delay parameter. Results are encouraging, more than 70% of the rides can be shared when passengers can accept a delay of up to 5 minutes. As expected, the benefit of ride sharing increases when the passengers are willing to take a higher discomfort, and the percentage of shareable rides is more than 90% when passengers can wait 30 minutes or more.

In this simplified analysis we considered the merging of two rides at a time, without taking into account the number of passengers in each ride. Since the average number of passengers per ride is 1.7, the number of saved rides could have been even higher by merging more than 2 rides at a time. On the other hand, in some cases, even the merging of two rides at a time might have resulted in overcrowding of the vehicle.

We model the high dynamic nature of the network by dividing the dataset into hourly snapshots, and creating $31 \times 24 = 744$ sub-networks. An illustration of one such sub-network is shown in Figure 3a. Intuitively we see that most of the nodes are highly connected, but a considerable number of nodes are connected to only one other node in the network.

We continue our analysis by considering each sub-network separately and averaging the results. Figure 2b (green solid line) shows the potential benefit of ride sharing which is first calculated for each sub-network separately and then averaged over all sub-networks. When averaging the sub-networks the benefits are constantly lower (around 10% less) than for the entire network. This reduction is expected as rides that start at the end of the hour have lower chances of

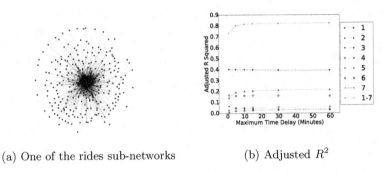

(a) One of the rides sub-networks   (b) Adjusted R^2

Fig. 3. Network analysis

being merged (i.e. they cannot be merged with rides that start at the beginning of the consecutive hour). Nevertheless, the curves have a similar trend in the two scenarios. This suggests that the effect of the quality of service is similar even on different networks.

Finally, for each one of the sub-networks, we extract a set of seven topological features — (1) the number of nodes, (2) the number of edges, (3) the averaged degree centrality score, (4) the averaged betweenness centrality score, (5) the averaged closeness centrality score, (6) the averaged eigenvector centrality score, and (7) the density score. Next, we use a linear regression model to fit the percentage of merged rides (dependent variable) to the various extracted features (independent variables) for the 744 instances. This process is then repeated multiple times, each time using a different maximum time delay. Figure 3b shows the obtained quality of fit as a function of the maximum time delay. As can be seen in the figure, the results are quite encouraging. When using all seven features as independent variables the quality of fit is remarkably high (R squared value close to 1). Clearly, when using each feature separately as the independent variable, the quality of fit drops.

4   Summary and Future Work

We aim at predicting the benefit of ride sharing expressed as the percentage of rides that can be shared with a limited discomfort for riders. We perform empirical analysis on a large-scale dataset, showing that we can predict the benefit of ride sharing based on the topological properties of the rides network only. In particular, we identify seven network topological features that combined can effectively predict the benefit.

Assessing the benefit of ride sharing is only a first step in devising a ride sharing solution. A direction for future work is to use the data-driven approach to study and incentivize participation in ride sharing. A complete solution should aim at guaranteeing benefits both for the riders (quality of service and low prices) and for the drivers (reducing the inconvenience of sharing the personal car) and at achieving the true potential of ride sharing by reducing the number of cars on the roads.

References

1. Santi, P., Resta, G., Szell, M., Sobolevsky, S., Strogatz, S.H., Ratti, C.: Quantifying the benefits of vehicle pooling with shareability networks. Proceedings of the National Academy of Sciences 111(37), 13290–13294 (2014)
2. Caragliu, A., Del Bo, C., Nijkamp, P.: Smart cities in europe. Journal of urban technology 18(2), 65–82 (2011)
3. Calabrese, F., Diao, M., Di Lorenzo, G., Ferreira Jr, J., Ratti, C.: Understanding individual mobility patterns from urban sensing data: A mobile phone trace example. Transportation research part C: emerging technologies 26, 301–313 (2013)
4. Noulas, A., Scellato, S., Lambiotte, R., Pontil, M., Mascolo, C.: A tale of many cities: universal patterns in human urban mobility. PloS one 7(5), e37027 (2012)
5. waze.com. http://www.waze.com (Online; accessed October 15, 2014)
6. Chin, R.: Solving transport headaches in the cities of 2050. BBC Future, June 2013 (Online; accessed October 15, 2014)
7. Altshuler, T., Shiftan, Y., Katoshevski, R., Oliver, N., Pentland, A.S., Altshuler, Y.: Mobile phones for on-demand public transportation. In: NetSci. (2014)
8. Cici, B., Markopoulou, A., Frias-Martinez, E., Laoutaris, N.: Assessing the potential of ride-sharing using mobile and social data: a tale of four cities. In: Proceedings of the 2014 International Conference on Pervasive and Ubiquitous Computing, pp. 201–211. ACM (2014)
9. Lazer, D., Pentland, A.S., Adamic, L., Aral, S., Barabási, A.L., Brewer, D., Christakis, N., Contractor, N., Fowler, J., Gutmann, M., Jebara, T., King, G., Macy, M., Roy, D., Van Alstyne, M.: Computational social science. Science 323(5915), 721–723 (2009)
10. Onnela, J.P., Saramäki, J., Hyvönen, J., Szabó, G., Lazer, D., Kaski, K., Kertész, J., Barabási, A.L.: Structure and tie strengths in mobile communication networks. Proceedings of the National Academy of Sciences 104(18), 7332–7336 (2007)
11. Altshuler, Y., Fire, M., Shmueli, E., Elovici, Y., Bruckstein, A., Pentland, A.S., Lazer, D.: The social amplifier–reaction of human communities to emergencies. Journal of Statistical Physics 152(3), 399–418 (2013)
12. Almaatouq, A., Alabdulkareem, A., Nouh, M., Shmueli, E., Alsaleh, M., Singh, V.K., Alarifi, A., Alfaris, A., Pentland, A.S.: Twitter: who gets caught? observed trends in social micro-blogging spam. In: Proceedings of the 2014 ACM Conference on Web Science. WebSci 2014, pp. 33–41. ACM (2014)
13. Pan, W., Aharony, N., Pentland, A.S.: Composite social network for predicting mobile apps installation. In: AAAI (2011)
14. Singh, V.K., Freeman, L., Lepri, B., Pentland, A.S.: Predicting spending behavior using socio-mobile features. In: Proceedings of the International Conference on Social Computing (SocialCom), pp. 174–179. IEEE (2013)
15. de Montjoye, Y.-A., Quoidbach, J., Robic, F., Pentland, A.S.: Predicting personality using novel mobile phone-based metrics. In: Greenberg, A.M., Kennedy, W.G., Bos, N.D. (eds.) SBP 2013. LNCS, vol. 7812, pp. 48–55. Springer, Heidelberg (2013)
16. NYC Open Data. http://data.ny.gov/ (Online; accessed September 01, 2014)

Complex Interactions
in Social and Event Network Analysis

Peter B. Walker[1(✉)], Sidney G. Fooshee[1], and Ian Davidson[2]

[1] United States Navy, Aerospace Experimental Psychology, Michigan, USA
peter.b.walker.mil@mail.mil, sidney.fooshee@navy.mil
[2] Department of Computer Science, University of California, Davis, USA
davidson@cs.ucdavis.edu

Abstract. Modern social network analytic techniques, such as centrality analysis, outlier detection, and/or segmentation, are limited in that they typically only identify interactions within the dataset occurring as a first-order effect. In our previous work, we illustrated how the use of tensor decomposition can be used to identify multi-way interactions in both sparse and dense data-sets. The primary aim of this paper will be to introduce innovative extensions to our tensor decomposition approach that target and/or identify second and third order effects.

Keywords: Tensor decomposition · Graph/Network modeling · Social network analysis

1 Introduction

Social and event network analyses have become more pervasive as a means for analyzing complex relationships in discrete datasets. These approaches attempt to analyze individuals/events or groups of individuals/events using a network-based view of the relationships and have widespread applications across a number of military domains. For example, military intelligence analysts often utilize the outputs of these results to make informed decisions about individuals such as friend or foe or determining an individual's sphere of influence.

In our previous work [2], we argued that both social and event networks naturally have two interpretations: 1) A tensor interpretation where the dimensions of space and time are used to record against event occurrence or absence and, 2) A graphical representation with each node being either a location or event-location combination with the behavior being encoded at the node as labels. We illustrated how the use of a tensor decomposition approach can be used to simplify these networks into much smaller and more meaningful sub-graphs [1, 8]. In the work we introduce in this paper, we focus on the graphical representation and describe several approaches to simplify these graphical representations of data.

Formally, the set of points in a network may be represented as a weighted undirected graph $G = (V, E)$, where the nodes are the set of points in a feature space and an edge is formed between each pair of nodes (see Figure 1). The weight (similarity) on each edge $w(i, j)$ is a function of the similarity between nodes i and j.

© Springer International Publishing Switzerland 2015
N. Agarwal et al. (Eds.): SBP 2015, LNCS 9021, pp. 440–445, 2015.
DOI: 10.1007/978-3-319-16268-3_56

To more effectively interpret the graph, grouping or clustering techniques may be applied that attempt to segment the graph into more similar sub graphs containing similar features. This may be accomplished by partitioning the graph into multiple disparate sets V_1, V_2,...,V_m, where some measure of similarity is high among vertices within set V_i but very low across different sets of vertices between sets V_i and V_j.

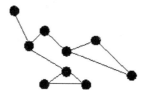

Fig. 1. Visualization of a graph or network. Black dots represent the nodes or vertices, and the lines connecting the dots represent the edges. Each node in the network may refer to a singular individual or event. The length of the edges connecting various nodes in the network suggests that the relationships among individuals or events in the network have different strengths.

The primary aim of our work here is to introduce two new segmentation approaches that have specific applications to social and network analysis. We describe in Section 2 an approach to segmentation that attempts to limit the likelihood of secondary or tertiary cuts of the graph. Section 3 describes a segmentation approach that differentiates sub graphs based on the similarity of post-cut node set weights and quantity. Specifically, this approach attempts to balance the resultant sub graphs in terms of the quantity and weight of the nodes in each sub graph. We conclude our discussion by describing potential applications for these techniques.

2   Minimizing Secondary Cuts in Graph Segmentation

Perhaps the most widely used graph segmentation approach, Min Cut attempts to segment the graph based on summed node weight. That is, an optimal *min cut* has been argued as one that minimizes the cut value; as the sum of the node weights included in the cut decreases, the cut value decreases as well. However, Wu and Leahy [9] suggested that an iterative optimal cut solution is one that attempts to partition the graph into several k-graphs so that the maximum cut across subgroups is minimized during each iteration. In other words, the Wu and Leahy segmentation approach attempts to find the minimum cuts that bisect the groups. However, this often results in the removal of outlier nodes whilst ignoring the impact that the removal of that node will have on the later segmentation of graphs.

Shi and Malik [3] attempted to resolve the min cut dilemma by constraining cuts such that each cut would be measured as a function of its edges *vol(A)* and *vol(B)*:

$$Ncut(A,B) = \frac{cut(A,B)}{vol(A_i)} + \frac{cut(A,B)}{vol(B_j)} \tag{1}$$

In contrast, Hagen and Kahng [4] introduced Ratiocut as an attempt to differentiate different groups in the graph by factoring in the number of vertices in each set $|A_i|$:

$$\text{Ratiocut}(A,B) = \frac{\text{cut}(A,B)}{|A_i|} + \frac{\text{cut}(A,B)}{|B_j|} \qquad (2)$$

Both algorithms attempt to weight the differences between each grouping as a function of the number of vertices or edge weights. Therefore, these algorithms eliminate the min cut problem raised earlier; however, each algorithm comes with their own set of problems. For example, Wagner and Wagner [7] identified both problems as being NP hard.

Still, both approaches modify the cost of cuts based on the similarity of the size of the edge weights in each resulting node set. Therefore, the cuts result in the lowest costs for cuts that yield node sets that have similar weight sums across nodes.

Both Ncut and Ratiocut might be viewed as appropriate in the analysis of social and event contexts where the phenomenon of interest is primarily driven by connection strengths, rather than connection quantity. In our example of a terrorist communication network, the use of a normed cut approach might suggest cuts that separate relatively few nodes from the network, if by doing so it is possible to isolate some of most important communicators from the rest of the network, if the combined value of the leaders so isolated by the cut in question (e.g., node set A) is roughly equivalent to the value of all the remaining terrorists in the network (e.g., node set B).

Similarly, this approach may yield cuts that have secondary effects on the sub graphs. For example, in Figure 2 below we see an imaginary network with two potential outlier nodes. Using an iterative (and recursive) segmentation approach, the first cut might result in the removal of the node in the upper right hand portion of the graph. However, upon the second iteration of segmentation, the second cut might result in the removal of the second outlier node in the same area of the graph. However, in many instances a more practical segmentation of this graph would be one in which only a single cut was performed that segmented the two outlier nodes into a single sub graph.

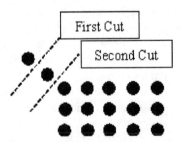

Fig. 2. Higher order effects on network segmentation. As can be seen from this Figure, the first cut may result in making a second cut and thereby isolate both nodes. A more practical segmentation in this case, might be one in which opts for the second cut so as to avoid isolating both nodes in the network.

To resolve this dilemma, we propose an alternative segmentation approach that attempts to minimize the likelihood of secondary cuts. Our 2cut approach (see Equation 2 below), attempts to minimize the nodes in sets A and B by maximizing the partitions of both $A(A_i \cup B_i)$ and $B(B_1 \cup B_2)$, respectively.

$$2\text{cut} = \underset{A,B,A_1,A_2,B_1,B_2}{\arg\min} \frac{\min cut(A,B)}{\min cut(A_1,A_2)\min cut(B_1,B_2)} \tag{3}$$

$$A_1 \cup A_2 = A$$

$$B_1 \cup B_2 = B$$

In the context of terrorist network analysis, a ratio cut approach would favor cuts that would bisect the size of the network, without respect to the number or strength of the connections of the members in one partition relative to the other. There may be instances where such an approach would be valuable, but more often than not, the members' connections (i.e., number and strength of edges) are too important an element to ignore in determining how best to disrupt the network.

3   Node Density Cuts: Graph Segmentation Based on Similarity of Post-Cut Node Set Weight and Quantity

An additional concern/criticism of both Ncut and Ratiocut segmentation approaches is that neither approach considers the resulting sub graphs. That is, both Ncut and Ratiocut segment a graph while considering only the cost of the cut relative to other cut costs. However, these algorithms ignore the quality of the resulting sub graphs. Therefore, in this section, we introduce an additional segmentation approach, which we term Pcut, that differentiates sub graphs based on the similarity of post-cut node set weights and quantity. That is, our Pcut algorithm selects a segmentation approach that attempts to "balance" the nodes and edges in each set after segmentation.

$$Pcut = \arg \min_{v} v'Lv + v'1 \tag{4}$$

As can be seen from Equation 5 above, Pcut seeks to minimize v (the set of vectors in the graph). This is accomplished because for each and every vector where $v'Lv$ is set to 1 (or -1), there is a separate vector where $v'1$ is 1 (or -1). In other words, this algorithm seeks to minimize the difference between two sets of vectors after segmentation.

In our example of analysis of a terrorist network, this approach would be likely to be the most fruitful. An approach favoring cuts that would seek to bisect both the number of nodes in each post-cut set as well as the combined weightings retained in each post-cut node set would be likely to do the most to disrupt the network's communications efficiency with each cut. This is ideal, since the goal of such analysis would be the identification of the single most valuable cut(s) for the purposes of network disruption to be used by military planners.

4   Future Directions

The primary focus of this paper is to illustrate the limitations of existing work in under-standing the relationships and dynamics of complex human behavior encoded as event networks. The research reported in this paper was applied to adversarial event data occur-ring in Iraq from 2003 through 2009 obtained from the World Incident Tracking System (WITS) database (wits.nctc.gov/). An example of this data can be observed in Table 1. These events are considered to be the result of human behavior. It is reasonable to con-clude that these events are not independent of each other but are instead related.

Our earlier work [2] focuses on analyzing this data directly as a tensor, here we suggest the application of both 2cut and Pcut to segment the nodes. This suggestion arises from the observation that in Social Networks, typical segmentation approaches such as Mincut result in a segmentation of the graph whereby several small networks are separated from much larger (and more connected) sub graphs. Such a result is often too difficult to interpret for the analyst and does not portray the relationship between subgraphs in an efficient manner. The approaches outlined above promise to provide significantly more information regarding how and why social networks (and specifically those social networks described in this paper) might be segmented so as to provide more information to the analyst.

In the work described in this paper we highlighted the need for the development of algorithms that explore higher order interactions within both social and event net-works. We introduced two separate algorithms in this paper: 1) 2cut which attempts to minimize the likelihood of secondary cuts through iterative trimming of the graph, and 2) Pcut which attempts to segment the graph such that there is a maximal amount of similarity in the sub-graphs in edge weight and node quantity. We believe that these algorithms represent the first concerted effort in the network analytic literature to focus on higher order effects within the subgraphs.

We have argued that social and event network is extremely complex and that inte-ractions within these datasets may ebb and flow over time. Because of this complexi-ty in the dataset, it remains imperative that algorithms such as the ones discussed in this paper are continued to be refined to allow for a thorough understanding of the interactions between social and cultural data.

Acknowledgements. The research reported in this paper was supported by Office of Naval Research Grant (NAVY 00014-11-1-0108). The opinions of the authors do not necessarily reflect those of the United States Navy or the University of California - Davis.

CDR Sidney Fooshee and LCDR Peter B Walker are military service members. This work was prepared as part of his official duties. Title 17 U.S.C. 101 defines U.S. Government work as a work prepared by a military service member or employee of the U.S. Government as part of that person's official duties.

References

1. Davidson, I.: Knowledge driven dimension reduction for clustering. IJCAI, 1034–1039 (2009)
2. Davidson, I., Gilpin, S., Walker, P. B.: Behavioral Event Data and Their Analysis. In: Data Mining & Knowledge Discovery, pp. 635–653 (2012)

3. Hagen, L., Kahng, A.: New spectral methods for ratio cut partitioning and clustering. IEEE Transactions in Computer-Aided Design **11**(9), 1074–1085 (1992)
4. Stoer, M., Wagner, F.: A simple min-cut algorithm. J. ACM **44**(4), 585–591 (1997)
5. Von Luxburg, U.: A Tutorial on Spectral Clustering. Statistics and Computing 17 (4) (2007)
6. Wagner, D., Wagner, F.: Between min cut and graph bisection. In: Proceedings of the 18th International Symposium on Mathematical Foundations of Computer Science (MFCS), pp. 744–750). Springer, London (1993)
7. Wang, X., Davidson, I.: Flexible constrained spectral clustering. In: KDD 2010, pp. 563–572 (2010)
8. Wu, Z., Leahy, R.: An Optimal Graph Theoretic Approach to Data Clustering: Theory and Its Application to Image Segmentation. IEEE Transactions on Pattern Analysis and Machine Intelligence 15 (11), 1,101–1,113 (1993)

Quantifying Tactical Risk: A Framework for Statistical Classification Using MECH

Xing Wang[1], Stephen George[2], Jason Lin[1], Jyh-Charn Liu[1]($\boxtimes$)

[1] Texas A and M University, College Station, TX 77840, USA
{xingwang,jason,liu}@cse.tamu.edu
[2] U.S. Department of Defense, Miami, USA
ticom.dev@gmail.com

Abstract. This paper presents a statistical classification framework for classification and prediction of asymmetric conflict (AC) locations. Different data normalization and feature reduction methods are coupled with supervised machine learning training algorithms to train classifiers. A set of 77 features derived from the MECH Model (Monitor, Emplacement, and Control in a Halo) were used to train the classifiers. The framework has been implemented and tested on real-world improvised explosive device and direct fire data collected from the conflict in Afghanistan in 2011-2012. Empirical results show that the classifiers achieve high accuracy, with human behavior-related features (visibility and population) exhibiting the most significant statistical differences. Performance of the classifiers is insensitive to the training algorithms. While performance is positively correlated to the training data size as expected, good performance is achieved with a fairly small amount of training data. Experiments based on cross-region training and prediction also show that classifiers are region dependent.

Keywords: Statistical pattern · Risk averse behavior · Asymmetric conflicts · Machine learning · Afghanistan

1 Introduction

Asymmetric conflict (AC) often pits a weaker attacker against a stronger target. To increase the probability of success, the attacker carefully assesses and optimizes attack-related tactical decisions. In [1], we propose a tactical behavior model for ACs called MECH that captures these decisions as features associated with common AC roles. A *Halo*, or annulus, surrounding the *Emplacement* site defines locations available for *Monitor* and *Control* activities. At the core of MECH is a set of 77 features that describe the local terrain, social/cultural features associated with nearby population centers, and AC actor positioning, under the assumption of a risk-averse attacker.

This work was supported in part by an ONR grant N00014-12-1-0531 and a National Defense Science and Engineering Graduate (NDSEG) fellowship. Opinions, findings and conclusions or recommendations expressed in this material are the author(s) and do not necessarily reflect those of the sponsors.

N. Agarwal et al. (Eds.): SBP 2015, LNCS 9021, pp. 446–451, 2015.
DOI: 10.1007/978-3-319-16268-3_57

In this work, we use supervised statistical machine learning (SML) techniques to predict high-threat locations for improvised explosive device (IED) and direct fire (DF) attacks. Raw measurements are transformed into features, with optional pre-processing steps of normalization and reduction for characteristic exploration. Here, stepwise feature selection (STP) and Principal Component Analysis (PCA) techniques are used for feature reduction. k-Nearest Neighbors (kNN), discriminant analysis (DA) and support vector machine (SVM) algorithms are used to train a binary (attack vs. no-attack) classifier. Using a dataset of roadside attacks in Afghanistan from 2011-2012, together with randomly selected non-incident locations, we develop a system for prediction of roadside attack locations. Experimental results show that features related to tactical risk-aversion and local population characteristics offer the strongest discriminant ability in comparison with generic geomorphometry features. Experiment results from cross-region training and evaluation suggest that the classifiers are region dependent.

The rest of the paper is organized as follows: Related Work in Section 2 is followed by Section 3 that presents features and their tactical meaning. Section 4 introduces the SML framework for road risk prediction and Section 5 presents analysis of experiments designed and executed using the framework.

2   Related Work

As an overview of the problem, Perry and Gordon [2] provide an intelligence-oriented view of asymmetric warfare. Existing efforts in the area of predictive analysis appear to have consistent shortcomings. First, the method is too simple to work in a real environment. Deitchman [3], for example, proposes a guerrilla warfare model based on a conventional Lanchester model, which only includes the attacker power factor. Second, the information output by the system is too coarse. Take SCARE-S2 [4] system as example. It is designed to identify regions (~100 sq km) containing high value targets but does not support more refined analysis. Third, systems like the two mentioned here depend on fixed rules and are difficult to adapt to changes.

An alternative approach uses SML to capture patterns of past events in optimized classifiers that can be used for prediction of future situations. An SML-based predictive framework overcomes the three problems mentioned above by including more features to describe the real situation, making adjustments to manage the problem of different scales and constraining the training data set temporally and spatially to make the predictor adaptive. SML has already been used to find the weapon cache location factors [2], CBRN usage prediction [5], and terrorist group growth prediction [6].

3   Road Risk Assessment Based on SML

We formalize the task as a binary classification problem which labels a random road location as useful or unfit for attack based on the classifier trained from the historical data. A binary classifier is generated by a linear process of feature extraction, data selection, feature normalization and reduction, classifier training, and finally evaluation of unlabeled data.

First, each potential AC event location is represented by 77 features using the MECH model [1] that captures the locational relationship of attackers and targets in an AC event. The first three groups of features in Table 1 are proposed based on tactical analysis. Relative visibility is a key component in the qualitative description of tactics. Viewshed and sparse viewshed derived from digital elevation maps (DEM) are key structures that store information about intervisibility between *Emplacement* locations and the neighboring region defined by the *Halo*. Next, general features based on the population and simple geomorphometric analysis are included. Given that available historical data does not contain full details of particular roadside attacks, it is infeasible to speculate optimal range parameters for each feature. Instead, the MECH behavior model uses the working experience of military experts to set a few representative ranges of roadside attack distances. As a result, each feature type is expanded into different ranges (shown in the title of each group of features) and each range is treated as a separate feature.

Table 1. Features and relevant behavior

G_1: Viewshed from the E location (window radius 100-350, 350, 500, 1K meters)	
Visibility Index (4)	A larger viewshed exposes a target to attackers for a longer period of time.
Shape Complexity (4)	A complex visible region is more difficult to assess and defend.
G_2: Sparse viewshed from the E location, # of directions (4,8,16,32,64)	
Local openness (5)	Open areas tend to offer the attacker a larger Emplacement while the target has less cover.
Distance to invisible region (min/mean/max) (15)	Nearby invisible regions offer cover for an attacker; more distant invisible regions are good for Monitor functions
Planimetric area (5)	Sparse viewshed version of visibility index.
Rugosity (5)	Sparse viewshed version of roughness within the visible range.
Shape Complexity 3D (5)	Shape complexity based on sparse viewshed.
G_3: From Monitor/Control view, route range (100, 250, 500, 1K, 3K meters)	
Route visibility near emplacement (min/median/max) (15)	Observability of a route from possible Monitor and Control locations.
G_4: General Terrain Features (window radius 50, 100, 350, 500,1000 meters)	
Elevation (1), Slope (1), Convexity(1) and Texture(1) Elevation range (5), Elevation roughness (5)	
G_5: Population related features (threshold of 1, 1K, 10K, 50K, 100K people)	
Minimum distance to city of at least certain size (5)	Populated areas act offer support for attackers.

After feature extraction, each sample is represented by a feature vector and label 1/0 indicating the sample location with/without event history. In our experiments, we ordered the event samples based on date and take the first 2/3 of the samples as training data, and the rest test data. The z-score technique is used for normalization. The first of the two dimension reduction methods is (unsupervised) principle component analysis (PCA) [7], which uses variances of high dimensional data to construct a low-dimensional, orthogonal projection. The second dimension reduction method is (supervised) stepwise feature selection [7] method, which is a regression-based iterative greedy algorithm. KNN, DA and SVM [7] were adopted for classifier training and performance evaluation.

4   Experiment, Result and Analysis

Our experiments are based on an AC event dataset containing IED and DF attacks collected in Afghanistan between February 1, 2011 and August 23, 2012 (with two gaps covering 09/12/2011-11/8/2011 and 02/14/2012-03/31/2012). The entire data set is used to estimate the performance and conduct comparative analysis of classifiers using various feature reduction methods and machine learning algorithms. The study assesses both general classification performance and the ranking of features based on their contributions to classification performance. To assess the location dependency of the classifiers, we also trained each classifier with data points from the different military command regions and then ran performance evaluations of each classifier with test data from other regions.

This analysis is constrained to roads and events along roads using road data collected and maintained by the Afghanistan Information Management Service. We discretized continuous roads to be a set R of locations according to the digital elevation map (DEM) from the ASTER Global Digital Elevation Model Version 2, which offers digital elevations with a horizontal resolution of approximately 30 meters.

The AC events dataset in the record format of [event type, location, time stamp] was provided by the ISAF-NATO RC South Civilian Integration Team. Among different event types, only IED and DF are considered in this work. Furthermore, we note that in this study we only considered events occurred at locations on or within 100 meters of known roads. Let T denote the full set of attack locations, and E the event set qualified for our experiments. The non-event class NE is the set of discrete road locations at least 250 meters from E locations. (This compensates for positional and estimation errors in the event data.) The resulting datasets include T_{IED}=3,224 events, T_{DF}=2,500 events and NE=3,326,250. The SML evaluation process is performed based on the average of 10 runs of experiments, where the NE is randomly down-sampled to the size of the E for each of the 10 runs.

We used machine learning algorithm with default parameters from the sklearn[1] package. KNN uses a k value of 3. SVM use C (box constraint) of 1 and an RBF kernel with γ set to 1/77. The cumulative variance threshold for PCA is 1, which give 5% accuracy improvement for SVM and DA methods compared with other thresholds. p-value threshold for STP is 10^{-5}. This p-value is the smallest value we tried and it yielded the best performance.

---

[1] http://scikit-learn.org/stable/

The average of classification accuracy is reported in Table 2, where the standard deviations of all cases are less than 0.0015. While the performance difference among all cases is small, classifiers based on feature reduction using STP method have better performance than classifiers based on feature reduction using PCA or the raw features. The kNN method with STP feature reduction offered the best performance.

Table 2. Accuracy for IED/DF risk prediction using kNN, DA and SVM

	IED			DF		
	kNN	DA	SVM	kNN	DA	SVM
None	0.9153	0.9695	0.9824	0.9272	0.9758	0.9862
STP	**0.9890**	0.9704	0.9858	**0.9903**	0.9760	0.9890
PCA	0.9114	0.9648	0.9788	0.9230	0.9715	0.9830

As for the contribution from each type of feature, the top ten ranked features are mostly from feature group G_3 and G_5 based on the STP feature selection process. Route visibility features G_3 alone achieve a classification accuracy of 90% or better. The second ranked feature is the population-related G_5, which achieves classification accuracy of ~80%. While it is expected that classification accuracy would increase with the size of the training dataset, it is interesting to assess the minimum dataset size required to produce usable classifiers (with acceptable classification accuracy.) For this study, we remove samples from the training data set gradually until a sharp decline of the classification performance is observed. In the end, with only about 100 samples we could produce classifiers with accuracies over 90%.

Using the NATO command regions to split the data into different groups, the accuracy loss of a classifier trained for one region ranges from 5% to 20% when it is used for tests in another region. This seems to support the argument that tactical behaviors of ACs differ across regions.

5   Conclusion

In this paper, we present a general pattern mining system based on a set of features derived from a tactical behavior model called MECH and some general demographic and geographic data. Experimental results show that the classifiers, together with their preprocessing steps, have good generalization abilities as defined in the machine learning (ML) community. Different feature reduction and ML methods lead to very similar performance, but the features have significant impact on the classification performance. It is interesting to observe that just a modest amount of training data is needed to produce good classifiers. The region-specific property of the classifier might also reveal the tactics difference across regions.

The study results derived from these experiments together its companion works in [1] and [8] are based on uncontrolled datasets obtained from the real world. They suggest that even though the data collection is not controlled, our results are highly indicative of the promising potential of computational behavior pattern extraction to support high level situational awareness analytics. How to develop new criteria to enable development and assessment of novel modeling and simulation techniques like MECH is a critical issue for the future generation of situational awareness tools that capture the abstractions of working military knowledge, behavior sciences, and geoscience. Our next efforts will fuse different techniques in order to support a broader range of experiments to explore the complex relation among human behaviors in low level conflicts, geolocations, and mining of their patterns to support field operations.

References

1. George, S., Wang, X., Liu, J.-C.: MECH: A model for predictive analysis of human choices in asymmetric conflicts. Presented at the International Conference on Social Computing, Behavior-Cultural Modeling and Prediction 2015, Washington, D.C. (2015)
2. Perry, W.L., Gordon, J.: Analytic Support to Intelligence in Counterinsurgencies. RAND National Defense Research Institute (2008)
3. Deitchman, S.J.: A lanchester model of guerrilla warfare. Operations Research **10**(6), 818–827 (1962)
4. Shakarian, P., Nagel, M., Schuetzle, B., Subrahmanian, V.: Abductive inference for combat: using SCARE-S2 to find high-value targets in Afghanistan. In: Proceedings of the 23rd Innovative Applications of Artificial Intelligence Conference, San Francisco, CA (2011)
5. Breiger, R.L., Ackerman, G.A., Asal, V., Melamed, D., Milward, H.B., Rethemeyer, R.K., Schoon, E.: Application of a profile similarity methodology for identifying terrorist groups that use or pursue CBRN weapons. In: Salerno, J., Yang, S.J., Nau, D., Chai, S.-K. (eds.) SBP 2011. LNCS, vol. 6589, pp. 26–33. Springer, Heidelberg (2011)
6. Bernica, T.W., Guarino, V.E., Han, A.J., Hennet, L.F., Mitchell, M.A., Gerber, M.S., Brown, D.E.: Analysis and prediction of insurgent influence for U.S. military strategy. In: 2013 IEEE Systems and Information Engineering Design Symposium (SIEDS), pp. 161–166 (2013)
7. Hastie, T., Tibshirani, R., Friedman, J.H.: The elements of statistical learning: data mining, inference, and prediction. Springer (2001)
8. Lin, J., Qu, B., Wang, X., George, S., Liu, J.-C.: Risk management in asymmetric conflict: using predictive route reconnaissance to assess and mitigate threats. Presented at the International Conference on Social Computing, Behavior-Cultural Modeling and Prediction 2015, Washington, D.C. (2015)

An Analysis of MOOC Discussion Forum Interactions from the Most Active Users

Jian-Syuan Wong, Bart Pursel, Anna Divinsky, and Bernard J. Jansen

The Pennsylvania State University, University Park, PA, USA
{jxw477,bkp10,axd289}@psu.edu, jjansen@acm.org

Abstract. Many massive open online courses (MOOCs) offer mainly video-based lectures, which limits the opportunity for interactions and communications among students and instructors. Thus, the discussion forums of MOOC become indispensable in providing a platform for facilitating interactions and communications. In this research, discussion forum users who continually and actively participate in the forum discussions throughout the course are identified. We then employ different measures for evaluating whether those active users have more influence on overall forum activities. We further analyze forum votes, both positive and negative, on posts and comments to verify if active users make positive contributions to the course conversations. Based the result of analysis, users who constantly participate in forum discussions are identified as statistically more influential users, and these users also produce a positive effect on the discussions. Implications for MOOC student engagement and retention are discussed.

Keywords: MOOC · Learning forum · Online learning · Distance education

1 Introduction

MOOCs have become a popular and significant source of distance-education due to the flexibility of course access and varied topics of courses. With these benefits, there are million students who enroll in one or more MOOCs. Since MOOCs are typically offered with only video-based lectures from are not co-located with the professor, students usually lack the opportunity to have interactions with other students and the instructor relative to traditional resident courses. Therefore, technology-based interactions, such as blogs and forums, are introduced as potential solutions. A discussion forum offers a platform for asynchronous communications that facilitates interactions and communications among students and instructors, and it also helps students build a learning community within the MOOC. Prior work has noted that discussions among peers help students improve their learning performance [1]. Additionally, discussions in a forum provide useful information for instructors to monitor course progress [2]. Therefore, discussion forums are widely adopted by MOOCs for interaction enhancement.

Due to the importance of the discussion forum to MOOCs, we conduct analyses to explore the interactions and communications in a MOOC forum. Forum users who

© Springer International Publishing Switzerland 2015
N. Agarwal et al. (Eds.): SBP 2015, LNCS 9021, pp. 452–457, 2015.
DOI: 10.1007/978-3-319-16268-3_58

actively participate in discussions are first identified based on their continuous participation, the number of replies they receive, and the number of responses they make. In order to verify whether active users are also influential users in the discussion forum, different measures of forum threads made by these active users are studied and compared with typical (i.e., non-active) forum users. Moreover, we analyze votes, both positive and negative, of posts and comments from active users to evaluate if active users make positive contributions to the MOOC forum.

2   Related Work

Anderson [3] discusses the factors of successful forums, as well as the measurements and limitations for online learning. Social network analysis is commonly applied to analyze the interactions on learning forums [4]. Dropout behavior is identified as relevant to sentiment of students' forum posts based on sentiment analysis of MOOC forums discussions [5]. Hauang et al. [6] studied the behavior of users who make great contributions across different MOOC forums offered by Coursera. The results indicate half of these 'superposters' are males, aged between 20 and 34. In addition, superposters commonly have better learning performance than the average forum users. Furthermore, these superposters generally have similar behaviors across different MOCCs that they take. However, there has been limited study of the effect of active posters on the learning environment of discussion forums.

3   MOOC Forum

We conceptualize a MOOC forum as composed of three types of hierarchical interactions, which are: A *thread* is created for initiating a new discussion. A *post* is a message for replying to a thread. A *comment* is a message used to reply to a post. In the 7-weeks course period for the MOOC used in this research, there are 8,169 threads, 21,434 posts, and 22,166 comments created by 7,389 forum users.

In order to provide insightful understanding of forum activities, we focus on user engagements. If a user initiates at least a new thread or makes a comment/post in a particular week, we consider this user to participate in the discussion for that week. Users' participations in each week are accumulated to study the number of weeks each user has engaged in the MOOC forum, and all of the forum users are divided into 7 groups based on weeks of involvements, from 1 to 7 weeks.

Table 1 denotes the number of weeks forum users participated in, and the average replies these users received and made during each week. For those users who participated in forum discussion for the entire course, the average replies they received and made are 80.31 and 113.00, respectively. So, these active users contributed more than others react to their contributions. Furthermore, the number of replies made is 1.41 times higher than replies received for these users, which implies that these users are dedicated to making responses to other users. For users who participate for 6 weeks, they also have the similar behavior but with fewer replies received and made.

Table 1. Number of users and average replies they receive/create based on weeks of participation

Weeks of Participation	# Users	Average Replies Receive	Average Replies Create
1	5,021	1.84	1.75
2	1,225	5.08	5.30
3	538	9.88	10.31
4	252	15.58	16.52
5	174	24.10	24.28
6	98	35.08	53.22
7	81	80.31	113.00

Therefore, users who participate in the forum discussion for 6 and 7 weeks can be considered as active users not only for the higher number of replies received and made but also for the continuous participation. Non-active users are classified as typical users who participated 5 weeks or less.

4   Research Question

In order to evaluate whether the active users actually play significant roles in forum discussions, the first research question is **do active users have more influence than typical users on the forum conversation?** In addition, it is important to discover whether these active users have a positive or negative effect on the forum discussion once they are identified as influential users. Therefore, the second research question is **do active users generally make a positive contribution to the MOOC forum?** By answering these two research question, we are able to assess whether actives users are central to the MOOC discussion forum.

5   Method

Analysis for Research Question 1. For evaluating whether active users are more influential in forum discussion than typical users, three analyses are conducted based on forum threads, since a thread initiates a new discussion topic among forum users. Three measures of forum threads are used for the comparisons. 1. Number of times a thread is read by users (*views of a thread*). 2. Amount of replies, both posts and comments, a thread receives (*replies of a thread*). 3. How long does the discussions continue within a thread (*duration of a thread*). The number of times a thread is viewed indicates whether a discussion topic is widely captivating among students. More views of a thread means the content/idea behind this thread can be broadly expressed, becoming more influential. The number of replies made for the discussions within a thread implies user involvement on this discussion topic. More replies created for a thread denotes that the content/idea of this thread is able to facilitate user participation. Lastly, by assessing how long a conversation lasts in a thread, whether an initiated topic is continuously discussed by users, can be identified. If the discussion in a thread continues for a long period, it could indicate an influential or engaging topic.

Analysis for Research Question 2. In order to answer whether active users generally make positive contributions to the MOOC forum, votes of posts and comments can be utilized for the evaluations. Forum users are allowed to give either a positive or negative vote to a post or comment based on the content. A post or comment that receives more positive votes implies the content is helpful and beneficial to other users; instead if a post or comment has more negative votes, it denotes that the content might be improper or irrelevant to the discussion topic. We can consider the posts or comments that receive more positive votes to have a positive contribution to the forum discussions, and the ones with more negative votes more likely to lead to negative impacts.

6   Result and Discussion

6.1   Are Active Users More Influential?

View of threads. For the 7 weeks of discussion, there were 828 threads created by active users and 7,370 threads created by typical users. The average views of threads created by active users and typical users are 126.14 and 53.93 times, respectively for overall discussions. As shown in Figure 1(a), the average views active users received for their threads are higher than typical users from week 1 through week 7. In addition, an increasing trend can be observed from week 1 to week 4 for both groups of users, and the only drops happened in week 5 and 7.

Reply of Thread. Average replies within a thread made by active users received for the overall course period is 12.96 compared to 5.61 for typical users. Figure 1(b) illustrates the average replies for threads made by both types of users during the 7 weeks. Figure 1(b) shows that threads made by active users received more replies than the ones created by typical users in all 7 weeks. Furthermore, a similar trend can be seen in Figure 1(a), with an increasing trend of replies from week 1 to week 4, but average replies decrease both on week 5 and 7. There are four weeks in which average replies of threads made by active users are twice that of typical users, which are weeks 1, 2, 5, and 6.

Duration of Thread. Duration of a thread is calculated by the time of last reply subtracted from the time a thread is initiated, with the difference rounded up to a day. The average duration of threads created by active users for the entire course is 3.89 days, which is 1.76 days more than typical users. In addition, Figure 1(c) shows the differences between active users and typical users for all 7 weeks for thread duration. Threads of active users have higher duration through the overall course, relative to typical users. The largest difference occurs for week 1 threads, which is 7.28 days for active users compared to 2.24 days for typical users. As shown in Figure 1(c), the duration of threads made by active users decrease weekly from 7.28 days in week 1 to 1.36 days in week 7. However, they are always longer than threads of typical users.

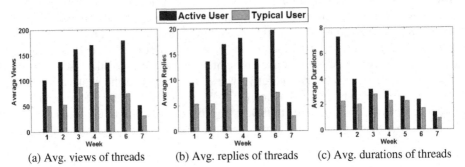

(a) Avg. views of threads   (b) Avg. replies of threads   (c) Avg. durations of threads

Fig. 1. Weekly analyses on three measures of forum threads created by active and typical users

6.2   Do Active Users Make Positive Course Contributions?

Average Votes of Post & Comment. The analysis for answering research question two is based on votes of posts and comments. The average votes of posts made by active and typical users for the overall course discussion is 1.74 and 1.25, respectively (the value of a positive vote is considered as +1, and a negative vote is -1). As shown in Figure 2(a), posts made by active users receive more positive votes than typical users in most of weeks, besides week 5 (1.91-1.92). In addition, the average votes of comments created by both groups of users are also studied. For the entire course period, active users' comments received an average 0.64 votes, which is 0.16 higher than typical users. For the 7-week discussion, active users received higher average votes in 5 of the weeks, but comments of typical users have more votes in week 5 and 6, which are 0.76-0.85, and 0.69-0.88 of active users-typical users, respectively.

Positive & Negative Votes in Posts & Comment. Beside examining average votes of posts and comments, the proportion of posts and comments that include the sum of votes greater or less than 0 is also studied (If the sum of votes in a post/comment is greater than 0, we consider it generally being a positive vote; otherwise it has a negative vote). For the analysis of posts with positive votes in the overall course period, 45.9% posts made by active users receive positive votes, which is 7.3% higher than the posts created by typical users. For the fraction of comments with positive votes in the overall course, active users also have the higher proportion than typical users, which are 37.7% and 26.6%, respectively. Moreover, as shown in Figure 2(b) comments made by active users have a higher proportion of positive votes than the comments created by typical users for all the weeks. Considering posts and comments with negative votes, active users have proportions of posts with negative votes as 1.5% for overall course, which is 1.1% higher than the posts of typical users. Additionally, the proportion of comments with negative votes made by active users for the entire course is 3.3% and 1% for typical users, which are both higher than negative votes of posts (both higher than the proportion of posts). Figure 2(c) shows the weekly analysis of posts and comments with negative votes. Active users have higher proportions of posts with negative vote in almost every week, with a tie in week 4. In addition, the fractions of comments made by active users having negative votes are higher than those of typical users in most of weeks, except week 1 and week 3.

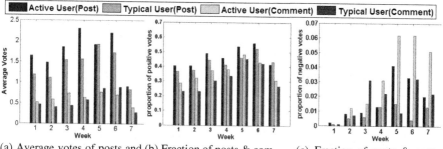

(a) Average votes of posts and (b) Fraction of posts & com- (c) Fraction of posts & com-
comments   ments with more positive votes   ments with more negative votes

Fig. 2. Weekly analysis on votes of forum posts and comments of active and typical users

7   Conclusion

Different analyses on the forum are conducted in this research for providing insightful understanding concerning different users participate in forum discussions. Three measures of threads are analyzed, including the *views*, *replies*, and *duration* of a thread to evaluate whether active users are also more influential users. Based on the analyses of forum threads, active users are also influential users. Furthermore, votes, both positive and negative, of posts and comments are evaluated for identifying whether active users generally make a positive contribution to the forum. Through the overall course period, posts and comments made by active users receive the highest number of votes, on average. For weekly analysis, posts and comments of active users have more votes than typical users in most weeks. However, posts and comments created by active users also have higher probability to receive negative vote than those from typical users. Based on the analysis of votes active, we can still consider that active users generally make a positive contribution to the forum discussion.

References

1. Smith, M.K., Wood, W.B., Adams, W.K., Wieman, C., Knight, J.K., Guild, N., Su, T.T.: Why peer discussion improves student performance on in-class concept questions. Science **323**, 122–124 (2009)
2. Stephens-Martinez, K., Hearst, M.A., Fox, A.: Monitoring moocs: which information sources do instructors value? In: Proceedings of the First ACM Conference on Learning@ Scale Conference, pp. 79–88. ACM (2014)
3. Andresen, M.A.: Asynchronous discussion forums: Success factors, outcomes, assessments, and limitations. Educational Technology & Society **12**, 249–257 (2009)
4. Zhu, E.: Interaction and cognitive engagement: An analysis of four asynchronous online discussions. Instructional Science **34**, 451–480 (2006)
5. Wen, M., Yang, D., Rosè, C.P.: Sentiment analysis in MOOC discussion forums: What does it tell us? In: Proceedings of Educational Data Mining (2014)
6. Huang, J., Dasgupta, A., Ghosh, A., Manning, J., Sanders, M.: Superposter behavior in mooc forums. In: Proceedings of the First ACM Conference on Learning@ Scale Conference, pp. 117–126. ACM (2014)

Inferring User Interests on Tumblr

Jiejun Xu$^{(\boxtimes)}$ and Tsai-Ching Lu

HRL Laboratories, LLC, Malibu, USA
{jxu,tlu}@hrl.com

Abstract. Inferring user interests is one of the core tasks for online social media services. It has direct impacts on personalization, recommendation and many other features for enhanced user experience. In this work, we proposed a novel bi-relational graph model to discover individual users' topics of interest from Tumblr, which is one of the most popular microblogging services. The proposed graph model contains two sub-graphs: one corresponds to users and the other corresponds to topics. Such a representation allows for effective exploitation of both user homophily relation and topic correlation simultaneously. This is in contrast with previous work where these two factors are considered in isolation. Subsequently, the user interest discovery problem is formulated as a multi-label learning problem on the bi-relational graph, with the goal to estimate the optimized associations between user nodes and topic nodes across the two sub-graphs. Our experiment is carried out with the complete data collected from Tumblr for a full month. To our knowledge, this work is the first attempt to conduct large-scale user interest inference on the platform.

Keywords: Tumblr · Social media · User interest modeling · Bi-relational graph · Multi-label learning

1 Introduction

The emergence and rapid development of social media and microblogging services provides users an excellent medium to interact with each other through their online presence. With more than 209 million user blogs and more than 69 million postings per day[1], Tumblr is one of the most popular microblogging services on the social web. In such a platform, users typically create and share contents based on their interests. Similar to Twitter, it has a social network aspect which allows users to interact with (e.g., *like*, *reblog*) each other. On the other hand, it favors multimedia content such as images and videos over traditional text. The combination of rich relational and content setting makes Tumblr an ideal platform for many different lines of research. In this paper, we focus on discovering the topics of interest for particular Tumblr users. Specifically, we aim to generate a ranked list of topics to characterize a particular user's interests.

---

[1] www.tumblr.com/about

© Springer International Publishing Switzerland 2015
N. Agarwal et al. (Eds.): SBP 2015, LNCS 9021, pp. 458–463, 2015.
DOI: 10.1007/978-3-319-16268-3_59

This is important as the research outcome allows for better clustering and search of Tumblr users, and it has direct impacts on personalization, recommendation and other aspects for improved online experience.

Our work focus on a class of approach which emphasizes on analyzing the graph topologies derived from social media platforms. The principle of homophily suggested that people with common characteristics are more likely to interact with one another, such as posting comments or re-publishing contents in the context of the social web. These actions serve as important cues to uncover implicit interests in social neighbors when modeled by graphs. For instance, a hypergraph structure is proposed in [3] to model the activities among users for music recommendation. Besides social homophily, correlation among topics is another important aspect for inference of user interests. Similarly, this can be captured by a graph-based framework. For instance, an entity graph is proposed in [5] to link concepts detected in tweets for user interest modeling. While social *homophily relation* and *topic correlation* are both relevant to user interest inference, they have been largely studied independently in existing literature. We believe that combining the two aspects in a principled manner will lead to substantial improvement on performance. To this end, we proposed a novel bi-relational graph model to discover individual users' topics of interest. The proposed graph model contains two sub-graphs: one corresponds to users and the other corresponds to topics. Subsequently, the user interest discovery problem is formulated as a multi-label learning problem on the bi-relational graph. The optimized associations between users and topics are learned by exploiting factors from both sub-graphs simultaneously.

2   Problem Formulation

Suppose that we have a collection of N users $\mathcal{U} = \{u_1, u_2, ..., u_N\}$ and K topics of interests $\mathcal{T} = \{t_1, t_2, ..., t_K\}$. Assuming that some of the users in U are (partially) labeled for their topics of interest, our goal is is to predict the topics of interest for the remaining unlabeled users u_i in the collection with a label subset $l_i \subseteq \mathcal{T}$.

The proposed graph-based multi-label learning[2] technique represents a transductive semi-supervised learning process that diffuses the label information from a small subset of nodes to the rest based on the intrinsic graph structure. The basic step in conventional graph-based learning is to construct a graph where vertices represent data instances and edge weights represent affinity between them. The key to graph-based multi-label learning is the prior assumption of consistency: nearby data instances or data instances that lie on the same structure are likely to share the same labels. Generally it is formulated in a regularization framework $\mathbf{F}^* = \mathbf{argmin}_F \{\Omega_{smooth}(\mathbf{F}) + \Omega_{prior}(\mathbf{F})\}$, where $\mathbf{F}$ is the to-be-learned matrix containing the label assignments of the graph nodes. The first term corresponds to a loss function which reflects the consistency assumption by imposing the smoothness constraint on the neighboring labels. The second

---

[2] Topics of interest and labels maybe be used interchangeably in the rest of the paper.

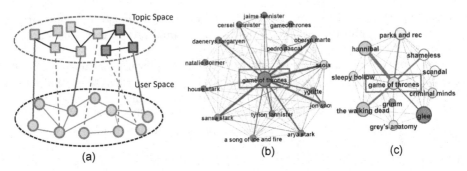

Fig. 1. (a) An illustration of the proposed bi-relational model. Solid lines within the two sub-graphs indicate affinity relationships among user nodes and topic nodes respectively. The solid red line across two sub-graphs denote the initial known user-topic assignments, and the dotted lines denote the topic assignments to be estimated. (b-c) A network of tags and a network of topics. The size and color of a node is proportional to its degree; the width of an edge is proportional to the co-occurence frequency of the correponding nodes.

term is a regularizer for the fitting constraint, which means that initial assigned labels should be changed as little as possible [7,8].

In our setting, data instances correspond to users, and their affinity can be characterized by the social interactions (e.g., *like,reblog*[3]) or computed based on any other similarity measures such as user demographic and geolocation. Note that the first term of the above regularization framework is also in accordance to the social homophily assumption. In addition to the user graph, we augment the conventional graph-based learning framework by introducing a new graph to emphasize the correlation among topics. In conjunction, the two graphs make up the bi-relational graph model as illustrated in Figure 1(a). Such a model allows for effective exploitation of the smoothness constraints on both sub-graphs as well as the interplay between them.

The construction of the user graph in this work is based on the *reblog* action on Tumblr. Specifically, we focus on *reblogs* that are reciprocated. In other words, a bidirectional edge is only introduced between user i and j if i *reblogs* j and j *reblogs* i at some point in time. The weight of an edge is determined based on the mutual *reblog* frequency. The construction of the topic graph is based on the co-occurrence among topics. However, unlike other microblogging platforms [4], topics are not explicitly defined in Tumblr. Alternatively we consider user tags as a medium to study Tumblr topics. This strategy has been studied in existing literature [6]. Figure 1(b) shows a snapshot of a co-occurrence network constructed from raw Tumblr tags. The tags in the network are related to a single coherent topic, which is the popular TV show "Game of Thrones". The nodes in the graph include character names, cast members, and variations of the show title. To reduce duplication and noise, we decided to aggregate and abstract raw tags to a more general level - clusters of semantically related tags, which we call topics. We detect these

---

[3] *reblog* is similar to *retweet* in Twitter

clusters by finding communities in the tag-based co-occurrence network. The Louvain community detection method [2] is applied to identify the topic clusters due to its computational efficiency. Figure 1(c) shows part of the resulting topic graph. Strong topic locality can be observed.

As mentioned earlier in this section, conventional graph-based learning framework minimizes a cost function with two terms. Introducing a new topic graph to the framework leads to the updated regularization framework regarding $\mathbf{F}$ as follows. Let $\mathbf{W}$ be a $N \times N$ affinity matrix denoting the data graph with N users, and $\mathbf{R}$ be a $K \times K$ affinity matrix denoting the label graph constructed for K topics. The frequency-based weights in $\mathbf{W}$ and $\mathbf{R}$ are normalized to the same dynamic range. Let $\mathbf{F} = (F_1, ..., F_N)^T = (C_1, ..., C_K)$ be a $N \times K$ matrix denoting the final association between every user topic pairs. $(C_1, ..., C_K)$ are the columns of $\mathbf{F}$, corresponding to the K labels. Similary, let $\mathbf{Y} = (Y_1, ..., Y_N)^T$ be an $N \times K$ matrix denoting the initial label assignments. Each $Y_{i,j}$ has 1 or 0 as the possible values: 1 if user i is labeled with topic j, 0 if it is unlabeled.

$$\Omega(\mathbf{F}) = \frac{1}{2}\eta \sum_{i,j}^{N} W_{ij} \left\| \frac{F_i}{\sqrt{D_{ii}}} - \frac{F_j}{\sqrt{D_{jj}}} \right\|^2 + \frac{1}{2}\mu \sum_{i,j}^{K} R_{ij} \left\| \frac{C_i}{\sqrt{D'_{ii}}} - \frac{C_j}{\sqrt{D'_{jj}}} \right\|^2 + \underbrace{\sum_{i}^{N} \|F_i - Y_i\|^2}_{\text{Prior constraint}}, \quad (1)$$

Smoothness on user graph (Homophily relation)   Smoothness on topic graph (Topic correlation)

where D and D' are both diagonal matrix whose (i, i) entries equal to the sum of the i-th row of $\mathbf{W}$ and $\mathbf{R}$, i.e., $D_{ii} = \sum_{j=1}^{N} W_{ij}$ and $D'_{ii} = \sum_{j=1}^{K} R_{ij}$.

The first term of the equation (1) is the smoothness constraint on the user graph. Minimizing it means neighboring vertices should share similar labels. For instance, if two users are close to each other based on their frequent mutual *reblog* activities, they will probably have common interests (thus with similar labels). The second term of (1) is the smoothness constraint on the label graph. Minimizing it means neighboring vertices should include similar users. For instance, if two topics are highly correlated with each other, then they are likely to be of interest to the same set of users. The third term indicates that the initially known user topic pairs should be changed as little as possible. η and μ are two constants controlling the trade-off of the regularization terms. If μ is set to zero, it means that we will ignore the correlation among topics, and the formulation is reduced to traditional multi-label learning on graph. Furthermore, the cost function in (1) can be written in a more concise matrix form as:

$$\Omega(\mathbf{F}) = \eta \mathrm{tr}(\mathbf{F}^T \mathbf{L}_g \mathbf{F}) + \mu \mathrm{tr}(\mathbf{F} \mathbf{L}_c \mathbf{F}^T) + \mathrm{tr}\left((\mathbf{F} - \mathbf{Y})^T (\mathbf{F} - \mathbf{Y})\right), \quad (2)$$

where $\mathbf{L}_g = \mathbf{I} - \mathbf{D}^{-1/2} \mathbf{W} \mathbf{D}^{-1/2}$ and $\mathbf{L}_c = \mathbf{I} - \mathbf{D}'^{-1/2} \mathbf{R} \mathbf{D}'^{-1/2}$. They are the *Normalized Laplacian* of the user graph and topic graph respectively. By applying the matrix properties: $\frac{\partial \mathrm{tr}(X^T A X)}{\partial X} = (A + A^T)X$ and $\frac{\partial \mathrm{tr}(X A X^T)}{\partial X} = X(A + A^T)$, we can differentiate equation (2) with respect to $\mathbf{F}$ as follows:

$$\frac{\partial \Omega(\mathbf{F})}{\partial \mathbf{F}} = \eta \mathbf{L}_g \mathbf{F} + \mu \mathbf{F} \mathbf{L}_c + (\mathbf{F} - \mathbf{Y}). \quad (3)$$

This is because both $\mathbf{L}_g$ and $\mathbf{L}_c$ are symmetric matrices. The solution of $\mathbf{F}$ can be obtained by requiring $\frac{\partial \Omega(\mathbf{F})}{\partial \mathbf{F}}$ to zero. Thus we have $(\eta \mathbf{L}_g + \mathbf{I})\,\mathbf{F} + \mu \mathbf{F} \mathbf{L}_c = \mathbf{Y}$, which is essentially a solvable matrix equation in the form of $AX + XB = C$. Note that F_{ij} is essentially a confidence value of user u_i being interested in topic t_j. Once it is determined, we can assign topics to users using simple thresholds.

3   Experiment

We now evaluate the effectiveness of the proposed bi-relational graph model for the task of inferring user interests. We obtain the Tumblr corpus over a 30 day period between June 1st, 2014 to June 30th, 2014 via GNIP[1] "firehose". During the period, every public activity on Tumblr is delivered to our system in real time. From the data, we construct a large user *reblog* graph and a tag co-occurrence graph. In order to maintain reliable social homophily relation, we only keep an edge between two users if their mutual *reblog* frequency is above 10. Similarly, we only keep an edge between two tags if they co-occur at least 10 times during the observed data window. After the filtering, we apply the Louvain community detection method to identify topics and subsequently construct the topic graph. The total number of topics obtained is 1086. Based on the mapping between tags and topics, each user now has a set of ground-truth topics of interest.

In the first experiment, we compare the proposed graph-based learning algorithm on bi-relational graph(BGML) with the standard graph-based learning techniques, namely Gaussian Filed and Harmonic Function(GFHF) [8] and Local and Global consistency method(LGC)[7]. Note that the user graph is large and disjoint, thus we only focus on the largest connected component to ensure the convergence of the label diffusion process. The parameters η and μ are determined empirically and fixed at 3 and 2 respectively . We randomly select 20% ground-truth as the initially known user topic assignments, and use the rest for testing. Figure 2(a) shows the MAP scores of the compared methods at different scopes (i.e., Top-k user-topic assignments by $\mathbf{F}$ confidence). The number suffixes for GFHF and LGC indicates the two best parameter sets tuned for the corresponding graph-based learning techniques. As can be seen, the proposed method consistently outperforms the other graph-based methods. The difference is more apparent toward the end of the scope range, with about $3 \sim 8\%$ improvement over the next best performing method.

In the second experiment, we investigate the graph-based learning performance with respect to different size of initial label assignments (i.e., known user topic pairs). We gradually increase the initial labeled size from 10% to 80%. For each size, we randomly select a set of user-topic pairs for initial labeling. MAP is computed at scope $= 40k$ for all size level. Intuitively, a better performance is expected with a larger initial set of labels. The results are shown on Figure 2(b). The MAP scores of all methods increase as the number of initial labels increase. However, our proposed method consistently outperforms all other methods. Substantial amount of improvement (up to 10%) is observed throughout the initial label size range.

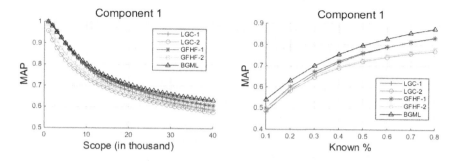

Fig. 2. (a) Performance comparison of different methods at various scopes. (b) Performance variations with respect to different number of initial labels.

4   Conclusion

This work addresses the problem of discovering user interests with a graph-based multi-label learning framework. Specifically, we proposed to model the social homophily relation and topic correlation via a unified bi-relational graph model. Such a representation allows for effective exploitation of the interplay between the two sub-graphs. Based on the proposed model, we then formulate the user topic inference problem as a semi-supervised label diffusion process. We conduct a large scale experiment on the dataset collected from the Tumlbr microblogging platform. Experimental results clearly demonstrate the effectiveness of the proposed approach.

References

1. Tumblr Data - Gnip (2014). https://gnip.com/sources/tumblr/
2. Blondel, V., Guillaume, J., Lambiotte, R., Mech, E.: Fast unfolding of communities in large networks. J. Stat. Mech., P10008 (2008)
3. Bu, J., Tan, S., Chen, C., Wang, C., Wu, H., Zhang, L., He, X.: Music recommendation by unified hypergraph: Combining social media information and music content. In: International Conference on Multimedia (MM) (2010)
4. Ottoni, R., Casas, D.B.L., Pesce, J.P., Meira Jr, W., Wilson, C., Mislove, A., Almeida, V.: Of pins and tweets: investigating how users behave across image- and text-based social networks. In: Proceedings of the Eighth International Conference on Weblogs and Social Media (ICWSM) (2014)
5. Shen, W., Wang, J., Luo, P., Wang, M.: Linking named entities in tweets with knowledge base via user interest modeling. In: Proceedings of the International Conference on Knowledge Discovery and Data Mining (KDD) (2013)
6. Yamaguchi, Y., Amagasa, T., Kitagawa, H.: Tag-based user topic discovery using twitter lists. In: International Conference on Advances in Social Networks Analysis and Mining (ASONAM), Kaohsiung, Taiwan, July 25–27 (2011)
7. Zhou, D., Bousquet, O., Lal, T.N., Weston, J., Schlkopf, B.: Learning with local and global consistency. In: NIPS. MIT Press (2004)
8. Zhu, X.: Semi-supervised learning literature survey. In University of Wisconsin Madison, Computer Sciences TR-1530 (2008)

Residential Mobility and Lung Cancer Risk: Data-Driven Exploration Using Internet Sources

Hong-Jun Yoon[1], Georgia Tourassi[1(✉)], and Songhua Xu[2]

[1] Health Data Sciences Institute, Oak Ridge National Laboratory,
Oak Ridge, TN 37831, USA
{yoonh,tourassig}@ornl.gov
[2] Department of Information Systems, College of Computing Sciences,
New Jersey Institute of Technology, University Heights, Newark, NJ 07102, USA
songhua.xu@njit.edu

Abstract. Frequent relocation has been linked to health decline, particularly with respect to emotional and psychological wellbeing. In this paper we investigate whether there is an association between frequent relocation and lung cancer risk. For the initial investigation we used web crawling and tailored text mining to collect cancer and control subjects from online data sources. One data source includes online obituaries. The second data source includes augmented LinkedIn profiles. For each data source, the subjects' spatiotemporal history is reconstructed from the available information provided in the obituaries and from the education and work experience provided in the LinkedIn profiles. The study shows that lung cancer subjects have higher mobility frequency than the control group. This trend is consistent for both data sources.

Keywords: Residential mobility · Lung cancer · Social media · Health data informatics

1 Introduction

There is rich literature in life-course epidemiology investigating the relationship between residential mobility and a person's health. The studies indicate an adverse effect and a fairly complex relationship, which includes both social and environmental factors. A systematic review of twenty-two studies [1] from the medical and social sciences literature reported that frequent residential change during childhood is a clin-

This manuscript has been authored by UT-Battelle, LLC under Contract No. DE-AC05-00OR22725 with the U.S. Department of Energy. The United States Government retains and the publisher, by accepting the article for publication, acknowledges that the United States Government retains a non-exclusive, paid-up, irrevocable, world-wide license to publish or reproduce the published form of this manuscript, or allow others to do so, for United States Government purposes. The Department of Energy will provide public access to these results of federally sponsored research in accordance with the DOE Public Access Plan (http://energy.gov/downloads/doe-public-access-plan).

N. Agarwal et al. (Eds.): SBP 2015, LNCS 9021, pp. 464–469, 2015.
DOI: 10.1007/978-3-319-16268-3_60

ical risk marker of behavioral and emotional health. Tønnessen et al [2] showed that frequent relocation during the early adolescent years is more detrimental than relocation in early childhood. Lin [3] studied the relationship between residential mobility history and self-rated health at midlife observing a similar association.

The association between residential mobility and cancer has been studied only in the context of radon exposure and lung cancer [e.g., 4-5]. The purpose of this paper is to explore the potential of a novel cyber-informatics approach to study a similar relationship, specifically if lifetime residential mobility is associated with lung cancer risk. Our study focuses on lung cancer because it is the leading cause of cancer death in the United States [6] both for males and females.

2   Methods

2.1   Data Sources

The typical retrospective case-control observational study involves first the collection of subjects with and without a specific disease (i.e., lung cancer) and then identification of each subject's exposure to the specific condition under investigation (i.e., lifetime residential mobility). We identified two different online data sources that contain the basic information needed for our study in a form that is relatively easily interpretable by computer. These sources include online obituaries and augmented LinkedIn social network profiles.

2.1.1   Online Obituaries

Online obituaries are widely available in newspaper sites, funeral homes' web pages, and web-based obituary archives. They have a largely similar format consisting of four sections; death announcement, biographical information, survivor information, and information about the funeral arrangements. Intrinsically, obituaries include basic information of the deceased that is essential for our study; name, age, birth date, cause of death, residence, and sometimes major life events (e.g., schools attended, military service, employments). Using an advanced web crawler developed in our laboratory [7], we searched the Internet for obituaries of people who died of lung cancer as well as non-cancer related obituaries. We applied the CoreNLP software package [8] to understand the obituaries' text content. Obituaries from which we could not identify or infer the essential information (i.e., age and gender) were excluded from this study.

- *Name and Age:* Typically a subject's name and age at death is explicitly stated in the first sentence (e.g.: "John Doe, 63, of Oak Ridge, TN passed away May 31, 2013."). In other cases age is inferred from the content by detecting dates at birth and at death (e.g.: "John Doe passed away Sunday, December 31, 2013... He was born January 1, 1950...").
- *Gender:* Gender is inferred by calculating the prevalence of male and female pronounces present in the obituary (e.g.: "She passed away at her residence...").
- *Cause of Death:* Lung cancer history was inferred from explicit statements (e.g.: "He passed away after a courageous battle with lung cancer..."). We applied

heuristic rules to filter out those obituaries that may contribute to false counts. For example, obituaries including sentences stating the family prefers monetary contributions to cancer research foundation rather than flowers (e.g.: "In lieu of flowers, please consider donations to lung cancer research.") were not considered cancer cases.

- *Locations of Residence:* Locations of residence of the deceased are stated in content, usually city and state. We collect locations of residence from the birthplace, residences, and the address where the funeral took place.

2.1.2 Augmented LinkedIn Dataset

LinkedIn is a business-oriented social network service which includes professional profiles (i.e., job history, work experience and education). Although LinkedIn profiles are a rich source of subjects with detailed spatiotemporal information during adulthood, few LinkedIn profiles contain the subjects' medical information, for example whether they have battled cancer. To leverage the advantages of the LinkedIn profiles while mitigating the limitations, we developed an additional cyber-informatics step.

Stories of lung cancer patients and survivors are abundant on the Internet, such as in open cancer survivor networks, lung cancer survivors' blogs, as well as national and local newspapers presenting lung cancer survivors' stories. First, we crawled such life stories of lung cancer patients and survivors using the advanced web crawler mentioned earlier. From the life stories of those candidate cancer subjects, we identified the subjects' names and other information that helped us search and match them with profiles available in LinkedIn. Since LinkedIn profiles have no direct indication of the subjects' gender and age, we developed tailored algorithms to infer this information. Gender was inferred from the first name of the LinkedIn profile utilizing the genderize.io API [9]. Education history (i.e., high school graduation year and college years) enabled us to estimate the subject's age. Profiles for which gender and age could not be inferred were excluded from further analysis. Furthermore, subjects with LinkedIn profiles with less 10 years location history were excluded from further analysis.

2.2   Number of Relocations in Lifetime

Collected residence locations from an obituary were aligned chronologically and then converted into geographical codes. Residence locations in LinkedIn profiles were collected from the "Experiences" and "Education" sections from which chronological residence locations were recomposed. We applied a simple rule to determine whether a particular geographical move was significant. If the distance of two consecutive locations was less than 50 miles, it was not considered a major move and was not included in the calculation of a person's mobility history.

3   Results

3.1   Online Obituary Dataset

Following the procedure described in Section 2.1 we formed a case group with lung cancer subjects and a control group of cancer-free subjects. To replicate a matched case-control study design, age and gender adjustment was achieved by selecting the same number of case and control subjects for each age and gender group.

The total number of lung cancer obituaries was 27,391. We found more male lung cancer subjects (16,129) than females (11,262). Table 1 shows the number of relocations per gender and age group. Statistical comparisons were made using the Student's t-test. Overall the average number of relocations for lung cancer subjects was 4.09 ± 3.16, which is significantly higher than that for the lung-cancer free group 3.01 ± 2.65. Significantly higher mobility was observed for the case group than the control group for all age groups and both genders consistently.

Table 1. Obituary Dataset: Number of obituaries of lung cancer diseased and cancer-free subjects by age and gender, average number of relocations (ARL), and their statistical comparison

GENDER	AGE GROUP	NO. OF SUBJECTS	ARL CASES	ARL CONTROLS	P-VALUE
Female	All	11,262	4.45	3.13	< 1e-5
	20~29	114	5.31	2.91	< 1e-5
	30~39	169	4.75	3.37	0.002
	40~49	580	3.74	2.83	< 1e-5
	50~59	1,686	3.77	2.73	< 1e-5
	60~69	2,886	4.08	3.08	< 1e-5
	70~79	3,471	4.46	3.20	< 1e-5
	80~89	2,048	5.30	3.41	< 1e-5
	90~	308	6.56	3.59	< 1e-5
Male	All	16,129	3.84	2.92	< 1e-5
	20~29	143	4.27	1.85	< 1e-5
	30~39	199	3.85	2.50	< 1e-5
	40~49	738	3.14	2.15	< 1e-5
	50~59	2,504	3.00	2.39	< 1e-5
	60~69	4,451	3.47	2.83	< 1e-5
	70~79	5,050	4.01	3.07	< 1e-5
	80~89	2,703	4.83	3.50	< 1e-5
	90~	341	6.00	3.60	< 1e-5
All	All	27,391	4.09	3.01	< 1e-5

Table 2. Obituary Dataset: Odds ratios (OR) and 95% confidence intervals (CI) between low mobility and high mobility groups, stratified by age and gender

GENDER	FEMALE		MALE	
AGE GROUP	NO. OF SUBJECTS	OR (95% CI)	NO. OF SUBJECTS	OR (95% CI)
All	11,262	1.75 (1.66~1.85)	16,129	1.47 (1.41~1.54)
20~29	114	3.03 (1.75~5.24)	143	3.65 (2.23~5.97)
30~39	169	2.17 (1.40~3.35)	199	2.17 (1.45~3.24)
40~49	580	1.47 (1.17~1.86)	738	1.72 (1.40~2.13)
50~59	1,686	1.74 (1.52~2.00)	2,504	1.40 (1.25~1.57)
60~69	2,886	1.57 (1.42~1.75)	4,451	1.38 (1.27~1.50)
70~79	3,471	1.74 (1.58~1.91)	5,050	1.47 (1.36~1.59)
80~89	2,048	2.00 (1.76~2.27)	2,703	1.55 (1.39~1.73)
90~	308	3.11 (2.20~4.38)	341	1.95 (1.43~2.66)
All	27,391		1.58 (1.53~1.64)	

We also measured the odds ratio (OR) between low mobility (0~2 times of relocation) and high mobility (3+ times of relocation) subjects (Table 2). We observed higher lung cancer incidence in high mobility subjects for all age groups and genders. Interestingly the OR is higher in younger groups for both genders.

3.2   Augmented LinkedIn Dataset

We repeated the same statistical analysis with subjects' profiles from the augmented LinkedIn dataset (Table 3). To replicate a matched case-control study design, we randomly selected an equal number of LinkedIn profiles with similar age and gender distribution from the remaining LinkedIn population.

Table 3. Augmented LinkedIn Dataset: Number of obituaries of lung cancer and cancer-free subjects by age and gender, average number of relocations (ARL), and statistical comparison

GENDER	AGE GROUP	NO. OF SUBJECTS	ARL CASES	ARL CONTROLS	P-VALUE
Female	**All**	**143**	**3.04**	**2.85**	**0.342**
	20~29	33	2.94	2.59	0.321
	30~39	45	2.87	2.99	0.732
	40~49	33	3.06	2.96	0.807
	50~59	26	3.23	2.90	0.522
	60~69	6	4.00	2.57	0.134
Male	**All**	**207**	**3.16**	**2.83**	**0.048**
	20~29	33	3.15	3.04	0.817
	30~39	66	3.29	2.61	0.019
	40~49	46	3.17	2.97	0.555
	50~59	33	3.09	3.08	0.983
	60~69	25	2.96	2.62	0.427
All	**All**	**350**	**3.11**	**2.84**	**0.032**

Table 4. Augmented LinkedIn Dataset: Odds ratios (OR) and 95% confidence intervals (CI) between low mobility and high mobility groups, stratified by age and gender

GENDER	FEMALE		MALE	
AGE GROUP	NO. OF SUBJECTS	OR (95% CI)	NO. OF SUBJECTS	OR (95% CI)
All	**143**	**1.06 (0.66~1.69)**	**207**	**1.10 (0.75~1.63)**
20~29	33	2.92 (1.06~8.03)	33	0.95 (0.36~2.51)
30~39	45	0.48 (0.20~1.12)	66	1.21 (0.60~2.41)
40~49	33	0.89 (0.34~2.33)	46	1.43 (0.62~3.26)
50~59	26	1.36 (0.46~4.05)	33	0.88 (0.33~2.34)
60~69	6	2.00 (0.19~20.61)	25	0.85 (0.28~2.58)
All	**350**		**1.08 (0.80~1.46)**	

It was observed that the average number of relocations in the case group was significantly higher than the average number of relocations in the control group (3.04±1.75 vs. 2.85±1.55). For each age and gender group, we observed mostly higher mobility in the cases than the controls. However, most differences were not statistically significant, possibly due to the smaller sample size of each subgroup. We also calculated the odds ratio of lung cancer risk between low mobility and high mobility groups observing similar trends but no significant differences within age and gender subgroups (Table 4).

4   Discussion

We presented a cyber-informatics approach to study the relationship between lifetime residential mobility frequency and lung cancer risk. The study utilized two distinct non-traditional data cybersources, namely obituaries and LinkedIn and replicated a matched case-control retrospective study design. Each data source shared its own strengths and limitations in terms of the sampling biases introduced. Regardless of the distinct differences between the two data sources, we observed consistently that frequent (and substantial) geographical relocation is linked to higher lung cancer risk. Furthermore, this study illustrated how non-traditional big data sources can be leveraged to execute cost-effective in silico epidemiological studies for knowledge discovery and hypotheses generation. However, issues of sampling bias and techniques to mitigate these challenges are still work in progress.

Acknowledgements. The study was funded by NIH/NCI (Grant #: 1R01CA170508-03).

References

1. Jelleyman, T., Spencer, N.: Residential Mobility in Childhood and Health Outcomes: A Systematic Review. J. Epidemiol Community Health **62**, 584–592 (2007)
2. Tønnessen, M., Telle, K., Syse, A.: Childhood Residential Mobility and Adult Outcomes. Statistics Norway Research Department. Discussion Papers (750) (2013)
3. Lin, K.C., Huang, H.C., Bai, Y.M., Kuo, P.C.: Lifetime residential mobility history and self-rated health at midlife. Journal of Epidemiology **22**(2), 113–122 (2012)
4. Warner, K.E., Mendez, D., Courant, P.N.: Toward a more realistic appraisal of the lung cancer risk from radon: the effects of residential mobility. American Journal of Public Health **86**(9), 1222–1227 (1996)
5. Krewski, D., et al.: Residential radon and risk of lung cancer: a combined analysis of 7 North American case-control studies. Epidemiology **16**(2), 137–145 (2005)
6. American Cancer Society: Cancer facts & figures (2014)
7. Xu, S., Yoon, H.J., Tourassi, G.D.: A user-oriented web crawler for selectively acquiring online content in e-health research. Bioinformatics **30**(1), 104–114 (2014)
8. Manning, C.D., Surdeanu, M., Bauer, J., Finkel, J., Bethard, S.J., McClosky, D.: The stanford corenlp natural language processing toolkit. In: Proceedings of 52nd Annual Meeting of the Association for Computational Linguistics: System Demonstrations, pp. 55–60 (2014)
9. Determine the Gender of a First Name. http://genderize.io/#overview

Author Index